Biomechanics of the Hand

Farid Amirouche • Alfonso Mejia
M. Daniel Wongworawat
Bassem Elhassan
Editors

Biomechanics of the Hand

Orthopedic and Clinical Perspectives

Springer

Editors
Farid Amirouche
Department of Orthopaedic Surgery
Orthopaedic and Spine Institute
Endeavor HealthSystem (Formerly
Northshore University HealthSystem)
An Affiliate of the University of Chicago
Pritzker School of Medicine
Skokie, IL, USA

Department of Orthopaedic Surgery
University of Illinois at Chicago
Chicago, IL, USA

M. Daniel Wongworawat
Department of Orthopedic Surgery
Loma Linda University, Orthopedic
Surgery
Loma Linda, CA, USA

Alfonso Mejia
Department of Orthopaedic Surgery
University of Illinois at Chicago
Chicago, IL, USA

Bassem Elhassan
Orthopedic Surgery, Co-Chief
Shoulder Service
Harvard Medical School, Massachusetts
General Hospital, Mass General West
Boston, MA, USA

ISBN 978-3-032-04687-1 ISBN 978-3-032-04688-8 (eBook)
https://doi.org/10.1007/978-3-032-04688-8

This Springer imprint is published by the registered company Springer Nature Switzerland AG
The registered company address is: Gewerbestrasse 11, 6330 Cham, Switzerland

Preface

The human hand represents a remarkable anatomical structure, functioning as an essential tool for expression, communication, and interaction with the surrounding environment. Its intricate anatomy and biomechanics provide an unparalleled range of motion and dexterity, enabling the execution of both delicate tasks and decisive actions. A thorough exploration of the biomechanics of the hand reveals the foundational principles that govern its functionality and potential for repair. This understanding is crucial in advancing our knowledge of both standard and pathological conditions affecting hand function.

This book, *Biomechanics of the Hand: Orthopedic and Clinical Perspectives*, is a collective effort from leading experts in the field, designed to bridge the gap between clinical practice and theoretical knowledge. By synthesizing the latest research and clinical insights, we aim to provide both a comprehensive resource and a practical guide for orthopedic surgeons, rehabilitation specialists, and students who are dedicated to improving hand health and function.

This comprehensive book explores the essential principles of hand and wrist anatomy and function. It examines the complex mechanics of the thumb, a vital component of manual dexterity. The book highlights the crucial role that advanced imaging techniques play in diagnosing and treating various conditions. Additionally, it emphasizes the significance of grip strength and the rehabilitation processes that can help restore functionality.

The text thoroughly investigates the biomechanics of hand tendons, illustrating their essential role in movement, and discusses the structural complexities of bone and cartilage, as well as the emerging field of biologics. Furthermore, it provides an in-depth analysis of the complexities involved in nerve repair and reconstruction, offering insights into the latest advancements in this area. The book also covers specialized techniques for repairing hand pulley function, ensuring a well-rounded understanding of the challenges and solutions in hand surgery and therapy.

Each chapter has been well written to explore various dimensions of hand biomechanics, from fundamental anatomy and kinematics to advanced treatment strategies for conditions such as tendon injuries and neuropathies. We also address contemporary topics, including the role of artificial intelligence and 3D printing in enhancing hand surgery. Our contributors bring a wealth of experience and knowledge, ensuring that readers gain insights into both established practices and cutting-edge innovations.

Some chapters focus on Artificial Intelligence and Machine Learning, including predictive analytics for outcomes in hand surgery, as well as AI-assisted imaging and diagnostics to enhance surgical precision. A topic of high interest is 3D Printing and Customization, which deals with personalized implants and prosthetics tailored to individual patients. Robotics in surgery is transforming the field, providing precision and control with advanced devices, which improves outcomes and enhances patient safety and recovery.

The Framework for the Future of Hand Surgery heavily emphasizes fostering partnerships between academic institutions and industry to create innovative solutions. It encourages multidisciplinary research to effectively address complex hand injuries and conditions.

Address the ethical implications of new technologies, particularly in AI and data usage. - Promoting equitable access to innovations in hand surgery for all patient demographics.

The future of hand surgery is poised for significant transformation driven by technological advancements and a focus on patient-centered care. By embracing these high-impact topics and collaborating within and across disciplines, we can improve outcomes and enhance the quality of life for patients with hand conditions. Continued dialogue and research will be crucial in navigating the challenges and opportunities that lie ahead in this dynamic field.

It is our hope that this book will not only serve as a crucial reference but also inspire ongoing dialogue and research in the biomechanics of the hand. We encourage readers to embrace the complexities of the hand's mechanics and to apply this knowledge to improve patient outcomes in their clinical work.

We want to extend our sincere gratitude to all the authors and collaborators whose contributions have made this comprehensive textbook possible. Your dedication and passion for the field are evident in every chapter. We also thank our readers for your enthusiasm and commitment to advancing the science and practice of hand biomechanics.

Chicago, IL, USA — Farid Amirouche
Chicago, IL, USA — Alfonso Mejia
Loma Linda, CA, USA — M. Daniel Wongworawat
Boston, MA, USA — Bassem Elhassan

Contents

Contributors

Nezar Abunnur Department of Orthopaedic Surgery, University of Illinois at Chicago, Chicago, IL, USA

John Alvarez Department of Orthopaedic Surgery, University of Illinois at Chicago, Chicago, IL, USA

Farid Amirouche Department of Orthopaedic Surgery, Orthopaedic and Spine Institute, Endeavor HealthSystem (Formerly Northshore University HealthSystem), an Affiliate of the University of Chicago Pritzker School of Medicine, Skokie, IL, USA

Department of Orthopaedic Surgery, University of Illinois at Chicago, Chicago, IL, USA

Diego Barragan Department of Orthopaedic Surgery, University of Illinois at Chicago, Chicago, IL, USA

Yvon Bogdonoff Department of Orthopaedic Surgery, University of Illinois at Chicago, Chicago, IL, USA

Hanny Chen Department of Orthopaedic Surgery, Loma Linda University, Orthopedic Surgery, Loma Linda, CA, USA

Bassem Elhassan Department of Orthopaedic Surgery, Massachusetts General Hospital, Boston, MA, USA

Bassem El Hassan Department of Orthopaedic Surgery, Massachusetts General Hospital, Boston, MA, USA

Anton J. Fakouri Mid America Orthopaedics, Palos Hills, IL, USA

Peter A. Falgiano Miami Orthopaedic Research Foundation, Miami, FL, USA

Mark H. Gonzalez Department of Orthopaedic Surgery, University of Illinois at Chicago, Chicago, IL, USA

Mark Gonzalez Department of Orthopaedic Surgery, University of Illinois at Chicago, Chicago, IL, USA

Steven Grindel Department of Orthopaedic Surgery, Medical College of Wisconsin, Milwaukee, WI, USA

Angela Hussain Department of Orthopaedic Surgery, University of Illinois College of Medicine at Chicago, Chicago, IL, USA

John D. King Department of Orthopaedic Surgery, Loma Linda University, Orthopedic Surgery, Loma Linda, CA, USA

Cadence Lee Department of Orthopaedic Surgery, University of Illinois College of Medicine at Chicago, Chicago, IL, USA

Ye Lin Department of Orthopaedic Surgery, University of Illinois at Chicago, Chicago, IL, USA

Gautam Malhotra Department of Orthopaedic Surgery, University of Illinois at Chicago, Chicago, IL, USA

Alfonso Mejia Department of Orthopaedic Surgery, University of Illinois at Chicago, Chicago, IL, USA

Barbara Mera Department of Orthopaedic Surgery, University of Illinois at Chicago, Chicago, IL, USA

Nirav Mungalpara Department of Orthopaedics, University of Illinois at Chicago College of Medicine, Chicago, IL, USA

Majd Mzeihem Department of Orthopaedic Surgery, University of Illinois at Chicago, Chicago, IL, USA

Kristi T. Nguyen Department of Orthopaedic Surgery, University of Illinois at Chicago, Chicago, IL, USA

Jorge L. Orbay Miami Bone and Joint Institute, Miami, FL, USA

Dmitriy Peresada Department of Orthopaedic Surgery, University of Illinois at Chicago, Chicago, IL, USA

Seth Roge Department of Orthopaedic Surgery, University of Illinois at Chicago, Chicago, IL, USA

Ali Rteil Department of Orthopaedic Surgery, Orthopaedic and Spine Institute, Northshore University Health System, An Affiliate of the University of Chicago Pritzker School of Medicine, Skokie, IL, USA

Jared Rubin Department of Orthopaedic Surgery, University of Illinois at Chicago, Chicago, IL, USA

Danil Rybalko Department of Orthopaedic Surgery, University of Illinois at Chicago, Chicago, IL, USA

Luis R. Scheker University of Louisville, Louisville, KY, USA

Christine M Kleinert Institute for Hand and Microsurgery, Louisville, KY, USA

Giovanni Solitro LSU Department of Orthopaedics, Shreveport, LA, USA

Aisha Suhaiba Department of Orthopaedic Surgery, University of Illinois at Chicago, Chicago, IL, USA

Isabella Trancoso Chicago Medical School, North Chicago, IL, USA

Yiru Wang Department of Orthopaedic Surgery, University of Illinois at Chicago, Chicago, IL, USA

Meagan Womack Department of Orthopaedic Surgery, Medical College of Wisconsin, Milwaukee, WI, USA

M. Daniel Wongworawat Department of Orthopaedic Surgery, Loma Linda University, Orthopedic Surgery, Loma Linda, CA, USA

Mohammad Zalzaleh Department of Orthopaedic Surgery, Medical College of Wisconsin, Milwaukee, WI, USA

Hand Biomechanics

1

Farid Amirouche and Bassem Elhassan

Introduction

Breakthroughs in cellular biology, particularly in bone morphogenic proteins, biologic scaffolding, and cartilage culturing, offer unprecedented opportunities to treat articular cartilage injuries effectively [1, 2]. This innovative approach includes fabricating cartilage autografts using a patient's cells, enhancing the likelihood of successful integration. Furthermore, ongoing investigations into the preservation and surgical implantation of cartilage and meniscal transplants will shed light on their long-term functionality, ensuring that patients receive the highest quality of care that closely mirrors the original donor characteristics [3].

Biomechanics is a cutting-edge field advancing our understanding of the intricate interactions among bones, soft tissues, biologics, and implants. This discipline, which focuses on the mechanics of the human body, encompasses both static and dynamic motions at the joints. It meticulously considers the various forces and moments that influence these movements, critical for maintaining joint function and overall mobility. This meticulous and reliable approach to research not only reassures us of the thoroughness and reliability of biomechanical studies but also stimulates our intellectual curiosity and pushes the boundaries of our knowledge.

Joint stability, a multifaceted concept, is a key area where biomechanics plays a crucial role. It relies on numerous mechanical factors, including ligaments' structural integrity, bones' alignment, and surrounding soft tissues' properties [4]. Modeling and understanding these interactions can be quite complex as they often involve mechanical behaviors that change under different conditions. The interplay between cellular-level mechanics and macro-level movements occurs in a multiphysics environment, indicating that factors like strain, stress, and fatigue can manifest in various forms, sometimes highlighted by injuries [5]. Biomechanics provides the tools and methodologies to dissect and understand these complex interactions, contributing significantly to our understanding of joint stability.

F. Amirouche (✉)
Department of Orthopaedic Surgery, Orthopaedic and Spine Institute, Endeavor HealthSystem (Formerly Northshore University HealthSystem), an Affiliate of the University of Chicago Pritzker School of Medicine, Skokie, IL, USA

Department of Orthopaedic Surgery, University of Illinois at Chicago, Chicago, IL, USA
e-mail: famirouche@northshore.org

B. Elhassan
Orthopedic Surgery, Co-Chief Shoulder Service, Harvard Medical School, Massachusetts General Hospital, Mass General West, Boston, MA, USA

F. Amirouche et al. (eds.), *Biomechanics of the Hand*,
https://doi.org/10.1007/978-3-032-04688-8_1

Biomechanics is inherently interdisciplinary. It draws on principles from biology and physics to understand the anatomical functions of human joints, ligaments, muscle fibers, and pathologies such as rheumatoid arthritis and cartilage degeneration. This practical and relevant interdisciplinary approach makes biomechanics a powerful tool that underpins the study of human movement. It also intersects with fields like engineering, computer science, and rehabilitation, demonstrating its broad scope and relevance to various disciplines. For instance, biomechanics principles are used in the design of prosthetics and orthotics [6], the development of sports equipment, and the optimization of rehabilitation protocols, showcasing its diverse applications.

The field of orthopedic biomechanics aims explicitly to transform the complexities associated with various anatomical structures, like the fine motor functions of hand digits, into simplified models. These models enhance our understanding and guide clinical applications and interventions to improve patient outcomes by providing a scientific basis for designing orthopedic implants, developing rehabilitation protocols, and understanding musculoskeletal pathologies.

Optimizing Performance

The study of sports biomechanics, specifically focusing on the hand and wrist, offers invaluable insights that are instrumental in optimizing the subject's performance [7]. This field meticulously examines movement efficiency before and after surgical procedures, providing a thorough understanding of how surgical interventions affect overall functionality. By exploring the dynamics of power, work, energy, resistance, and fatigue, researchers can identify the best physical therapy exercises that are beneficial for diagnosis and critical for rehabilitation.

Advanced technologies and methodologies are the backbone of hand biomechanics research. Motion capture systems record the precise movements of the hand and wrist, enabling detailed analysis of kinematic variables such as speed, trajectory, and angle of movement [8, 9]. Sophisticated sensors gather kinetic data, capturing forces and torques exerted during different activities. High-resolution imaging techniques, such as MRI or ultrasound, provide visual insights into soft tissue structures and joint mechanics [10, 11], further enhancing the understanding of biomechanical interactions. This technological prowess is the cornerstone of our research, offering a promising future for patient care and rehabilitation.

Electromyography (EMG) offers another layer of depth by measuring the electrical activity of muscles [8], revealing how muscle coordination and activation patterns change in response to injury or rehabilitation. Collectively, these data collection methods enable researchers to establish key performance indicators—critical metrics such as joint range of motion, movement repeatability, and force production—which can be meticulously analyzed to improve patient outcomes [12].

By integrating this wealth of data, clinicians can develop tailored rehabilitation strategies that address specific deficiencies, promoting effective recovery and functionality. Ultimately, this comprehensive approach improves patients' physical capabilities and elevates their overall quality of life as they regain strength and dexterity in their hands and wrists [13].

Injury Prevention

Hand biomechanics plays a vital role in injury prevention. By closely examining specific movement patterns while monitoring the forces applied to the hand during various tasks, researchers can pinpoint potential risk factors linked to musculoskeletal injuries. This proactive approach, often complemented by cadaveric experiments, provides valuable guidance for surgeons and healthcare professionals, enabling them to implement targeted interventions that can significantly lower the risk of injuries in patients. Understanding the biomechanics of the hand can help in designing ergonomic tools and workstations, developing effective rehabilitation strategies, and guiding surgical interventions, all of which contribute to injury prevention and improved patient outcomes.

Hand clinical biomechanics is a multidisciplinary field that focuses on the intricate mechanics of the hand. It combines advanced diagnostic imaging techniques like MRI and ultrasound with innovative tissue bioengineering and regenerative medicine approaches to understand and optimize hand function [14]. Using finite element analysis allows stress distribution simulations, while extensive mechanics databases provide invaluable hand anatomy and performance data.

Furthermore, modern modeling techniques assist in analyzing complex hand functions and movements, leading to better insights into injuries and diseases. Incorporating cutting-edge artificial intelligence enhances our ability to refine diagnostic accuracy and predict surgical outcomes [15]. By utilizing a comprehensive set of quantitative descriptive parameters, this discipline fundamentally elucidates the complex functions of the hand, enabling healthcare professionals to develop more effective treatment strategies and improve patient care. This remarkable transformation is driven by skilled surgeons who have masterfully woven together the latest advancements in research, clinical practices, and comprehensive analyses of measurable outcomes.

Hand surgeons are now pioneering a revolutionary approach that prioritizes patient needs, integrating various innovative techniques and state-of-the-art technologies. Advanced biomaterials, meticulously designed implants, and 3D computer modeling have opened new avenues for surgical interventions [16–18]. Furthermore, virtual testing is vital in refining these procedures, paving the way for more effective and precise treatments tailored to individual patients [19]. This convergence of science and art in hand surgery, facilitated by biomechanics, is genuinely reshaping the landscape of surgical possibilities. By optimizing surgical procedures through biomechanical analysis, surgeons can ensure better patient outcomes, reduced recovery times, and improved quality of care, thereby shaping the future of orthopedics [19, 20].

The Mechanics of the Human Hand

The human hand is an astonishingly intricate and versatile mechanism that functions on both broad and detailed levels, showcasing remarkable dexterity and control. With its 27 bones, numerous muscles, and complex network of tendons and ligaments, the hand allows us to perform various tasks, from delicate manipulations to powerful grips (Fig. 1.1). Our hands are not just tools for action; they also serve as conduits for our thoughts and emotions. We express our ideas, desires, and feelings through gestures and touch, illustrating how deeply interconnected our hands are with our cognitive and emotional experiences. Our hands are pivotal in shaping our interactions with the world around us, influencing how we think and feel.

To fully grasp the typical characteristics of the human hand, one must appreciate its intricate sensory and mechanical features. The hand is a marvel of engineering, consisting of twenty-four distinct muscle groups that work harmoniously, each controlled by a complex network of motor and sensory nerve pathways. This rich tapestry of connections, combined with the hand's sophisticated arrangement of bones and joints, empowers it to execute a stunning variety of movements, from the delicate precision required to thread a needle to the powerful grip needed to lift heavy objects.

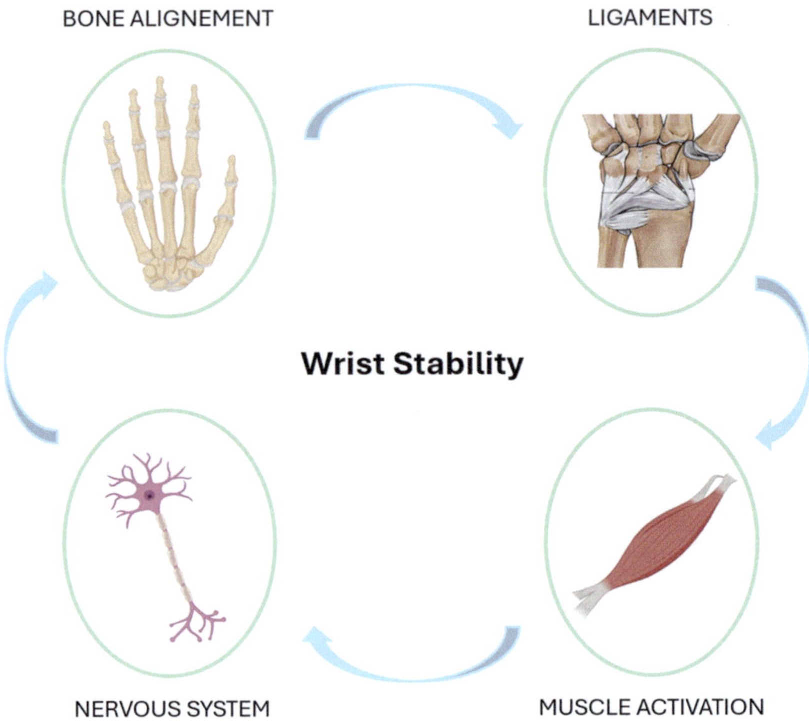

Fig. 1.1 Wrist stability: the vital alliance of muscles, bone ligaments, and nervous system coordination

The Musculoskeletal Structure

The human hand is a complex structure composed of various bones that can be categorized into two main groups: the carpus and the digits. The carpus consists of eight distinct bones that form the intricate framework of the wrist, serving as the critical connection between the forearm and the hand itself. These carpal bones are organized into two rows; the proximal row articulates with the radius and ulna, the two long bones of the forearm, facilitating movement and stability [21].

The intercarpal articulation between the two rows of carpal bones is vital in providing mobility for hand function. This remarkable arrangement of bones and their ligament attachments establishes a strong structure that effectively restricts lateral (side-to-side) and dorsal-volar (front-to-back) movements. Furthermore, this configuration allows for a broad range of wrist motions, significantly enhancing the dexterity and overall functionality of the hand. The synergy between the carpal bones and the interconnected metacarpal and phalangeal segments of the fingers showcases the intricate design that enables the hand to perform various tasks with remarkable precision [22].

The anatomical structure of each digit is fundamentally similar, except for the thumb. The metacarpals II–V are closely connected to the adjacent carpal bones in the distal row. While they can perform some flexion and extension, their independence of movement is quite limited. The shafts of the metacarpals are arched, forming the palm of the hand, and the distal ends are almost hemispherical, designed to fit the concave curvature of the proximal ends of the first phalanges [21].

The metacarpophalangeal joint displays a pattern similar to the interphalangeal joints. The center of rotation is located approximately at the center of curvature of the distal end of the proximal bone. The lateral sides of the joint surfaces are narrower and tightly bound by ligaments, resulting in minimal lateral rotation in the metacarpophalangeal joints and a complete absence of lateral movement in the phalangeal joints. Thus, the phalangeal joints function as typical hinge joints. Additionally, the thumb is unique compared to the other fingers for two main reasons: the absence of the second phalanx and the increased mobility of the carpometacarpal joint [21].

Muscles and Tendons

The intricate mechanics of hand and wrist movement are governed mainly by a complex network of muscles primarily located in the forearm. These muscles extend into tendons that weave through the wrist, ultimately anchoring themselves to the hand's various bony and ligamentous structures. The flexor muscles, responsible for bending the hand and fingers, typically originate from the medial epicondyle of the humerus or the adjacent volar surfaces of the radius and ulna. These muscles run along the inner side of the forearm, and in addition to their flexing function, they also contribute to the supination of the forearm [23].

Conversely, the extensor muscles, which facilitate the straightening of the wrist and fingers, emerge from the lateral epicondyle and extend from parts of the ulna. These muscles travel down the dorsal side of the forearm, playing a critical role in pronation—the rotation of the forearm so that the palm faces downward [24]. With its unique functionality, the thumb follows a comparable flexor-extensor structure; however, its extensors and abductors originate from the mid and distal regions of the radius and ulna, highlighting its remarkable versatility [25].

The tendons of the wrist and hand navigate through a series of bony and ligamentous guide systems that ensure smooth and efficient movement. The flexor tendons traverse a confined "tunnel" bounded on the dorsal side by the carpal bones, laterally by the greater multangular and the projection of the hamate, and on the volar side by a robust structure known as the transverse carpal ligament [23]. Similarly, the extensor tendons are directed by the dorsal carpal ligament [24]. At the same time, a sophisticated network of sheaths expertly guides both the flexor and extensor tendons through the intricate metacarpal and phalangeal regions.

In addition to these muscles, the intrinsic muscles of the hand, which possess both their origin and insertion points within the confines of the wrist and hand (except for the abductors of the thumb and little finger), are specifically adapted to facilitate the delicate movements of the fingers. These specialized muscles are crucial for abducting the digits and performing intricate opposition movements, enabling tasks like making a fist or executing a spherical grasp, among other vital functions [26, 27].

The Static Hand

The resting hand assumes a characteristic posture, a feature easily seen when the hand hangs loosely at the side. The resting wrist takes a mid-position in which the extended forearm axis is dorsiflexed 35°. It is worth noting that this is the position of the most significant prehensile force. The mid position for radial or ulnar flexion appears to be such that the metacarpophalangeal joint center of digit III lies in the extended sagittal plane of the wrist [28].

Typically, the conformation of fingers and thumb is similar to that shown for palmar prehension, the fingers being more flexed from the index to the little finger. The relations between thumb, palm, and fingers permit the grasp of a 1.75 in. cylinder crossing the palm at about 45° to the radioulnar axis [28]. Bunnell considers this "an ancestral position ready for grasping limbs, weapons, or other creatures" [29].

In thrusting or striking actions, the hand may assume fixed and rigid postures while functioning with the arm in support. These represent nonspecialized functions in which the hand serves merely as an adapted "end of the arm." The various forms include the flat of the hand, the clenched fist, the knuckle and digital support postures, and so on [28].

Hand in Motion

The second law of physics states that the acceleration of an object is directly proportional to the net force acting upon it and inversely proportional to its mass. It can be summarized by the equation $F = ma$, where F is the resultant force applied to the mass, m is the mass, and a is the acceleration. This equation assumes the hand is a mass similar to a particle mass. By definition, it only has translation and is not a rigid body where we need to account for the rotational motion. If the hand is modeled or viewed as a rigid body, we need to use $M = I$. Alpha, where M is the sum of all the moments (torques) applied to the rigid mass, I is the moment of inertia associated with the geometry of the hand being used as a model, and Alpha denotes the angular acceleration.

In the biomechanics study of the hand, several key principles can be observed:

1. Force generation: The muscles of the hand and forearm generate moving forces. For example, when you grasp an object, the muscles contract, creating a force that acts on the tendons attached to the bones in the fingers. The more force generated by the muscles (for instance, by increasing contraction), the greater the acceleration of the fingers, allowing for the precise movement needed to grasp or manipulate different objects.
2. Mass and control: Each finger and part of the hand has its mass, which must be accounted for when applying force. For instance, if you are trying to lift a heavy object, the mass of the fingers and hand contributes to how much force needs to be applied to achieve the desired acceleration upward. The heavier the object, the greater the force required to lift it, following Newton's second law.
3. Manipulation of objects: The hand can apply varying forces to different objects, considering their mass and the need for control and precision. For instance, when threading a needle, only a small force is needed to move the needle forward, but careful acceleration control is required to ensure accuracy. In contrast, picking up a heavy bag would require a more potent force and potentially faster acceleration to lift it off a surface.
4. Movement dynamics: The hand's ability to perform intricate movements, such as those needed for playing a musical instrument or typing, also illustrates the application of the second law. As fingers strike keys or strings, their forces accelerate each movement, producing the desired sound or action. The mass of each finger and the object being manipulated will affect the speed and timing of these movements.

In summary, the second law of physics, which combines translation and rotation, helps explain how generalized forces, moments, mass, and acceleration interplay in the hand's muscle movements. This allows for the wide range of motions and functions we perform every day. The hand muscles, actions, and contractile forces act all together or individually, depending on the task.

References

1. Wang M, Wu Y, Li G, et al. Articular cartilage repair biomaterials: strategies and applications. Mater Today Bio. 2024;24:100948. https://doi.org/10.1016/j.mtbio.2024.100948.
2. Liang J, Liu P, Yang X, et al. Biomaterial-based scaffolds in promotion of cartilage regeneration: recent advances and emerging applications. J Orthop Transl. 2023;41:54–62. https://doi.org/10.1016/j.jot.2023.08.006.
3. Gersoff WK. Meniscal restoration: scaffolds, transplantation, and implants. J Cartil Jt Preserv. 2023;3(1) https://doi.org/10.1016/j.jcjp.2023.100108.
4. Eschweiler J, Li J, Quack V, et al. Anatomy, biomechanics, and loads of the wrist joint. Life Basel Switz. 2022;12(2):188. https://doi.org/10.3390/life12020188.
5. Edwards WB. Modeling overuse injuries in sport as a mechanical fatigue phenomenon. Exerc Sport Sci Rev. 2018;46(4):224–31. https://doi.org/10.1249/JES.0000000000000163.
6. Bedotto RA. Biomechanical assessment and treatment in lower extremity prosthetics and orthotics: a clinical perspective. Phys Med Rehabil Clin N Am. 2006;17(1):203–43. https://doi.org/10.1016/j.pmr.2005.10.007.
7. Busuttil NA, Reid M, Connolly M, Dascombe BJ, Middleton KJ. A kinematic analysis of the upper limb during the topspin double-handed backhand stroke in

tennis. Sports Biomech. 2022;21(9):1046–64. https://doi.org/10.1080/14763141.2020.1726994.
8. Israel JS, Loushin SR, Tetzloff SU, Ellenbecker T, Kaufman KR, Kakar S. Wrist motion assessment in tennis players using three-dimensional motion capture and dynamic electromyography. J Wrist Surg. 2024;13(3):264–71. https://doi.org/10.1055/s-0043-1777024.
9. Shibata S, Kageyama M, Inaba Y, Yoshioka S, Fukashiro S. Kinetic analysis of the wrist and fingers during fastball and curveball pitches. Eur J Sport Sci. 2022;22(2):136–45. https://doi.org/10.1080/17461391.2020.1866080.
10. Padilla-Magaña JF, Peña-Pitarch E, Sánchez-Suarez I, Ticó-Falguera N. Hand motion analysis during the execution of the action research arm test using multiple sensors. Sensors. 2022;22(9):3276. https://doi.org/10.3390/s22093276.
11. Connell D, Page P, Wright W, Hoy G. Magnetic resonance imaging of the wrist ligaments. Australas Radiol. 2001;45(4):411–22. https://doi.org/10.1046/j.1440-1673.2001.00948.x.
12. Al-Ayyad M, Owida HA, De Fazio R, Al-Naami B, Visconti P. Electromyography monitoring systems in rehabilitation: a review of clinical applications, wearable devices and signal acquisition methodologies. Electronics. 2023;12(7):1520. https://doi.org/10.3390/electronics12071520.
13. Duncan SFM, Saracevic CE, Kakinoki R. Biomechanics of the hand. Hand Clin. 2013;29(4):483–92. https://doi.org/10.1016/j.hcl.2013.08.003.
14. Kim W, Ferguson VL, Borden M, Neu CP. Application of elastography for the noninvasive assessment of biomechanics in engineered biomaterials and tissues. Ann Biomed Eng. 2016;44(3):705–24. https://doi.org/10.1007/s10439-015-1542-x.
15. Shaikh HJF, Hasan SS, Woo JJ, Lavoie-Gagne O, Long WJ, Ramkumar PN. Exposure to extended reality and artificial intelligence-based manifestations: a primer on the future of hip and knee arthroplasty. J Arthroplast. 2023;38(10):2096–104. https://doi.org/10.1016/j.arth.2023.05.015.
16. Keller M, Guebeli A, Thieringer F, Honigmann P. Overview of in-hospital 3D printing and practical applications in hand surgery. BioMed Res Int. 2021;2021(1):4650245. https://doi.org/10.1155/2021/4650245.
17. Chen L, Zhou C, Jiang C, et al. Translation of nanotechnology-based implants for orthopedic applications: current barriers and future perspective. Front Bioeng Biotechnol. 2023;11:1206806. https://doi.org/10.3389/fbioe.2023.1206806.
18. Ledet EH, Liddle B, Kradinova K, Harper S. Smart implants in orthopedic surgery, improving patient outcomes: a review. Innov Entrep Health. 2018;5:41–51. https://doi.org/10.2147/IEH.S133518.
19. Longo UG, De Salvatore S, Candela V, et al. Augmented reality, virtual reality and artificial intelligence in orthopedic surgery: a systematic review. Appl Sci. 2021;11(7):3253. https://doi.org/10.3390/app11073253.
20. Verhey JT, Haglin JM, Verhey EM, Hartigan DE. Virtual, augmented, and mixed reality applications in orthopedic surgery. Int J Med Robot. 2020;16(2):e2067. https://doi.org/10.1002/rcs.2067.
21. Panchal-Kildare S, Malone K. Skeletal anatomy of the hand. Hand Clin. 2013;29(4):459–71. https://doi.org/10.1016/j.hcl.2013.08.001.
22. Kauer JM. Functional anatomy of the wrist. Clin Orthop. 1980;149:9–20.
23. Bogumill GP. Functional anatomy of the flexor tendon system of the hand. Hand Surg. 2002;7(1):33–46. https://doi.org/10.1142/s0218810402000790.
24. Wehbé MA. Anatomy of the extensor mechanism of the hand and wrist. Hand Clin. 1995;11(3):361–6.
25. von Schroeder HP, Botte MJ. Anatomy and functional significance of the long extensors to the fingers and thumb. Clin Orthop. 2001;383:74–83. https://doi.org/10.1097/00003086-200102000-00010.
26. Gupta S, Michelsen-Jost H. Anatomy and function of the thenar muscles. Hand Clin. 2012;28(1):1–7. https://doi.org/10.1016/j.hcl.2011.09.006.
27. Pasquella JA, Levine P. Anatomy and function of the hypothenar muscles. Hand Clin. 2012;28(1):19–25. https://doi.org/10.1016/j.hcl.2011.09.003.
28. Taylor CL, Schwarz RJ. The anatomy and mechanics of the human hand. Artif Limbs. 1955;2(2):22–35.
29. Bunnell S. Reconstructive surgery of the hand. Surgery. 1946;19:867.

Anatomy of the Hand

2

Isabella Trancoso and Farid Amirouche

Introduction

The hand is a remarkable structure intricately designed to perform various movements and functions essential for daily activities. As a pivotal instrument in human interaction with the environment, its anatomy comprises an elaborate network of bones, joints, muscles, tendons, ligaments, nerves, skin, and fascia. Each component is critical in ensuring mobility, dexterity, strength, and precision in grasping and manipulation. This chapter is divided into several sections, starting with the intricate anatomy of the hand and beginning with the distinct characteristics of its skin and fascia, which provide both protection and flexibility. We will explore the bones and joints that form the skeletal framework of the hand, facilitating complex movements. By understanding the anatomy and biomechanics of the hand, we can appreciate how these interconnected parts allow for the extraordinary range of functions we often take for granted.

I. Trancoso
Chicago Medical School, North Chicago, IL, USA

F. Amirouche (✉)
Department of Orthopaedic Surgery, Orthopaedic and Spine Institute, Endeavor HealthSystem (Formerly Northshore University HealthSystem), an Affiliate of the University of Chicago Pritzker School of Medicine, Skokie, IL, USA

Department of Orthopaedic Surgery, University of Illinois at Chicago, Chicago, IL, USA
e-mail: famirouche@northshore.org

Skin and Fascia

Like the skin on the rest of the body, skin on the hand and wrist is a protective barrier and has unique characteristics to ensure optimum function. The skin on the hand's palmar (or volar) side differs from the dorsal. The skin on the dorsum of the hand is thin and pliable, whereas the volar side is thickened with a heavily keratinized surface [28]. The thick skin on the palm and volar aspect of the fingers allows for grip and protection against friction [42]. The skin of the hand must also be flexible to allow for movement; creases in the volar skin, also known as flexion creases, ensure the hand can flex and extend without the skin moving too much and being damaged [7].

Another crucial component of hand anatomy is the palmar and dorsal fascia. As a continuation of the antebrachial fascia, the hand fascia is divided into separate components and compartments. The fascial components of the volar side of the hand are the flexor retinaculum, the palmar aponeurosis, and the digital sheaths. The flexor retinaculum is the most proximal component and is a dense band of connective tissue that covers the carpal bones and forms the carpal tun-

F. Amirouche et al. (eds.), *Biomechanics of the Hand*,
https://doi.org/10.1007/978-3-032-04688-8_2

nel. The carpal tunnel contains nine flexor tendons as well as the median nerve. The flexor retinaculum is anchored medially to the pisiform and hook of the hamate and laterally to the trapezium and scaphoid [29]. Continued from the flexor retinaculum is the palmar aponeurosis, a triangular-shaped fibrous membrane of the deep fascia that protects the underlying tendons, nerves, and vasculature [12]. The palmar aponeurosis is also involved in proprioception and grip ability [16]. Extending distally from the palmar aponeurosis are the fibrous digital sheaths, and these are collagenous connective tissue tunnels that enclose and protect tendons inserted on the digits [29].

The fascial components of the dorsal hand are the extensor retinaculum and the dorsal digital expansions. The extensor retinaculum is a connective tissue band that covers the dorsum of the wrist and protects the underlying extensor tendons. Due to pulley system damage, bowstringing is also prevented when the tendon deviates from its typical path [21]. The dorsal distal expansions, or extensor expansions, are triangular aponeuroses that cover the dorsum of the digits and attach the extensor tendons to the distal phalanges [29]. The extensor expansions are also the site for lumbrical and interossei insertion [41]. The fascial compartments of the hand can be divided into the thenar, hypothenar, central, interosseous, and adductor compartments. The thenar compartment holds three muscles that act on the thumb (digit 1), and the hypothenar compartment contains three muscles that act on the little finger (digit 5). The central compartment is between these two compartments and holds flexor tendons and the lumbricals [11]. The interosseous compartment can be found between the metacarpals and carries the palmar and dorsal interossei. Finally, the adductor compartment contains the adductor pollicis muscle [11].

The dorsal wrist has six fascial compartments, each separated by a longitudinal vertical septum [36]. The first compartment contains the abductor pollicis longus and extensor pollicis brevis muscles. Compartment two holds the extensor carpi radialis longus and brevis muscles, and compartment three contains the extensor pollicis longus muscles. The fourth compartment contains the extensor digitorum and extensor indicis muscles. Compartment five contains the extensor digiti minimi muscles, and compartment six carries the extensor carpi ulnaris muscles [38]. The sixth dorsal wrist compartment is more complex than the other compartments and is unique in that it has a fibro-osseous tunnel [26]. This tunnel enclosing the extensor carpi ulnaris tendon is formed superiorly and inferiorly by the two layers of the extensor retinaculum: the supra tendinous and infra tendinous [36]. Its lateral border is the sixth septum, and the medial border is a bundle of collagen fibers called the linea jugata, which inserts onto the interosseous membrane [8] Fig. 2.1.

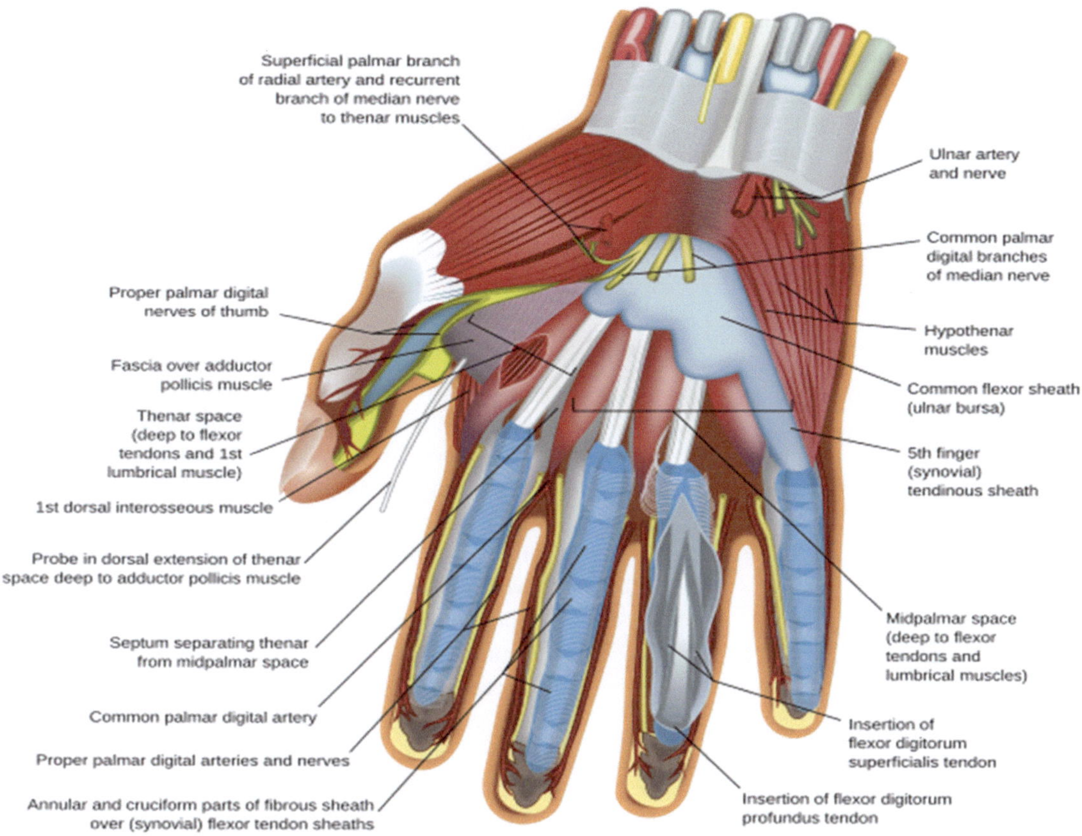

Fig. 2.1 Wrist and hand deep palm dissection. The image depicts a deep dissection of the volar aspect of the right hand. The hand fascia can show prominent muscles, arteries, nerves, and tendons. (*Wrist and hand deeper palmar dissection*, by Wilfredo Hernandez, 2008, Wikimedia (https://commons.wikimedia.org/wiki/File:Wrist_and_hand_deeper_palmar_dissection-en.svg). CC by Attribution 3.0 Unported)

Bones and Joints

The bones of the hand include the carpals, metacarpals, and phalanges. There are eight carpal bones, the small ones that comprise the wrist. They can be divided into proximal and distal rows. Moving from lateral to medial, the proximal row consists of the scaphoid, lunate, triquetrum, and pisiform, and the distal row contains the trapezium, trapezoid, capitate, and hamate [37].

The metacarpals are the five bones of the palm, and each is associated with one digit. The metacarpals articulate at their proximal end with the carpal bones and the proximal phalanges at their distal end. The phalanges are the bones that make up the digits; there are 14 of these bones. Digits 2–5 have a proximal, middle, and distal phalange, whereas the thumb has a proximal and distal phalange [29] Fig. 2.2.

Joints can be divided into joints of the wrist and the hand. The wrist joints consist of the radiocarpal, ulnocarpal, distal radioulnar, and carpocarpal joints. The radiocarpal joint articulates the distal radius and the scaphoid, lunate, and triquetrum [19]. It is an ellipsoid synovial joint and allows for flexion, extension, and radial and ulnar deviation [14]. The ulnocarpal joint is where the ulna joins with the lunate and triquetrum [15]. It aids forearm rotation while serving as a critical joint in wrist function [45]. The distal radioulnar joint (DRUJ) is the articulation between the ulnar head and the ulnar notch of the radius. Its primary function is to stabilize the movement of the radius around the stationary ulna during supination and pronation [4]. The carpocarpal/intercarpal joints are synovial joints that connect carpal bones, and while each carpocarpal joint contributes to total wrist motion, some serve a larger contribution [35].

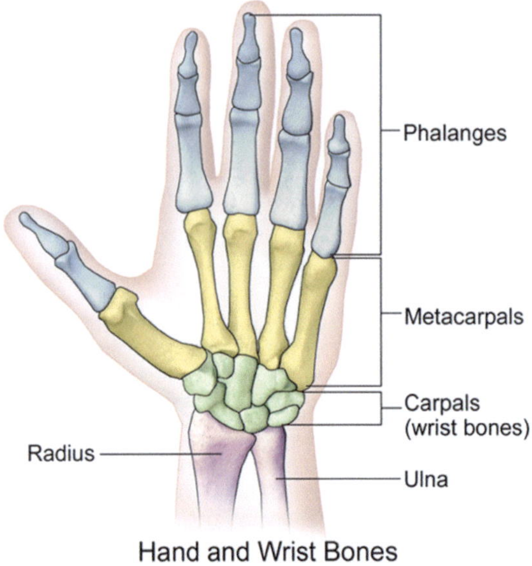

Fig. 2.2 Bones of the hand and wrist. This is a color-coded illustration of the bones of the hand and wrist. It is a view of the left hand's volar side, where all eight carpal bones are visible (the pisiform is not visible on the dorsal side). (Hand Bones, From Medical Gallery of Blausen Medical, 2014, *Wikijournal of Medicine* https://doi.org/10.15347/WJM/2014.010. CC by Attribution 3.0 Unported)

The hand joints consist of the carpometacarpal (CMC), metacarpophalangeal (MCP), proximal interphalangeal (PIP), and distal interphalangeal (DIP) joints. The CMC joint is the articulation between the distal carpals and the metacarpals. The CMC joint of the thumb is unique from the other joints as it is a saddle joint, which allows for a much greater range of motion [18]. The MCP joint is the articulation between the distal aspect of the metacarpal and each proximal phalanx, otherwise known as the "knuckles." The metacarpal head has a convex surface shape and is positioned in the concave shape of the proximal phalanx. The MCP allows for flexion, extension, abduction, adduction, and restricted circumduction, and the thumb is unique with more range of motion as a hinge MCP joint [6]. The interphalangeal joints are synovial hinge joints between the phalanges and allow for flexion and extension of the digits [29]. Digits 2–5 have two interphalangeal joints: the PIP joint is between the proximal and middle phalanges, and the DIP joint is between the middle and distal phalanges. The thumb, however, has one interphalangeal joint as there are only two phalanges, so the joint is between the proximal and distal phalanx [20] Fig. 2.3.

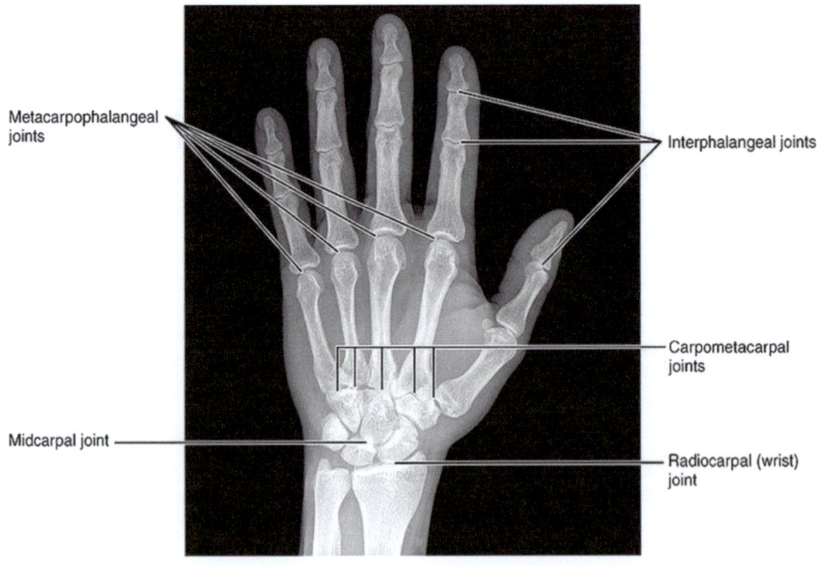

Fig. 2.3 Radiograph of the hand. This image is a radiograph of the hand with labeled joints within the wrist and hand. This is a view of the volar side of the right hand. (*Radiograph of Hand*, From Openstax by Rice University, 2013, Wikimedia https://commons.wikimedia.org/wiki/File:814_Radiograph_of_Hand.jpg. CC by Attribution 3.0 Unported)

Muscles and Tendons

Many muscles contribute to wrist and hand movement. These can be categorized into the extrinsic muscles, which originate from the forearm and control crude movements of the wrist and hand, and the intrinsic muscles, which originate and insert within the hand and are responsible for precise and fine motor movements [22]. The extrinsic muscles of the wrist and hand can be further subdivided into their actions of flexors and extensors. The flexors of the wrist include the flexor carpi ulnaris (FCU), flexor carpi radialis (FCR), and palmaris longus (PL), and all three muscles originate from the medial epicondyle of the humerus [24]. The FCU also originates from the ulnar head, which is inserted on the pisiform, hook of the hamate, and volar side of the fifth metacarpal to flex and adduct the wrist. The FCR inserts on the volar side of the second metacarpal, and the PL tendon inserts on the palmar aponeurosis and flexor retinaculum [27]. The extensors of the wrist include the extensor carpi radialis longus (ECRL), extensor carpi radialis brevis (ECRB), and extensor carpi ulnaris (ECU). The ECRL originates from the lateral supracondylar ridge of the humerus and inserts on the dorsal side of the second metacarpal [23]. The ECRB and ECU both originate from the lateral epicondyle of the humerus, and the ECRB inserts on the dorsal side of the third metacarpal, whereas the ECU inserts on the dorsum of the fifth metacarpal [23]. The FCR and ECRL also deviate the wrist radially, while the FCU and ECU are responsible for ulnar deviation [24].

The flexors of the hand (digits) include the flexor digitorum superficialis (FDS) and the flexor digitorum profundus (FDP), both of which originate from the medial epicondyle of the humerus [31]. The FDS inserts on the middle phalanx of digits 2–5 and flexes the PIP joint, and the FDP inserts on the distal phalanx of digits 2–5 and flexes the DIP joint [31]. The extensors of the digits include the extensor digitorum communis (EDC), extensor indices propius (EIP), and extensor digiti minimi (EDM). The EDC originates from the lateral epicondyle of the humerus and inserts into the extensor hood (covering of MCP joint) of digits 2–5 [44]. The EIP originates from the posterior surface of the ulna and inserts on the proximal phalange of the second digit; its primary function is to extend the index finger [24]. The EDM originates from the lateral epicondyle of the humerus and inserts into the extensor hood of the fifth digit, functioning to extend the little finger [44].

The movers of the thumb include the flexor pollicis longus (FPL), extensor pollicis longus (EPL), extensor pollicis brevis (EPB), and abductor pollicis longus (APL). The FPL originates from the radius, ulna, and interosseous membrane in the forearm and is inserted on the distal phalanx of the thumb to flex the IP joint [1]. The EPL, EPB, and APL originate from the dorsal ulna [24]. The EPL inserts on the distal phalanx of the thumb, and the EPB inserts on the proximal phalanx of the thumb; both work to extend the thumb [1]. The APL inserts on the base of the first metacarpal and produces abduction of the thumb [10].

Four muscle groups comprise the intrinsic hand muscles: the thenar, hypothenar, lumbrical, and interossei muscles. The thenar muscles include the opponents pollicis (OP), abductor pollicis brevis (APB), and flexor pollicis brevis (FPB), and all three muscles originate from the volar side of the scaphoid and trapezium as well as the flexor retinaculum [24]. The OP inserts the distal aspect of the first metacarpal and moves the thumb across the palm toward the fifth digit, or opposition [30]. The APB and FPB both insert on the proximal phalanx of the thumb, and the APB is responsible for abduction at the CMC and MCP while the FPB flexes the thumb at the MCP joint [9]. It should be noted that the adductor pollicis is responsible for thumb adduction but is not part of the thenar eminence [1].

The hypothenar muscles consist of the opponens digit minimi (ODM), flexor digiti minimi brevis (FDMB), and abductor digiti minimi (ADM). These muscles originate from the pisiform, hamate, and flexor retinaculum and are inserted on the proximal phalanx of the fifth digit to move the little finger [24]. The ODM allows for movement of the little finger across the palm, the FDMB flexes it at the MCP joint, and the ADM is responsible for slight finger abduction [9].

The lumbricals are worm-like muscles that aid in fine control of finger movements and there are four in each hand. They originate from the FDP tendons and attach to the extensor expansions distally [41]. The lumbricals flex the digits at the MCP joint and extend at the PIP and DIP joints [24]. The interosseous muscles are located between the metacarpals, with three palmar interossei and four dorsal interossei. The palmar interossei are responsible for finger adduction, while the dorsal interossei abduct the fingers [41]. Like the lumbricals, the interossei flex the digits at the MCP joint and extend at the IP joints [40] Fig. 2.4.

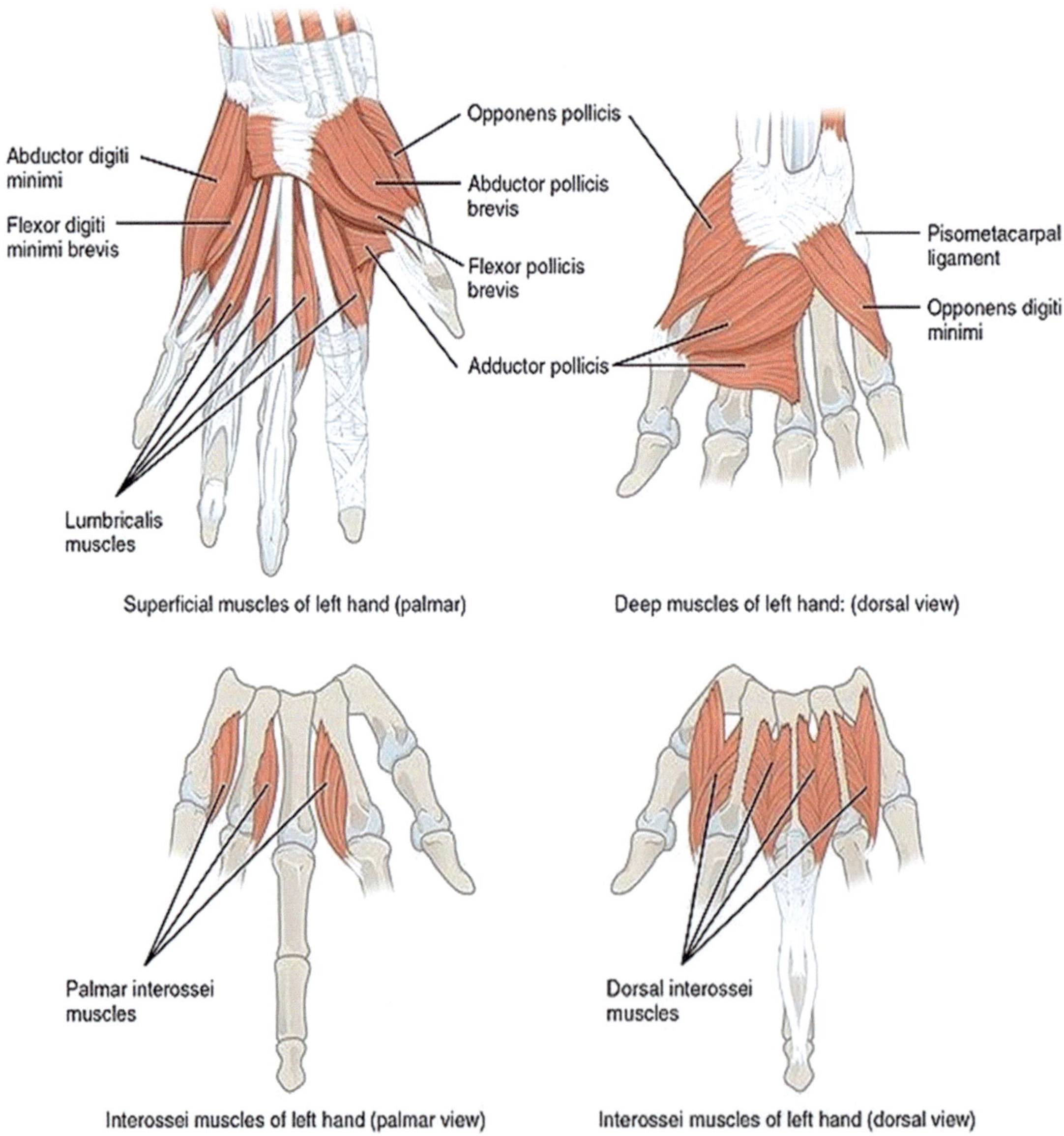

Fig. 2.4 Intrinsic hand muscles. The image depicts all four intrinsic hand muscle groups. The left illustrations are from a volar view of the left hand, and the right is from a dorsal view of the left hand. (*Intrinsic Muscles of the Hand*, From Openstax by Rice University, 2016, Wikimedia https://commons.wikimedia.org/wiki/File:1121_Intrinsic_Muscles_of_the_Hand.jpg. CC by 4.0)

Ligaments

While ligaments and tendons are both fibrous connective tissue crucial to movement, the difference is that ligaments connect bone to bone, and tendons attach muscle to bones or other structures [3].

Many ligaments of the hand and wrist function to stabilize, protect, and allow for range of motion. The collateral ligaments of the hand can be found on the sides of the MCP joints, originating from the sides of the metacarpal head and inserted into the proximal phalanx [2]. The volar plate is a thick ligament in the hand that connects the proximal and middle phalanges, preventing hyperextension of the digits [32].

The ligaments of the wrist can be divided into intrinsic (originate and insert only on carpal bones) and extrinsic (can attach to non-carpal bones). An extrinsic ligament is the collateral ligament, which consists of the radial collateral ligament (RCL) and ulnar collateral ligament (UCL). These ligaments support and stabilize the medial and lateral aspects of the wrist and connect to the forearm [39]. Other extrinsic ligaments would be the volar and dorsal wrist ligaments. The volar ligaments include the volar radiocarpal ligaments (supports supination), the volar radioulnar ligament (stabilizes the DRUJ), and the volar carpal ligaments (support the flexor tendons) [25]. The dorsal radiocarpal ligament is essential for stability of the dorsal wrist and movement, especially pronation [43]. The principal intrinsic wrist ligaments are the lunotriquetral interosseous ligament (LTIL) and the scapholunate interosseous ligament (SLIL). The LTIL connects the volar and dorsal surfaces of the lunate and triquetrum, and the SLIL is a C-shaped ligament attaching the volar, proximal, and dorsal aspects of the scaphoid and lunate bones [33].

Nerves

There are three primary nerves of the hand and wrist: the ulnar nerve, the radial nerve, and the median nerve. All three have a sensory and motor component. The ulnar nerve enters the hand via Guyon's Canal, which is formed by the pisiform and hook of the hamate [5]. It then divides into the superficial and deep branches, the superficial branch acting primarily as sensory innervation and the deep branch supplying mainly motor innervation to the hypothenar, lumbrical, and interosseous muscles [34]. The deep branch of the radial nerve innervates the hand and wrist extensors, and the superficial branch is purely sensory [17]. Finally, the median nerve enters the hand via the carpal tunnel and then divides into the recurrent branch and the palmar digital nerves [13]. The motor component innervates the flexor and pronators, except for the FCU and part of the FDP [34]. The sensory innervation to the skin by the ulnar, radial, and median nerves is seen in Fig. 2.5.

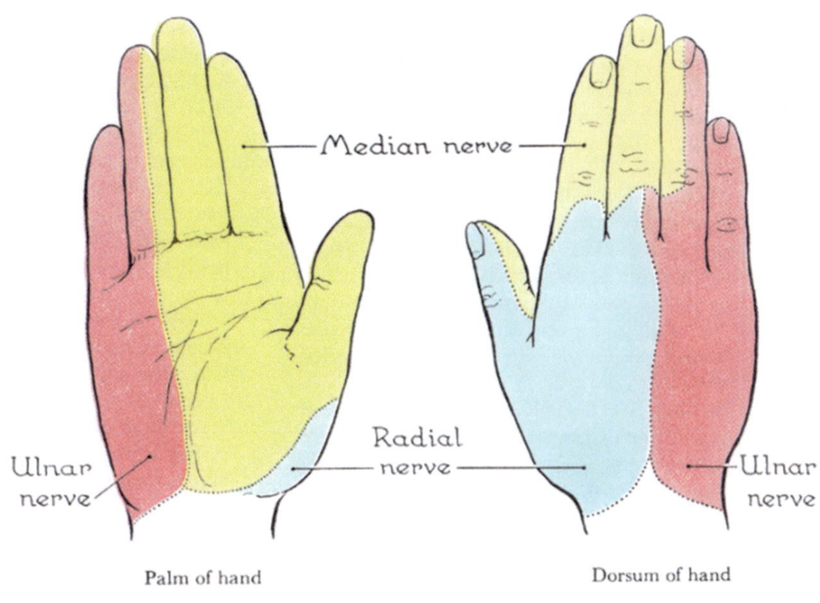

Fig. 2.5 Cutaneous innervation of the hand. Cutaneous innervation of the right hand. The blue region denotes sensory innervation by the radial nerve, yellow by the median nerve, and red by the ulnar nerve. (Reproduced from *Atlas of Anatomy* by J.C.B. Grant, 1962. Public domain image via Wikimedia Commons (https://commons.wikimedia.org/wiki/File:Grant_1962_84.png). No changes were made

Conclusion

Understanding the fundamentals of hand anatomy fosters an appreciation for how various components work in harmony to enable movement and function. The intricate system formed by the interconnectedness of the skin, fascia, bones, muscles, ligaments, and nerves in the hand is often overlooked or poorly understood. Each element plays a distinct role, contributing to our hands' overall functionality and capabilities. It is crucial to advance clinical research on hand mechanics and conduct further studies at anatomical and biomechanical levels. This approach will deepen our understanding of the diverse range of hand movements, ultimately leading to improved diagnosis and treatment options.

References

1. Acosta JR, Graefe SB, Varacallo MA. Anatomy, shoulder and upper limb, hand adductor pollicis. In: StatPearls. StatPearls Publishing; 2023. http://www.ncbi.nlm.nih.gov/books/NBK526059/.
2. Allison DM. Anatomy of the collateral ligaments of the proximal interphalangeal joint. J Hand Surg Am. 2005;30(5):1026–31. https://doi.org/10.1016/j.jhsa.2005.05.015.
3. Amiel D, Frank C, Harwood F, Fronek J, Akeson W. Tendons and ligaments: a morphological and biochemical comparison. J Orthop Res. 1983;1(3):257–65. https://doi.org/10.1002/jor.1100010305.
4. Arias DG, Black AC, Varacallo MA. Anatomy, shoulder and upper limb, distal radio-ulnar joint. In: StatPearls. StatPearls Publishing; 2023. http://www.ncbi.nlm.nih.gov/books/NBK547720/.
5. Becker RE, Manna B. Anatomy, shoulder and upper limb, ulnar nerve. In: StatPearls. StatPearls Publishing; 2023. http://www.ncbi.nlm.nih.gov/books/NBK499892/.
6. Benson DC, Graefe SB, Varacallo MA. Anatomy, shoulder and upper limb, metacarpophalangeal joints. In: StatPearls. StatPearls Publishing; 2023. http://www.ncbi.nlm.nih.gov/books/NBK538428/.
7. Bugbee WD, Botte MJ. Surface anatomy of the hand: the relationships between palmar skin creases and osseous anatomy. Clin Orthop Relat Res. 1993;1976–2007(296):122.
8. Byrd JN, Sasor SE, Chung KC. Extensor carpi ulnaris subluxation. Hand Clin. 2021;37(4):487–91. https://doi.org/10.1016/j.hcl.2021.06.005.
9. Dawson-Amoah K, Varacallo MA. Anatomy, shoulder and upper limb, hand intrinsic muscles. In: StatPearls. StatPearls Publishing; 2023. http://www.ncbi.nlm.nih.gov/books/NBK539810/.
10. Deivasigamani S, Azad A, Yang SS. The variable insertional anatomy of the abductor pollicis longus: functional relevance and relationship to adjacent thumb extensors. Hand (New York, NY). 2023;18(1):145–52. https://doi.org/10.1177/1558944721999734.
11. DiFelice A, Seiler JG, Whitesides TE. The compartments of the hand: an anatomic study. J Hand Surg Am. 1998;23(4):682–6. https://doi.org/10.1016/S0363-5023(98)80055-5.
12. Doyle JR. Anatomy and function of the palmar aponeurosis pulley. J Hand Surg. 1990;15(1):78–82. https://doi.org/10.1016/S0363-5023(09)91109-1.
13. Dydyk AM, Negrete G, Sarwan G, Cascella M. Median nerve injury. In: StatPearls. StatPearls Publishing; 2023. http://www.ncbi.nlm.nih.gov/books/NBK553109/.
14. Erwin J, Varacallo MA. Anatomy, shoulder and upper limb, wrist joint. In: StatPearls [Internet]. StatPearls Publishing; 2023. https://www.ncbi.nlm.nih.gov/sites/books/NBK534779/.
15. Eschweiler J, Li J, Quack V, Rath B, Baroncini A, Hildebrand F, Migliorini F. Anatomy, biomechanics, and loads of the wrist joint. Life. 2022;12(2):2. https://doi.org/10.3390/life12020188.
16. García-Martínez I, García-Mesa Y, García-Piqueras J, Martínez-Pubil A, Cobo JL, Feito J, García-Suárez O, Vega JA. Sensory innervation of the human palmar aponeurosis in healthy individuals and patients with palmar fibromatosis. J Anat. 2022;240(5):972–84. https://doi.org/10.1111/joa.13609.
17. Glover NM, Black AC, Murphy PB. Anatomy, shoulder and upper limb, radial nerve. In: StatPearls. StatPearls Publishing; 2023. http://www.ncbi.nlm.nih.gov/books/NBK534840/.
18. Gunther SF. The carpometacarpal joints. Orthop Clin N Am. 1984;15(2):259–77. https://doi.org/10.1016/S0030-5898(20)31314-6.
19. Helms JT, Maldonado KA, Burns B. Anatomy, shoulder and upper limb, hand radiocarpal joint. In: StatPearls [internet]. StatPearls Publishing; 2023. https://www.ncbi.nlm.nih.gov/sites/books/NBK539744/.
20. Kamnerdnakta S, Huetteman HE, Chung KC. Complications of proximal interphalangeal joint injuries: prevention and treatment. Hand Clin. 2018;34(2):267–88. https://doi.org/10.1016/j.hcl.2017.12.014.
21. Komutrattananont P, Gumpangseth T, Omokawa S, Mahakkanukrauh P. Cadaveric study of extensor retinaculum of the wrist. J Anat Soc India. 2019;68(4):299. https://doi.org/10.4103/JASI.JASI_84_19.
22. Li ZM, Zatsiorsky VM, Latash ML. Contribution of the extrinsic and intrinsic hand muscles to the moments in finger joints. Clin Biomech. 2000;15(3):203–11. https://doi.org/10.1016/S0268-0033(99)00058-3.

23. Lieber RL, Fazeli BM, Botte MJ. Architecture of selected wrist flexor and extensor muscles. J Hand Surg Am. 1990;15(2):244–50. https://doi.org/10.1016/0363-5023(90)90103-x.
24. Lifchez SD, Cho BH, Brunicardi F, Andersen DK, Billiar TR, Dunn DL, Kao LS, Hunter JG, Matthews JB, Pollock RE, editors. Surgery of the hand and wrist. Schwartz's principles of surgery, vol. 11e. McGraw-Hill Education; 2019. https://accessmedicine-mhmedical-com.rosalindfranklin.idm.oclc.org/content.aspx?bookid=2576&sectionid=216217370
25. Logan SE, Nowak MD. Intrinsic and extrinsic wrist ligaments: biomechanical and functional differences. ISA Trans. 1988;27(3):37–41.
26. Lui TH, Lau AYC. Endoscopic ganglionectomy and release of the sixth extensor compartment. Arthrosc Tech. 2019;8(2):e111–5. https://doi.org/10.1016/j.eats.2018.10.002.
27. Mathew AJ, Sukumaran TT, Joseph S. Versatile but temperamental: a morphological study of palmaris longus in the cadaver. J Clin Diagn Res : JCDR. 2015;9(2):AC01–AC03. https://doi.org/10.7860/JCDR/2015/11212.5542.
28. McLean WHI, Epithelial Genetics Group. Genetic disorders of palm skin and nail. J Anat. 2003;202(1):133–41. https://doi.org/10.1046/j.1469-7580.2003.00141.x.
29. Morton DA, Foreman K, Albertine KH, editors. The big picture: gross anatomy, medical course & step 1 review. 2nd ed. McGraw-Hill Education. Hand; 2018. https://accessmedicine-mhmedical.com.rosalindfranklin.idm.oclc.org/content.aspx?bookid=2478&sectionid=202021641
30. Nguyen JD, Duong H. Anatomy, shoulder and upper limb, hand opponens pollicis muscle. In: StatPearls. StatPearls Publishing; 2023. http://www.ncbi.nlm.nih.gov/books/NBK546603/.
31. Okafor L, Varacallo MA. Anatomy, shoulder and upper limb, hand flexor digitorum superficialis muscle. In: StatPearls. StatPearls Publishing; 2022. http://www.ncbi.nlm.nih.gov/books/NBK539723/.
32. Pattni A, Jones M, Gujral S. Volar plate avulsion injury. Eplasty. 2016;16:ic22.
33. Rajan PV, Day CS. Scapholunate interosseous ligament anatomy and biomechanics. J Hand Surg Am. 2015;40(8):1692–702. https://doi.org/10.1016/j.jhsa.2015.03.032.
34. Rapp FA, Soos MP. Anatomy, shoulder and upper limb, hand cutaneous innervation. In: StatPearls. StatPearls Publishing; 2022. http://www.ncbi.nlm.nih.gov/books/NBK544247/.
35. Seradge H, Owens W, Seradge E. The effect of intercarpal joint motion on wrist motion: are there key joints? An in vitro study. Orthopedics. 1995;18(8):727–32. https://doi.org/10.3928/0147-7447-19950801-07.
36. Taleisnik J, Gelberman RH, Miller BW, Szabo RM. The extensor retinaculum of the wrist. J Hand Surg Am. 1984;9(4):495–501. https://doi.org/10.1016/s0363-5023(84)80100-8.
37. Tang A, Varacallo M. Anatomy, shoulder and upper limb, hand carpal bones. In: StatPearls. Treasure Island: StatPearls Publishing; 2023.
38. Thompson ST, Wehbé MA. Extensor physiology in the hand and wrist. Hand Clin. 1995;11(3):367–71.
39. Türker T, Sheppard JE, Klauser AS, Johnston SS, Amerongen H, Taljanovic MS. The radial and ulnar collateral ligaments of the wrist are true ligaments. Diagn Int Radiol (Ankara, Turkey). 2019;25(6):473–9. https://doi.org/10.5152/dir.2019.19036.
40. Valenzuela M, Bordoni B. Anatomy, shoulder and upper limb, hand dorsal interossei muscle. In: StatPearls. StatPearls Publishing; 2023. http://www.ncbi.nlm.nih.gov/books/NBK536922/.
41. Valenzuela M, Launico MV, Varacallo MA. Anatomy, shoulder and upper limb, hand lumbrical muscles. In: StatPearls. StatPearls Publishing; 2023. http://www.ncbi.nlm.nih.gov/books/NBK534876/.
42. Valle M, Zamorani MP. Skin and subcutaneous tissue. In: Bianchi S, Martinoli C, editors. Ultrasound of the musculoskeletal system. Springer; 2007. p. 19–43. https://doi.org/10.1007/978-3-540-28163-4_2.
43. Viegas SF, Yamaguchi S, Boyd NL, Patterson RM. The dorsal ligaments of the wrist: anatomy, mechanical properties, and function. J Hand Surg Am. 1999;24(3):456–68. https://doi.org/10.1053/jhsu.1999.0456.
44. von Schroeder HP, Botte MJ. Functional anatomy of the extensor tendons of the digits. Hand Clin. 1997;13(1):51–62.
45. Williams CS, Jupiter JB. The painful ulnocarpal joint. diagnosis and therapy. Der Orthopade. 1993;22(1):36–45.

Hand and Wrist Biomechanics Basics

3

Kristi T. Nguyen, Mark H. Gonzalez, and Farid Amirouche

Wrist Anatomy

Bones of the Wrist and Hand

The wrist and the hand are collectively composed of 27 bones. Of these 27 bones, 8 carpal bones make up the wrist (carpus); 5 metacarpal bones that make up the palm (metacarpus), distal to the wrist; and 14 phalanges that make up the fingers and thumb, distal to the palm.

In the wrist, the eight bones are organized into two rows: the proximal carpal row (PCR), including the scaphoid, lunate, triquetrum, and pisiform, and the distal carpal row (DCR), including trapezium, trapezoid, capitate, and hamate. These two rows are functionally distinct and operate as independent functional units. In the DCR, the bones are tightly connected, and there is little movement between them. In contrast, the bones of the PCR have considerably more movement between bones, which characterizes its function as a unit between the radius and DCR during wrist motion [1]. These unique anatomic considerations of the carpus form the basis of the biomechanics of wrist and hand movement.

Joints of the Wrist and Hand

The wrist consists of three joints:

1. The radiocarpal joint connects the PCR to the forearm's radius.
2. The ulnocarpal joint connects the lunate and triquetrum to the forearm's ulna.
3. The distal radioulnar joint is an articulation between the radius and ulna of the forearm.

In the palm, the metacarpus has five bones that articulate proximally with the carpus, comprising the carpometacarpal (CMC) joints, and distally with the phalanges, comprising the metacarpophalangeal (MCP) joints.

There are two joints per finger in the fingers, except for the thumb, which has one joint. The thumb only has one joint in the fingers: the proximal interphalangeal (PIP) joint. The remaining fingers have PIP joints and distal interphalangeal (DIP) joints. Compared to the remaining fingers, another anatomic difference of the thumb

K. T. Nguyen · M. H. Gonzalez
Department of Orthopaedic Surgery, University of Illinois at Chicago, Chicago, IL, USA

F. Amirouche (✉)
Department of Orthopaedic Surgery, Orthopaedic and Spine Institute, Endeavor HealthSystem (Formerly Northshore University HealthSystem), an Affiliate of the University of Chicago Pritzker School of Medicine, Skokie, IL, USA

Department of Orthopaedic Surgery, University of Illinois at Chicago, Chicago, IL, USA
e-mail: famirouche@northshore.org

F. Amirouche et al. (eds.), *Biomechanics of the Hand*,
https://doi.org/10.1007/978-3-032-04688-8_3

includes its resting position, which lays in a different plane than the other fingers, being situated laterally and anteriorly [2].

Soft Tissue That Stabilizes the Wrist and Hand

While the wrist's stability relies heavily on the independent morphologies of its carpal bones and their specific articulations, the extrinsic and intrinsic ligaments play a significant role. These ligaments, crucial components of the wrist anatomy, provide stability at a wide range of loads, underscoring their immense significance in the biomechanics of the wrist and hand.

The intrinsic ligaments of the wrist are internal stabilizers that stabilize the wrist between carpal bones. Two important intrinsic ligaments include the scapholunate interosseous (SL) and the lunotriquetral (LT) interosseous ligaments. Disruptions to these ligaments may lead to carpal instability, with SL dissociation being the most common pattern. Anatomically, carpal instability often leads to widening the space between the carpal bones, thereby altering the wrist's biomechanical stability and ability to transmit forces and bear load. Injury to these stabilizing ligaments may transmit more force to the cartilage of the joints in the hand and wrist. These ligament injuries may thus predispose individuals to arthritic changes in the hand and wrist [3]. Clinically, ligament injuries typically manifest as wrist sprains, which compromise joint stability and thus lead to uneven weight-bearing on the cartilage, increasing the risk of rapid arthritic changes.

The extrinsic ligaments course between the carpus and the radius or metacarpals. Their precise kinematic and biomechanical functions are poorly understood [4]. However, the different positions of the various extrinsic ligaments are strained differently depending on the planes of wrist motion.

Biomechanics of the Wrist

Range of Motion, Axes of Movement, and Force Transmission

Wrist motion depends on articulation and movement between the carpal bones, primarily the PCR and the radius. PCR motion is elicited by mechanical forces between their articulations with the radius and DCR, as there are no muscles directly attached to the PCR that cause wrist motion. Thus, the motion of the carpal bones is passive and a result of mechanical forces that balance both motion and load [5]. During wrist motion, there is appreciable movement between the PCR and radius. The PCR and DCR also experience movement, albeit less than that of the former. Lastly, there is minimal to no movement between the DCR and PCR.

The wrist primarily operates in two planes of motion, which allows for flexion and extension of the wrist in the flexion-extension axis (FE axis) and radial and ulnar deviation of the wrist in the radioulnar deviation axis (RUD axis) [6]. These two planes of motion are combined to allow for circumduction of the wrist or "circular" movements of the wrist. Circumduction provides for the more complex movements of the wrist, which are necessary for activities of daily living [7]. A study in a healthy wrist joint concluded that carpal bone movement is the same regardless of active or passive wrist movement [8].

While wrist motion is initiated by contracture of muscles in the forearm, force transmission across the carpal bones informs the movement patterns that the wrist can make. Many theories outline the mechanism of intercarpal and wrist motion, including two main theories called the row and column theories [4]. The row theory follows the anatomic arrangement of the carpus into PCR and DCR, as aforementioned. In contrast, the column theory describes three longitudinal columns created by the bones of the carpus. With these theories in mind, force transmission shows that most force is loaded onto the radiocarpal joint at 80%, and the remaining 20% is transmitted to the ulnocarpal joint [9]. As most force transmission occurs at the radiocarpal and ulnocarpal joints, distal radius fractures (DRFs) significantly alter the standard transmission of forces, particularly to the radiocarpal joint. With DRFs, the force increases the load through the ulna and, consequently, the ulnocarpal joint [10].

The wrist also plays a vital role in load transmission from the forearm to the hand. The scaphoid and lunate bones in the wrist concentrate force during gripping and weight-bearing activations. In extension, weight is shifted radially, which concentrates and is transmitted through the scaphoid [11, 12]. In fact, in neutral position and flexion, the scaphoid bears more weight than the lunate, though force is almost equal between the two [4] Fig 3.1.

The hand requires a harmonious interplay between muscles of the forearm, tendons, bones, joints of the wrist, palm, and hand, and nerves to allow for a wide range of maneuvers, including manipulation via gripping and grasping, sensation, and communication. Like the wrist, the hand joints are load-bearing joints, which transmit force from the hand during gripping and grasping to the radius and ulna of the forearm. The two significant categories of hand maneuvers include the grasp and pinch maneuvers. Grasping involves transmitting force via the entire hand, using the whole hand to enclose an object. On the contrary, pinching involves precision movements between the fingers and thumb.

There are multiple pinch maneuvers, including the precision pinch, oppositional pinch, and key pinch. As for grasp maneuvers, there are chuck grip, hook grip, power grasp, and span grasp [13]. These different maneuvers differ in the spatial organization of the fingers and thumb and for flexion and extension of specific interphalangeal joints (Fig. 3.2). For example, the precision pinch transmits force to the distal radius by flexing the thumb and index finger PIP joints. In contrast, the opposition pinch maintains force on the thumb by extending both the thumb and index finger PIP joints. Notably, stability in hand biomechanics is typically maintained by thenar muscles and not through articulations, unlike the stability achieved through the joints in the wrist.

In the hand, torque movements are initiated by the intrinsic and extrinsic muscles of the hands. The extrinsic muscles, including the flexor digitorum profundus (FDP) and flexor digitorum superficialis (FDS), generate gross torque movements that allow the hand to grasp objects. On the other hand, the intrinsic muscles, including the lumbrical and interossei muscles, are involved in precision movements when grasping. Tendon injury and subsequent repair can alter torque movements by affecting the tension–length relationship. When a tendon is repaired, scar tissue can reduce the ability of the tendon to glide and thus impact the tension–length relationship. Changes to the tension–length relationship will reduce the torque extrinsic hand muscles can produce, leading to decreased finger flexion and gripping ability.

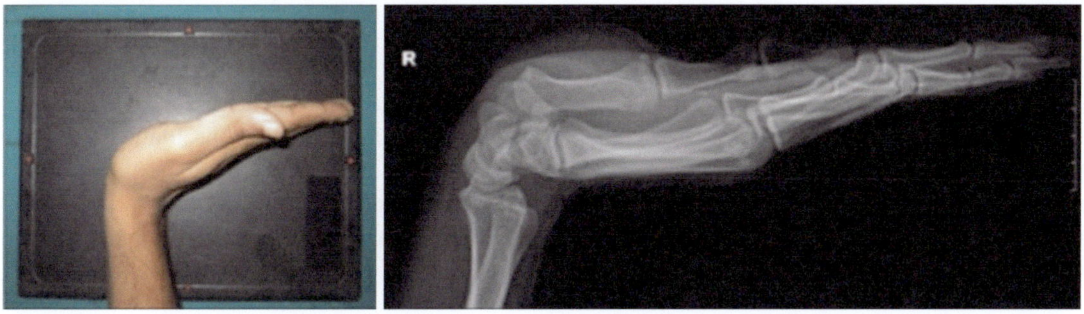

Fig. 3.1 Wrist extension concentrates force onto the scaphoid bone [12]

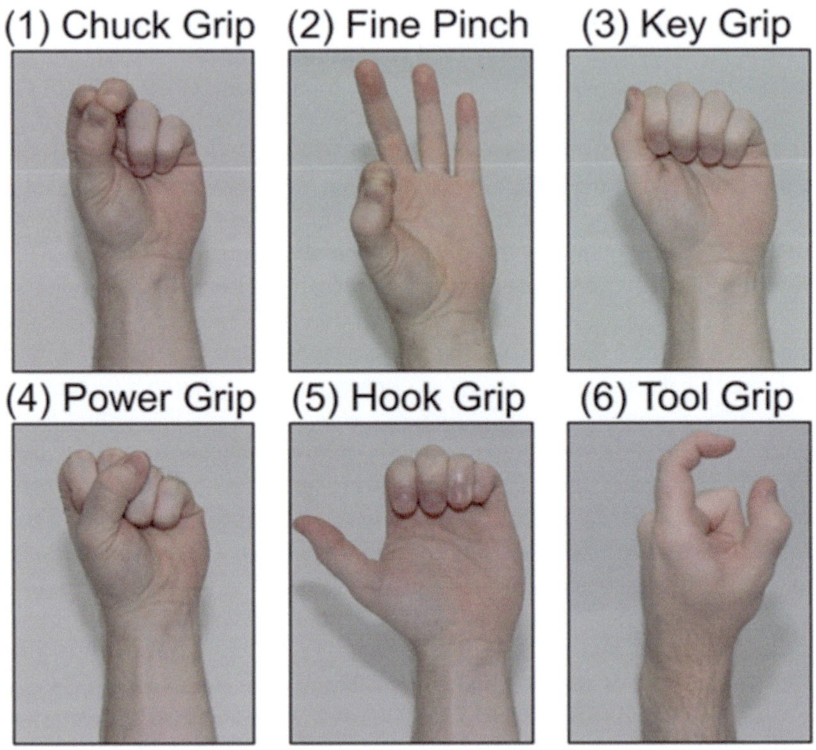

Fig. 3.2 Pinch maneuvers are used in daily living activities [14]

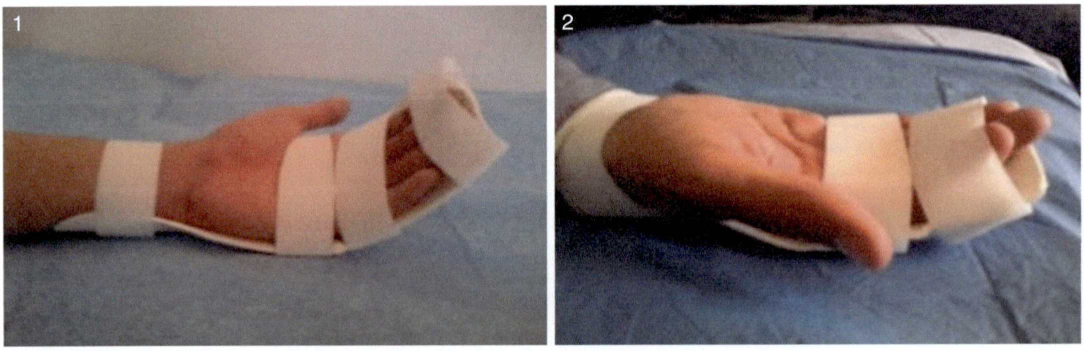

Fig 3.3 A wrist brace that immobilizes both the wrist and fingers [15]

Similarly, injuries and conditions requiring finger splinting can decrease finger range of motion, gripping, and grasping abilities. In many splints, both the wrist and hands are immobilized (Fig. 3.3). In finger splints specifically, the MCP joints are typically flexed to minimize movement at the PIP and DIP joints, thereby reducing tension on tendons. However, prolonged splinting may result in flexion contractures of the flexed joints, causing permanent torque alteration generated by the movements of these joints.

Wrist and Hand Biomechanical Models to Optimize Clinical Outcomes

External Modulators of Hand and Wrist Biomechanics

External modulators impact hand and wrist biomechanics by altering load transmission and joint alignment. External modulators include hand tools, like those used in construction or daily tasks, and sports equipment, like a bat or a racket, or even sports balls, like the handling of a football or basketball. These external modulators often transmit additional forces to the force generated in the hand and wrist to execute motion. In ball-handling sports like basketball and handball, the fingers take much of the force and rely on maximal grip strength. Increasing finger length is associated with maximal handgrip strength [16].

Load and Stabilization Restoration with Hand/Wrist Surgery and Rehab

The interplay of wrist and hand anatomy and biomechanics is integral for effectively treating and rehabilitating wrist and hand injuries and pathologies. Biomechanical models of the wrist and hand can be used explicitly as clinical aids that guide appropriate conservative or surgical treatment of wrist injuries, specifically to optimize load bearing and stabilization during hand and wrist movement. For example, biomechanical properties of force and movement can inform the development of braces and splints to restore physiologic anatomy and force distribution in acute and chronic hand and wrist conditions. Similarly, recovery and rehabilitation can be improved with assistive devices that restore anatomy or physiologic movement by concentrating forces on specific bones or joints in the hand and wrist.

Thus, the underlying goals of the biomechanical wrist and hand models are to inform surgical and nonsurgical treatments so that biomechanical motions are restored, and patients can have optimal hand use. A stable wrist is needed to bear load to perform grip and grasp maneuvers in the hand. Similarly, a stable wrist is necessary to maintain smooth and controlled wrist movements. By restoring anatomy and thus biomechanical forces distributed across the wrist and hand, surgeons can maintain optimal function in patients with wrist and hand pathologies.

Biomechanical Applications That Aid Clinical Outcomes

To improve clinical outcomes of wrist and hand pathologies, we can leverage biomechanical models and tools to assess for physiologic restoration and rehabilitation. These tools include grip dynamometers and pressure sensors, which measure grip strength and force distribution. These can be utilized perioperatively to assess status before surgical intervention and progress postoperatively. These tools can also guide safe physical therapy progression and prevent force overload during different parts of the healing stage. These tools will provide personalized patient care and improve hand and wrist outcomes.

Conclusion

The wrist and hand comprise a complex network of bones, ligaments, and muscles that facilitate precise movements and transmit large forces. From grasping objects to undertaking complex motions, the function of the hand and wrist is critical for daily activities. Whether it be injury or external modulation, such as sports equipment use, hand and wrist disruption can significantly impact performance and quality of life.

Understanding wrist and hand biomechanics allows for more effective clinical interventions, guiding surgical and nonsurgical approaches that restore stability, load distribution, and, ultimately, function. External modulators like braces and splints can be individualized for each patient to optimize force transmission and rehabilitation. Other assistive tools like grip dynamometers enable clinicians to track patient progress and customize treatment strategies.

With biomechanical models, healthcare professionals can enhance patient care, ensuring conservative management and surgical reconstruction align with the body's natural mechanics. Integrating these models into clinical decision-making improves outcomes and advances the precision of rehabilitation and recovery in individuals with wrist and hand conditions.

References

1. Wolfe SW, Garcia-Elias M, Kitay A. Carpal instability nondissociative. J Am Acad Orthop Surg. 2012;20(9):575–85. https://doi.org/10.5435/JAAOS-20-09-575.
2. Nichols DS, Oberhofer HM, Chim H. Anatomy and biomechanics of the thumb carpometacarpal joint. Hand Clin. 2022;38(2):129–39. https://doi.org/10.1016/j.hcl.2021.11.001.
3. Kim S, Nyaaba W, Mzeihem M, Mejia A, Gonzalez M, Amirouche F. Wrist kinetics post-Scapholunate dissociation: experimental and computational analysis of Scapholunate interosseous ligament injury. J Hand Surg Global Online. 2024;7 https://doi.org/10.1016/j.jhsg.2024.11.011.
4. Eschweiler J, Li J, Quack V, et al. Anatomy, biomechanics, and loads of the wrist joint. Life (Basel). 2022;12(2):188. Published 2022 Jan 27. https://doi.org/10.3390/life12020188.
5. Kamal RN, Starr A, Akelman E. Carpal kinematics and kinetics. J Hand Surg Am. 2016;41(10):1011–8. https://doi.org/10.1016/j.jhsa.2016.07.105.
6. Rainbow MJ, Wolff AL, Crisco JJ, Wolfe SW. Functional kinematics of the wrist. J Hand Surg Eur. 2016;41(1):7–21. https://doi.org/10.1177/1753193415616939.
7. Palmer AK, Werner FW, Murphy D, Glisson R. Functional wrist motion: a biomechanical study. J Hand Surg Am. 1985;10(1):39–46. https://doi.org/10.1016/s0363-5023(85)80246-x.
8. Patterson RM, Williams L, Andersen CR, Koh S, Viegas SF. Carpal kinematics during simulated active and passive motion of the wrist. J Hand Surg Am. 2007;32(7):1013–9. https://doi.org/10.1016/j.jhsa.2007.05.004.
9. Berger RA. The anatomy and basic biomechanics of the wrist joint. J Hand Ther. 1996;9(2):84–93. https://doi.org/10.1016/s0894-1130(96)80066-4.
10. Short WH, Palmer AK, Werner FW, Murphy DJ. A biomechanical study of distal radial fractures. J Hand Surg Am. 1987;12(4):529–34. https://doi.org/10.1016/s0363-5023(87)80202-2.
11. Majima M, Horii E, Matsuki H, Hirata H, Genda E. Load transmission through the wrist in the extended position. J Hand Surg Am. 2008;33(2):182–8. https://doi.org/10.1016/j.jhsa.2007.10.018.
12. Bhat AK, Kumar B, Acharya A. Radiographic imaging of the wrist. Indian J Plast Surg. 2011;44(2):186–96. https://doi.org/10.4103/0970-0358.85339.
13. Duncan SF, Saracevic CE, Kakinoki R. Biomechanics of the hand. Hand Clin. 2013;29(4):483–92. https://doi.org/10.1016/j.hcl.2013.08.003.
14. Earley EJ, Hargrove LJ, Kuiken TA. Dual window pattern recognition classifier for improved partial-hand prosthesis control. Front Neurosci. 2016;10:58. https://doi.org/10.3389/fnins.2016.00058. PMID: 26941599; PMCID: PMC4763088.
15. Rrecaj S, Martinaj M, Murtezani A, Ibrahimi-Kaçuri D, Haxhiu B, Zatriqi V. Physical therapy and splinting after flexor tendon repair in zone II. Med Arch. 2014;68(2):128–31. https://doi.org/10.5455/medarh.2014.68.128-131. PMID: 24937939; PMCID: PMC4272500.
16. Visnapuu M, Jürimäe T. Handgrip strength and hand dimensions in young handball and basketball players. J Strength Cond Res. 2007;21(3):923–9. https://doi.org/10.1519/1533-4287(2007)21[923:HSAHDI]2.0.CO;2.

4 Kinematics-Based Diagnosis and Treatment of Hand–Finger Joint Diseases

Majd Mzeihem, Farid Amirouche,
and Bassem Elhassan

Introduction

The human hand has remarkable artistry in biomechanics and understands various tasks, from precise precision policies to the mightiest of gears [77]. Their general applicability and fine motor control are involved in tasks ranging from writing and typing to lifting and picking up objects [64]. This complexity has been realized due to the complexity of the kinematic system of hand and finger joints [61]. The theory that offers fundamental information about the movements of the hand and fingers is kinematics, which in mechanics deals with the movement of an object without reference to the forces that produce that movement [56]. Knowledge of relative movements of a joint allows physicians to diagnose and manage various diseases of the hand and finger joints.

Kinematics can involve any area, including range of motion (ROM), joint stability, and movement patterns [22]. These elements define physiological and pathological conditions [18]. For example, reaching angular movement or ROM modifications may suggest pathologies such as arthritis, ligament injury, or tendon pathology [4]. Assessing all these parameters provides better insight into joint biomechanics and the formulation of unique therapy approaches [83].

This chapter briefly discusses the relevance of kinematics to hand–finger joint health. It lays out the biomechanics of normally functioning hands, analyzes the effects of different disorders on joint movement, and considers the application and possibilities of kinematic studies as a diagnostic and therapeutic tool. However, the chapter also provides information on innovative methods of kinematic assessment by establishing the usage of three-dimensional movement analysis, dynamic imaging, and wearable sensors.

Principles of Kinematics in Hand and Finger Joints

Normal Kinematics

The work of the dominant human hand is based on articulation, with a focus on the metacarpophalangeal (MCP), proximal interphalangeal

M. Mzeihem
Department of Orthopaedic Surgery, University of Illinois at Chicago, Chicago, IL, USA

F. Amirouche
Department of Orthopaedic Surgery, Orthopaedic and Spine Institute, Endeavor HealthSystem (Formerly Northshore University HealthSystem), an Affiliate of the University of Chicago Pritzker School of Medicine, Skokie, IL, USA

Department of Orthopaedic Surgery, University of Illinois at Chicago, Chicago, IL, USA

B. Elhassan (✉)
Orthopedic Surgery, Co-Chief Shoulder Service, Harvard Medical School, Massachusetts General Hospital, Mass General West, Boston, MA, USA
e-mail: BELHASSAN@MGH.HARVARD.EDU

F. Amirouche et al. (eds.), *Biomechanics of the Hand*,
https://doi.org/10.1007/978-3-032-04688-8_4

(PIP), and distal interphalangeal (DIP) [109]. These joints allow many movements like flexion, extension, abduction, adduction, and circumduction [108]. For example, the MCP joints are formed for flexion-extension and abduction-adduction; thus, they significantly assist in hand flexibility [73]. The PIP and DIP joints mainly consist of flexing and extending movements needed in the form of grip and release of work [17].

Ligaments and tendons are as important as cartilage in contributing to joint stability and direction of movement [9]. Collateral ligaments play a significant role in offering lateral stability to the joints and guaranteeing easy and frictionless motion of joints [88]. On the other hand, volar plates counter hyperextension to minimize the stress on the joints [45]. Flexor and extensor tendons control the fingers in tasks that require accurate motor activity [57].

Knowledge of these standard kinematic patterns is necessary to determine possible pathologic from normal ones. For instance, a breakdown of the involuntarily coordinated relationship between flexor and extensor tendons leads to the pathology of movement, which may limit the hand's functional versatility in several tasks [14, 24]. Normal kinematics are not only compared with pathologic kinematics but also help understand the complex working of the hand and its versatility [112].

Kinematics measurements of the hand's finger joints include assessing joint angles (flexion, extension, abduction, and adduction) and their ROM at the MCP, PIP, and DIP joints. Angular velocity and acceleration provide insight into the speed and dynamics of joint movement and could be measured in degrees per second [78].

Altered Kinematics and Tools of Measurement

Abnormal pathologies affect the kinematics of hand and finger joints, vastly obliterating functional coverage [80]. These changes are evidenced by such disturbances as changes in movement coordination, reduced ROM, and impaired joint stability [94]. First, it is possible to mention that tissue damage due to underlying diseases such as osteoarthritis or rheumatoid arthritis may lead to abnormal rotational movement or angular deviations [74]. Like in the earlier case, post-traumatic stiffening or establishing scarring caused by Dupytren's contracture also affects the extension and flexion of the hand, thus reducing its functionality [59].

Another frequent occurrence is instability when the parts of a joint—ligaments or capsules—weaken [38]. This can lead to such blastic movements that they are out of control, which worsens the extent of the injuries and functional disability [46]. For instance, wrist scapholunate ligament injuries lead to changes in motion that limit grip strength and coordination [76].

These changes in the kinematics of locomotion are crucial to diagnose using several diagnostic tools. One of these tools is a goniometer, a device that allows physicians to give objective measurements of joint angles, thus allowing the physician to assess passive and active ROM definitively [33]. Numerous applications have been developed to enable physicians to use the goniometer function in smartphones (Fig. 4.1). Keugler et al. showed that goniometer applications were reliable and valid for measuring angles as long as the smartphone's adequate positioning was maintained while measuring [53].

Another method for measuring kinematics involves using specialized gloves (Powerglove) with sensors placed on patients' fingers. These gloves work in tandem with an Optotrak system, which tracks the movement of markers. The captured data calculates joint angles and ROM (Fig. 4.2). Before the calculations, a calibration process defines the neutral positions of the hand and fingers. Once calibrated, the recorded data is transmitted to specialized software, which processes and calculates the degrees of flexion and extension [6].

Computerized systems such as three-dimensional (3D) motion analysis systems pro-

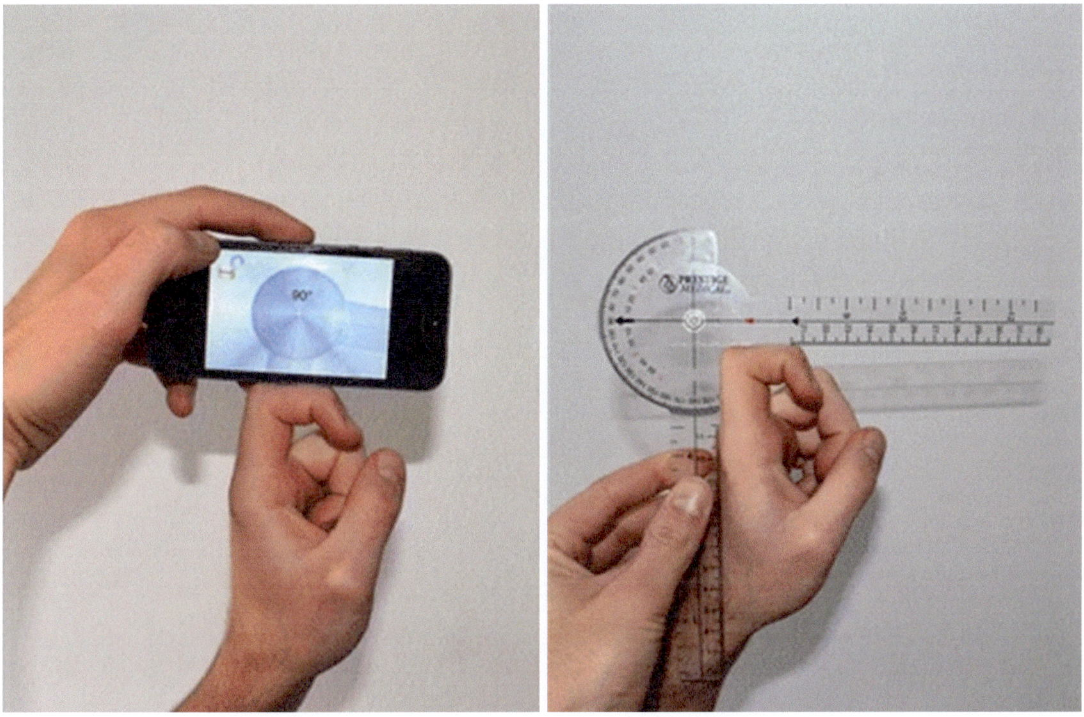

Fig. 4.1 An example demonstrating how goniometer applications measure the PIP joint angle compared to a mechanical goniometer. (All rights reserved to Keugler et al. [53])

vide quantitative and qualitative descriptions of evolved dynamic traits, which are minuscule but evident and might not be detected by the naked eye during physical assessment [79]. For instance, 3D motion analysis may indicate compensatory patterns in patients with thumb CMC osteoarthritis that can be utilized during therapy development [48, 65] (Fig. 4.3).

Techniques like fluoroscopy and magnetic resonance imaging (MRI) also add diagnostic depth to the method since they can view joint and soft tissue functionality in real time [29]. While dynamic instability, such as scapholunate advanced collapse (SLAC) III, is better demonstrated by fluoroscopy because of the change that occurs in the wrist during motion, MRI gives detailed information about soft tissue structures, for example, ligament integrity, tendon tear, or synovitis [96].

Finally, hand kinematics can be assessed as an overall measure of hand functionality in performing activities of daily living (ADL). Functional tests involving simulated ADL objectively measure hand dexterity and performance [62, 87, 91]. Among these, the Sollerman Hand Function Test (SHIFT) is widely recognized for its comprehensive and representative evaluation of ADL performance [91]. Recent studies have demonstrated that monitoring hand kinematics during these functional tests offers an objective and accurate assessment of hand functionality, enabling more informed therapeutic and motor strategy applications [35]. However, these tests are rarely used in clinical practice due to their time-consuming nature and are often replaced by patient-completed questionnaires assessing ADL skills [68]. A promising alternative is the BE-UJI test (Biomechanics and Ergonomics Group of the Universitat Jaume I), developed by Jarque-Bou et al. in a cross-sectional study. This reduced set of ADL tasks provides comparable kinematic information to the SHIFT test regarding

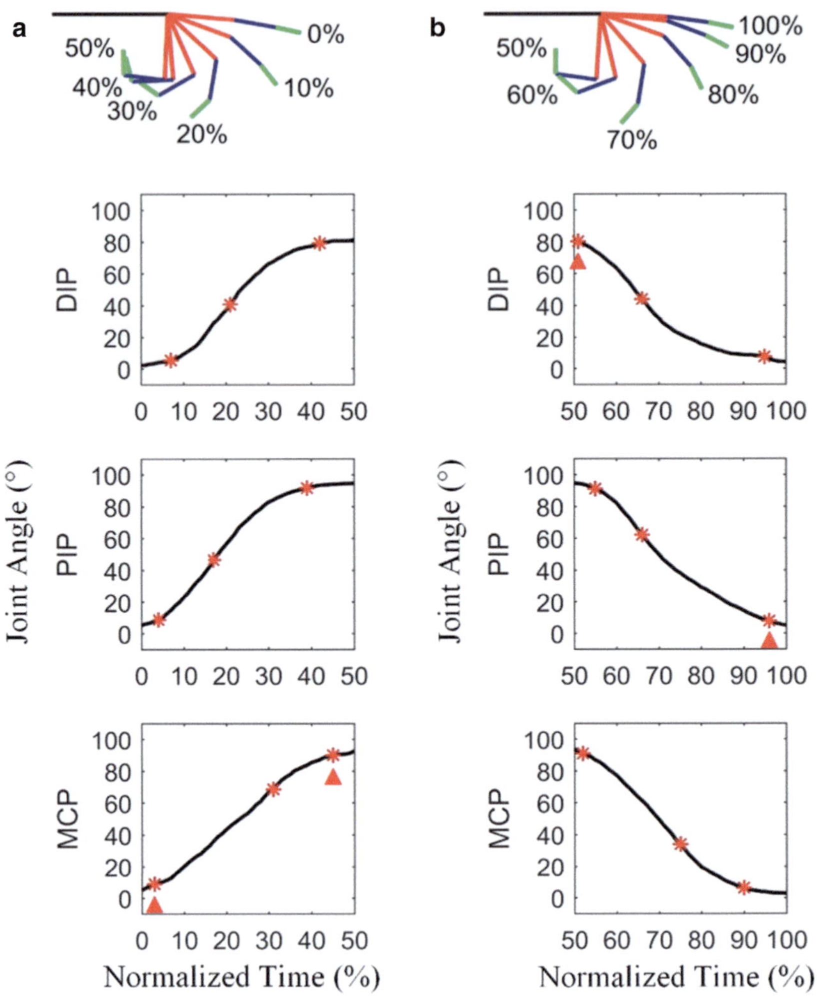

Fig. 4.2 Example of middle finger motion and angular curves for three joints: (**a**) flexion and (**b**) extension. The top row shows finger motion with green, blue, red, and black lines representing the distal, middle, proximal phalanges, and the metacarpal, respectively. The bottom rows display joint angle changes. Red asterisks mark joint motion points (initiation, maximum velocity, completion), while red triangles indicate finger motion points (initiation, completion). (All rights reserved to Li et al. [61])

synergies, ROM, and velocities. Its shorter administration time makes it a practical option for clinical settings, potentially facilitating more widespread adoption of objective hand function assessments [43] (Fig. 4.4).

Nonetheless, incorporating these diagnostic tools enhances the precision of kinematic evaluations and strengthens the design of specific interventions. The results uncover specific kinematic changes linked to the different pathologies, allowing therapists to prescribe treatments that tackle the root causes of dysfunction, thus enhancing the patients' lives.

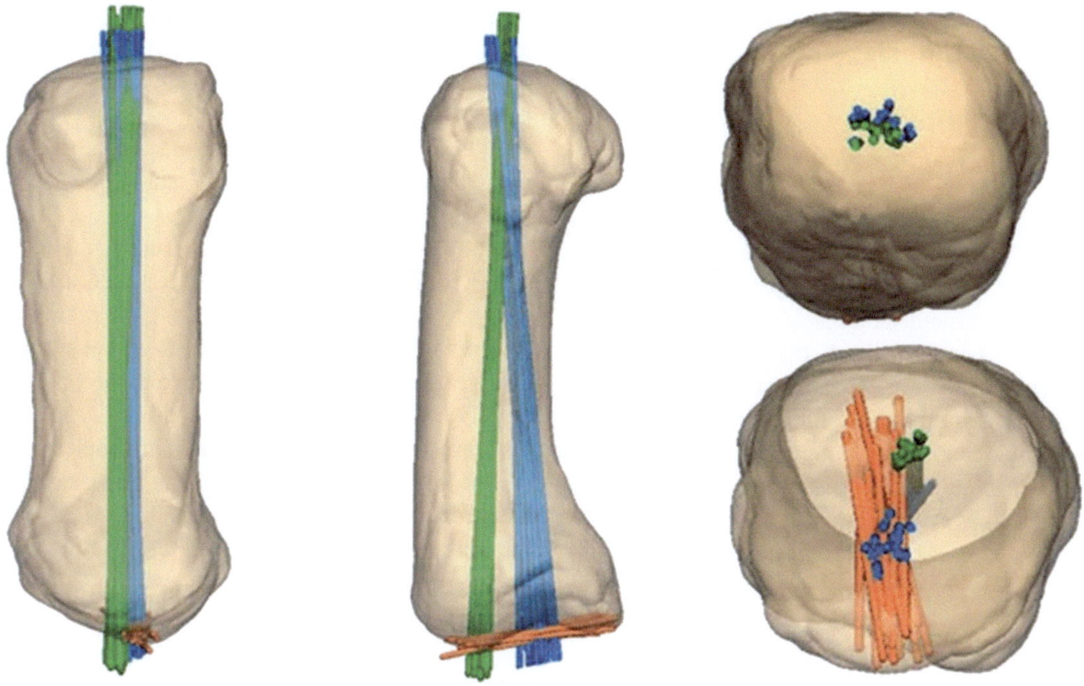

Fig. 4.3 Anatomical axis of MC I combined = green; mechanical axis of MC I combined = blue; proximal joint axis of MC I combined = orange. (All rights reserved to Keller et al. [48])

Activities of daily living considered in the Sollerman Hand Function Test
1. Pick coins up from flat surface, put into purses mounted on wall
2. Open/close zip
3. Pick up coins from purses
4. Lift wooden cubes over edge 5 cm in height
5. Lift iron over edge 5 cm in height
6. Turn screw with screwdriver
7. Pick up nuts and turn them until completely screwed onto bolts
8. Put key into Yale lock, turn 90°
9. Turn door-handle 30°
10. Put key into Yale lock, turn 90°
11. Do up buttons
12. Put Tubigrip stocking on the other hand
13. Cut Play-Doh with a knife and fork
14. Write with pen
15. Fold paper, put into envelope
16. Put paper-clip on envelope
17. Lift telephone receiver, put to ear
18. Pour water from Pure-Pak
19. Pour water from a jar
20. Pour water from a cup

BE-UJI reduced set	
Grasp	*Description*
1. Pulp pinch	Pour water from a cup
2. Transvers volar grip	Pour water from a jar
3. Five-finger pinch	Unscrew lid of jars
4. Pulp pinch	Pour water from Pure-Pak
5. Diagonal volar grip	Lift telephone receiver, put to ear
6. Diagonal volar grip	Cut play-doh with a knife and fork
7. Tripod pinch	Write with pen
8. Lateral pinch	Pick up nuts and turn them until completely screwed onto bolts
9. Five-finger pinch	Lift wooden cubes over edge 5 cm in height
10. Lateral pinch	Pick up coins from purses

Fig. 4.4 Comparison of the components tested by the SHIFT and BE-UJI tests. (Adapted with modifications; all rights reserved to Jarque-Bou et al. [43])

Common Disorders and Kinematics-Based Diagnosis

Osteoarthritis (OA)

OA is best defined by the progressive erosions and progressive narrowing of joint space, formation of osteophytes, and sclerosis of underlying subchondral bone [26]. OA primarily affects the thumb's carpometacarpal (CMC) joint and the fingers' proximal and distal interphalangeal joints [60]. These anatomical sites form focal contact areas that bear high loads and undergo repetitive motions, making them prone to degenerative alterations [54]. The use of kinematic analysis has revealed that OA is associated with decreased ROM, compromised joint biomechanics, and compromised hand use [3]. This provides crucial insights for diagnosis and treatment, empowering medical professionals to make informed decisions.

Thumb CMC OA patients typically have weak abduction and extension. Therefore, they have significant challenges in pinching, gripping, and grasping [100]. For instance, decreased thumb opposition capability means they cannot hold objects between the thumb and the index finger, so they cannot write, paint, or use small utensils [67]. In any individual with a motor disorder, these functional deficits can be accurately represented and enumerated by computing kinematic features like joint angles, velocities, and activity coordination during tasks [49]. These objective data points allow clinicians to make management decisions with a high degree of precision, ensuring the most effective treatment for each patient.

For example, a recent experiment published in 2022 demonstrated that active ROM measurements can serve as reliable indicators for the early diagnosis of OA. Patients with OA exhibited reduced functional ranges, diminished thumb opposition, and impaired gross dexterity, irrespective of the severity of their condition. These findings highlight that reduced active ROM could serve as an early marker of OA, potentially minimizing the need for imaging in diagnosis. Importantly, rehabilitation strategies focusing on strengthening the thumb, enhancing precision, and improving task completion times have shown promise in effectively addressing these deficits [35], offering hope for patients.

While radiographic assessments are beneficial in determining non-negligible anatomic alterations in certain rheumatic diseases about joint space and osteophyte formation alone, they are less helpful in detecting the dysfunction and adaptations of the patients [95]. For example, patients who experience pain in their joints may change movement activity to avoid the painful area, which results in other conditions, such as muscle incoordination or injuries in another part of the structure [5]. Organizing assessments based on the kinematic approach allows for a better evaluation of the effects of OA and aids in closing the gap between lesion detection and the functional outcome.

Rheumatoid Arthritis (RA)

Rheumatoid arthritis, a chronic inflammatory disease, often leads to joint deterioration and functional disability [42]. The hand is a commonly affected area, with characteristic deformities such as ulnar drift, boutonniere deformity, and swan neck deformity, each presenting unique challenges to hand function [97].

- *Ulnar drift*: This deformity includes bending the fingers toward the ulnar side owing to the synovium's inflammation and subsequent joint laxity [70]. Kinematic evaluations show decreased force exerted by hands and worsening finger movements since the angle changes the efficiency of flexor and extensor muscles [110].
- *Boutonniere deformity*: This deformity, represented by flexion at the PIP and hyperextension of the DIP joint, prevents finger extension almost completely [105]. Static and dynamic kinematic analyses show that patients have less finger extension during tasks that involve fine motor skills or when a steady force is required to extend the fingers, for instance, when pushing something [36].

- *Swan-neck deformity*: This congenital is characterized by the PIP joint being locked in hyperextension while the DIP joint is flexed and inhibits tasks that require finger flexion, particularly in gripping [23]. Social and physical functional groups may worsen as many patients exhibit severe impairment in carrying out basic tasks like buttoning shirts or opening the tops of jars [23].

Real-time ultrasound and motion capture systems play a central role in evaluating the increased severity of these deformities [31]. These tools help obtain the multibody system mechanics of the joint, tendon gliding, and muscle activities to help develop patients' specific management plans [31].

Trigger Finger

Trigger finger, or stenosing tenosynovitis, is a condition in which the flexor tendons of the affected finger become thickened and stuck within the tendon sheath ([89]. This condition most often affects the thumb and the ring finger, resulting from gripping activities or secondary to diabetes and RA [44].

Conventional kinematic analyses have demonstrated that the trigger finger leads to the staccato motion profiles of the fingers measured in terms of the rapid jerk or locking at the flexion or extension phase [98]. Some other frameworks of ultrasonography are employed to show the tendon's thickness and evaluate the condition's magnitude by examining the tendon's movement while conducting movements [16].

Tung et al. demonstrated that different grades of trigger fingers, as classified by Froimson's system, exhibit distinct kinematic patterns. Patients with grade II trigger fingers had a significantly larger workspace than those with higher grades (32.61 cm^2 vs. 24.42 cm^2 and 23.78 cm^2 for grades III and IV, respectively; $P = 0.02$). Higher-grade patients also achieved less finger flexion. Additionally, grades III and IV were associated with a smaller ROM at the PIP joint compared to grade II (94.24° vs. 84.32° and 80.70°, respectively; $P = 0.005$). These findings suggest that the PIP joint is the first joint significantly affected by tendon catching at the A1 pulley [98].

Such assessments are crucial for determining the most appropriate treatment plan. For mild to moderate cases, splinting and corticosteroid injections for 3 months coupled with physiotherapy for optimal results are viable treatments, and in some instances, surgery may be required [13]. In severe forms, the exploration with the release of the tendon sheath and incisions of the A1 pulley improve tendon gliding mechanics [13].

Dupuytren's Contracture

Dupuytren's contracture is a fibroproliferative disease of palms and hands in which the palmar fascia forms cords, and the fingers contract with time [27]. It manifests in the ring and little fingers and can lead to almost complete loss of hand use [86]. Thus, the disease progresses through nodular thickening and fibrous cords, which place the affected fingers into a flexed position [102].

Kinematic investigations aim to measure pliability's contour and its effect on specific manipulations with objects, for instance, grasping, typing, or holding [8]. Active ROM is determined using a goniometer to assess the angle of flexion deformity. In addition, motion analysis systems display dynamic limitations during hand movements [55]. For example, patients may have some constraints of finger extension, in which they support their body weight through the radial aspect of their hand with the thumb on the volar side using fingers 2–5. They may have increased use of the unaffected fingers.

These kinematic assessments are essential for providing surgical strategies appropriate for needle aponeurotomy or fasciectomy [39]. Kinematic monitoring is then used in postoperative rehabilitation treatment plans to articulate the changes in ROM and functional capacity and minimize reoccurrence risk [101].

Scapholunate Ligament Injury

As a part of the wrist joint, the scapholunate ligament is also responsible for maintaining the proper coupling between two carpal bones, scaphoid and lunate [7]. Injury to this ligament poses normal carpal kinematics to abnormality, which results in strange motion patterns, shifts in load, and eventual degeneration [104]. Consequently, a frequent consequence of nonsurgical undertreatment of scapholunate ligament injuries is scapholunate advanced collapse (SLAC), defined as arthritis and instability of the joint [66].

Initial imaging kinematics evaluations using 3D CT scan motion analysis and real-time MRI, a technique that provides dynamic images of the wrist in motion, are critical for decision-making, such as performing ligament repair, reconstruction, or wrist stabilizing operations [28]. Kinematic follow-up can also guarantee that wrist motion function returns to normal and detect potential complications, including residual wrist instability or stiffness.

Kinematics-Based Treatment Strategies

The kinematics of the hand is crucial in developing effective treatment strategies because it provides a comprehensive understanding of how the hand moves, distributes forces, and performs complex tasks. This knowledge is instrumental in identifying and addressing specific functional impairments, leveraging your expertise to tailor interventions that restore or enhance motion and optimize surgical or rehabilitative outcomes. By analyzing hand kinematics, clinicians can address issues such as grip strength, joint stability, and coordination, ultimately improving patient quality of life and the ability to perform daily activities.

Nonsurgical Approaches

Nonsurgical hand and finger joint treatment methods use simple techniques to help people recover functionality and reduce symptoms. Healthcare providers start with these methods because they cause minor damage to the body and pose fewer risks than surgical treatment [107].

Splinting and Dynamic Bracing

Hand and finger joint disorder treatment relies on custom-made splints and dynamic braces to support proper joint positioning and reduce pain while stopping deformations [81]. Different types of splints work best for distinct injuries. For example, resting splints can help RA patients prevent ulnar drift and provide trigger finger victims with extension splints to manage tendon discomfort [21].

The medical field has recently adopted dynamic support bracing to help people recover joint mobility after traumatic events. According to scientific research, long-term dynamic brace use results in better joint mobility, reduced discomfort, and improved functionality [50]. After ligament injuries, patients wear specialized braces designed for safe joint movement that help avoid joint stiffness and muscle loss [50].

ROM and Flexibility Exercises

People need specific exercises to enhance joint mobility while reducing stiffness and increasing ROM [93]. ROM exercises work best for easing stiffness from trauma injuries plus help treat injuries stemming from Dupuytren's contracture and early OA [2]. Scissor tension exercises work muscles near the joint, while simple stretches help lengthen the joint area [51].

Stretching programs work best when therapists add manual physical treatment to the regimen [75]. Medical guidelines recommend gradually stepping up exercise routines to keep existing health problems from worsening [84].

Strengthening Programs

Regular muscle-strengthening workouts build support structures around joints to stabilize movements and slow down future joint problems [37]. Grip-strengthening programs use hand grippers, stress balls, and resistance bands to build strength in the hand's muscle groups that lie within the hand and beyond it [99]. Studies show that specific exercises help OA patients reduce pain and maintain joint stability to do everyday tasks more comfortably [52].

Corticosteroid Injections

Healthcare providers use corticosteroid injections as a pain-free treatment method for both trigger finger and carpal tunnel syndrome without surgery [20]. Injections of anti-inflammatory medications at the source offer quick pain relief and let joints function more easily [10].

Medical professionals urge patients to exercise extra caution when receiving multiple corticosteroid injections because they can damage tissue and cause joint problems [20]. Results improve most when patients receive corticosteroid injections alongside physical therapy to reduce inflammation and correct structural abnormalities [20].

Surgical Approaches

Patients need surgery if other joint treatment methods fail to restore normal movement and performance. These operations improve joint structure, helping people enjoy less discomfort and lead better daily lives.

Arthroplasty

Healthcare providers use CMC arthroplasty to replace joints and give patients back their range of motion while easing their pain from advanced osteoarthritis [40]. New prosthetic systems copy how joints move in the body to produce better patient recovery results [72]. Medical studies show patients with CMC arthroplasty surgery gain notable enhancement in hand grip power, joint mobility, and pain management [72]. A recent study investigating the optimal center of rotation for the ball and socket of CMC arthroplasty has shown that when the offset was 4.5 mm, a neck rotation angle of 130.6° from the radial side to the palmar side of the first metacarpal and an implant height off 43.6 mm from the distal end of the metacarpal, normal kinematics of the thumb could be restored [71].

Modern arthroplasty includes new methods like pyrocarbon implant arthroplasty, which shows great potential for patient recovery [40]. These implants stand out for their safe interaction with human tissue and ability to effectively replace MCP and PIP joints throughout time [40].

Arthrodesis

Doctors may perform arthrodesis surgery when patients develop severe arthritis or post-traumatic arthritis because keeping the joint intact is no longer possible [19]. Joint fusion surgery cuts pain through joint stabilization but limits movement functions [63]. The procedure works well to keep joints stable even though it reduces movement.

Arthrodesis succeeds better today thanks to new fixation methods such as locking plates and bioabsorbable screws [47]. After surgery, we train nearby joints to perform better so patients can move without their fused joints [47]. Thus, kinematics-based physical rehabilitation is essential for long-term results.

Tendon Transfers and Releases

Through tendon transfer surgeries, medical professionals take healthy tendons from one part of the body and use them to fix another damaged tendon joint [69]. The procedure helps restore function when tendons break or extreme deformities affect the patient, like advanced rheumatoid arthritis does [69].

Surgeons often perform tendon release operations to correct trigger finger problems [32]. Surgery releases the limited tendon space to let tendons move freely, ending the sharp pain when moving the finger [32]. Doctors use improved percutaneous release procedures to speed recovery and boost patient satisfaction [92].

Postoperative Rehabilitation

Physical therapy after surgery helps make operations successful [85]. Starting movement therapy early helps patients avoid joint stiffness and achieve better healing [85]. Treatment plans customize rehabilitation exercises based on specific surgery type, injury status, and target movement abilities [1].

People recovering from tendon transfer surgery perform unique movements to teach their new tendon its proper use [69]. After arthroplasty, people need ROM exercises to stop scar tissue building up and keep their joints flexible [2].

Modern motion-tracking systems and kinematics-based measurement tools enable doctors and therapists to effectively monitor patients' recovery progress. These advanced tools provide real-time insights into patients' movements, allowing healthcare providers to promptly identify areas for improvement and adjust treatment plans as needed [12]. Wearable sensor technology helps identify inefficient compensation patterns, which lets providers focus therapy on fixing them [12]. This scientific kinematic-based monitoring system improves patient healthcare results and speeds up healing processes.

Advanced Technologies in Kinematics-Based Hand Assessment

Machine Learning and Artificial Intelligence

Modern technologies such as machine learning (ML) and artificial intelligence (AI) now power the way physicians diagnose medical conditions using patient movement patterns [58]. ML technologies examine extensive movement patterns to identify minor health problems before clinical detection [90]. Healthcare providers use these predictive systems to monitor RA and OA patients before disabilities appear and provide early treatments [90, 106].

AI systems with neural networks show exceptional performance in finding minor kinematic irregularities [58]. Researchers created AI systems to examine wrist movement patterns and spot potential damage in fingers through early motion sensors [58]. The system improves medical detection abilities and lets healthcare teams plan tailored therapies based on what each patient responds best to.

Virtual Reality

Users can perform assessments and receive rehabilitation using virtual reality (VR) and augmented reality (AR) systems that deliver complete digital experiences [11]. VR rehabilitation programs let patients safely work on real-world activities using virtual simulation technology [11]. Digital therapeutic simulations draw patient attention and help them better follow treatment requirements [82]. Numerous immersive VR systems allow patients to be evaluated at home or independently [25]. Several studies have compared upper limb kinematics in adults with stroke to healthy controls using VR. The findings indicate that VR-based kinematic evaluations are highly discriminative and can effectively quantify movement and movement quality [30, 34, 41]. Therefore, this approach could make it easier for patients to adhere to physical therapy by providing an engaging and interactive experience in a playful environment while objectively measuring the patient's progress (Fig. 4.5).

Fig. 4.5 A participant wearing VR glasses and performing target-to-target tasks. (All rights reserved to Hussain et al. [41])

Wearable Sensors

Hand activity tracking occurs through wearable technology using glove sensors with accelerometers and gyroscopes [15]. These sensors record actual activities to measure how people move their hands and determine their grip strength while tracking all of their joints [15].

Wearable devices help patients track their recovery and create better therapy plans that fit their needs [111]. Wearable sensors can detect tiny differences in joint movements that show when treatment starts working or when problems appear [103]. By tracking patient progress, these data help doctors adjust therapy for better results.

The development of stretchable electronic sensors will upgrade what wearable devices can do for users [103]. Next-generation sensors provide better precision and ease alongside long-term clinical monitoring options for various patients with hand motions [111].

References

1. Adogwa O, Elsamadicy AA, Fialkoff J, Vuong VD, Mehta AI, Vasquez RA, Cheng J, Karikari IO, Bagley CA. Assessing the effectiveness of routine use of post-operative in-patient physical therapy services. J Spine Surg. 2017;3(2):149–54. https://doi.org/10.21037/jss.2017.04.03.
2. Afonso J, Ramirez-Campillo R, Moscão J, Rocha T, Zacca R, Martins A, Milheiro AA, Ferreira J, Sarmento H, Clemente FM. Strength training versus stretching for improving range of motion: a systematic review and meta-analysis. Healthcare. 2021;9(4):427. https://doi.org/10.3390/healthcare9040427.
3. Al-Mahmood MR, Uddin MT, Islam MT, Fuad SM, Rahman Shah T. Correlation between goniometric measurements of range of motion and radiographic scores in osteoarthritis knee: an observational study among females. Medicine. 2022;101(32):e29995. https://doi.org/10.1097/md.0000000000029995.
4. Almansoof HS, Nuhmani S, Muaidi Q. The role of the kinetic chain in sports performance and injury risk is a narrative review. J Med Life. 2023;16(11):1591–6. https://doi.org/10.25122/jml-2023-0087.
5. Alnahdi AH, Zeni JA, Snyder-Mackler L. Muscle impairments in patients with knee osteoarthritis. Sports Health. 2012;4(4):284–92. https://doi.org/10.1177/1941738112445726.
6. An den Noort JC, Kortier HG, van Beek N, Veeger DH, Veltink PH. Measuring 3D hand and finger kinematics-A comparison between inertial sensing and an opto-electronic marker system. PLoS One. 2016;11(11):e0164889. https://doi.org/10.1371/journal.pone.0164889.
7. Andersson JK. Treatment of scapholunate ligament injury. EFORT Open Rev. 2017;2(9):382–93. https://doi.org/10.1302/2058-5241.2.170016.
8. Ansuini C, Cavallo A, Koul A, Jacono M, Yang Y, Becchio C. Predicting object size from hand kinematics: a temporal perspective. PLoS One. 2015;10(3):e0120432. https://doi.org/10.1371/journal.pone.0120432.
9. Asahara H, Inui M, Lotz MK. Tendons and ligaments: connecting developmental biology to musculoskeletal disease pathogenesis. J Bone Miner Res. 2017;32(9):1773–82. https://doi.org/10.1002/jbmr.3199.
10. Baker JF, Olave M, Marshall Leach W, Doherty CR, Gillcrist RL, White DK, Ogdie A, England BR, Wysham KD, Quinones M, Xiao R, Neogi T, Scanzello CR. Corticosteroid injections for symptomatic treatment of osteoarthritis of the knee: a pilot blinded randomized trial. ACR Open Rheumatol. 2023;5(10):529–35. https://doi.org/10.1002/acr2.11596.
11. Bateni H, Carruthers J, Mohan R, Pishva S. Use of virtual reality in physical therapy as an intervention and diagnostic tool. Rehabil Res Pract. 2024;2024:1–9. https://doi.org/10.1155/2024/1122286.
12. Becker ML, Hurkmans HL, Verhaar JAN, Bussmann JBJ. Monitoring postures and motions of hospitalized patients using sensor technology: a scoping review. Ann Med. 2024;56(1):2399963. https://doi.org/10.1080/07853890.2024.2399963.
13. Belloti JC, Sato ES, Faloppa F. Tratamento do Dedo em Gatilho. Rev Bras Ortop. 2020; https://doi.org/10.1055/s-0040-1713765.
14. Beutel BG, Gutowski KS, Marappa-Ganeshan R. Hand extensor tendon lacerations. StatPearls Publishing; 2024. https://www.ncbi.nlm.nih.gov/sites/books/NBK554431/.
15. Bian S, Liu M, Zhou B, Lukowicz P. The state-of-the-art sensing techniques in human activity recognition: a survey. Sensors. 2022;22(12):4596. https://doi.org/10.3390/s22124596.
16. Bianchi S, Gitto S, Draghi F. Ultrasound features of trigger finger: review of the literature. J Ultrasound Med. 2019;38(12):3141–54. https://doi.org/10.1002/jum.15025.
17. Bindra RR, Foster BJ. Management of proximal interphalangeal joint dislocations in athletes. Hand Clin. 2009;25(3):423–35. https://doi.org/10.1016/j.hcl.2009.05.008.
18. Bloomer C, Wang S, Kontson K. Kinematic analysis of motor learning in upper limb body-powered bypass prosthesis training. PLoS One.

2020;15(1):e0226563. https://doi.org/10.1371/journal.pone.0226563.
19. Brumat P, Kunšič O, Novak S, Slokar U, Pšenica J, Topolovec M, Mihalič R, Trebše R. The surgical treatment of osteoarthritis. Life. 2022;12(982):982. https://doi.org/10.3390/life12070982.
20. Cage ES, Beyer JJ, Ebraheim NA. Injections for treatment of carpal tunnel syndrome: a narrative review of the literature. J Orthop. 2023;37:81–5. https://doi.org/10.1016/j.jor.2023.02.011.
21. Carruthers KH, O'Reilly O, Skie M, Walters J, Siparsky P. Casting and splinting management for hand injuries in the in-season contact sport athlete. Sports Health. 2017;9(4):364–71. https://doi.org/10.1177/1941738117700133.
22. Cerfoglio S, Capodaglio P, Rossi P, Conforti I, D'Angeli V, Milani E, Galli M, Cimolin V. Evaluation of upper body and lower limbs kinematics through an IMU-based medical system: a comparative study with the optoelectronic system. Sensors. 2023;23(13):6156. https://doi.org/10.3390/s23136156.
23. Chetta M, Burns PB, Kim HM, Burke FD, Wilgis EFS, Fox DA, Chung KC. The effect of swan neck and boutonniere deformities on the outcome of silicone metacarpophalangeal joint arthroplasty in rheumatoid arthritis. Plast Reconstr Surg. 2013;132(3):597–603. https://doi.org/10.1097/prs.0b013e31829ad1c1.
24. Chinchalkar SJ, Gan BS, Mcfarlane RM, King GJ, Roth JH. Extensor quadriga: pathomechanics and treatment. Can J Plast Surg. 2004;12(4):174–8. https://doi.org/10.1177/229255030401200405.
25. Clay F, Howett D, FitzGerald J, Fletcher P, Chan D, Price A. Use of immersive virtual reality in the assessment and treatment of Alzheimer's disease: a systematic review. J Alzheimer's Dis JAD. 2020;75(1):23–43. https://doi.org/10.3233/JAD-191218.
26. Coaccioli S, Sarzi-Puttini P, Zis P, Rinonapoli G, Varrassi G. Osteoarthritis: new insight on its pathophysiology. J Clin Med. 2022;11(20):6013. https://doi.org/10.3390/jcm11206013.
27. Dutta A, Jayasinghe G, Deore S, Wahed K, Bhan K, Bakti N, Singh B. Dupuytren's contracture – current concepts. J Clin Orthop Trauma. 2020;11(4):590–6. https://doi.org/10.1016/j.jcot.2020.03.026.
28. Eftekhar P, Li MH, Semple MJ, Richardson D, Brooks D, Mochizuki G, Dutta T. Investigation of the kinematic upper-limb movement assessment (KUMA): a pilot study. Physiother Can. 2022;74(3):316–23. https://doi.org/10.3138/ptc-2019-0023.
29. El-Najjar AR, Abu-Elsoaud AM, Mohammed HT, Shawky KM. Diagnostic potential of magnetic resonance imaging (MRI) of the first carpometacarpal joint in hand osteoarthritis. Egypt Rheumatol. 2021;43(1):59–64. https://doi.org/10.1016/j.ejr.2020.07.010.
30. Everard G, Boivin S, Boulay G, Duchemin R, Batcho CS. Immersive virtual reality to assess arm kinematics among older adults with and without Major neurocognitive disorder – an exploratory cross-sectional study. Neuroscience. 2024;537:47–57. https://doi.org/10.1016/j.neuroscience.2023.10.024.
31. Ezzati F, Pezeshk P. Radiographic findings of inflammatory arthritis and mimics in the hands. Diagnostics. 2022;12(9):2134. https://doi.org/10.3390/diagnostics12092134.
32. Fiorini HJ, Tamaoki MJ, Lenza M, Gomes dos Santos JB, Faloppa F, Belloti J, carlos. Surgery for trigger finger. Cochrane Database Syst Rev. 2018; https://doi.org/10.1002/14651858.cd009860.pub2.
33. Gandbhir VN, Cunha B. Goniometer. PubMed. StatPearls Publishing; 2020. https://www.ncbi.nlm.nih.gov/books/NBK558985/
34. Gilliaux M, Lejeune TM, Detrembleur C, Sapin J, Dehez B, Selves C, Stoquart G. Using the robotic device REAplan as a valid, reliable, and sensitive tool to quantify upper limb impairments in stroke patients. J Rehabil Med. 2014;46(2):117–25. https://doi.org/10.2340/16501977-1245.
35. Gracia-Ibáñez V, Rodríguez-Cervantes PJ, Bayarri-Porcar V, Granell P, Vergara M, Sancho-Bru JL. Using sensorized gloves and dimensional reduction for hand function assessment of patients with osteoarthritis. Sensors (Basel, Switzerland). 2021;21(23):7897. https://doi.org/10.3390/s21237897.
36. Grau L, Baydoun H, Chen K, Solitro G, Gonzalez MH, Sankary S. Biomechanics of the acute boutonniere deformity. J Hand Surg. 2018;37(8):10. https://doi.org/10.1016/s0363-5023(12)60014-8.
37. Hameed I, Farooq N, Haq A, Aimen I, Shanley J. Role of strengthening exercises in management and prevention of overuse sports injuries of the lower extremity: a systematic review. J Sports Med Phys Fitness. 2024; https://doi.org/10.23736/S0022-4707.23.15470-3.
38. Hauser, R. (2024). The joint instability diagnosis. Caringmedical.com. https://caringmedical.com/prolotherapy-news/the-joint-instability-diagnosis/
39. Herrera FA, Mitchell S, Elzik M, Roostaeian J, Benhaim P. Modified percutaneous needle aponeurotomy for the treatment of dupuytren's contracture: early results and complications. Hand (New York, NY). 2015;10(3):433–7. https://doi.org/10.1007/s11552-015-9740-4.
40. Hidajat NN, Aditya R, Arsa W, Malik R. Outcome comparison between carpometacarpal arthroplasty and trapeziectomy in patients with carpometacarpal osteoarthritis: a systematic review. Orthop Rev. 2024;16 https://doi.org/10.52965/001c.117651.
41. Hussain N, Alt Murphy M, Sunnerhagen KS. Upper limb kinematics in stroke and healthy controls using target-to-target task in virtual reality. Front Neurol. 2018;9:300. https://doi.org/10.3389/fneur.2018.00300.
42. Jahid M, Khan KU, Ul-Haq R, Ahmed RS. Overview of rheumatoid arthritis and scientific understanding of the disease. Mediterr J Rheumatol.

2023;34(3):284–91. https://doi.org/10.31138/mjr.20230801.oo.

43. Jarque-Bou NJ, Gracia-Ibáñez V, Vergara M, Sancho-Bru JL. The BE-UJI hand function activity set: a reduced set of activities for the evaluation of the healthy and pathological hand. J Neuroeng Rehabil. 2023;20(1):122. https://doi.org/10.1186/s12984-023-01245-1.
44. Jeanmonod R, Harberger S, Waseem M. Trigger finger. PubMed. StatPearls Publishing; 2021. https://www.ncbi.nlm.nih.gov/books/NBK459310/
45. Kamnerdnakta S, Huetteman HE, Chung KC. Complications of proximal interphalangeal joint injuries. Hand Clin. 2018;34(2):267–88. https://doi.org/10.1016/j.hcl.2017.12.014.
46. Kawabata S, Murata K, Nakao K, Sonoo M, Morishita Y, Oka Y, Kubota K, Kuroo-Nakajima A, Kita S, Nakagaki S, Arakawa K, Kokubun T, Kanemura N. Effects of exercise therapy on joint instability in patients with osteoarthritis of the knee: a systematic review. Osteoarthr Cartil Open. 2020;2(4):100114. https://doi.org/10.1016/j.ocarto.2020.100114.
47. Kazmers NH, Hippensteel KJ, Calfee RP, Wall LB, Boyer MI, Goldfarb CA, Gelberman RH, Osei DA. Locking plate arthrodesis compares favorably with LRTI for thumb trapeziometacarpal arthrosis: early outcomes from a longitudinal cohort study. HSS J. 2017;13(1):54–60. https://doi.org/10.1007/s11420-016-9527-3.
48. Keller M, Rueegg J, Haefeli M, Honigmann P. Three-dimensional analysis of the first metacarpal axes in healthy individuals and early-stage thumb carpometacarpal osteoarthritis patients—potential implication on first metacarpal corrective osteotomy. J Clin Med. 2024;13(18):5513. https://doi.org/10.3390/jcm13185513.
49. Kerkhof FD, Brugman E, D'Agostino P, Dourthe B, van Lenthe GH, Stockmans F, Jonkers I, Vereecke EE. Quantifying thumb opposition kinematics using dynamic computed tomography. J Biomech. 2016;49(9):1994–9. https://doi.org/10.1016/j.jbiomech.2016.05.008.
50. Kim D-G, Choi S, Park EJ. Effectiveness of a novel finger range-of-motion brace for extensor tendon injury: a report of 10 patients. BMC Musculoskelet Disord. 2024;25(1) https://doi.org/10.1186/s12891-024-07899-9.
51. Koizumi K, Sasao K, Koike Y, Okino A, Takeda K, Hamaguchi T. Usefulness of Scissors with a power-support mechanism to assist thumb movement: an observational study. Appl Sci. 2021;11(16):7756. https://doi.org/10.3390/app11167756.
52. Kong H, Wang X-Q, Zhang X-A. Exercise for osteoarthritis: a literature review of pathology and mechanism. Front Aging Neurosci. 2022;14(14) https://doi.org/10.3389/fnagi.2022.854026.
53. Kuegler P, Wurzer P, Tuca A, et al. Goniometer-apps in hand surgery and their applicability in daily clinical practice. Saf Health. 2015;1:11. https://doi.org/10.1186/s40886-015-0003-4.
54. Ladd AL, Weiss A-PC, Crisco JJ, Hagert E, Wolf JM, Glickel SZ, Yao J. The thumb carpometacarpal joint: anatomy, hormones, and biomechanics. Instr Course Lect. 2014;62:165. https://pmc.ncbi.nlm.nih.gov/articles/PMC3935621/
55. Lapresa M, Zollo L, Cordella F. A user-friendly automatic toolbox for hand kinematic analysis, clinical assessment and postural synergies extraction. Front Bioeng Biotechnol. 2022;10 https://doi.org/10.3389/fbioe.2022.1010073.
56. Lapresa M, Corradini V, Iacca A, Scotto di Luzio F, Zollo L, Cordella F. A comprehensive analysis of task-specific hand kinematic, muscle and force synergies. Biocybern Biomed Eng. 2024;44(1):218–30. https://doi.org/10.1016/j.bbe.2024.01.006.
57. Lee SW, Qiu D, Fischer HC, Conrad MO, Kamper DG. Modulation of finger muscle activation patterns across postures is coordinated across all muscle groups. J Neurophysiol. 2020;124(2):330–41. https://doi.org/10.1152/jn.00088.2020.
58. Lee HJ, Jin SM, Kim SJ, Kim JH, Kim H, Bae E, Yoo SK, Kim JH. Development and validation of an artificial intelligence-based motion analysis system for upper extremity rehabilitation exercises in patients with spinal cord injury: a randomized controlled trial. Healthcare. 2023;12(1):7–7. https://doi.org/10.3390/healthcare12010007.
59. Lee J, Choi YJ, Seo BF, Jung S, Choi JY. Dupuytren contracture after acute traumatic hand injury in an adolescent: a case report. Joint Dis Relat Surg. 2023;34(3):737–40. https://doi.org/10.52312/jdrs.2023.1168.
60. Li YK, White CP. Carpometacarpal osteoarthritis of the thumb. Can Med Assoc J. 2012;185(2):149. https://doi.org/10.1503/cmaj.111444.
61. Li X, Wen R, Duanmu D, Huang W, Wan K, Hu Y. Finger kinematics during human hand grip and release. Biomimetics. 2023;8(2):244. https://doi.org/10.3390/biomimetics8020244.
62. Light CM, Chappell PH, Kyberd PJ. Establishing a standardized clinical assessment tool of pathologic and prosthetic hand function: normative data, reliability, and validity. Arch Phys Med Rehabil. 2002;83(6):776–83. https://doi.org/10.1053/apmr.2002.32737.
63. Lisanti M, Rosati M, Spagnolli G, Luppichini G. Trapeziometacarpal joint arthrodesis for osteoarthritis: results of power staple fixation. J Hand Surg: Br Eur. 2005;22(5):576–9. https://doi.org/10.1016/S0266-7681(97)80348-8.
64. Liu Y, Jiang L, Liu H, Ming D. A systematic analysis of hand movement functionality: qualitative classification and quantitative investigation of hand grasp behavior. Front Neurorobot. 2021;15 https://doi.org/10.3389/fnbot.2021.658075.
65. Luker KR, Aguinaldo A, Kenney D, Cahill-Rowley K, Ladd AL. Functional task kinematics of the thumb carpometacarpal joint. Clin Orthop Relat Res. 2014;472(4):1123–9. https://doi.org/10.1007/s11999-013-2964-0.

66. Maris S, Apergis E, Apostolopoulos A, Melissaridou D, Koulouvaris P, Papagelopoulos PJ, Savvidou O. Scapholunate advanced collapse (SLAC) and scaphoid nonunion advanced collapse (SNAC): a review of treatment options for stage II. Cureus. 2022;16(4):e59014. https://doi.org/10.7759/cureus.59014.
67. Marquardt TL, Nataraj R, Evans PJ, Seitz WH, Li Z-M. Carpal tunnel syndrome impairs thumb opposition and circumduction motion. Clin Orthop Relat Res. 2014;472(8):2526–33. https://doi.org/10.1007/s11999-014-3651-5.
68. Marshall SC, Gray D, Wilson KG, Yelle JD, Hébert P, O'Rourke K, Jamshidi M. A prospective study to validate an impairment questionnaire for major trauma survivors. Am J Phys Med Rehabil. 2007;86(2):114–24. https://doi.org/10.1097/01.phm.0000253256.11686.23.
69. Monir JG, Wagner ER. Tendon transfers in the setting of shoulder arthroplasty. JSES Rev Rep Tech. 2024;4(3):607–14. https://doi.org/10.1016/j.xrrt.2024.03.007.
70. Morco S, Bowden A. Ulnar drift in rheumatoid arthritis: a review of biomechanical etiology. J Biomech. 2015;48(4):725–8. https://doi.org/10.1016/j.jbiomech.2014.12.052.
71. Murai A, Kurosawa A, Tada K, Tachiya H, Tamai A, Akahane M, Matsuta M, Nakamura Y, Kawashima H, Tsuchiya H. Optimal center of rotation for ball-and-socket thumb carpometacarpal arthroplasty identified using three-dimensional kinematic analysis: a pilot study. Front Bioeng Biotechnol. 2022;10(868):150. https://doi.org/10.3389/fbioe.2022.868150.
72. Newton A, Talwalkar S. Arthroplasty in thumb trapeziometacarpal (CMC joint) osteoarthritis: an alternative to excision arthroplasty. J Orthop. 2022; https://doi.org/10.1016/j.jor.2022.11.011.
73. Okafor L, Sinkler MA, Varacallo M. Anatomy, shoulder and upper limb, hand metacarpal phalangeal joint. PubMed: StatPearls Publishing; 2022. https://www.ncbi.nlm.nih.gov/books/NBK538343/
74. Ouyang Z, Dong L, Feng Y, Wang K, Chen Y, Li S, Zhou R, Zhao Y, Hu W. Cartilage-related collagens in osteoarthritis and rheumatoid arthritis: from pathogenesis to therapeutics. Int J Mol Sci. 2023;24(12):9841. https://doi.org/10.3390/ijms24129841.
75. Page P. Current concepts in muscle stretching for exercise and rehabilitation. Int J Sports Phys Ther. 2012;7(1):109. https://pmc.ncbi.nlm.nih.gov/articles/PMC3273886/
76. Pappou IP, Basel J, Deal DN. Scapholunate ligament injuries: a review of current concepts. Hand (New York, NY). 2013;8(2):146–56. https://doi.org/10.1007/s11552-013-9499-4.
77. Pilacinski A, Vandenberghe A, Andrietta G, Vannuscorps G. Humans underestimate the movement range of their own hands. Commun Psychol. 2024;2(1) https://doi.org/10.1038/s44271-024-00153-x.
78. Prosise JF, Hendrix CM, Ebner TJ. Joint angles and angular velocities and relevance of eigenvectors during prehension in the monkey. Exp Brain Res. 2015;233(2):339–50. https://doi.org/10.1007/s00221-014-4117-4.
79. Provini P, Camp AL, Crandell KE. Emerging biological insights enabled by high-resolution 3D motion data: promises, perspectives and pitfalls. J Exp Biol. 2023;226(Suppl_1) https://doi.org/10.1242/jeb.245138.
80. Rath S. Hand kinematics: application in clinical practice. Indian J Plast Surg. 2011;44(2):178. https://doi.org/10.4103/0970-0358.85338.
81. Richards T, Clement R, Russell I, Newington D. Acute hand injury splinting—the good, the bad and the ugly. Ann R Coll Surg Engl. 2018;100(2):92–6. https://doi.org/10.1308/rcsann.2017.0195.
82. Rodríguez-Almagro D, Achalandabaso-Ochoa A, Ibáñez-Vera AJ, Góngora-Rodríguez J, Rodríguez-Huguet M. Effectiveness of virtual reality therapy on balance and gait in the elderly: a systematic review. Healthcare. 2024;12(2):158. https://doi.org/10.3390/healthcare12020158.
83. Roelker SA, Schmitt LC, Chaudhari AMW, Siston RA. Discover your potential: the influence of kinematics on a muscle's ability to contribute to the sit-to-stand transfer. PLoS One. 2022;17(3):e0264080. https://doi.org/10.1371/journal.pone.0264080.
84. Rooney D, Gilmartin E, Heron N. Prescribing exercise and physical activity to treat and manage health conditions. Ulster Med J. 2023;92(1):9. https://pmc.ncbi.nlm.nih.gov/articles/PMC9899030/
85. Saito T, Nakamichi R, Nakahara R, Nishida K, Ozaki T. The effectiveness of rehabilitation after open surgical release for trigger finger: a prospective, randomized, controlled study. J Clin Med. 2023;12(22):7187. https://doi.org/10.3390/jcm12227187.
86. Sarkar M, Kashyap N, Madabhavi I. Dupuytren's disease: a review. Chin J Plast Reconstr Surg. 2024; https://doi.org/10.1016/j.cjprs.2024.08.002.
87. Sears ED, Chung KC. Validity and responsiveness of the Jebsen-Taylor hand function test. J Hand Surg. 2010;35(1):30–7. https://doi.org/10.1016/j.jhsa.2009.09.008.
88. Shetty A, Prabhath S, Alappatt K, Kn LK, Bhat N, Sumalatha S. Lateral collateral ligament and anterolateral ligament of the knee—a morphological analysis with orthopedic significance. Knee. 2021;28:202–6. https://doi.org/10.1016/j.knee.2020.12.002.
89. Shohda E, Sheta RA. Misconceptions about trigger finger: a scoping review. Definition, pathophysiology, site of lesion, etiology. Trigger finger solving a maze. Adv Rheumatol. 2024;64(1) https://doi.org/10.1186/s42358-024-00379-7.
90. Simeoli R, Rega A, Cerasuolo M, Nappo R, Marocco D. Using machine learning for motion analysis to early detect autism spectrum disorder: a systematic

review. Rev J Autism Dev Disord. 2024; https://doi.org/10.1007/s40489-024-00435-4.
91. Sollerman C, Ejeskär A. Sollerman hand function test. A standardised method and its use in tetraplegic patients. Scand J Plast Reconstr Surg Hand Surg. 1995;29(2):167–76. https://doi.org/10.3109/02844319509034334.
92. Steinman S. Surgical treatment of trigger thumb and trigger finger. J Pediatr Orthop Soc North Am. 2021;3(1) https://doi.org/10.55275/jposna-2021-229.
93. Suzuki Y, Iijima H, Tashiro Y, Kajiwara Y, Zeidan H, Shimoura K, Nishida Y, Bito T, Nakai K, Tatsumi M, Yoshimi S, Tsuboyama T, Aoyama T. Home exercise therapy to improve muscle strength and joint flexibility effectively treats pre-radiographic knee OA in community-dwelling elderly: a randomized controlled trial. Clin Rheumatol. 2018;38(1):133–41. https://doi.org/10.1007/s10067-018-4263-3.
94. Tanashi A, Szekeres M, MacDermid J, Lalone EA. Comparison of finger kinematics between patients with hand osteoarthritis and healthy participants with and without joint protection programs. J Hand Ther. 2020; https://doi.org/10.1016/j.jht.2020.10.010.
95. Tins BJ, Butler R. Imaging in rheumatology: reconciling radiology and rheumatology. Insights Imaging. 2013;4(6):799–810. https://doi.org/10.1007/s13244-013-0293-1.
96. Tischler BT, Diaz LE, Murakami AM, Roemer FW, Goud AR, Arndt WF, Guermazi A. Scapholunate advanced collapse: a pictorial review. Insights Imaging. 2014;5(4):407–17. https://doi.org/10.1007/s13244-014-0337-1.
97. Toyama S, Tokunaga D, Tsuchida S, Kushida R, Oda R, Kawahito Y, Takahashi K. Comprehensive assessment of alterations in hand deformities over 11 years in patients with rheumatoid arthritis using cluster analysis and analysis of covariance. Arthritis Res Ther. 2021;23(1) https://doi.org/10.1186/s13075-021-02448-4.
98. Tung W-L, Kuo L-C, Lai K-Y, Jou I-M, Sun Y-N, Su F-C. Quantitative evidence of kinematics and functional differences in different graded trigger fingers. Clin Biomech. 2010;25(6):535–40. https://doi.org/10.1016/j.clinbiomech.2010.02.009.
99. Vaishya R, Misra A, Vaish A, Ursino N, D'Ambrosi R. Hand grip strength as a proposed new vital sign of health: a narrative review of evidences. J Health Popul Nutr. 2024;43(1) https://doi.org/10.1186/s41043-024-00500-y.
100. Villafañe JH, Valdes K. Combined thumb abduction and index finger extension strength: a comparison of older adults with and without thumb carpometacarpal osteoarthritis. J Manipulative Physiol Ther. 2013;36(4):238–44. https://doi.org/10.1016/j.jmpt.2013.05.004.
101. Vorro J, Bush TR, Rutledge B, Li M. Kinematic measures during a clinical diagnostic technique for human neck disorder: inter- and intraexaminer comparisons. Biomed Res Int. 2013;2013 https://doi.org/10.1155/2013/950719.
102. Walthall J, Anand P, Rehman UH. Dupuytrens contracture. PubMed. StatPearls Publishing; 2020. https://www.ncbi.nlm.nih.gov/books/NBK526074/
103. Wang X, Yu H, Kold S, Rahbek O, Bai S. Wearable sensors for activity monitoring and motion control: a review. Biomim Intell Robot. 2023;3(1):100089. https://doi.org/10.1016/j.birob.2023.100089.
104. Werner FW, Wang H, Short WH, Sutton LG, Rosenbaum PF. Identifying scapholunate ligamentous injury. J Orthop Res. 2008;27(3):394–9. https://doi.org/10.1002/jor.20749.
105. Williams K, Terrono AL. Treatment of boutonniere finger deformity in rheumatoid arthritis. J Hand Surg. 2011;36(8):1388–93. https://doi.org/10.1016/j.jhsa.2011.05.029.
106. Woo H-C, White P, Ng H-K, Lai CWK. Development of kinematic graphs of median nerve during active finger motion: implications of smartphone use. PLoS One. 2016;11(7):e0158455. https://doi.org/10.1371/journal.pone.0158455.
107. Xie J, Huang X, Gao M, Wei L, Wang R, Chen J, Zeng Y, Ji B, Liu T, Wang J, Wu H, Wang Y, Qin L, Wang Y, Zheng Z, Xue J, Wu J, Chen X, Zheng Z, Li X. Surgical pharmacy for optimizing medication therapy management services within enhanced recovery after surgery (ERAS®) programs. J Clin Med. 2023;12(2):631. https://doi.org/10.3390/jcm12020631.
108. Yang G, McGlinn EP, Chung KC. Management of the stiff finger. Clin Plast Surg. 2014;41(3):501–12. https://doi.org/10.1016/j.cps.2014.03.011.
109. Yang T-H, Lu S-C, Lin W-J, Zhao K, Zhao C, An K-N, Jou I-M, Lee P-Y, Kuo L-C, Su F-C. Assessing finger joint biomechanics by applying equal force to flexor tendons in vitro using a novel simultaneous approach. PLoS One. 2016;11(8):e0160301. https://doi.org/10.1371/journal.pone.0160301.
110. Yayama T, Kobayashi S, Kokubo Y, Inukai T, Mizukami Y, Kubota M, Ishikawa J, Baba H, Minami A. Motion analysis of the wrist joints in patients with rheumatoid arthritis. Mod Rheumatol. 2007;17(4):322–6. https://doi.org/10.1007/s10165-007-0590-y.
111. Ye X, Sakurai K, Nair N-KC, Wang K, I-Kai. Machine learning techniques for sensor-based human activity recognition with data heterogeneity—a review. Sensors. 2024;24(24):7975. https://doi.org/10.3390/s24247975.
112. Zaidi KF, Harris-Love M. Upper extremity kinematics: development of a quantitative measure of impairment severity and dissimilarity after stroke. Peer J. 2023;11:–e16374. https://doi.org/10.7717/peerj.16374.

5 Mechanics of the Thumb

Hanny Chen, John D. King, and M. Daniel Wongworawat

What It Is

The thumb, referred to as the *anti-hand* by Hippocrates, has long been cited as an element that separates humans from other animals as it allows for prehension, which is the act of using the hand to grasp an object. Prehension is a complex skill that involves multiple coordinated components [1, 2].

The thumb with its four joints is not in the same plane as the other four digits in a resting position. Sitting at a right angle to the other four digits at rest, this allows the thumb to perform opposition as well as flattening [3].

Anatomy

The thumb is the shortest digit and is composed of the trapezium, first metacarpal, the proximal, and the distal phalanx.

Within the thumb, there are four articulations from proximal to distal being the scaphotrapeziotrapezoid (STT), carpometacarpal (CMC), metacarpophalangeal (MCP), and the interphalangeal (IP) joint. The thumb sits in a plane of 60–80° to the metacarpal arch of the hand that can allow opposition to each of the four fingers, powered by forearm and hand muscles [3, 4].

Osteology

Trapezium The trapezium has six faces. On the palmar face, the FCR passes through the anterolateral groove onto its insertion at the volar base of the second metacarpal [1]. A part of CMC joint stability, the trapezial recess, is near this aspect of the trapezium. The dorsal aspect involves the dorsal ligaments attached to the trapezial tubercle, providing stability to the CMC joint. The proximal face articulates with the scaphoid and trapezoid forming the STT joint and the distal face articulates with the thumb metacarpal. The abductor pollicis brevis (ABP), opponens pollicis (OP), and flexor pollicis brevis (FPB) muscles originate on the radial face. The last face, the ulnar face, involves the articular surface abutting the trapezoid proximally and second metacarpal distally [3].

Thumb Metacarpal The shortest metacarpal, it has a concavoconxex saddle-shaped *surface* that extends palmar, forming a volar beak [3]. This surface is convex front to back and concave from inside out, creating a ridge not congruent with any other joint surface. Four muscles are along its shaft: the abductor polli-

H. Chen · J. D. King · M. D. Wongworawat (✉)
Department of Orthopaedic Surgery, Loma Linda University, Orthopedic Surgery,
Loma Linda, CA, USA
e-mail: dwongworawat@llu.edu

F. Amirouche et al. (eds.), *Biomechanics of the Hand*,
https://doi.org/10.1007/978-3-032-04688-8_5

cis longus (APL) dorsally, opponens pollicis (OP) on the volar/radial surface, first dorsal interosseous on the dorsal/medial surface, and a small portion of the flexor pollicis brevis (FPB), yet this not may be in all individuals [3].

Proximal and Distal Phalanges

The thumb is comprised of two phalanges, proximal and distal, whereas the other four digits have three. Phalanges are tubular bones consisting of a body and two extremities on the proximal and distal ends [5].

The body is flat with a concavity on the volar surface and convexity on the dorsal. Rough areas arise on the bony surface to allow for attachment tendon sheaths. The proximal extremity is larger than the distal. The distal extremity ends in two condyles covered in articular cartilage as part of the interphalangeal (IP) joint [6].

The distal phalanx is flat on the volar surface with a roughened and elevated surface to support the pulp of the finger. The distal tip of the distal phalanx has flat and wide expansions called apical tufts, which support the fingertip and nailbed. Specific to the thumb, there is an asymmetrically large insertion for the flexor pollicis longus (FPL) radially, and ungal fossa, and ungal spines. Asymmetry is needed to ensure the thumb pulp always faces the pulps of the other digits [7].

Sesamoids

The thumb often has two sesamoid bones at the distal volar surface of the metacarpal. These are within the tendons of the adductor pollicis and flexor pollicis brevis.

The STT Joint

The triscaphoid joint is comprised of the scaphoid, trapezium, and trapezoid.

STT Ligaments

The main stabilizer of the STT joint is the radio-palmar scapho-trazpezial ligament, one of the four ligaments in the STT ligament complex. The other three are the palmar scapho-trapezial-trapezoidal capsule, dorsal scapho-trapezial-trapezoidal capsule, and the scapho-capitate ligament.

The CMC Joint

This joint, also known as the trapeziometacarpal joint, provides the ability for the human thumb to diametrically oppose the other digits [3].

The geometry of this joint has been described as a "biconcavoconvex universal joint, both concave and convex on each side of the joint, and there are two reciprocally interlocking saddle shapes that oppose one another. The trapezium itself articulates with the thumb metacarpal, index metacarpal, scaphoid, and trapezoid, with a deep groove on the volar ulnar side for the FCR tendon" [1].

This biconcave double saddle joint with a permissively loose capsule that is incongruous and relaxed in the resting position allows for flexion, extension, adduction, and abduction through various blends, resulting in circumduction, and more importantly, during the opposition screw home torque rotation allows the joint to turn incongruous to congruous and becoming a stable joint, allowing for power grip, power pinch, and the precision pinch [1].

The mean amount of joint distraction can be up to 3 mm [1]. The articular surface sacrifices stability for mobility [7]. More specifically, there are three degrees of motion: (1) flexion/extension; (2) abduction/adduction; and (3) medial rotation/lateral rotation. When the metacarpal is abducted, it locks into lateral rotation, and when the metacarpal is adducted, it locks into medial rotation, positioning the thumb for pulp-to-pulp opposition of thumb to fingers [8]. Another way to view it is relative to the thumb's tip, pronation can turn the thumb's distal end to oppose itself to the tips of other fingers (internal rotation) and supination can turn the thumbs tip in the other direction (external rotation) [3].

Technically, the CMC joint has two axes of rotation with flexion/extension and abduction/adduction, and summation of these two axes allow for various movements. Additional movements are due to the CMC joint hinge offset from anatomical planes. Movement in varus and valgus as well as pronation and supination can occur, therefore the motions of flexion/extension, abduction/adduction, and rotation that occur in three planes can occur with only two axes of control this acts different from joints such as the glenohumeral joint that has independent movement in three planes of motion with three axes of control [7].

At the neutral position with the central ridge as the axis of the trapezium, sits 60° from a line through the heads of the second and third metacarpal heads [8].

Isolated movements of abduction and adduction occur in the sagittal plane wherein abduction moves the thumb further from the second metacarpal and adduction closer. Flexion and extension occur in the AP plane wherein flexion moves the tip of the thumb to the palm and extension moves it away. There is also specific terminology for the CMC joint that arise from combined joint motion. Antepulsion and retropulsion occur around an axis passing through the trapezium. During antepulsion, the thumb is brought to touch the base of the fifth metacarpal, with the nail of the thumb parallel to the palm. In this position, the CMC is maximally rotated and flexed. Retropulsion brings the CMC into full abduction and extension so that the first metacarpal sits dorsal to the plane of the rest of the metacarpals. Circumduction is a combined movement of the above that occurs around the axis at the center of the trapezial articular surface. Lastly, opposition is the motion of bringing the thumb tip into contact with the tip of one of the other four digits. This is a combination of antepulsion (CMC), adduction (CMC), flexion (MCP), and pronation. At terminal opposition, the lax CMC joint becomes stable through the screw home torque rotation mechanism, which will be further expanded on below [9].

Thumb position nomenclature has been confusing, and some advocate for clearer terminology, with simplified terminology as resting, hitchhiker, opposition without screw home and opposition with screw home torque rotation. In the resting position, the CMC joint is lax, and in hitchhiker position, the thumb metacarpal is retracted by the extrinsic APL, EPB, and EPL, so the volar beak is not engaged with the trapezial recess. It is only in this position the volar beak ligament is taut, the dorsal ligament complex (DLC) is lax, and the volar beak is disengaged from the recess, allowing for an unstable CMC joint.

Edmunds Screw Home Description

The CMC joint at rest is lax but forms a dynamic force couple through the articular congruence of the volar beak on the first MC base and the trapezium through the action of the dorsal ligament complex. The DLC is the strongest and most important ligament of the CMC joint arising from the trapezium and inserting on the radial base of the thumb MC or sometimes beneath the APL tendon insertion. When the DLC tightens, the CMC joint is compressed leading the volar beak on the thumb MC to recess into the trapezium.

The intrinsic APB abducts the thumb, the OP rotates the thumb, and the FB, AP, and FPL then compress the TM joint into stability. The DLC shortens under tension and becomes oblique as the CMC joint space is compressed. The recess in the trapezium is the pivot area, and the DLC shortens, allowing more powerful stability [1].

Diagram of the CMC (Fig. 5.1)

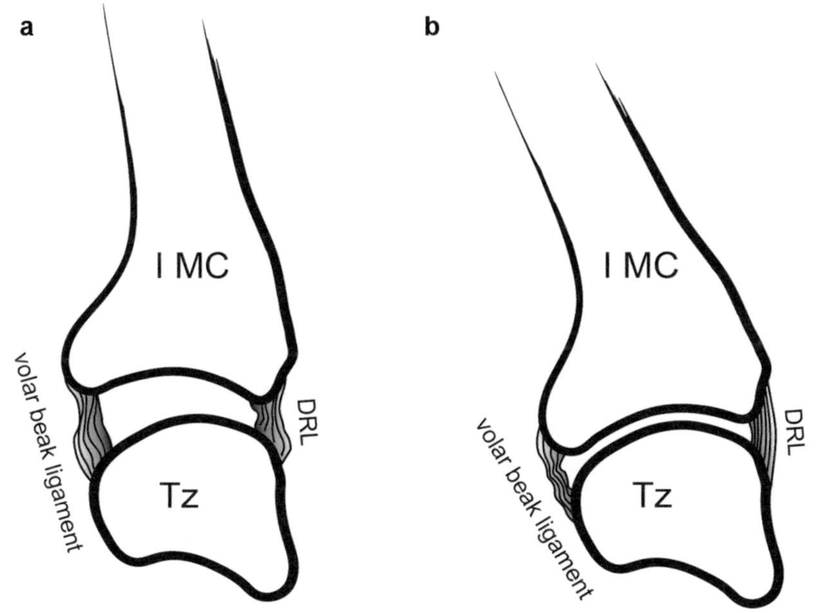

Fig. 5.1 Dynamic force couple. (**a**) The TM joint in the static resting position. The volar beak of the thumb metacarpal is disengaged from its recess in the trapezium, the TM joint space is large, and both the volar beak ligament and the dorsal ligament complex are lax. (**b**) The dynamic TM position of power pinch or power grip with screw home torque rotation, in which the dorsal ligament complex tightens, the volar beak ligament becomes even more lax, the TM joint is compressed, and the volar beak of the thumb metacarpal is compressed into its recess in the trapezium. This forms a dynamic force couple that changes the TM joint from incongruity and laxity to congruity and rigid stability (Drawings by M. D. Wongworawat, adapted from Edmunds JO. Traumatic dislocations and instability of the trapeziometacarpal joint of the thumb. Hand Clinics 2006;22:365–392)

CMC Ligaments

There have been several previous authors who have studied the various ligaments and stabilizers of the CMC. Adding to the confusion are the differing names that refer to the same structure. Bettinger et al. [10] conducted an anatomic cadaveric study to provide a detailed and comprehensive anatomic description of the ligaments stabilizing the trapezium and TMC.

Diagram of Main CMC Ligaments (Fig. 5.2)

Dorsally, the ligaments consist of

1. Dorsoradial (DRL)
2. Posterior oblique (POL)
3. Dorsal trapeziotrapezoid (DTT)
4. Dorsal trapezio-second metacarpal (DT-II MC)
5. Dorsal intermetacarpal (DIML)
6. Radial scaphotrapezial (RST)

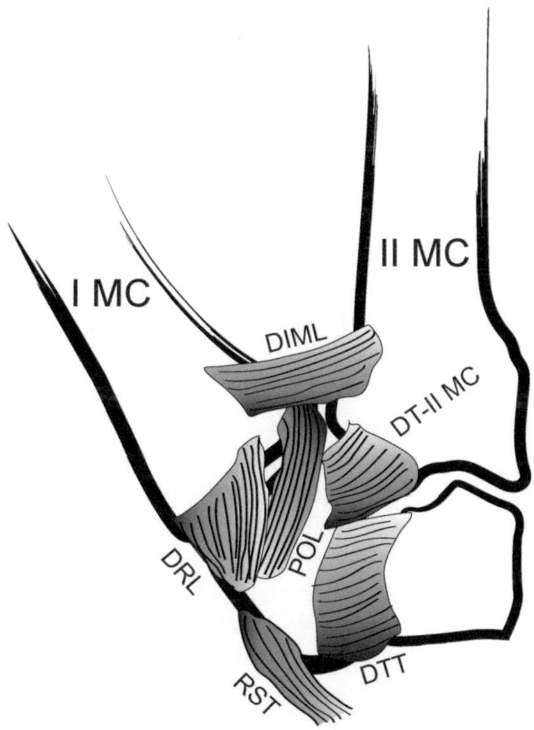

Fig. 5.2 The dorsal thumb CMC ligaments from a right hand showing the dorsal deltoid ligament complex consisting of the dorsal radial ligament (DRL), dorsal central ligament (DCL), and posterior oblique ligament (POL), all emanating from the dorsal tubercle of the trapezium. Also seen are the dorsal bases of metacarpal 1 and 2 (I MC, II MC) Other structures include the DIML, DT-II MC, DTT, and RST. (Drawing by M. D. Wongworawat, adapted from Katarincic KJ. Thumb kinematics and their relevance to function. Hand Clinics 2001;17:169–174)

Volarly, the ligaments are

1. Superficial anterior oblique (SAOL)
2. Ulnar collateral (UCL)
3. Transverse carpal ligament TCL
4. Intermetacarpal (IML)
5. Volar scaphotrapezial (VST)
6. Trapeziocapitate (TC)
7. Volar trapezio second metacarpal (VT-II MC)
8. Trapezio third metacarpal (T-III MC)
9. Volar trapeziotrapezoid
10. Deep anterior oblique ligament aka volar beak ligament

Although there are 16 distinct ligaments, the three most important ligaments: the dorsal ligament complex, the volar beak ligament, and the intermetacarpal ligament [1].

The DLC has also been referred to as the deltoid ligament complex [11]. The dorsal ligament complex is comprised of the dorsal radial ligament (DRL), dorsal central ligament (DCL), and posterior oblique ligament (POL), which all originate from the dorsal tubercle of the trapezium and insert onto the dorsal thumb metacarpal base.

DRL: This originates on the radial side of the dorsal trapezial tubercle and inserts on the dorsal radial base of the thumb metacarpal. This is deep and medial to the insertion of the APL tendon that inserts onto the thumb metacarpal, and the APL reinforces this tendon [7]. This ligament is taut in abduction/flexion and relaxed in antepulsion/opposition [3].

DCL: This is the shortest, thicket ligament that spans the joint [7] and it inserts on the central dorsal edge of the first metacarpal, originating on the center of the trapezium dorsal tubercle.

POL: This originates on the dorsal ulnar edge of the trapezium and inserts onto the dorsal ulnar base of the first metacarpal, "this is the longest ligament and broadest insertion onto the first metacarpal" [7]. This ligament provides stability in flexion, opposition, and pronation. The POL is taut in extension and relaxed in retropulsion/constraining supination [1, 3].

The volar beak ligament (VBL) is often also referred to as either the palmar beak ligament [1, 12], the anterior oblique ligament [13], the ulnar ligament [14], or the palmar ligament [15]. The VBL is comprised of two ligaments, the superficial anterior oblique ligament (SAOL) and the deep anterior oblique ligament (DAOL). The VBL ligament is an intra-articular ligament and spans the trapezium to the ulnar volar base of the thumb metacarpal. This is the closest ligament to the center of the joint and in opposition, the AOL/VBL is a pivot point for pronation [16, 17].

Earlier studies concluded that the anterior ligaments were the primary stabilizers with the dorsal ligaments not playing an important role [4, 10, 12]. However, Bettinger and colleague's cadaveric studies found the anterior ligaments to be weaker compared to the dorsal ligaments and changed their view [12, 18]. The volar beak ligament arises from the volar beak of the thumb metacarpal and inserts on the volar trapezium. In addition, the volar beak ligament is only under tension when the thumb is placed in the hitchhiker position, and it has no part in stabilizing the CMC during power pinch and grasp.

More recent studies [10, 11, 17, 19] concluded the dorsal ligaments restrict the joint dorsally during the screw home phase of opposition, preventing disengagement of the first metacarpal head and dorsal subluxation. However, if the dorsal ligament complex is compromised and power grip is performed, the CMC joint becomes unstable, and the volar beak dislocates [3, 20]. Ladd and Kuhlman further delineated additional dorsal ligaments in cadaveric studies [11, 21] involved in joint stability.

The intermetacarpal ligament rests between the thumb and index metacarpal. If both the DLC and beak ligament are cut, the CMC can dislocate but remains in contact with the index metacarpal [1]. It is the function of the intermetacarpal ligament that is mimicked following trapeziectomy and reconstruction with the FCR, APL, or suspensionplasty. This ligament is firm in abduction, opposition, and supination as it holds the first and second metacarpal tightly together [1]. The volar 1–2 intermetacarpal ligament spans the volar ulnar base of the thumb metacarpal to the radio-volar base of the second metacarpal, whereas the dorsal 1–2 intermetacarpal ligament spans from the ulnar to the dorsoradial tubercle of the second metacarpal adjacent to the volar side of the ECRL [7].

The CMC Muscles

There are four extrinsic muscles (three on the dorsal side), four intrinsic muscles (all on volar side) [1]. The first dorsal interosseous muscle does not cross the joint but stabilizes the joint with contraction and is one of the muscles that is addressed during rehab for CMC joint arthritis [3, 21, 22].

The dorsal extrinsic muscles are the abductor pollicis longus (APL), extensor pollicis longus (EPL), and brevis (EPB). The APL abducts the thumb and supports opposition. It has a superficial and deep aspect, with the deep inserting at the thumb MC at the CMC [23]. The EPL extends and adducts the CMC joint (Chim) [3]. The EPB is the primary extensor of the CMC joint with some contribution from the APL [24]. The FPL is the only volar extrinsic muscle and is responsible for thumb CMC and IP flexion. It also functions during thumb opposition [25].

The intrinsic muscles from superficial to deep in reference to the FPL consist of the abductor pollicis brevis (APB), flexor pollicis brevis (FPB), and opponens pollicis (OP). These make up the thenar eminence (Chim) [3]. The adductor pollicis (AP) forms the bulk of the webspace between the first and second metacarpals. The recurrent branch of the median nerve innervates the superficial thenar muscles (APB, OP, FPB superficial head) while the deep branch of the ulnar nerve innervates the deep thenar muscles (AP, FPB deep head) [22].

The APB is radial to the FPB, and it inserts on the radial side of the proximal phalanx and originating from the TCL, as well as the scaphoid tubercle and trapezium. This is primarily responsible for abduction and opposition. This muscle allows for flexion and abduction of the first metacarpal, and the start of opposition is the act of pulling the thumb away from the palm at a right angle, "this muscle also functions to extend the thumb IP joint through its EPL insertion and ulnarly deviates the MCP joint [6, 26–28].

The FPB has a superficial and deep head. The superficial portion inserts on the base of the thumb proximal phalanx and originates from the tubercle of the trapezium as well as the TCL. The deep head inserts on the radial sesamoid and at the proximal phalanx base and originates on the trapezoid, capitate, and the distal carpal row. The superficial head is lateral and is innervated by the recurrent branch of the median nerve, whereas the deep head is medial and is commonly innervated by the ulnar nerve deep motor branch. The FPB flexes the MCP joint, extends the distal phalanx, and pronates the thumb metacarpal [14, 27–30].

The OP is deep to the APB and inserts on the volar radial surface of the first metacarpal. This also originates from the TCL, tubercle of the trapezium, as well as the CMC joint capsule. This is specifically designed for opposition. This allows for flexion and pronation of the thumb metacarpal and assists the opposition of the APB [14, 31].

The AP has an oblique head and a transverse head. The oblique portion unites with the tendons of the FPB deep head and originates from the second and third metacarpals, capitate, the volar metacarpal ligaments, and can originate as well from the FCR tendon sheath. The transverse head inserts on the ulnar side of the proximal phalanx base as well as the extensor apparatus, and fibers can converge to the FPB deep head. The AP and first dorsal interosseous muscle make the bulk of the first web space, the FDIM is dorsal to the AP. The AP adducts the first metacarpal and can assist with thumb IP joint extension. The ulnar nerve deep motor branch innervates the AP.

The MCP Joint

The metacarpophalangeal joints (MCP) and interphalangeal joints (IP) contribute to flexion and extension.

The MCP joint is a condyloid joint composed of a biconvex metacarpal head and a concave base of the proximal phalanx, with flexion/extension and abduction/adduction [4]. There is a small amount of rotation through this joint, but it is extremely limited by the capsular ligamentous complex [32].

The MCP joint is unique as there is no mechanical equivalent in the human body as it is a ball and socket joint with theoretically three degrees of freedom [4, 20]. The thumb MCP joint is different from the fingers' MCP joints as the thumbs have less range of motion in flexion/extension and abduction/adduction, and the radial articular prominence of the first metacarpal head is larger than the ulnar side [8].

The MCP Ligaments

The stout MCP joint has strong proper and accessory collateral ligaments as well as a strong volar plate acting as static stabilizers. The volar plate acts to prevent hyperextension as well as dorsal subluxation of the joint during pinch.

These proper collateral ligaments originate on the metacarpal head and insert at the base of the proximal phalanx [33]. The accessory collateral ligaments are proximal and volar to the proper collateral ligaments and insert onto the volar plate. However, flexion turns the MCP joint into a hinge joint, as the metacarpophalangeal portion is taut in flexion and looser in extension, and in extension the fanlike collateral ligaments to the palmar plate is taut; therefore, adduction and abduction are limited [8]. When the MCP joint is in extension, pronation but no supination is allowed, and the joint will lock into secure position for secure grasping in supination [8].

The MCP Muscles

The intrinsic adductor brevis augments the ulnar side of the MCP joint and the intrinsic abductor pollicis brevis on the radial side of the joint. The extrinsic FPL stabilizes the MCP joint in flexion, and compressive and shear loads that translate the proximal phalanx dorsally are stabilized by the intrinsic flexor pollicis brevis, as it pulls on the volar plate and base of the proximal phalanx [34]. The FPB pulling on the sesamoids increases the moment arm at the MCP joint, allowing for a more effective FPL [35].

The two sesamoid bones, one lateral to the FPL and one medial, are the last hand bones to ossify, with the medial/adductor sesamoid appearing before the lateral/flexor sesamoid. The volar plate regularly incorporates the two sesamoid bones, and there is a fibrous linkage that is the thick palmar capsule between the two bones that provide a channel for the FPL. These sesamoids are articular in nature and the sesamoids act as insertion sites for the radially located FPB insertion and the ulnar-sided insertion of the AP [8].

The EPB and EPL, attached to the extensor expansion, along with the FPL, can have adduction at the MCP joint, helping the adductor pollicis in providing stability and to balance flexion with pinch and grasp [4, 20].

The IP Joint

The IP joint is a bicondylar joint comprised of the base of the distal phalanx and the condyle of the proximal phalanx. The IP joint range of motion is from 15° hyperextension to 80° of flexion [32].

The IP joint static stabilizers are the volar plate, accessory, and proper collateral ligaments and dorsal capsule, and these resist rotation and lateral movement during flexion and extension [32, 36]. Flexion is solely from the flexor pollicis longus (FPL), yet extension is from the extensor pollicis longus (EPL), as well as the intrinsic adductor pollicis longus (APL) and abductor pollicis brevis (APB) [4].

The IP joint does have one degree of freedom with flexion and extension, but the tip rotates into 5–10° of pronation due to shapes of the articular surfaces of the proximal and distal phalanges. There is a larger circumferential arc for rotation for rotation of the distal phalanx, and this design is helpful for stable pulp to pulp pinching [8]. During pinch, applied force has been measured to be 3 kg of flexor tendon force and 0.5 kg of extensor tendon support [20]. Extension has dynamic resistance of the FPL as well as static support of the volar plate [4]. Degenerative arthritis is most common at the IP joint due to the high compressive forces during use [37].

Function of the Human Thumb

John Napier once proclaimed, "The hand without a thumb is at worst nothing but an animated spatula and at best a pair of forceps whose points don't meet properly" [38]. The thumb is integral in providing precise, powerful, and coordinated hand movements that enabled our early ancestors to develop the ability to make and use tools with heighted manual dexterity [39]. The thumb has also become an instrument of communication as various forms of "body language," whether that be for hitchhiking or to convey approval or disapproval through a thumbs up or thumbs down gesture.

Whether it is prehensile pinch or grasp, the loss of the thumb can be devastating, and the loss of the thumb underscores how important its function is for the human to interact with his or her environment, as this loss of thumb function accounts for 40–50% loss of entire hand function [40–42].

In thumb injuries, there has been a push to focus not just on loss of anatomy to evaluate post injury recovery and ability, but also to focus on patient-reported outcomes in addressing more complex elements of these thumb injuries [43]. In traumatic settings, multiple reconstruction methods have been described and compared with no reconstruction, any procedure is of benefit [44].

Thumb and hand function can be broadly grouped into pinch and grasp maneuvers.

Duncan defined various types of pinches and grasp as detailed below [31]:

Pinch:

1. Precision pinch: Terminal pinch involves thumb and index interphalangeal (IP) joint flexion to pick up small items
2. Oppositional: Subterminal pinch brings pulp of thumb to pulp of index with IP and DIP in extension that provides more stability than precision. The first dorsal interossei as well as index flexor digitorum profundus are contracting
3. Key pinch: Thumb adduction so that pulp of thumb contacts the radial aspect of the middle phalanx of the index
4. Chuck pinch: Three-finger pinch involves the thumb in opposition against the index and third

Intrinsic muscles responsible for precision and key pinch are the adductor pollicis, first dorsal interossei, and flexor pollicis brevis [45].

Biomechanical studies reveal that during pinch both the intrinsic and extrinsic tendons of the thumb sustained 10–30 kg of force. During simple pinch, joint contact forces at the IP, MCP, and CMC joint measured 3 kg, 5.4 kg, and 10 kg, respectively [20].

Grasp:

1. Chuck grasp aka directional grasp allows hand to hold cylindrical object and apply rotational force
2. Hook grasp is also wrapping fingers around object but requires greater finger flexion and does not require the thumb—for example, a door handle
3. Power grasp involves finger flexion with thumb involvement directly opposing other digits to support grip
4. Span grasp used when object completely fills the palm like picking up a ball

Thumb opposition is needed for all prehension that relies on the actions of the APB, OP, and superficial head of the FPL (sFPL). The APB provides the major force generation, with the OP and sFPL acting as secondary motors. Compressive forces up to 2× and 3× of applied load occur at the IP and MCP during grasp and compressive forces of up to 120 kg at the CMC during strong grasp [20, 46].

Alterations of Normal Kinematics

CMC Pathology

1. CMC arthritis.
   (i) CMC joint osteoarthritis (OA) pathogenesis is not due to CMC ligament instability or volar beak ligament degeneration like previously thought [47]. There is stress causing erosions between the volar beak and the trapezium recess. During the final phase of opposition which occurs during grip and power pinch, compressive rotational shear forces are concentrated in the pivot area in the trapezium recess, next to the volar beak. The arthritic changes start at the highest shear contact area where the volar beak engages in the trapezium recess. These compressive shear forces over time can result in joint arthritis. The volar surface of the CMC joint has a significantly higher OA stage and significantly thinner cartilage [48]. The compressive forces at the TM joint can be 12 times the force in lateral pinch at the tips of the thumb and index finger, and the shear stresses are 2.5 times the applied force. Power grasp can also generate up to 120 kilograms of force in the TM joint. The DLC is the key ligament stabilizing these powerful dorsal subluxating forces in power pinch and power grip.
   (ii) There has been interest in the findings of relaxin, a member of the insulin-like growth factor family, that is involved in changing ligament compliance, and there has been a proposed role in relaxin involving attenuation of the ligaments or inhibition of repair of the CMC ligaments [49, 50].
2. The FCR passes through the volar ulnar trapezial groove as it inserts on the second metacarpal base, and this is a frequent site of tendonitis. Ossification can occur over the tendon, which places this at risk during trapezium removal.

MCP Pathology

1. Instability
   (a) Ulnar collateral ligament. The UCL originates from the TCL and trapezial ridge and inserts on the first metacarpal ulnar base. The AOL is deep and radial to the UCL [7]. Also known as a gamekeeper's thumb or skier's thumb, injury to the thumb UCL can lead to decreased effectiveness in thumb pinch and grasp. This is usually caused by a radially directed force, causing a hyper-abduction moment across the MCP joint. The avulsed UCL ligament can come to rest superficial to the adductor aponeurosis, which is referred to as a Stener lesion. The overlying abductor aponeurosis prevents healing of the UCL and generally necessitates surgical repair [51].
   (b) Hyperextension deformity. This deformity is often seen in tandem with CMC arthritis. The MCP joint assumes an extended moment secondary to CMC adduction and pull of the extensor pollicis brevis. These deformities together cause a significant reduction in pinch and grasp strength [52].
2. Arthritis
   (a) Etiologies include primary, post-traumatic, and inflammatory. Osteoarthritis uncommonly affects the thumb MCP joint. In inflammatory conditions such as rheumatoid, the MCP joint can be affected, leading to characteristic deformities, for example, Boutonniere and swan neck deformities [53].

IP Pathology

The DIP joints of the hands are subject to the highest joint forces in the hand and thus undergo more wear and tear [54–56]. Primary osteoarthritis at these joints can present as Heberden's nodules, mucous cysts, and angular deformities causing pain and stiffness. The IP are also involved in many inflammatory arthritic conditions and can lead to erosive osteoarthritis. Limitation in thumb IP joint motion alters joint stability, grasp and pinch strength, and function.

Management of IP joint pathology includes intra-articular injections, implant arthroplasty, and joint fusion.

Summary

The importance of the thumb in human evolution to everyday life cannot be understated. The thumb's unique abilities stem from its unique anatomy as demonstrated mostly by the CMC joint. The thumb sits in a different plane from the rest of the four digits that allow for circumduction to oppose the other digits and prehension. The joints within the thumb sustain great joint reactive forces during use that are stabilized by various ligamentous complexes and tendons. Loss of hand function leads to profound impairment, with most reconstructive efforts aimed at mitigating functional loss.

References

1. Edmunds JO. Current concepts of the anatomy of the thumb Trapeziometacarpal joint. J Hand Surg Am. 2011;36(1):170–82. https://doi.org/10.1016/j.jhsa.2010.10.029.
2. https://pubmed.ncbi.nlm.nih.gov/15781342/.
3. Nichols DS, Oberhofer HM, Chim H. Anatomy and biomechanics of the thumb carpometacarpal joint. Hand Clin. 2022;38(2):129–39. https://doi.org/10.1016/j.hcl.2021.11.001.
4. Imaeda T, An KN, Cooney WP. Functional anatomy and biomechanics of the thumb. Hand Clin. 1992;8(1):9–15.
5. Singh V. Textbook of anatomy upper limb and thorax, vol. 1. 2nd ed. Elsevier Health Sciences; 2014. p. 32. ISBN 978-8-131-23625-3.
6. Gray H. Gray's anatomy. Arcturus Publishing; 2008. p. 192–3. ISBN 978-1-841-93958-2.
7. Komatsu I, Lubahn JD. Anatomy and biomechanics of the thumb carpometacarpal joint. Oper Tech Orthop. 2018;28(1):1–5. https://doi.org/10.1053/j.oto.2017.12.002.
8. Chase R, Hentz V. Hand surgery: a clinical atlas. W. B. Saunders; 2001. ISBN 0721655327.
9. Fontaine C, D'Agostino P, Maes-Clavier C, Boutan M, Sturbois-Nachef N. Anatomy and biomechanics of healthy and arthritic trapeziometacarpal joints. Hand Surg Rehabil. 2021;40S:S3–S14. https://doi.org/10.1016/j.hansur.2020.09.014. Epub 2021 Jun 9. PMID: 34118467.
10. Bettinger PC, Linscheid RL, Berger RA, Cooney WP III, An KN. An anatomic study of the stabilizing ligaments of the trapezium and trapeziometacarpal joint. J Hand Surg Am. 1999;24(4):786–98.
11. Ladd AL, Lee J, Hagert E. Macroscopic and microscopic analysis of the thumb carpometacarpal ligaments: a cadaveric study of ligament anatomy and histology. J Bone Joint Surg Am. 2012;94(16):1468–77. https://doi.org/10.2106/JBJS.K.00329. PMID: 22992815; PMCID: PMC3412634.
12. Pellegrini VD Jr, Olcott CW, Hollenberg G. Contact patterns in the trapeziometacarpal joint: the role of the palmar beak ligament. J Hand Surg Am. 1993;18(2):238–44. https://doi.org/10.1016/0363-5023(93)90354-6. PMID: 8463587.
13. Napier JR. The form and function for the carpometacarpal joint of the thumb. J Anat. 1955;89:362–9.
14. Spinner M, editor. Kaplan's functional and surgical anatomy of the hand, 3rd ed. Philadelphia: J. B. Lippincott; 1984. 446 pp.
15. Zancolli E, Cozzi E. Atlas of surgical anatomy of the hand. Churchill Livingstone; 1992. ISBN 0-443-08176-X.
16. Colman M, Mass DP, Draganich LF. Effects of the deep anterior oblique and Dorsoradial ligaments on Trapeziometacarpal joint stability. J Hand Surg Am. 2007;32(3):310–7. https://doi.org/10.1016/j.jhsa.2006.12.002.
17. Strauch RJ, Behrman MJ, Rosenwasser MP. Acute dislocation of the carpometacarpal joint of the thumb: an anatomic and cadaver study. J Hand Surg Am. 1994;19(1):93–8. https://doi.org/10.1016/0363-5023(94)90229-1.
18. Bettinger PC, Berger RA. Functional ligamentous anatomy of the trapezium and trapeziometacarpal joint (gross and arthroscopic). Hand Clin. 2001;17(2):151–68. https://doi.org/10.1016/S0749-0712(21)00236-5.
19. Van Brenk B, Richards RR, Mackay MB, Boynton EL. A biomechanical assessment of ligaments preventing dorsoradial subluxation of the trapeziometacarpal joint. J Hand Surg Am. 1998;23(4):607–11. https://doi.org/10.1016/s0363-5023(98)80045-2.
20. Cooney WP 3rd, Chao EY. Biomechanical analysis of static forces in the thumb during hand function. JBJS. 1977;59(1) https://journals.lww.com/jbjsjournal/fulltext/1977/59010/biomechanical_analysis_of_static_forces_in_the.4.aspx
21. Smutz WP, et al. Mechanical advantage of the thumb muscles. J Biomech. 1998;31(6):565–70.
22. Kerkhof FD, Deleu G, D'Agostino P, Vereecke EE. Subject-specific thumb muscle activity during functional tasks of daily life. J Electromyogr

Kinesiol. 2016;30:131–6. https://doi.org/10.1016/j.jelekin.2016.06.009.
23. Panchal-Kildare S, Malone K. Skeletal anatomy of the hand. Hand Clin. 2013;29(4):459–71. https://doi.org/10.1016/j.hcl.2013.08.001.
24. Kijima Y, Viegas SF. Wrist anatomy and biomechanics. J Hand Surg Am. 2009;34(8):1555–63. https://doi.org/10.1016/j.jhsa.2009.07.019.
25. Pellegrini VDJ. Osteoarthritis and injury at the base of the human thumb: survival of the fittest? Clin Orthop Relat Res. 2005;438. https://journals.lww.com/clinorthop/fulltext/2005/09000/the_abjs_2005_nicolas_andry_award__osteoarthritis.43.aspx.
26. Doyle JR, Botte MJ. Surgical Anatomy of the Hand and Upper Extremity. 1st ed. Lippincott Williams & Wilkins (LWW); 2002. ISBN: 978-0-39-751725-1.
27. Leversedge FJ. Anatomy and pathomechanics of the thumb. Hand Clin. 2008;24(3):219–29.
28. Leversedge F, Goldfarb C, Boyer M. Chapter 1, Hand; Chapter 5, Neuroanatomy. In: A pocket manual of hand and upper extremity primus manus. Philadelphia: Wolters Kluwer Lippincott Williams & Wilkins; 2010. p. 24–9.
29. Morris's human anatomy, part II: a complete systematic treatise by English and American authors. 4th ed. Philadelphia: Blakiston's Son & Co; 1907. p. 395–8; *Book digitized by Google from the library of Harvard University and uploaded to the Internet Archive.
30. Loréa P, Berthe JV, De Mey A, Coessens BC, Rooze M, Foucher G. Thenar branch of the median nerve and innervation of the trapeziometacarpal joint. Chir Main. 2001;20(4):257–62. https://doi.org/10.1016/S1297-3203(01)00044-0.
31. Duncan SF, Saracevic CE, Kakinoki R. Biomechanics of the hand. Hand Clin. 2013;29(4):483–92. https://doi.org/10.1016/j.hcl.2013.08.003. Epub 2013 Oct 15. PMID: 24209947.
32. Katarincic JA. Thumb kinematics and their relevance to function. Hand Clin. 2001;17(2):169–74. PMID: 11478039.
33. Eaton RG, Littler JW. A study of the basal joint of the thumb. J Bone Joint Surg. 1969;51(A):661–8.
34. Chao EYS, An KN, Cooney WP, et al. Abnormal hand joint forces. Biomechanics of the hand: a basic research study, chap 6. Singapore, World Scientific.
35. Imaeda T, An KN, Cooney WP et al. Anatomy and biomechanics of the trapeziometacarpal (TMC) ligaments. Presented at the Annual Meeting of the ASSH, Orlando, FL, 1991.
36. Imaeda T, An KN, Cooney WP, et al. Anatomy of the trapeziometacarpal ligaments. K Hand Surg [Am]. 1993;18:226.
37. Radin EL, Parker HG, Paul IL. Patterns of degenerative arthritis: preferential involvement of the distal finger joints. Presented at the Annual Meeting of the ASSH, Washington, DC, 1972.
38. Napier J. Hands, vol. 55. Rev ed. Princeton: Princeton University Press; 1993. [Google Scholar].
39. Karakostis FA, Haeufle D, Anastopoulou I, et al. Biomechanics of the human thumb and the evolution of dexterity. Curr Biol. 2021;31(6):1317–1325.e8. https://doi.org/10.1016/j.cub.2020.12.041.
40. Flatt AE. Our thumbs. Proc (Bayl Univ Med Cent). 2002;15(4):380–7. https://doi.org/10.1080/08998280.2002.11927870. PMID: 16333469; PMCID: PMC1276642.
41. Adani R, Woo SH. Microsurgical thumb repair and reconstruction. J Hand Surg (European Volume). 2017;42(8):771–88. https://doi.org/10.1177/1753193417723310.
42. Slocum DB, Pratt DR. Disability evaluation for the hand. JBJS. 1946;28(3) https://journals.lww.com/jbjsjournal/fulltext/1946/28030/disability_evaluation_for_the_hand.9.aspx.
43. Giladi AM, McGlinn EP, Shauver MJ, Voice TP, Chung KC. Measuring outcomes and determining long-term disability after revision amputation for treatment of traumatic finger and thumb amputation injuries. Plast Reconstr Surg. 2014;134(5):746e–55e. https://doi.org/10.1097/PRS.0000000000000591.
44. Graham DJ, Venkatramani H, Sabapathy SR. Current reconstruction options for traumatic thumb loss. J Hand Surg Am. 2016;41(12):1159–69. https://doi.org/10.1016/j.jhsa.2016.09.012. PMID: 27916148.
45. Lee SK, Wisser JR. Restoration of pinch in intrinsic muscles of the hand. Hand Clin. 2012;28(1):45–51. https://doi.org/10.1016/j.hcl.2011.10.002.
46. Toft R, Berme N. A biomechanical analysis of the joints of the thumb. J Biomech. 1980;13(4):353–60. https://doi.org/10.1016/0021-9290(80)90015-9. PMID: 7400164.
47. Jónsson H, Valtýsdóttir ST, Kjartansson O, Brekkan A. Hypermobility associated with osteoarthritis of the thumb base: a clinical and radiological subset of hand osteoarthritis. Ann Rheum Dis. 1996;55(8):540–3. https://doi.org/10.1136/ard.55.8.540.
48. Xu L, Strauch RJ, Ateshian GA, et al. Topography of the osteoarthritic thumb carpometacarpal joint and its variations with regard to gender, age, site, and osteoarthritic stage. J Hand Surg Am. 1998;23(3):454–64.
49. Lubahn J, Ivance D, Konieczko E, Cooney T. Immunohistochemical detection of relaxin binding to the volar oblique ligament. J Hand Surg Am. 2006;31(1):80–4. https://doi.org/10.1016/j.jhsa.2005.09.012.
50. Tregear GW, Bathgate RAD, Hossain MA, Lin F, Zhang S, Shabanpoor F, Scott DJ, Ma S, Gundlach AL, Samuel CS, Wade JD. Structure and activity in the relaxin family of peptides. Ann N Y Acad Sci. 2009;1160:5–10. https://doi.org/10.1111/j.1749-6632.2009.03955.x.
51. Green DP. Green's operative hand surgery. 7th ed. (Wolfe SW, Hotchkiss RN, Pederson WC, Kozin SH, Cohen MS, eds.). Elsevier; 2017.
52. Armbruster EJ, Tan V. Carpometacarpal joint disease: addressing the metacarpophalangeal joint defor-

mity. Hand Clin. 2008;24(3):295–9, vii. https://doi.org/10.1016/j.hcl.2008.03.013. PMID: 18675721.
53. Earp B, Cefalu C, Blazar P. Thumb metacarpophalangeal joint arthritis. J Am Acad Orthop Surg. 2019;27(23) https://journals.lww.com/jaaos/fulltext/2019/12010/thumb_metacarpophalangeal_joint_arthritis.5.aspx.
54. Dahaghin S, Bierma-Zeinstra SM, Ginai AZ, et al. Prevalence and pattern of radiographic hand osteoarthritis and association with pain and disability (the Rotterdam study). Ann Rheum Dis. 2005;64:682–7.
55. Kalichman L, Cohen Z, Kobyliansky E, et al. Patterns of joint distribution in hand osteoarthritis: contribution of age, sex, and handedness. Am J Hum Biol. 2004;16:125–34.
56. Wilder FV, Barrett JP, Farina EJ. Joint-specific prevalence of osteoarthritis of the hand. Osteoarthr Cartil. 2006;14:953–7.

6 Imaging Techniques, Diagnostics, and Evaluations of the Hand

Aisha Suhaiba, Alfonso Mejia, and Farid Amirouche

Introduction

With its intricate network of bones, numerous muscles, complex neurovasculature, and diverse connective tissues, the human hand stands as one of humanity's most vital tools. Its anatomical design allows for a remarkable range of motion and dexterity, making it integral to countless activities, from fine motor tasks to powerful grips [74]. However, this complexity also makes the hand particularly vulnerable to injury, which can be profoundly debilitating.

Accurately identifying the various pathologies that can compromise hand biomechanics, such as fractures, ligament injuries, tendon ruptures, arthritis, and nerve compression syndromes, is crucial for maintaining optimal hand function [73]. For instance, a fracture in the metacarpal bones can severely limit grip strength, while ligament injuries could lead to instability in joint function [31]. A comprehensive approach is necessary to diagnose these conditions effectively [47]. Gathering a detailed patient history allows clinicians to understand the context of the injury, including any previous traumas, existing medical conditions, and the specific mechanisms of the injury [6, 38, 52]. A thorough physical examination is essential to evaluate swelling, range of motion, and any signs of nerve involvement or circulation issues [14, 34].

Imaging modalities, such as X-rays, MRIs, and ultrasounds, play a pivotal role in visualizing the underlying structures of the hand. X-rays are often the first step in assessing fractures, while MRI and ultrasound can provide more detailed views of soft tissue injuries, such as tendon ruptures or ligament damage [5, 71, 73]. In addition to these methods, practicality must also be taken into account. This includes ensuring patient safety during examinations, the availability of diagnostic resources, and the cost-effectiveness of the chosen imaging techniques [1]. By balancing these elements, healthcare providers can maximize their effectiveness in diagnosing and treating hand injuries, ultimately preserving the function and utility of this indispensable tool [4].

A. Suhaiba · A. Mejia
Department of Orthopaedic Surgery, University of Illinois at Chicago, Chicago, IL, USA

F. Amirouche (✉)
Department of Orthopaedic Surgery, Orthopaedic and Spine Institute, Endeavor HealthSystem (Formerly Northshore University HealthSystem), an Affiliate of the University of Chicago Pritzker School of Medicine, Skokie, IL, USA

Department of Orthopaedic Surgery, University of Illinois at Chicago, Chicago, IL, USA
e-mail: famirouche@northshore.org; amirouch@uic.edu

F. Amirouche et al. (eds.), *Biomechanics of the Hand*,
https://doi.org/10.1007/978-3-032-04688-8_6

X-Ray and Computed Tomography (CT)

X-Ray

Clinical Applications

X-ray, as the primary imaging modality for the hand, plays a crucial role in the initial assessment of fractures, dislocations, and joint pathologies [27, 73]. It is rare for a patient to undergo hand surgery without plain radiographs as they aid in assessing bone alignment preoperatively, intraoperatively (fluoroscopy), and postoperatively [73] (Fig. 6.1). Fluoroscopy, a real-time X-ray imaging technique, is commonly used intraoperatively to determine the reduction of dislocations and fractures at multiple time points throughout operative procedures [33]. This method allows for serial multi-plane viewing of the hand and any hardware introduced during the procedure [33]. Similarly, fluoroscopic images are often used in the emergency department during and after the nonsurgical reduction of a fracture to ensure adequate alignment [73]. This is especially important in the assessment of fractures that have the potential to be managed nonoperatively [73]. X-rays also play a key role in monitoring postoperative healing and identifying complications, such as hardware loosening or malunion [30]. In uncomplicated recovery from surgery, an X-ray is often the only imaging modality used to assess recovery in conjunction with the physical examination [30].

Key Considerations

When ordering plain films of the hand, it is important to determine proper hand positioning and plane selection for specific conditions (e.g., PA, lateral, and oblique views) [73] (Fig. 6.2). Although X-rays do not allow for direct visualization of soft tissue structures, stress radiographs are often utilized to evaluate ligamentous instability [41]. For example, increased distance between the scaphoid and lunate on X-ray supports a diagnosis of scapholunate ligament injury [57]. Although it has been standard practice to evaluate the hand using orthogonal views, many recent studies have sought to identify the minimum number of radiographic planes required for adequate diagnosis. For example, emerging evidence suggests that dorsopalmar views alone may be sufficient to diagnose hand osteoarthritis [68, 71]. However, multiple views remain helpful for academic purposes and in the event of a poor image in one plane of a series [20].

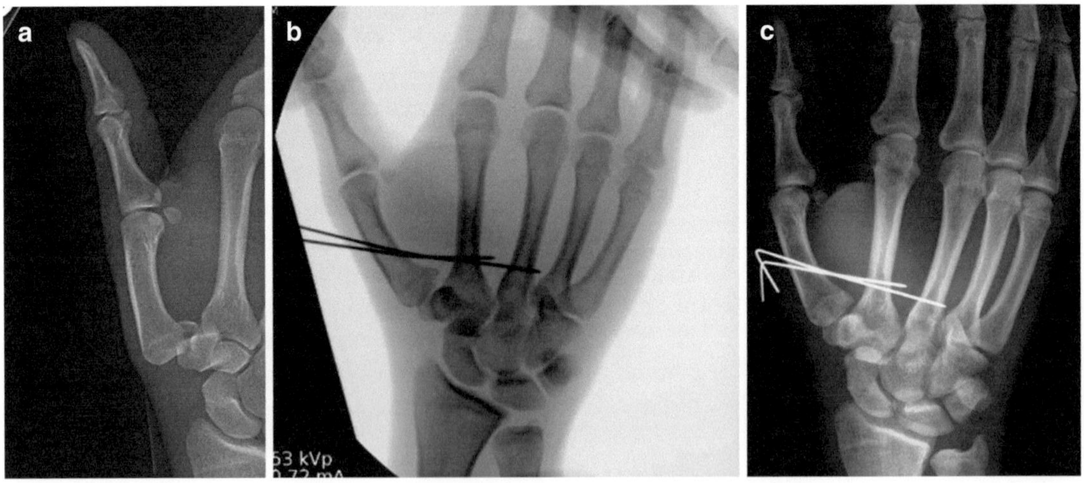

Fig. 6.1 Right first metacarpal (**a**) X-ray before, (**b**) fluoroscopic image during, and (**c**) X-ray after closed reduction and percutaneous fixation of dislocated CMC joint

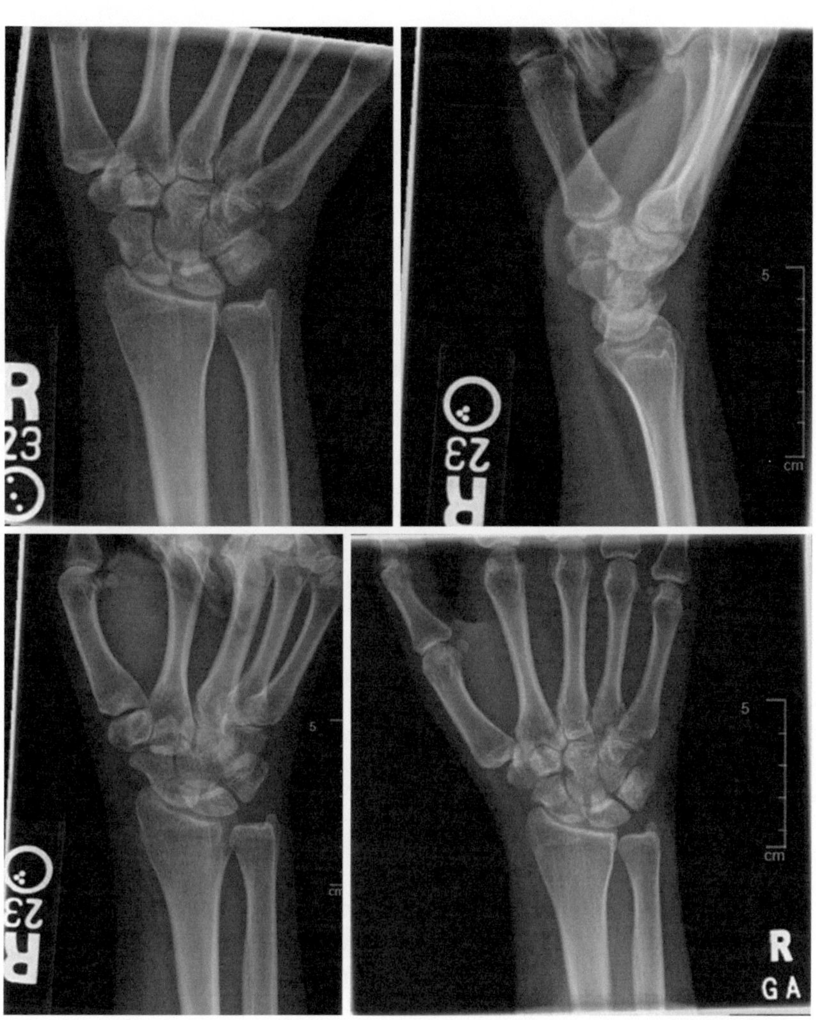

Fig. 6.2 Scaphoid series radiographs of right scaphoid fracture

New Technology

Traditional radiographs have long been the primary imaging modality in clinical practice, serving as a first-line tool for assessing hand pathology. However, they are inherently limited by projection bias, lack of depth information, and an inability to reliably measure anatomical structures and their interrelationships [44]. In recent years, significant advancements have been made in imaging techniques, mainly through the application of three-dimensional (3D) reconstruction from biplanar radiographs. This emerging technology, building on established methods used in the lower extremity, enhances the accuracy of hand visualization by offering a more precise and objective representation of bone morphology [44, 59, 63]. Moreover, it holds great promise for improving surgical planning and enabling earlier detection of ligamentous pathology by providing more reliable measurements [44]. In terms of accuracy, this technique is now approaching the gold standard set by segmented computed tomography (CT) scans, making it an invaluable asset for the detailed assessment of hand biomechanics [44].

Computed Tomography (CT)

Advantages over X-Ray

Computed tomography (CT) utilizes the power of X-rays to produce intricate, 3D cross-sectional

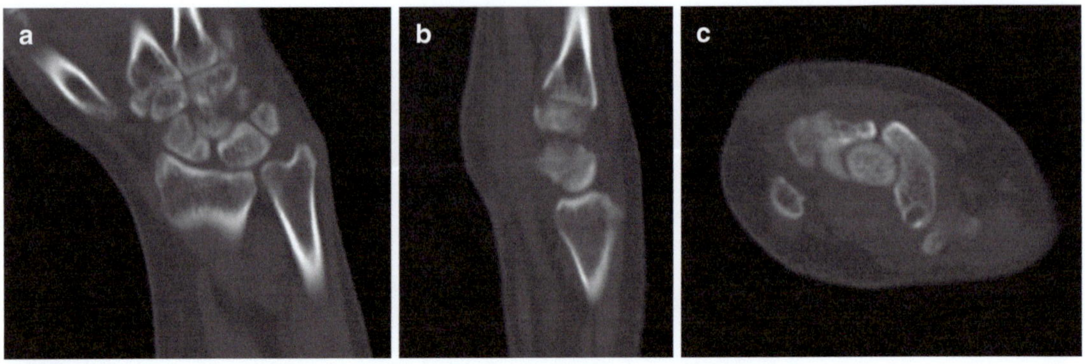

Fig. 6.3 (**a**) Coronal, (**b**) sagittal, and (**c**) axial views of a subtle scaphoid fracture on CT

images that reveal the complexities of injuries and fractures. This imaging modality excels in identifying hidden fractures, such as scaphoid (Fig. 6.3) and stress fractures, that may not be evident on conventional X-ray images [49]. Furthermore, CT is invaluable in evaluating the healing process of fractures, mainly when there is concern for delayed recovery, allowing for a comprehensive assessment of bone integrity and alignment [16].

Clinical Uses

In a clinical setting, CT is commonly used for the detailed evaluation of intra-articular fractures, intra-articular loose bodies, and comminuted fractures, finding frequent employment in traumatic crush injuries [2]. CT scans are often converted to a composite 3D image, typically via proprietary software, to assist with preoperative planning for complex reconstructions [65]. In recent years, these virtual three-dimensional models have been 3D printed to help with accurate visualization and diagnosis when managing complex problems [16]. CT may also evaluate bone when pathologies such as osteoporosis, metabolic bone conditions, or osteomyelitis are suspected [10, 19, 24]. However, imaging techniques with better sensitivity and specificity for these conditions exist [22].

New Technology

Cone-beam computed tomography (CBCT), while still an emerging technique in orthopedics, has become a mainstay in maxillofacial imaging [9, 53]. Unlike traditional CT, CBCT uses a cone-shaped X-ray beam that irradiates the entire volume of interest simultaneously, rather than thin slices. This difference confers several advantages: CBCT systems are typically smaller, lighter, and more affordable, while offering higher spatial resolution, reduced radiation exposure, and faster imaging times [18, 26, 53]. These benefits enable quicker and more precise high-resolution 3D reconstructions of the area being examined [18, 26, 53]. In the context of hand evaluation, CBCT is particularly valuable for revealing intricate fracture patterns, assessing changes in articular alignment, detecting degeneration, and identifying foreign bodies [26, 51, 53]. As a result, CBCT is emerging as a promising modality for both initial assessments and ongoing management, including surgical planning and post-operative follow-up [26, 54].

Magnetic Resonance Imaging (MRI)

Clinical Applications

Magnetic resonance imaging (MRI), harnessing the power of radio waves and magnetic fields, excels at generating strikingly detailed images of soft tissue structures within the body. This includes intricate components such as tendons, ligaments, cartilage, synovial membranes, and nerves [17, 67]. One of the standout features of MRI is its remarkable ability to reveal subtle

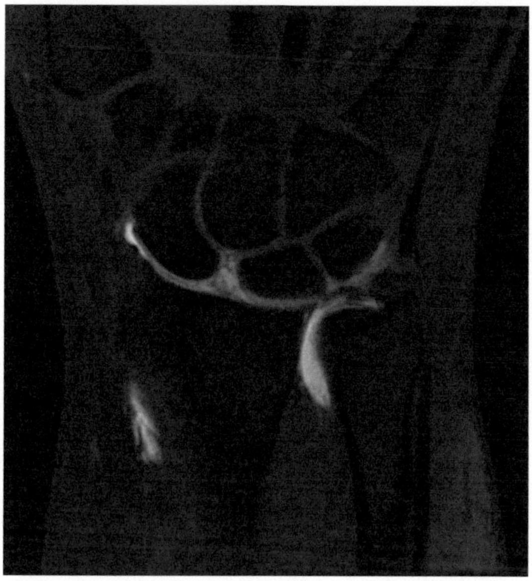

Fig. 6.4 Coronal T1-weighted fat-suppressed view of scapholunate ligament tear and TFCC tear on MRI of the right wrist

pathologies that may elude detection during a physical examination. This capability is invaluable in diagnosing conditions such as tears in the triangular fibrocartilage complex (TFCC) and the scapholunate ligament (Fig. 6.4), which are vividly visualized through this advanced imaging technique [13, 56].

Benefits in Specific Scenarios

The high resolution of the images produced by MRI is particularly beneficial in delineating lesion boundaries, such as ganglion cysts, for surgical planning, even when history and physical examination are sufficient for a confident diagnosis [11, 15, 28]. MRI's unique ability to readily identify occult soft tissue injuries when physical examination findings and radiographs are equivocal is a significant advantage [20, 73]. While MRI cannot be performed as quickly as CT, it does not involve ionizing radiation and is generally favored during pregnancy and in children [12, 22, 40].

In the case of ulnar neuropathy, especially one that results in rapid hypothenar eminence atrophy, MRI is functional for visualizing the ulnar nerve as it passes through the Guyon canal to determine whether a release is required at the canal or more proximally in the cubital tunnel [3].

Scaphoid fractures are missed on 5–20% of initial scaphoid series radiographs, and due to inconsistencies in both imaging protocol and technical acumen, scaphoid series radiographs are often subpar [66, 76]. This is where the expertise of radiologists comes into play. Various algorithms dictate that the series should be repeated if no scaphoid fracture is identified on initial imaging [73]. MRI, CT, or bone scan should be performed within 3–10 days of initial imaging [73]. Inadequate union is the most common complication seen in these injuries, followed by osteonecrosis, typically occurring at the proximal pole due to the scaphoid's retrograde arterial supply [42]. Malunion and osteonecrosis are more readily identified on MRI and CT scans than on X-rays, highlighting the crucial role of radiologists in patient care.

Emerging Techniques

Mechanical dysfunction of the wrist, resulting in weakness and pain, is often referred to as "dynamic instability" and is considered a diagnosis of exclusion when conventional imaging cannot identify clinically significant symptoms of carpal instability. The development of real-time MRI has allowed for dynamic wrist and hand evaluation, leading to accurate diagnosis of hand and wrist loading and movement dysfunction, such as dynamic carpal instability [64, 73].

Previously, electromyography was the primary means of obtaining functional information regarding peripheral nerve injuries, which was used in conjunction with the structural information obtained via MRI for a comprehensive assessment of these injuries. Functional MRI techniques, such as diffusion-weighted imaging (DWI) and diffusion tensor imaging (DTI), have recently been applied to functional magnetic resonance neurography. This technique can provide morphologic and functional information about peripheral nerve injuries [46, 73].

Ultrasound

Advantages

Ultrasound is a cost-effective and noninvasive means of real-time imaging of soft tissue structures. Because it utilizes sound to produce images, there is no radiation risk, and it is preferred in pregnancy and children when ionizing radiation carries a greater-than-average risk of harm [62, 75]. Additionally, ultrasound machines are portable and accessible for point-of-care diagnostics, a recent iteration of which is as simple as a probe plugged into a smartphone or tablet for visualization on the device's screen [60, 75]. With the volume seen in many emergency departments, ultrasound can be a handy tool, minimizing delays in patient care [60]. It is also easy to adjust and repeat the imaging for optimization at the point of care [60].

Clinical Applications

There are many practical applications of ultrasound: tendons can be assessed statically and dynamically for gliding, tears, and tenosynovitis disruptions [36, 50]. The etiology of various neuropathies, such as cubital tunnel syndrome and carpal tunnel syndrome, can be elucidated, assisting with surgical planning [62]. Real-time visualization is helpful for injections into the small spaces in the hand and wrist, as is commonly done for CMC joint osteoarthritis and Dupuytren's contracture [43, 72]. Cystic structures can be identified, and lesions with serous, mucoid, or purulent content are readily aspirated with ultrasound guidance [7]. In patients who present unable to cooperate during physical examination, as is often the case with acute, painful injuries in the emergency department or the young or unconscious patient, gross disruptions to structures in the hand may be identified by ultrasound [37, 73]. Literature also states that metacarpal fractures can be diagnosed with 92.5% sensitivity and 98.3% specificity with X-ray as the gold standard [35], a helpful application in global low-resource catchment areas [23, 25, 69].

Limitations

Although ultrasound imaging may offer lower resolution compared to other advanced imaging modalities such as MRI or CT scans, its effectiveness is heavily contingent upon the operator's proficiency and experience [32]. This reliance on skill emphasizes the importance of training in maximizing the technique's potential. Despite inherent limitations in resolution, ultrasound provides numerous advantages, including rapid accessibility and cost-effectiveness, making it a preferred option in various clinical scenarios [60, 75].

The ability to produce immediate results with relatively low financial investment underscores ultrasound as an essential tool within the medical field. It plays a critical role in initial diagnoses, real-time monitoring of patients, and guiding interventions, thus facilitating timely decision-making and improved patient outcomes [60, 75]. Furthermore, its noninvasive nature and lack of ionizing radiation contribute to its status as a low-risk examination option, further solidifying its value in everyday medical practice.

Nuclear Imaging

Clinical Applications

A bone scan, also known as skeletal scintigraphy, is a nuclear imaging technique that involves injecting a radiopharmaceutical into a vein, which then accumulates in the bone in varying concentrations according to the metabolic activity of the bone. It can be used to detect occult fractures (e.g., stress fractures, scaphoid fractures), osteomyelitis, malignancy, and avascular necrosis, and can play a role in evaluating bone healing and identifying prosthetic loosening [8, 21, 39, 45, 58].

Single-photon emission computed tomography-CT, or SPECT-CT, is a hybrid technique that combines SPECT, which utilizes a radiotracer to create 3D images, with CT. It is an increasingly common choice when the etiology of postoperative pain is unable to be identified using more conventional techniques [29, 48, 55]. It can assess cartilage and ligament injury using arthrography and shows more cortical detail with less sensitivity to metal artifacts when compared to MRI [29, 61, 70].

Conclusion

When evaluating a patient for conditions that compromise the integrity of the hand and wrist, it is essential to carefully select imaging modalities and planes optimized for identifying the specific anatomical structures most likely to be affected. This selection process must balance several critical factors: ensuring that imaging is feasible and can be completed in a reasonable timeframe, minimizing any potential risks to the patient, aligning with the clinical context and practice standards, and maintaining cost-effectiveness to facilitate access to care.

With the advent of new imaging technologies, dynamic imaging of the hand and wrist has become increasingly feasible and practical. These state-of-the-art techniques enhance our diagnostic capabilities by providing real-time visualization of the joint mechanics and soft tissue interactions during movement. Such advancements are crucial in accurately diagnosing ligament injuries, tendon pathologies, and arthropathies.

Incorporating these innovative imaging modalities into clinical practice is essential as we strive to preserve and restore optimal hand function. By employing dynamic imaging, healthcare providers can significantly improve treatment precision, tailor interventions to the individual needs of each patient, and ultimately enhance functional outcomes. The integration of these technologies represents a vital step forward in the comprehensive management of hand and wrist disorders.

References

1. Adler RS. What is the place of ultrasound in MSK imaging? Skeletal Radiol. 2024;53(1699):1709. https://doi.org/10.1007/s00256-024-04642-2.
2. Asran Y, Mutungi T, Shirodkar K, et al. Musculoskeletal computed tomography: how to add value when reporting adult lower limb trauma. J Comput Assist Tomogr. 2025;49(313):326. https://doi.org/10.1097/RCT.0000000000001675.
3. Baal JD, Yoon D, Patel RP, et al. Advanced imaging of the peripheral nerves, from the AJR "how we do it" special series. AJR Am J Roentgenol. 2024;223:e2430826. https://doi.org/10.2214/AJR.24.30826.
4. Balzer A, Patel A, Palisch A. Wrist and hand trauma imaging. Clin Sports Med. 2021;40:625–39. https://doi.org/10.1016/j.csm.2021.05.003.
5. Bohndorf K, Kilcoyne RF. Traumatic injuries: imaging of peripheral musculoskeletal injuries. Eur Radiol. 2002;12:1605–16. https://doi.org/10.1007/s00330-002-1461-8.
6. Bowden G, McNally MA, Thomas SRYW, et al., editors. Orthopaedic history and examination. In: Oxford handbook of orthopaedics and trauma. Oxford University Press; 2010. p 0.
7. Bratsman A, Cano M, Phillips T, et al. Surgical excision versus ultrasound-guided treatment of symptomatic wrist ganglia in pediatric patients. J Pediatr Orthop Soc North America. 2023;5:519. https://doi.org/10.55275/JPOSNA-2023-519.
8. Brenner AI, Koshy J, Morey J, et al. The bone scan. Semin Nucl Med. 2012;42(11):26. https://doi.org/10.1053/j.semnuclmed.2011.07.005.
9. Bromberg N, Brizuela M. Dental cone beam computed tomography. In: StatPearls. Treasure Island: StatPearls Publishing; 2025.
10. Burghardt AJ, Link TM, Majumdar S. High-resolution computed tomography for clinical imaging of bone microarchitecture. Clin Orthop Relat Res. 2011;469:2179–93. https://doi.org/10.1007/s11999-010-1766-x.
11. Chhabra A, Soldatos T, Thawait GK, et al. Current perspectives on the advantages of 3-T MR imaging of the wrist. Radiographics. 2012;32:879–96. https://doi.org/10.1148/rg.323115741.
12. Copel J, El-Sayed Y, Heine RP, Wharton KR. Guidelines for diagnostic imaging during pregnancy and lactation. 2017. https://www.acog.org/clinical/clinical-guidance/committee-opinion/articles/2017/10/guidelines-for-diagnostic-imaging-during-pregnancy-and-lactation. Accessed 12 Mar 2025.
13. Daunt N, Couzens GB, Cutbush K, et al. Accuracy of magnetic resonance imaging of the wrist for

clinically important lesions of the major interosseous ligaments and triangular fibrocartilage complex; correlation with radiocarpal arthroscopy. Skeletal Radiol. 2021;50:1605–16. https://doi.org/10.1007/s00256-020-03701-8.
14. Day CS, Wu WK, Smith CC. Examination of the hand and wrist. N Engl J Med. 2019;380 https://doi.org/10.1056/NEJMvcm1407111.
15. DeFrance MJ, Lutsky KF, Ilyas AM, Matzon JL. MRI utility in hand surgery. Hand (NY). 2021;16(298):302. https://doi.org/10.1177/1558944719861725.
16. Demehri S, Baffour FI, Klein JG, et al. Musculoskeletal CT imaging: state-of-the-art advancements and future directions. Radiology. 2023;308:e230344. https://doi.org/10.1148/radiol.230344.
17. Dewan AK, Chhabra AB, Khanna AJ, et al. MRI of the elbow: techniques and spectrum of disease. AAOS exhibit selection. J Bone Joint Surg Am. 2013;95:e99 1–13. https://doi.org/10.2106/JBJS.L.01621.
18. Doan MK, Long JR, Verhey E, et al. Cone-beam CT of the extremities in clinical practice. Radiographics. 2024;44:e230143. https://doi.org/10.1148/rg.230143.
19. Expert Panel on Musculoskeletal Imaging, Pierce JL, Perry MT, et al. ACR appropriateness criteria® suspected osteomyelitis, septic arthritis, or soft tissue infection (excluding spine and diabetic foot): 2022 update. J Am Coll Radiol. 2022;19:S473–87. https://doi.org/10.1016/j.jacr.2022.09.013.
20. Expert Panel on Musculoskeletal Imaging: Rubin DA, Roberts CC, et al. ACR appropriateness criteria® chronic wrist pain. J Am Coll Radiol. 2018;15:S39–55. https://doi.org/10.1016/j.jacr.2018.03.021.
21. Expert Panel on Pediatric Imaging, Shet NS, Iyer RS, et al. ACR appropriateness criteria® osteomyelitis or septic arthritis-child (excluding axial skeleton). J Am Coll Radiol. 2022;19:S121–36. https://doi.org/10.1016/j.jacr.2022.02.017.
22. Florkow MC, Willemsen K, Mascarenhas VV, et al. Magnetic resonance imaging versus computed tomography for three-dimensional bone imaging of musculoskeletal pathologies: a review. J Magn Reson Imaging. 2022;56(11):34. https://doi.org/10.1002/jmri.28067.
23. Frija G, Blažić I, Frush DP, et al. How to improve access to medical imaging in low- and middle-income countries? eClinicalMedicine. 2021;38 https://doi.org/10.1016/j.eclinm.2021.101034.
24. Gausden EB, Nwachukwu BU, Schreiber JJ, et al. Opportunistic use of CT imaging for osteoporosis screening and bone density assessment: a qualitative systematic review. J Bone Joint Surg Am. 2017;99:1580–90. https://doi.org/10.2106/JBJS.16.00749.
25. Ginsburg AS, Liddy Z, Khazaneh PT, et al. A survey of barriers and facilitators to ultrasound use in low- and middle-income countries. Sci Rep. 2023;13:3322. https://doi.org/10.1038/s41598-023-30454-w.
26. Grunz J-P, Pennig L, Fieber T, et al. Twin robotic x-ray system in small bone and joint trauma: impact of cone-beam computed tomography on treatment decisions. Eur Radiol. 2021;31:3600–9. https://doi.org/10.1007/s00330-020-07563-5.
27. Ha AS, Chang EY, Bartolotta RJ, et al. ACR appropriateness criteria® osteonecrosis: 2022 update. J Am Coll Radiol. 2022;19:S409–16. https://doi.org/10.1016/j.jacr.2022.09.009.
28. Hayter CL, Gold SL, Potter HG. Magnetic resonance imaging of the wrist: bone and cartilage injury. J Magn Reson Imaging. 2013;37(1005):1019. https://doi.org/10.1002/jmri.23845.
29. Huellner MW, Strobel K. Clinical applications of SPECT/CT in imaging the extremities. Eur J Nucl Med Mol Imaging. 2014;41(Suppl 1):S50–8. https://doi.org/10.1007/s00259-013-2533-5.
30. Johnson SP, Chung KC, Zhong L, et al. Use of postoperative radiographs following operative fixation of distal radius fractures. Plast Reconstr Surg. 2016;138:1255–63. https://doi.org/10.1097/PRS.0000000000002746.
31. Jones NF, Jupiter JB, Lalonde DH. Common fractures and dislocations of the hand. Plast Reconstr Surg. 2012;130(722e):736e. https://doi.org/10.1097/PRS.0b013e318267d67a.
32. Joshua F. Ultrasound applications for the practicing rheumatologist. Best Pract Res Clin Rheumatol. 2012;26:853–67. https://doi.org/10.1016/j.berh.2012.10.002.
33. Keil H, Trapp O. Fluoroscopic imaging: new advances. Injury. 2022;53(Suppl 3):S8–S15. https://doi.org/10.1016/j.injury.2022.05.035.
34. Kenney RJ, Hammert WC. Physical examination of the hand. J Hand Surg Am. 2014;39:2324–34. https://doi.org/10.1016/j.jhsa.2014.04.026.
35. Kocaoğlu S, Özhasenekler A, İçme F, et al. The role of ultrasonography in the diagnosis of metacarpal fractures. Am J Emerg Med. 2016;34:1868–71. https://doi.org/10.1016/j.ajem.2016.06.083.
36. Koppikar S, Diaz P, Kaeley GS, Eder L. Seeing is believing: smart use of musculoskeletal ultrasound in rheumatology practice. Best Pract Res Clin Rheumatol. 2023;37:101850. https://doi.org/10.1016/j.berh.2023.101850.
37. Kransdorf MJ, Murphey MD, Wessell DE, et al. ACR appropriateness criteria® soft-tissue masses. J Am Coll Radiol. 2018;15:S189–97. https://doi.org/10.1016/j.jacr.2018.03.012.
38. Krastman P, Mathijssen NMC, Bierma-Zeinstra SMA, et al. Diagnostic accuracy of history taking, physical examination, and imaging for non-chronic finger, hand, and wrist ligament and tendon injuries: a systematic review update. BMJ Open. 2020;10:e037810. https://doi.org/10.1136/bmjopen-2020-037810.
39. Krumme JW, Lauer MF, Stowell JT, et al. Bone scintigraphy: a review of technical aspects and applications in orthopedic surgery. Orthopedics. 2019;42:e14–24. https://doi.org/10.3928/01477447-20181120-05.
40. Kutanzi KR, Lumen A, Koturbash I, Miousse IR. Pediatric exposures to ionizing radiation: carcinogenic considerations. Int J Environ Res Public

Health. 2016;13:1057. https://doi.org/10.3390/ijerph13111057.
41. Lee SK, Desai H, Silver B, et al. Comparison of radiographic stress views for scapholunate dynamic instability in a cadaver model. J Hand Surg Am. 2011;36(1149):1157. https://doi.org/10.1016/j.jhsa.2011.05.009.
42. Li NY, Dennison DG, Shin AY, Pulos NA. Update to management of acute scaphoid fractures. J Am Acad Orthop Surg. 2023;31:e550–60. https://doi.org/10.5435/JAAOS-D-22-01210.
43. Li X, Yi PH, Curry EJ, Murakami AM. Ultrasonography as a diagnostic, therapeutic, and research tool in orthopaedic surgery. J Am Acad Orthop Surg. 2018;26:187–96. https://doi.org/10.5435/JAAOS-D-16-00221.
44. Loisel F, Durand S, Goubier J-N, et al. Three-dimensional reconstruction of the hand from biplanar x-rays: assessment of accuracy and reliability. Orthop Traumatol Surg Res. 2023;109:103403. https://doi.org/10.1016/j.otsr.2022.103403.
45. Love C, Din AS, Tomas MB, et al. Radionuclide bone imaging: an illustrative review. Radiographics. 2003;23:341–58. https://doi.org/10.1148/rg.232025103.
46. Martín Noguerol T, Barousse R, Gómez Cabrera M, et al. Functional MR neurography in evaluation of peripheral nerve trauma and postsurgical assessment. Radiographics. 2019;39(427):446. https://doi.org/10.1148/rg.2019180112.
47. McDonald AP, Lourie GM. Complex surgical conditions of the hand: avoiding the pitfalls. Clin Orthop Relat Res. 2005;65–71 https://doi.org/10.1097/01.blo.0000159825.48072.16.
48. Mokoala KMG, Ndlovu H, Lawal I, Sathekge MM. PET/CT and SPECT/CT for infection in joints and bones: an overview and future directions. Semin Nucl Med. 2024;54:394–408. https://doi.org/10.1053/j.semnuclmed.2023.10.007.
49. Moran DS, Evans RK, Hadad E. Imaging of lower extremity stress fracture injuries. Sports Med. 2008;38:345–56. https://doi.org/10.2165/00007256-200838040-00005.
50. Motamedi D, Bauer AH, Patel R, Morgan TA. Problem solved: integral applications of musculoskeletal ultrasound. J Ultrasound Med. 2021;40:1693–704. https://doi.org/10.1002/jum.15551.
51. Mys K, Stockmans F, Vereecke E, van Lenthe GH. Quantification of bone microstructure in the wrist using cone-beam computed tomography. Bone. 2018;114(206):214. https://doi.org/10.1016/j.bone.2018.06.006.
52. O'Donnell EA, Best MJ, Price MD. The preparticipation orthopedic history and physical examination for american football players. HSS J. 2023;19(299):301. https://doi.org/10.1177/15563316231175648.
53. Pallaver A, Honigmann P. The role of cone-beam computed tomography (CBCT) scan for detection and follow-up of traumatic wrist pathologies. J Hand Surg Am. 2019;44(1081):1087. https://doi.org/10.1016/j.jhsa.2019.07.014.
54. Patzer TS, Grunz J-P, Huflage H, et al. Combining gantry-free cone-beam computed tomography with iterative metal artefact reduction for surgical follow-up imaging of the appendicular skeleton. Eur J Radiol. 2022;155:110465. https://doi.org/10.1016/j.ejrad.2022.110465.
55. Pinski JM, Chen AF, Estok DM, Kavolus JJ. Nuclear medicine scans in total joint replacement. J Bone Joint Surg Am. 2021;103:359–72. https://doi.org/10.2106/JBJS.20.00301.
56. Potter HG, Asnis-Ernberg L, Weiland AJ, et al. The utility of high-resolution magnetic resonance imaging in the evaluation of the triangular fibrocartilage complex of the wrist. J Bone Joint Surg Am. 1997;79:1675–84. https://doi.org/10.2106/00004623-199711000-00009.
57. Rachunek K, Springer F, Barczak M, et al. An algorithmic diagnostic approach to scapholunate ligament injuries based on comparison of x-ray examinations and arthroscopy in 414 patients. J Plast Reconstr Aesthet Surg. 2022;75(3293):3303. https://doi.org/10.1016/j.bjps.2022.04.083.
58. Ralston SH, Corral-Gudino L, Cooper C, et al. Diagnosis and management of Paget's disease of bone in adults: a clinical guideline. J Bone Miner Res. 2019;34:579–604. https://doi.org/10.1002/jbmr.3657.
59. Rohan P-Y, Perrier A, Ramanoudjame M, et al. Three-dimensional reconstruction of foot in the weightbearing position from biplanar radiographs: evaluation of accuracy and reliability. J Foot Ankle Surg. 2018;57(931):937. https://doi.org/10.1053/j.jfas.2018.03.014.
60. Royer DF. Seeing with sound: how ultrasound is changing the way we look at anatomy. Adv Exp Med Biol. 2019;1138(47):56. https://doi.org/10.1007/978-3-030-14227-8_4.
61. Saha S, Burke C, Desai A, et al. SPECT-CT: applications in musculoskeletal radiology. Br J Radiol. 2013;86:20120519. https://doi.org/10.1259/bjr.20120519.
62. Scholtz LC, Rosenberg J, Robbins MS, et al. Ultrasonography in neurology: a comprehensive analysis and review. J Neuroimaging. 2023;33:511–20. https://doi.org/10.1111/jon.13124.
63. Setliff JC, Anderst WJ. A scoping review of human skeletal kinematics research using biplane radiography. J Orthop Res. 2024;42:915–22. https://doi.org/10.1002/jor.25806.
64. Shaw CB, Foster BH, Borgese M, et al. Real-time three-dimensional MRI for the assessment of dynamic carpal instability. PLoS One. 2019;14:e0222704. https://doi.org/10.1371/journal.pone.0222704.
65. Sivagurunathan G, Shirodkar K, Hegde G, et al. Musculoskeletal computed tomography: how to add value when reporting adult upper limb trauma. J Comput Assist Tomogr. 2023;47(264):276. https://doi.org/10.1097/RCT.0000000000001417.
66. Smith M, Bain GI, Turner PC, Watts AC. Review of imaging of scaphoid frac-

tures. ANZ J Surg. 2010;80(82):90. https://doi.org/10.1111/j.1445-2197.2009.05204.x.
67. Sneag DB, Abel F, Potter HG, et al. MRI advancements in musculoskeletal clinical and research practice. Radiology. 2023;308:e230531. https://doi.org/10.1148/radiol.230531.
68. Staats K, Sunk I-G, Weidekamm C, et al. Hand x-ray examination in two planes is not required for radiographic assessment of hand osteoarthritis. Ther Adv Musculoskelet Dis. 2020;12:1759720X20934934. https://doi.org/10.1177/1759720X20934934.
69. Stewart KA, Navarro SM, Kambala S, et al. Trends in ultrasound use in low and middle-income countries: a systematic review. Int J MCH AIDS. 2020;9(103):120. https://doi.org/10.21106/ijma.294.
70. Strobel K, van der Bruggen W, Hug U, et al. SPECT/CT in postoperative hand and wrist pain. Semin Nucl Med. 2018;48:396–409. https://doi.org/10.1053/j.semnuclmed.2018.05.001.
71. Tiegs-Heiden CA, Howe BM. Imaging of the hand and wrist. Clin Sports Med. 2020;39:223–45. https://doi.org/10.1016/j.csm.2019.10.003.
72. To P, McClary KN, Sinclair MK, et al. The accuracy of common hand injections with and without ultrasound: an anatomical study. Hand (NY). 2017;12(591):596. https://doi.org/10.1177/1558944717692086.
73. Torabi M, Lenchik L, Beaman FD, et al. ACR appropriateness criteria® acute hand and wrist trauma. J Am Coll Radiol. 2019;16:S7–S17. https://doi.org/10.1016/j.jacr.2019.02.029.
74. Valero-Cuevas FJ. An integrative approach to the biomechanical function and neuromuscular control of the fingers. J Biomech. 2005;38:673–84. https://doi.org/10.1016/j.jbiomech.2004.04.006.
75. Wang S, Hossack JA, Klibanov AL. From anatomy to functional and molecular biomarker imaging and therapy: ultrasound is safe, ultrafast, portable, and inexpensive. Investig Radiol. 2020;55:559–72. https://doi.org/10.1097/RLI.0000000000000675.
76. Yin Z-G, Zhang J-B, Kan S-L, Wang X-G. Diagnosing suspected scaphoid fractures: a systematic review and meta-analysis. Clin Orthop Relat Res. 2010;468:723–34. https://doi.org/10.1007/s11999-009-1081-6.

7 Motion Analysis of Hand Movement

Yiru Wang and Farid Amirouche

Introduction

In this chapter, we delve into motion's intricacies, as shown in Fig. 7.1, produced by the human hand, focusing on the biomechanical principles that govern its remarkable versatility. Motion analysis of the hand explores how its anatomy—comprising bones, joints, muscles, tendons, and ligaments—enables a range of movements, from precise fine motor tasks to powerful grips. The insights gained from studying hand motion are also pivotal in sports, where performance, injury prevention, and rehabilitation often rely on a deep understanding of hand biomechanics. This understanding is critical for diagnosing and treating hand-related pathological conditions, optimizing athletic performance, developing tailored rehabilitation programs for sports injuries, and designing advanced technologies such as prosthetics, robotic systems, and virtual reality. By examining the kinetics, kinematics, and clinical applications of hand motion, this chapter aims to highlight the pivotal role of motion analysis in advancing healthcare, sports science, and engineering solutions.

The original version of the chapter has been revised. A correction to this chapter can be found at https://doi.org/10.1007/978-3-032-04688-8_29

Y. Wang
Department of Orthopaedic Surgery, University of Illinois at Chicago, Chicago, IL, USA

F. Amirouche (✉)
Department of Orthopaedic Surgery, Orthopaedic and Spine Institute, Endeavor HealthSystem (Formerly Northshore University HealthSystem), an Affiliate of the University of Chicago Pritzker School of Medicine, Skokie, IL, USA

Department of Orthopaedic Surgery, University of Illinois at Chicago, Chicago, IL, USA
e-mail: famirouche@northshore.org

Foundations of Motion in Biomechanics

What Is Motion?

Motion is a fundamental concept that intertwines two essential variables: time and space. It describes how an object's position alters over time in a phenomenon classified as being in motion. This captivating process can be segmented into three primary categories: linear motion, where an object moves along a straight path; rotational motion, where an object pivots around a fixed axis; and coupled movements, also known as general movements, which involve a combination of both linear and rotational motion. Each of these movement types is governed by the Newton–Euler laws of physics, which provide a structured understanding of how objects navigate the physical world [2].

To fully understand the intricacies of hand motion, two critical biomechanical concepts

F. Amirouche et al. (eds.), *Biomechanics of the Hand*,
https://doi.org/10.1007/978-3-032-04688-8_7

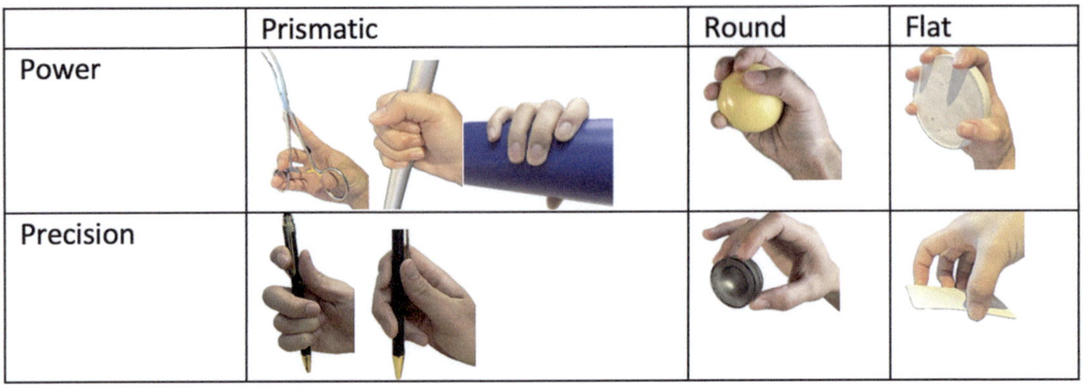

Fig. 7.1 Nine example grasp types based on functionality (power and precision) and object shape (prismatic, round, and flat)

come to the forefront: kinematics and kinetics. Kinematics is the branch of mechanics that examines the motion of the hand without regard to the forces that cause this motion. It looks at the hand's position, velocity, and acceleration, focusing on how these quantities change over time. Conversely, kinetics deals with the forces and torques influencing hand movements, revealing how these factors generate motion. Together, kinematics and kinetics offer a comprehensive perspective on the complex dynamics of hand movements, and often, principles of dynamics such as work and energy are used to quantify the effort needed for an object to move from one position to another. This is illustrated by how various forces, angular displacements, and velocities interplay to create free movements—like waving a hand—and constrained movements, such as gripping an object. This nuanced understanding of hand motion is crucial for biomechanics, robotics, and rehabilitation.

Types of Motion

Linear Motion (Translational Motion)

Linear motion occurs when all parts of an object move the same distance, in the same direction, and at the same time along a straight or curved path. In hand biomechanics, this type of motion refers to movements without rotation or angular displacement of the hand or its segments. Based on the nature of the path, linear motion can be further classified into *rectilinear motion*, which involves movement along a straight line, and *curvilinear motion*, characterized by movement along a curved path. For example, performing a chest pass in basketball exemplifies rectilinear motion, while executing a forehand stroke in tennis illustrates curvilinear motion.

Rotational (Angular) Motion

Rotational motion involves movement around a fixed axis or pivot point, where different hand parts move through varying distances along a circular path. Typical examples are flexion and extension of the fingers at the metacarpophalangeal joints and the rotational movements of the forearm during pronation and supination, such as the wrist's rotation during a golf swing and the intricate finger rotation in badminton.

Coupled Motion

Coupled motion is the coordinated combination of linear and rotational movements, enabling complex and multidimensional hand functions. Actions such as reaching to grasp an object, where the hand moves linearly toward the target while simultaneously rotating to orient the fingers appropriately, are coupled motions. In sports, an example is a tennis player reaching to hit a serve, where the hand moves linearly upward to meet the ball while simultaneously rotating to apply spin and control the racket's angle.

Understanding hand motion is crucial in biomechanics, and it is guided by two key biochemical principles: kinematics and kinetics. *Kinematics*, in particular, is a cornerstone in this field as it focuses on motion's spatial and temporal characteristics. It provides vital insights into how hand movements are executed without delving into the forces involved. The analysis of displacement, velocity, and acceleration of the hand and its components in space is a fundamental aspect of kinematics. The concept of a *degree of freedom (DOF)*, which represents an independent movement a joint or system can perform, is a key element in defining the motion capabilities across specific planes or axes. Specific reference terminologies and spatial systems are used to standardize and communicate kinematic analyses, laying the groundwork for further biomechanical studies.

Joint Movement Terminology

The motion of hand joints is typically described using standardized anatomical terms. These terms define the direction and plane of movement, ensuring clarity in biomechanical analysis [20].

Term	Definition	Example
Flexion	Decreasing the angle between two bones or surfaces at a joint	Bending the fingers to form a fist
Extension	Increasing the angle between two bones or surfaces at a joint	Straightening the fingers
Abduction	Movement away from the midline of the hand or body	Spreading the fingers apart
Adduction	Movement toward the midline of the hand or body	Bringing the fingers back together
Opposition	Movement of the thumb across the palm to touch another fingertip	Touching the thumb to the index finger
Pronation	Rotational movement turning the palm downward	Turning the palm to face the ground
Supination	Rotational movement, turning the palm upward	Turning the palm to face upward

Spatial Reference Systems

Kinematic analyses of hand motion often use a three-dimensional spatial reference system to quantify and describe movements relative to defined axes and planes.

Component	Definition	Application in hand kinematics
Axes		
X-axis (mediolateral)	Horizontal axis running side to side	Describes lateral movements, such as wrist ulnar deviation
Y-axis (vertical)	Vertical axis running up and down	Tracks elevation changes, such as wrist extension
Z-axis (anteroposterior)	Horizontal axis running front to back	Measures movements in depth, such as forward reach
Planes		
Sagittal plane	Divides the body into left and right halves	Flexion and extension of the fingers occur in this plane
Frontal plane	Divides the body into front and back halves	Abduction and adduction of fingers occur in this plane
Transverse plane	Divide the body into top and bottom halves	Pronation and supination of the forearm occur in this plane

Kinetics is the branch of biomechanics that studies the forces and moments (torques) responsible for producing motion. Unlike kinematics, which describes how the hand moves, kinetics explains why the hand moves in a certain way by analyzing the forces behind the motion. In hand motion analysis, kinetics commonly measures the following parameters, each with practical applications that engage and interest the audience:

Joint Moments (Torques) The rotational effect of forces around a joint, such as the torque at the wrist during supination or pronation. A prime example is the torque at the wrist during a baseball pitch, where wrist pronation generates spin on the ball to influence its speed and trajectory.

Muscle Forces The internal forces generated by muscle contractions to move or stabilize the hand, such as those involved in gripping or pinching. In rock climbing, the intense muscle contractions required for gripping holds and stabilizing the hand against the rock surface exemplify the importance of studying muscle-generated forces.

Pressure Distribution How forces are distributed across the hand's surface during contact with objects is relevant for analyzing ergonomic design or prosthetic fit.

Other kinetic parameters, such as inertia, center of gravity, and impulse, can also be measured by hand. In boxing, analyzing the pressure distribution across the hand and knuckles during a punch helps optimize glove design and minimize injury risks.

Building upon the basic concepts in kinetics, *mechanical loads* play a crucial role in understanding hand biomechanics by describing the forces and stresses acting on the hand during various activities. These loads are inherently *vector quantities*, meaning they have both magnitude and direction, influencing their biomechanical effects. For instance, *work*, the force applied over a distance, is evident during gripping or weight-bearing tasks, such as a powerlifter gripping a barbell, where *compression* stabilizes bones and joints by applying forces along specific axes. Similarly, mechanical energy is expended and transferred during movements like finger flexion and extension, such as when a tennis player grips and swings a racket, where tendons and ligaments experience tension through pulling forces oriented along their structure.

Unlike other body parts, the hand contains relatively few muscles, relying heavily on tendons and ligaments to transmit forces and generate motion. This unique characteristic makes the hand a fascinating and challenging subject of kinetic studies. The forces generated in the forearm muscles are transferred through tendons, such as the flexor and extensor tendons, to facilitate precise and powerful hand movements. These structures are subjected to significant *loads* during activities, such as *shear* forces when sliding objects or torsion during twisting motions like opening a jar. However, because tendons and ligaments lack direct contractile properties and are challenging to measure in vivo, quantifying the precise forces and stresses they endure presents a significant obstacle in kinetic analysis. Advanced technologies such as motion capture, pressure sensors, and computational modeling have become crucial tools for addressing these challenges.

The effects of these vector forces and the associated energy expenditure depend on their magnitude, direction, and duration. Appropriate loading and energy use foster tissue adaptation, such as increased tendon stiffness or bone density. At the same time, excessive or repetitive work may lead to injuries like tendonitis, ligament sprains, or cartilage wear. Recognizing the interplay of forces, work, energy, and the unique biomechanics of tendons and ligaments is essential for analyzing hand function, optimizing rehabilitation strategies, improving ergonomic designs, and advancing surgical interventions.

Anatomy and Biomechanics of the Hand

Structural Overview

The *human hand*, with roughly *23 total degrees of freedom (DOF)* [44], consists of 27 bones, 29 joints, and over 120 ligaments; it provides a foundation for diverse functional tasks. The *carpal bones* (eight in total) form the wrist, which has *two DOFs*, allowing flexion/extension and radial/ulnar deviation—commonly measured in wrist motion analysis. The *metacarpals* (five bones) serve as levers during tasks like gripping and pushing, transmitting forces to the *phalanges* (14 bones). Key joints include the *MCP joints* with *two DOF* (flexion/extension and abduction/

adduction) and the *PIP and DIP joints*, each with *one DOF* (flexion/extension), often evaluated in studies of grasp and release patterns. With its three DOFs, the thumb carpometacarpal (CMC) joint enables complex motions such as opposition, a critical feature in precision grip analysis. Ligaments, including the *collateral ligaments* and the *volar plate*, provide stability and limit excessive motion, ensuring controlled and functional movements. This comprehensive range of motion makes the hand a primary focus for motion analysis in clinical and research settings [13].

Lever Systems

The *biomechanical functions* of the hand are optimized for both *precision* and *strength*, enabling a wide range of tasks from delicate manipulations to powerful grips. The hand operates through a system of *lever mechanisms*, where the *bones* act as levers, the *joints* serve as fulcrums, and the *muscles and tendons* provide the forces necessary for movement. There are mainly three classes of a lever based on the location of the fulcrum and the length of the effort and weight arm [21]. *First-class levers*, with the fulcrum positioned between the force and the load, are less common in hand but can be observed in wrist extension during sports like archery, where the wrist acts as the fulcrum to stabilize the bowstring pull; *second-class levers* where the load is between the fulcrum and force, can be seen during a rock climber's grip; and *third-class levers*, the most common in the body, have the force applied between the fulcrum and the load (e.g., biceps lifting the forearm), favoring speed and range of motion. A good example is the rapid finger flexion during a basketball shot or a baseball pitch, where the metacarpophalangeal (MCP) joints function as third-class levers to control and guide the ball. The coordinated force transmission across the tendons and joints also ensures efficient load distribution, particularly during activities like volleyball spikes or a boxer's punch. The *ligaments*, including the *collateral ligaments* and *volar plates*, stabilize the joints and prevent excessive motion, contributing to the hand's structural integrity. These biomechanical functions are central to the hand's ability to perform complex motions, making it a focus of clinical and research analyses in motion biomechanics.

Common Hand Movements and Motion Patterns

Functional Motions

Human hand movements can be classified based on the type of grip/grasp employed, each tailored to specific tasks and object geometries. These movements fall into power grips, precision grips, and pinch grips. Depending on the fingers involved, the location of the contact points, or the object's shape, a diverse range of grasp types can be utilized to accommodate various activities of daily living. *Power grips* maximize the contact area and stability between the hand and the object, engaging the palm and fingers for force-intensive tasks. Examples include the cylindrical grip used for holding a hammer and the spherical grip used to grasp a ball [11]. *Precision grips* emphasize fine motor control and minimal contact, allowing delicate manipulation of objects. Examples include the tip-to-tip grip for threading a needle and the palmar pinch for picking up small items. These grips depend heavily on thumb opposition and refined inter-joint coordination [14]. *Pinch grips* are a specialized subset of precision grips that involve pressing an object between the thumb and another finger, typically without affecting the palm. Common forms include the lateral pinch, used to hold flat objects like cards or keys, and the tripod pinch, where the thumb, index, and middle fingers stabilize an object. These grips are essential for writing, turning a key, or handling coins [14, 57]. The GRASP taxonomy, a comprehensive classification system encompassing 33 unique grasp types, provides a framework for categorizing these actions based on thumb position, finger roles, and object geometry [14]. Figure 7.1 illustrates a few selected examples.

Dexterity represents the hand's ability to execute precise, adaptable, and coordinated movements, making it the cornerstone of fine motor skills. This capability enables tasks requiring rapid, controlled motions and precise force application, such as playing a musical instrument or assembling small components. The concept of opposition is central to both gripping and dexterity, where different parts of the hand work together to stabilize and manipulate objects. *Pad opposition* involves thumb-to-finger pad contact, ideal for delicate and precise tasks. *Palm opposition* engages the palm and fingers to secure a grip on larger or heavier objects. Lastly, *side opposition* features the thumb opposing the side of the index finger, commonly used in lateral grips like holding a key [14].

Motion variability highlights the diverse range of hand movements influenced by individual differences and task-specific adaptations. Age, injury, and pathology significantly shape hand motion, with older adults often experiencing reduced dexterity and grip strength due to degenerative changes. In contrast, injuries may lead to compensatory movement patterns—additionally, specific tasks like sports or musical performance demand unique adaptations in hand motion to optimize performance. For example, athletes develop specialized grips for better control and stability, while musicians refine finger coordination and speed for complex compositions.

Pathological conditions affecting the hand, such as rheumatoid arthritis (RA), can profoundly impair hand function and motion. RA is a chronic autoimmune disorder that primarily targets the joints, including the metacarpophalangeal (MCP) and interphalangeal (IP) joints, causing inflammation, swelling, pain, and eventual deformity [6]. Common hand deformities associated with RA include ulnar drift, buttonhole deformity, and swan-neck deformity. Among these, ulnar drift, where the fingers deviate toward the ulnar side, is the most prevalent, affecting 44% of patients within the first 10 years of diagnosis [26]. The biomechanical etiology of deformity remains an active area of research [38]. Dupuytren's contracture is another example. It involves the thickening and shortening of the palmar fascia, leading to the progressive flexion of the fingers, particularly the ring and little fingers. As the disease progresses, the fingers become increasingly fixed in a bent position, limiting the ability to perform basic tasks such as gripping or holding objects [23]. Dupuytren's contracture causes a significant alteration in gripping biomechanics, as the contracture prevents the hand from fully opening, thereby affecting both power grips and precision grips.

Sports-Related Hand Movements and Motion Patterns

Hand movements in sports are highly specialized and vary depending on the demands of the activity, involving unique combinations of power, precision, and pinch grips to achieve optimal performance. Power grips are essential in sports like weightlifting, where athletes rely on a firm cylindrical grip to stabilize and lift heavy bars, and rock climbing, where the spherical grip is crucial for holding rounded holds on climbing walls. Precision grips play a significant role in sports requiring fine motor skills, such as archery, where a tip-to-tip grip allows the athlete to delicately hold and release the bowstring, and golf, where a refined palmar pinch ensures precise control over the club's angle and orientation. Pinch grips are critical in activities like football, where a lateral pinch is used to grip and spin the ball during a throw, and billiards, where a tripod pinch stabilizes the cue for accurate shots. These motions are often accompanied by rapid changes in force, direction, and coordination, highlighting the complexity of hand biomechanics in sports. The repetitive and high-impact nature of these movements also exposes athletes to a higher risk of hand injuries, such as ligament sprains, tendonitis, or even fractures, emphasizing the importance of understanding and optimizing motion patterns to enhance performance and prevent injury.

Tools and Techniques for Hand Motion Analysis

Hand motion analysis is a multidisciplinary field that integrates biomechanics, neuroscience, engineering, and medicine to understand and measure the complexities of hand movement. Various tools and techniques are utilized to quantify, monitor, and evaluate hand motion. Below is a detailed exploration of these tools and techniques, grounded in scientific evidence and supported by current research.

Quantitative motion capture systems play a pivotal role in studying the dynamic range of hand movements by transforming physical actions into precise, measurable data, thereby offering critical insights into motion mechanics and biomechanics. *Optical systems* are extensively used in motion capture, leveraging cameras to track reflective markers or operate with markerless technologies [4]. Despite significant advancements in wearable sensors, these systems remain the gold standard in motion analysis due to their unmatched precision and reliability. Marker-based systems, such as Vicon and OptiTrack, utilize markers strategically placed on the skin over bony landmarks to capture hand motion accurately [7, 25, 38]. Figure 7.2 shows an example of a setup using reflective markers [45]. These markers are categorized into two types: active markers, which emit infrared light to enhance signal clarity and minimize interference [58], and passive markers, which consist of reflective surfaces offering a more straightforward and more cost-effective solution [30].

Among those systems, both 2D and 3D data can be collected, 3D Motion Analysis captures complex hand movements in three-dimensional space, enabling researchers to study intricate interactions between hand joints and tendons. Protocols and systems optimized for specific applications such as swimming, which requires underwater camera setups and overcoming challenges of refraction, are also an area of interest [37]. Applications include virtual reality, prosthetic development, and surgical simulations [16].

Force and torque, two important aspects of motion kinetics, are commonly measured by dynamometers with direct gauges quantifying the force exerted during a hand squeeze and are widely used in clinical settings and sports science to assess muscle strength and endurance [12]. With a demand for dynamic force and torque measures, instrumented gloves and wearable sensors are also used with increasing DOF for real-time monitoring in various research settings [50].

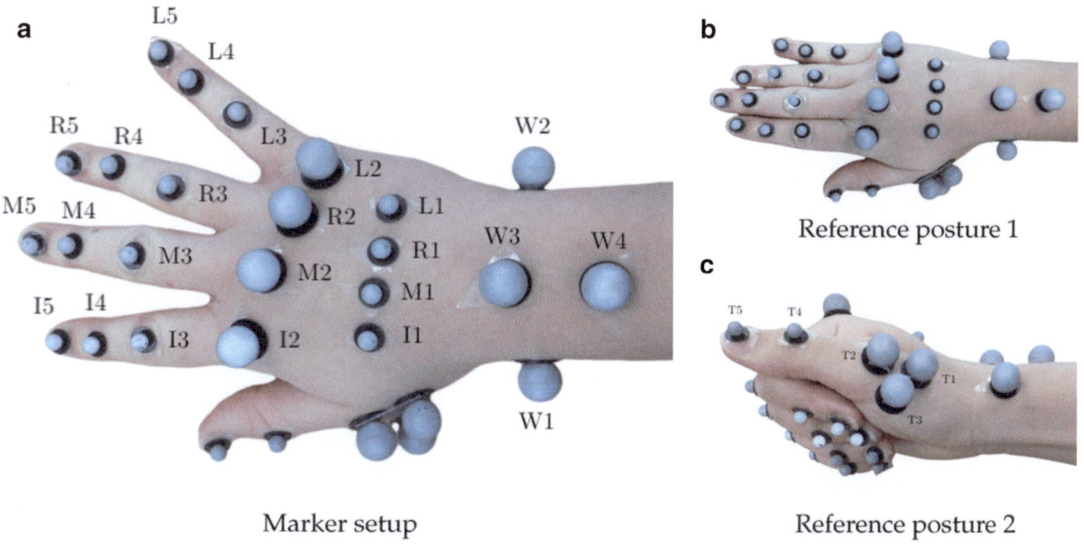

Fig. 7.2 29 marker setup (**a**) with two reference postures in (**b**) and (**c**) [45]

Electromyography (EMG) is a standard technique for analyzing muscle electrical activity. EMG sensors detect the activation patterns of hand muscles during various tasks. The method can distinguish between voluntary and reflexive movements, providing insight into neuromuscular control [24]. It is commonly used to design prosthetics that mimic natural muscle activation patterns and diagnose neuromuscular disorders.

Other *advanced imaging techniques* allow for visualizing soft tissue and joint mechanics. Magnetic resonance imaging (MRI) provides detailed images of joints, tendons, and muscles, enabling the assessment of structural integrity and dynamic function. Ultrasound offers a real-time view of tissue motion, particularly useful for analyzing tendons during hand movement [1].

Each of the tools mentioned above can generate large datasets with varying dimensions. As a result, *computational models* that integrate data from multiple sources to simulate hand mechanics are now widely employed, including machine learning models designed to train robots. Biomechanical models and simulations use finite element analysis to study stress distribution and joint loading. Figure 7.3 shows an example of finite element analysis of the bones of the hand. Simulations help study pathomechanics and the changes resulting from the disease progression or predict the outcomes of surgical interventions and assist in designing ergonomic tools [3]. Machine learning algorithms analyze hand motion data to identify patterns associated with specific tasks or disorders. These models have been widely used in surgical task simulation and automation [54].

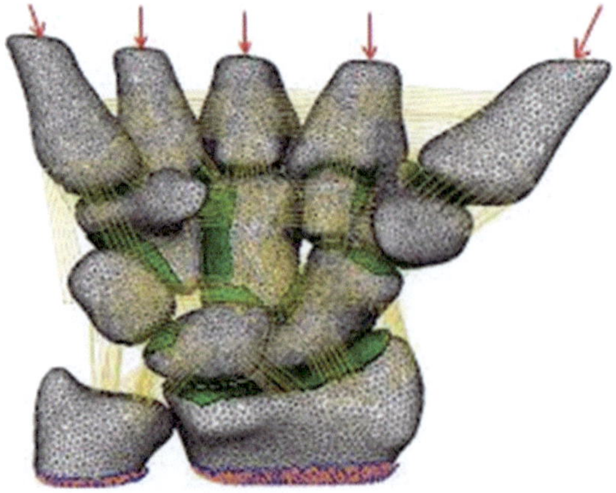

Fig. 7.3 Finite element analysis of the hand bones. Yellow marks the ligaments, green marks the cartilage, and red arrows show the force applied

Physical examination is foundational for assessing hand biomechanics, providing direct insights into structure and function. Key components include visual inspection for deformities, swelling, and skin changes, as well as range of motion (ROM) assessment through active and passive movements to evaluate joint flexibility. Strength testing uses clinical tools like hand dynamometers to evaluate grip and pinch strength. Palpation identifies tenderness, swelling, or structural abnormalities, while functional tests, such as opposition tasks or object manipulation, assess coordination and dexterity. As shown in Fig. 7.4, clinical tools are used to diagnose nerve dysfunctions (e.g., carpal tunnel syndrome) [29]. This examination provides a comprehensive foundation for further biomechanical or imaging studies. It involves evaluating joint mobility, grip strength, and range of motion through clinical tests.

Clinical Applications of Hand Motion Analysis

Hand motion analysis is critical in various clinical settings, providing invaluable data for understanding, diagnosing, and treating motion-related disorders. The following sections delve into the significant clinical applications, emphasizing their biomechanical implications.

In *rehabilitation and physical therapy*, Hand motion analysis informs the development of individualized rehabilitation plans for patients recovering from injuries, surgeries, or neurological conditions. By identifying deficiencies in motion patterns, therapists can design targeted therapy regimens to improve range of motion, coordination, and strength. In addition, hand rehabilitation robots have been developed and used since

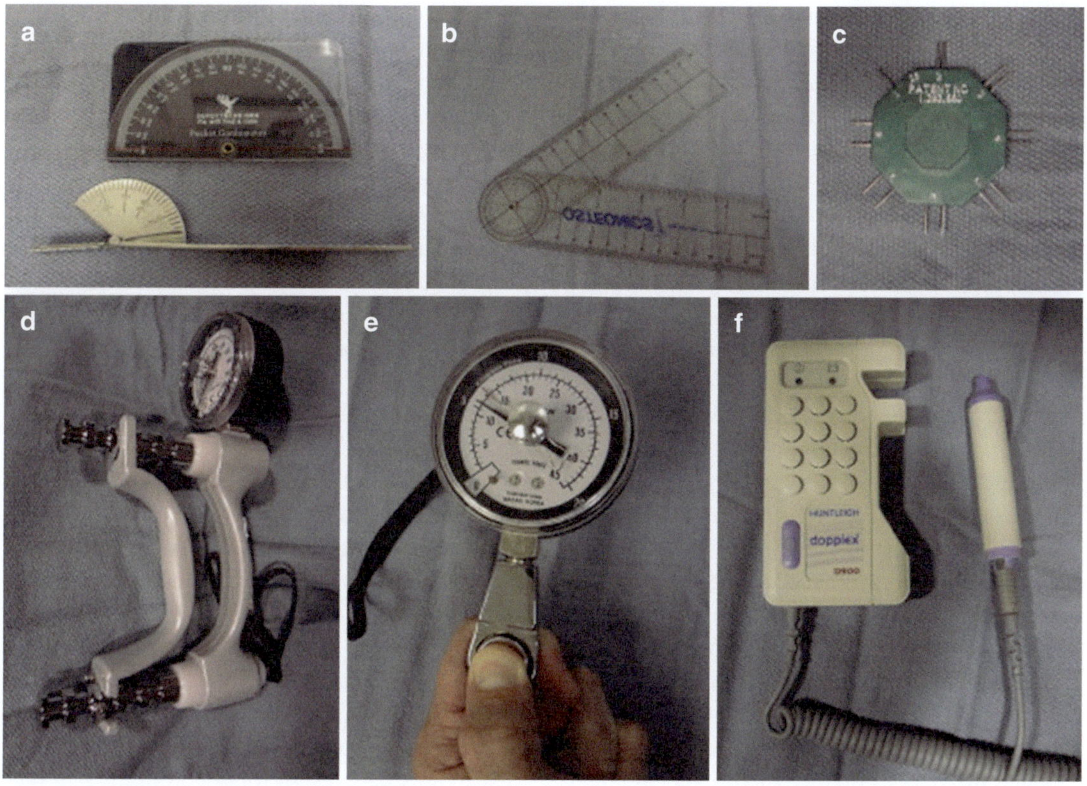

Fig. 7.4 Standard instruments are used in the physical examination of the hand [29]. (**a**) Two goniometers, one digital and one analog, used for measuring angles. (**b**) A clear plastic goniometer with measurement markings. (**c**) A circular device with protruding metal pins labeled "PATIENT TAG 1234-56." (**d**) A hand dynamometer with a dial gauge for measuring grip strength. (**e**) A close-up of a pressure gauge held in a hand, displaying measurements from 0 to 35. (**f**) A handheld Doppler device labeled "HUNTLEIGH dopplex D900" with a probe attached by a coiled cable

the 1990s in cases such as poststroke patients [50, 55]. Functional Recovery Tracking uses quantitative analysis tools, such as motion capture systems and grip strength measurements, to monitor progress over time and adjust interventions accordingly [5].

Motion analysis is integral to both preoperative and postoperative phases of *surgical care*, particularly for joint reconstruction and hand deformity corrections. Identifying dysfunctional joints or tendons using physical examination or imaging tools in the preoperative phase allows surgeons to plan precise interventions, improving surgical outcomes. Tracking functional gains after procedures provides objective data to measure success and guide further recovery efforts. In tendon transfer surgeries, motion analysis assists in validating new surgery techniques and their efficacy [35].

Advances in hand motion analysis have profoundly influenced the development of assistive devices, including *prosthetics and orthotics*. Through sophisticated analysis of natural motion patterns using advanced sensor technologies, researchers have been able to design prosthetic hands that replicate human movement with remarkable precision and functionality [27]. In rehabilitation, orthoses are widely utilized to support recovery and improve patient outcomes. By integrating detailed motion analysis, these devices can be tailored to patients' specific biomechanical and functional needs, enabling more effective rehabilitation protocols to restore mobility and optimize recovery following injury or surgery [22]. This personalized approach enhances the therapeutic process and better aligns with the patient's unique motion requirements.

Hand biomechanics analysis is a cornerstone for the early detection, management, and rehabilitation of *neurological disorders* such as Parkinson's, ALS, and stroke. In Parkinson's, it provides precise metrics on tremor frequency, grip strength, and finger dexterity, enabling clinicians to monitor disease progression, evaluate treatment efficacy, and tailor interventions [40]. For ALS, motion analysis detects early muscle weakness, reduced range of motion, and compensatory movement patterns, facilitating timely interventions to slow functional decline and maintain independence [15]. In post-stroke patients, motion analysis identifies specific impairments such as spasticity, asymmetrical hand use, and impaired grasp-release cycles, informing targeted therapies like robotic-assisted exercises, constraint-induced movement therapy, and neuroplasticity-focused strategies [10]. By integrating advanced motion analysis technologies, clinicians can develop personalized, data-driven rehabilitation protocols that improve motor function and patient outcomes and enhance the overall quality of life and independence for individuals with these debilitating conditions.

Hand motion analysis is crucial in *sports science*, enhancing performance, preventing injuries, and optimizing training. It provides insights into biomechanics during activities such as tennis serves or baseball pitches, helping athletes refine techniques to improve precision, reduce energy expenditure, and maximize output [32]. In injury prevention and rehabilitation, motion analysis identifies risk factors like repetitive stress in rock climbing grips, enabling targeted corrective exercises and tracking recovery progress [53]. It also informs sports equipment design, such as improving grip ergonomics for golf clubs or distributing impact forces in boxing gloves to reduce injury risk [36, 51]. Additionally, sport-specific adaptations, such as controlled grip strength during basketball free throws, are analyzed, allowing athletes to optimize performance [56]. Hand motion analysis bridges the gap between athletic potential and biomechanics by leveraging these applications, driving improved performance, injury prevention, and tailored training strategies across various sports.

Building on its applications in neurological disorders, hand motion analysis is also pivotal in addressing motor challenges in *pediatric and aging* populations. In pediatrics, it is used to assess developmental delays, neuromuscular disorders, and conditions such as cerebral palsy, providing critical insights into impairments in grip strength, coordination, and fine motor skills [47]. For aging populations, hand motion analysis helps detect early signs of motor decline, arthritis, and neurodegenerative diseases like Parkinson's and Alzheimer's, as mentioned in previous sections. Tracking changes in dexterity, grip strength, and coordination informs interventions such as targeted strength training, the design of assistive devices, and strategies for fall prevention.

Hand motion analysis is a cornerstone in developing *robotic surgery systems*, enhancing precision, ergonomics, and training effectiveness by extending its utility into surgical innovation. Researchers can design robotic systems that replicate and augment human dexterity by analyzing the biomechanics of hand movements during surgical tasks. For instance, the da Vinci Surgical System, guided by motion analysis data, translates a surgeon's hand movements into precise micro-movements of robotic instruments, reducing tremors and improving accuracy in minimally invasive procedures [17]. Studies have shown that robotic-assisted surgeries, including those performed using the da Vinci system, result in superior surgical outcomes, such as reduced blood loss, lower infection rates, and shorter hospital stays than traditional open or laparoscopic procedures [48]. Additionally, motion tracking optimizes haptic feedback systems, allowing surgeons to experience realistic resistance and texture during remote operations, enhancing precision [43]. In training, hand motion analysis evaluates surgeon performance by measuring metrics like motion efficiency, tool path accuracy, and hand stability, which can guide skill development in surgical simulators [33]. By leveraging these advancements, hand motion analysis drives the evolution of robotic surgical systems and ensures safer surgeries, quicker recoveries, and improved patient outcomes, revolutionizing the standard of care in operating rooms.

Integration of Motion Analysis with Emerging Technologies

The integration of motion analysis with emerging technologies has revolutionized its applications across diverse fields, including diagnostics, rehabilitation, training, and regenerative medicine. By combining advanced tools like artificial intelligence (AI), wearable technology, virtual reality (VR), and bioengineering, motion analysis has become more precise, predictive, and accessible.

AI and machine learning transform motion analysis by enabling pattern recognition and predictive diagnostics. For example, AI algorithms analyze hand motion data to detect and monitor neurological disorders such as Parkinson's or Alzheimer's, predicting progression with remarkable accuracy for effective management [42]. In sports, AI-driven motion analysis improves player performance by analyzing finger movements to optimize spin delivery or monitoring a basketball player's shooting technique to enhance free-throw accuracy [9]. In robotic surgery, machine learning models enhance automation, improve the precision of robotic instruments, and provide real-time assistance to surgeons [8].

Wearable devices equipped with sensors and real-time feedback systems have made motion analysis more accessible to patients, clinicians, and researchers. Bright gloves like the Rapael Smart Glove provide continuous monitoring of hand movements with game-based VR and deliver instant feedback during rehabilitation exercises and motivation, helping patients regain motor function after strokes or injuries [49]. In sports, similar wearable technologies are used to monitor and enhance athletic performance, such as smart gloves in baseball that track grip pressure and throwing mechanics to optimize a pitcher's technique or prevent overuse injuries [32]. Recent innovations, such as AI-powered wearables, further enhance these devices by predicting recovery trajectories and optimizing therapy [18].

Integrating *virtual reality* (VR) with motion analysis has revolutionized rehabilitation and skill training by creating immersive, interactive environments. VR-based systems like the Neofect Smart Rehab system combine motion tracking with gamified exercises to engage patients during therapy [41]. In surgical training, VR simulators use real-time hand motion analysis to evaluate performance metrics such as precision, stability, and efficiency. This allows surgeons to refine their skills in a risk-free environment [34]. These systems also leverage motion tracking to simulate complex tasks and provide detailed feedback, accelerating skill acquisition and improving outcomes.

There has been a growing integration of motion analysis sensors with stem cell and regenerative therapies. Besides assessing stem cell transplant outcomes and functional validation, advanced systems can be used to record hand function to determine the efficacy of engineered tissues, guiding further optimization. For example, a microhand constructed from muscle tissue responds to neuronal signals, enabling voluntary movement through electromyogram (sEMG) control. Motion analysis ensures precise evaluation of muscle contraction, movement accuracy, and functionality, driving prosthetic design innovations and promising solutions for restoring lost extremity functions [28]. Emerging technologies such as bio-integrated sensors embedded in regenerative scaffolds allow for continuous recovery monitoring, providing invaluable data for tailoring therapies [31].

Future Directions in Hand Motion Biomechanics

As the field of hand biomechanics continues to advance, several emerging trends and innovations are poised to shape the future. These directions not only aim to deepen our understanding of hand motion but also hold the potential to revolutionize patient care and technological applications.

Personalized Medicine

Integrating biomechanics with motion analysis offers exciting possibilities for tailored rehabilitation and treatment. For instance, real-time motion capture technologies, combined with biomechanical modeling, enable the development of customized therapy plans for patients with conditions such as carpal tunnel syndrome or post-stroke motor impairments. Tools like Vicon motion analysis systems or OpenSim biomechanical modeling software have been instrumental in advancing this domain. A study using OpenSim highlighted its ability to simulate subject-specific pathological tremors and provide insights into optimal therapeutic interventions for improved motor recovery [46].

Interdisciplinary Collaboration

Bridging biomechanics with fields such as engineering, regenerative medicine, and clinical practice fosters innovation in both research and application. For example, collaborations between mechanical engineers and clinical practitioners have developed soft robotic gloves that assist individuals with impaired grip strength. Similarly, partnerships in regenerative medicine have explored the use of tissue-engineered scaffolds to repair damaged tendons, combining biomechanical principles with cutting-edge biological materials [52].

Advancements in Robotic and Prosthetic Design

Integrating biomechanics into robotics has ushered in a new era of prosthetic innovation. Modern hand prosthetics, such as the Bebionic hand, leverage biomechanical principles to replicate natural motion and provide enhanced functionality. These devices utilize electromyography (EMG) signals to interpret muscle activity, allowing users to perform intricate tasks like typing or gripping delicate objects [19]. Furthermore, robotics research has made strides in developing surgical robots capable of precise hand surgeries, such as the da Vinci Surgical System. This system incorporates biomechanical insights to enhance dexterity and minimize invasiveness, transforming how surgeries are performed.

Big Data in Biomechanics

The growing availability of large datasets has transformed biomechanics research, allowing for the refinement of hand motion models through advanced computational techniques. Databases like the Carnegie Mellon Mocap Library provide extensive kinematic data researchers use to analyze complex hand movements. Machine learning algorithms applied to such datasets can identify patterns and optimize rehabilitation protocols.

Conclusion

Hand motion analysis has emerged as a vital field, unlocking new possibilities in understanding the mechanics of human movement and translating these insights into tangible advancements. Beyond current applications, the future of hand biomechanics lies in the seamless integration of motion data with predictive technologies, enabling early diagnosis of disorders and precise interventions. Innovations such as bio-integrated sensors and regenerative medicine hold the potential to redefine rehabilitation and prosthetics, offering unprecedented levels of personalization. Hand motion analysis is revolutionizing performance optimization, injury prevention, and recovery in sports, with technologies like AI-powered wearables helping athletes refine techniques. Furthermore, interdisciplinary collaborations will likely fuel robotics, surgical precision, and ergonomic design breakthroughs. By continuing to explore the intersection of biomechanics, data science, and engineering, hand motion analysis will address existing challenges and chart new healthcare paths, enhancing functionality and quality of life for diverse populations.

References

1. Akhlaghi N, Baker CA, Lahlou M, Zafar H, Murthy KG, Rangwala HS, Kosecka J, Joiner WM, Pancrazio JJ, Sikdar S. Real-time classification of hand motions using ultrasound imaging of forearm muscles. IEEE Trans Biomed Eng. 2016;63(8):1687–98. https://doi.org/10.1109/TBME.2015.2498124.
2. Amirouche F. Fundamentals of multibody dynamics: theory and applications. 1st ed. Springer Nature; 2007. https://doi.org/10.1007/b137682.
3. Bajuri MN, Kadir MRA, Raman MM, Kamarul T. Mechanical and functional assessment of the wrist affected by rheumatoid arthritis: a finite element analysis. Med Eng Phys. 2012;34(9):1294–302. https://doi.org/10.1016/j.medengphy.2011.12.020.
4. Barca JC, Rumantir G, Koon Li R. A new illuminated contour-based marker system for optical motion capture. Paper presented at the 1–5. 2006. https://doi.org/10.1109/INNOVATIONS.2006.301980.
5. Bhatt E, Nagpal A, Greer KH, Grunewald TK, Steele JL, Wiemiller JW, Lewis SM, Carey JR. Effect of finger tracking combined with electrical stimulation on brain reorganization and hand function in subjects with stroke. Exp Brain Res. 2007;182(4):435–47. https://doi.org/10.1007/s00221-007-1001-5.
6. Bielefeld T, Neumann DA. The unstable metacarpophalangeal joint in rheumatoid arthritis: anatomy, pathomechanics, and physical rehabilitation considerations. J Orthop Sports Phys Ther. 2005;35(8):502–20. https://doi.org/10.2519/jospt.2005.35.8.502.
7. Casile A, Fregna G, Boarini V, Paoluzzi C, Manfredini F, Lamberti N, Baroni A, Straudi S. Quantitative comparison of hand kinematics measured with a markerless commercial head-mounted display and a marker-based motion capture system in stroke survivors. Sensors (Basel, Switzerland). 2023;23(18):7906. https://doi.org/10.3390/s23187906.
8. Chadebecq F, Lovat LB, Stoyanov D. Artificial intelligence and automation in endoscopy and surgery. Nat Rev Gastroenterol Hepatol. 2023;20(3):171–82. https://doi.org/10.1038/s41575-022-00701-y.
9. Cheng Y, Liang X, Xu Y, Kuang X. Artificial intelligence technology in basketball training action recognition. Front Neurorobot. 2022;16:819784. https://doi.org/10.3389/fnbot.2022.819784.
10. Choi H, Park D, Rha D, Nam HS, Jo YJ, Kim DY. Kinematic analysis of movement patterns during a reach-and-grasp task in stroke patients. Front Neurol. 2023;14:1225425. https://doi.org/10.3389/fneur.2023.1225425.
11. Cutkosky MR. On grasp choice, grasp models, and the design of hands for manufacturing tasks. IEEE Trans Robot Autom. 1989;5(3):269–79. https://doi.org/10.1109/70.34763.
12. Dodrill CB. The hand dynamometer is a neuropsychological measure. J Consult Clin Psychol. 1978;46(6):1432–5. https://doi.org/10.1037/0022-006X.46.6.1432.
13. Duncan SFM, Saracevic CE, Kakinoki R. Biomechanics of the hand. Hand Clin. 2013;29(4):483–92. https://doi.org/10.1016/j.hcl.2013.08.003.
14. Feix T, Romero J, Schmiedmayer H, Dollar AM, Kragic D. The GRASP taxonomy of human grasp types. IEEE Trans Hum-Mach Syst. 2016;46(1):66–77. https://doi.org/10.1109/THMS.2015.2470657.
15. Fernandes APM, de Holanda LJ, de Lucena LC, da Silva KER, Lopes ACSM, Borges DT, Nagem DAP, de M Valentim RA, Bougrain L, Rodrigues Lindquist AR, Egwuonwu VA. Electromyography as a tool to motion analysis for people with amyotrophic lateral sclerosis: a protocol for a systematic review. PloS One. 2024;19(5):e0302479. https://doi.org/10.1371/journal.pone.0302479.
16. Fonk R, Schneeweiss S, Simon U, Engelhardt L. Hand motion capture from a 3D leap motion controller for a musculoskeletal dynamic simulation. Sensors (Basel, Switzerland). 2021;21(4):1199. https://doi.org/10.3390/s21041199.
17. Freschi C, Ferrari V, Melfi F, Ferrari M, Mosca F, Cuschieri A. Technical review of the da Vinci surgical telemanipulator. Int J Med Robot Comp Assisted Surg. 2013;9(4):396–406. https://doi.org/10.1002/rcs.1468.
18. Goto R, Aharari A, Mehdipour F, Nakamatsu K, Kountchev R, Patnaik S, Nakamatsu K, Kountchev R, Patnaik S. Development of an AI-powered interactive hand rehabilitation system. Springer; 2024. p. 429–41. https://doi.org/10.1007/978-981-99-9018-4_31.
19. Guo K, Lu J, Wu Y, Hu X, Yang H. The latest research progress on bionic artificial hands: a systematic review. Micromachines (Basel). 2024;15(7):891. https://doi.org/10.3390/mi15070891.
20. Hall S. Basic biomechanics. Mcgraw-Hill US Higher Ed; 2019.
21. Hazari A, Maiya AG, Nagda TV. Conceptual biomechanics and kinesiology. 1st ed. Springer Singapore Pte. Limited; 2021. https://doi.org/10.1007/978-981-16-4991-2.
22. Hirth MJ, Howell JW, O'Brien L. Two case reports–use of relative motion orthoses to manage extensor tendon zones III and IV and sagittal band injuries in adjacent fingers. J Hand Ther. 2017;30(4):546–57. https://doi.org/10.1016/j.jht.2017.04.006.

23. Hurst LC, Badalamente MA, Komatsu DE. Dupuytren's disease: a scientific review. Elsevier; 2024.
24. Jaramillo-Yánez A, Benalcázar ME, Mena-Maldonado E. Real-time hand gesture recognition using surface electromyography and machine learning: a systematic literature review. Sensors (Basel, Switzerland). 2020;20(9):2467. https://doi.org/10.3390/s20092467.
25. Jiang Y, Ye Y, Gopinath D, Won J, Winkler AW, Liu CK, Jung SK, Lee J, Bargteil A. Transformer inertial poser: real-time human motion reconstruction from sparse IMUs with simultaneous terrain generation. Paper presented at the 1–9. 2022. https://doi.org/10.1145/3550469.3555428.
26. Johnsson PM, Eberhardt K. Hand deformities are important signs of disease severity in patients with early rheumatoid arthritis. Rheumatology (Oxford). 2009;48(11):1398–401. https://doi.org/10.1093/rheumatology/kep253.
27. Ju Z, Liu H. Human hand motion analysis with multisensory information. IEEE/ASME Trans Mechatron. 2014;19(2):456–66. https://doi.org/10.1109/TMECH.2013.2240312.
28. Kabumoto K, Hoshino T, Akiyama Y, Morishima K. Voluntary movement controlled by the surface EMG signal for tissue-engineered skeletal muscle on a gripping tool. Tissue Eng Part A. 2013;19(15-16):1695–703. https://doi.org/10.1089/ten.tea.2012.0421.
29. Kenney RJ, Hammert WC. Physical examination of the hand. J Hand Surg (American Ed). 2014;39(11):2324–34. https://doi.org/10.1016/j.jhsa.2014.04.026.
30. Lee Y, Yoo H. Low-cost 3D motion capture system using passive optical markers and monocular vision. Optik (Stuttgart). 2017;130:1397–407. https://doi.org/10.1016/j.ijleo.2016.11.174.
31. Liu M, Zhang Y, Tao TH. Recent progress in bio-integrated intelligent sensing system. Adv Intell Syst. 2022;4(6):n/a. https://doi.org/10.1002/aisy.202100280.
32. Loushin SR, Verhoeven M, Christoffer DJ, Camp CL, Kaufman KR. Are 4D motion sensors valid and reliable for studying baseball pitching? Am J Sports Med. 2023;51(6):1608–14. https://doi.org/10.1177/03635465231166423.
33. Mackenzie CF, Yang S, Garofalo E, Hu PF, Watts D, Patel R, Puche A, Hagegeorge G, Shalin V, Pugh K, Granite G, Stansbury LG, Shackelford S, Tisherman S. Enhanced training benefits of video recording surgery with automated hand motion analysis. World J Surg. 2021;45(4):981–7. https://doi.org/10.1007/s00268-020-05916-1.
34. Mao RQ, Lan L, Kay J, Lohre R, Ayeni OR, Goel DP, SA, D. d. Immersive virtual reality for surgical training: a systematic review. J Surg Res. 2021;268:40–58. https://doi.org/10.1016/j.jss.2021.06.045.
35. Mardula KL, Balasubramanian R, Allan CH. Implanted passive engineering mechanism improves hand function after tendon transfer surgery: a cadaver-based study. Hand (New York, NY). 2015;10(1):116–22. https://doi.org/10.1007/s11552-014-9676-0.
36. Menzel T, Potthast W. Validation of a unique boxing monitoring system. Sensors (Basel, Switzerland). 2021;21(21):6947. https://doi.org/10.3390/s21216947.
37. Monnet T, Samson M, Bernard A, David L, Lacouture P. Measurement of three-dimensional hand kinematics during swimming with a motion capture system: a feasibility study. Sports Eng. 2014;17(3):171–81. https://doi.org/10.1007/s12283-014-0152-4.
38. Morco S, Bowden A. Ulnar drift in rheumatoid arthritis: a review of biomechanical etiology. J Biomech. 2015;48(4):725–8. https://doi.org/10.1016/j.jbiomech.2014.12.052.
39. Muschter E, Schulz J, Tömösközi M, Herbst L, Küssel L, Sefunç M, Holtzhausen S, Speidel S, Fitzek FHP, Li S. Coming in handy: CeTI-age – a comprehensive database of kinematic hand movements across the lifespan. Sci Data. 2023;10(1):826. https://doi.org/10.1038/s41597-023-02738-3.
40. Pang Y, Christenson J, Jiang F, Lei T, Rhoades R, Kern D, Thompson JA, Liu C. Automatic detection and quantification of hand movements toward the development of an objective assessment of tremor and bradykinesia in Parkinson's disease. J Neurosci Methods. 2020;333:108576. https://doi.org/10.1016/j.jneumeth.2019.108576.
41. Parikh A, Legault C, Flavin K, Lansberg M. Neofect glove: virtual reality device for home therapy in stroke survivors (P5.007). Neurology. 2018;90(15) https://doi.org/10.1212/WNL.90.15_supplement.P5.007.
42. Park KW, Mirian MS, McKeown MJ. Artificial intelligence-based video monitoring of movement disorders in the elderly: a review on current and future landscapes. Singapore Med J. 2024;65(3):141–9. https://doi.org/10.4103/singaporemedj.SMJ-2023-189.
43. Patel RV, Atashzar SF, Tavakoli M. Haptic feedback and force-based teleoperation in surgical robotics. Proc IEEE. 2022;110(7):1012–27. https://doi.org/10.1109/JPROC.2022.3180052.
44. Peña-Pitarch E, Falguera NT, Yang J. Virtual human hand: model and kinematics. Comput Methods Biomech Biomed Eng. 2014;17(5):568–79. https://doi.org/10.1080/10255842.2012.702864.
45. Phutane U, Liphardt A, Bräunig J, Penner J, Klebl M, Tascilar K, Vossiek M, Kleyer A, Schett G, Leyendecker S. Evaluation of optical and radar-based motion capturing technologies for characterizing

hand movement in the rheumatoid arthritis-A pilot study. Sensors (Basel, Switzerland). 2021;21(4):1208. https://doi.org/10.3390/s21041208.

46. Pinheiro WC, Ferraz HB, Castro MCF, Menegaldo LL. An OpenSim-based closed-loop biomechanical wrist model for subject-specific pathological tremor simulation. IEEE Trans Neural Syst Rehabil Eng. 2024;32:1. https://doi.org/10.1109/TNSRE.2024.3373433.
47. Ramey SL, DeLuca S, Stevenson RD, Case-Smith J, Darragh A, Conaway M. Children with hemiparesis arm and movement project (CHAMP): protocol for a multisite comparative efficacy trial of paediatric constraint-induced movement therapy (CIMT) testing effects of dosage and type of constraint for children with hemiparetic cerebral palsy. BMJ Open. 2019;9(1):e023285. https://doi.org/10.1136/bmjopen-2018-023285.
48. Ricciardi R, Seshadri-Kreaden U, Yankovsky A, Dahl D, Auchincloss H, Patel NM, Hebert AE, Wright V. The COMPARE study: comparing perioperative outcomes of oncologic minimally invasive laparoscopic, da Vinci robotic, and open procedures: a systematic review and meta-analysis of the evidence. Ann Surg. 2024; https://doi.org/10.1097/SLA.0000000000006572.
49. Shin S, Lee H, Chang WH, Ko SH, Shin Y, Kim Y. A smart glove digital system promotes restoration of upper limb motor function and enhances cortical hemodynamic changes in subacute stroke patients with mild to moderate weakness: a randomized controlled trial. J Clin Med. 2022;11(24):7343. https://doi.org/10.3390/jcm11247343.
50. Stienen AHA, Moulton TS, Miller LC, Dewald JPA. Wrist and finger torque sensor for the quantification of upper limb motor impairments following brain injury. IEEE Int Conf Rehabil Robot. 2011;2011:1–5. https://doi.org/10.1109/ICORR.2011.5975464.
51. Tsunoda M, Bours R, Hasegawa H. Three-dimensional motion analysis and inverse dynamic modelling of the human golf swing, SAE technical aper 2004-01-2163. 2004. https://doi.org/10.4271/2004-01-2163.
52. Varaganti P, Seo S. Recent advances in biomimetics for the development of bio-inspired prosthetic limbs. Biomimetics (Basel, Switzerland). 2024;9(5):273. https://doi.org/10.3390/biomimetics9050273.
53. Walsh A, Seifert L, Button C, Vial S, Croft J. The effect of fatigue on climbing fluidity and hand movements. Sports Biomech. 2023:1–13. https://doi.org/10.1080/14763141.2023.2182703.
54. Watson RA. Use of a machine learning algorithm to classify expertise: analysis of hand motion patterns during a simulated surgical task. Acad Med. 2014;89(8):1163–7. https://doi.org/10.1097/ACM.0000000000000316.
55. Yue Z, Wang J, Zhang X, Hu X. Hand rehabilitation robotics on poststroke motor recovery. Behav Neurol. 2017;2017(2017):1–20. https://doi.org/10.1155/2017/3908135.
56. Zhang K, Guo B, Yang M, Jia Y, Zhang K, Wang L. The assessment of sports performance by grip pressure using flexible piezoresistive pressure sensors in seven sports events. Sci Rep. 2024;14(1):31750–12. https://doi.org/10.1038/s41598-024-82274-1.
57. Zheng W, Guo N, Zhang B, Zhou J, Tian G, Xiong Y. Human grasp mechanism understanding, human-inspired grasp control, and robotic grasping planning for agricultural robots. Sensors (Basel, Switzerland). 2022;22(14):5240. https://doi.org/10.3390/s22145240.
58. Zunino A, Cavazza J, Volpi R, Morerio P, Cavallo A, Becchio C, Murino V. Predicting intentions from motion: the subject-adversarial adaptation approach. Int J Comput Vis. 2020;128(1):220–39. https://doi.org/10.1007/s11263-019-01234-9.

8 The Significance of Gripping, Forces, and Rehabilitation in Hand Biomechanics for Orthopedics and Rehabilitation

Barbara Mera, Anton J. Fakouri, and Farid Amirouche

Introduction

Hand biomechanics, particularly in grasping and gripping functions, involves a complex interplay between intrinsic and extrinsic muscles, neuromuscular coordination, and sensory feedback.

The extrinsic muscles, such as the flexor digitorum profundus (FDP) and flexor digitorum superficialis (FDS), are primarily responsible for generating the force required for power grips [1, 2]. These muscles originate in the forearm and are inserted into the fingers, enabling strong flexion. In contrast, the intrinsic muscles—including the thenar muscles (abductor pollicis brevis, flexor pollicis brevis, opponens pollicis), hypothenar muscles, interossei, and lumbricals—are essential for fine motor control and stabilization during precision grips [1, 2]. These muscles originate and insert within the hand, providing the nuanced control needed for delicate tasks [1, 2].

Neuromuscular coordination is crucial for effective gripping and involves the integration of signals from the motor cortex, the corticospinal tract, and the reticular system. In power grips, a significant reduction in intracortical inhibition facilitates greater force generation [3]. Conversely, precision grips demand more refined motor control, characterized by stronger interactions within frontoparietal networks to meet the demands of fine motor activity [3].

Sensory feedback, particularly from cutaneous receptors, plays a pivotal role in modulating grip strength. It provides essential information about object characteristics such as texture, weight, and slip, allowing for real-time adjustment of grip forces to maintain secure handling. This feedback loop supports the dynamic adaptation of grip strength and coordination, ensuring effective manipulation of objects under varying conditions. Sensory deficits—such as those observed in conditions like diabetes mellitus—can impair this feedback mechanism, leading to diminished grip strength and reduced dexterity. Understanding and managing these sensory aspects is crucial in our work as clinicians, therapists, and researchers in orthopedics and rehabilitation.

B. Mera
Department of Orthopaedic Surgery, University of Illinois at Chicago, Chicago, IL, USA

A. J. Fakouri
Mid America Orthopaedics, Palos Hills, IL, USA

F. Amirouche (✉)
Department of Orthopaedic Surgery, Orthopaedic and Spine Institute, Endeavor HealthSystem (Formerly Northshore University HealthSystem), an Affiliate of the University of Chicago Pritzker School of Medicine, Skokie, IL, USA

Department of Orthopaedic Surgery, University of Illinois at Chicago, Chicago, IL, USA
e-mail: famirouche@northshore.org; amirouch@uic.edu

F. Amirouche et al. (eds.), *Biomechanics of the Hand*,
https://doi.org/10.1007/978-3-032-04688-8_8

The Practical Implications of Hand Gripping Biomechanics: Hand gripping biomechanics represent a sophisticated integration of muscular function, neuromuscular coordination, and sensory feedback. A thorough understanding of these muscle groups and their functional range of motion is essential for diagnosing hand pathologies, planning surgical interventions, and designing effective rehabilitation protocols.

Anatomy and Kinematics of the Hand Grip

The Interdisciplinary Nature of Hand Gripping Biomechanics: Gripping is the human hand's fundamental and intricate function, necessitating a seamless collaboration between extrinsic and intrinsic muscle groups. The primary extrinsic muscles, namely the flexor digitorum profundus (FDP) and the flexor digitorum superficialis (FDS), are primarily tasked with the flexion of the fingers. The FDP plays a particularly pivotal role in enabling the formation of a composite fist, allowing for a robust and secure grip. In contrast, the FDS, alongside the various intrinsic muscles, contributes to grip strength and adapts its involvement based on the specific demands of the task [4].

The intrinsic muscles, which include the finely tuned thenar muscles (comprising the abductor pollicis brevis, flexor pollicis brevis, and opponens pollicis) and the hypothenar muscles (including the abductor digiti minimi, flexor digiti minimi brevis, and opponens digiti minimi), are crucial for executing precise movements and stabilizing the fingers during gripping activities. Additionally, the interossei and lumbricals—small yet powerful muscles located in the hand—provide essential support in fine motor control, enhancing dexterity and ensuring that the fingers work harmoniously to create and maintain a firm grip [5].

The intricate functionality of the human hand's gripping mechanism relies on the seamless interaction between various functional units [6]. A delicate balance must be maintained between the hand's extrinsic and intrinsic systems to achieve adequate finger flexion.

The extrinsic system comprises the long flexor and long extensor tendons that span multiple finger phalanges [7]. These tendons are vital for producing the range of motion necessary for grasping and manipulating objects, with the flexors facilitating bending and the extensors enabling straightening of the fingers [7].

In contrast, the intrinsic system plays a crucial role in coordinating and fine-tuning the movements initiated by the extrinsic system [8]. This system comprises the lumbrical muscles, which originate from the flexor digitorum profundus tendons and attach to the extensor expansions, enabling complex finger movements [8]. Additionally, the two interossei muscles—dorsal and palmar—assist in abduction and adduction of the fingers, further enhancing dexterity.

These intrinsic muscles are crucial for executing precise movements and maintaining fine motor control. For example, the first dorsal interosseous (FDI) and abductor pollicis brevis (APB) are heavily engaged during precision grips, such as holding a pen or picking up small objects [9]. These muscles enable the fine adjustments required for accurate object manipulation. Studies have demonstrated that dynamic coordination between the APB and FDI is critical for maintaining grip stability and adapting to varying load conditions [9]. Intrinsic muscles also play a significant role in stabilizing the fingers during gripping tasks. In single-finger isometric force tasks, intrinsic muscles display individuation, wherein the agonistic and antagonistic muscles of the instructed fingers show the highest activation. This activation pattern supports finger stabilization, enabling fingers to maintain their position and exert force effectively during gripping [10]. Additionally, spinal premotor interneurons (PreM-INs) have been shown to facilitate the coactivation of intrinsic hand muscles, further enhancing the stability and coordination essential for precision grips [11].

The Landsmeer ligament, also known as the oblique retinacular ligament (ORL), is a crucial component that connects the flexor and extensor systems, enabling the efficient transfer of forces during finger motion. This intricate interplay among both the extrinsic and intrinsic components is fundamental to the hand's ability to perform precise gripping and manipulation tasks [12–14].

The hand's unique anatomy enables the functional range of motion (ROM) necessary for gripping, with ROM varying across the finger joints. For most activities of daily living, the metacarpophalangeal (MCP) joints require a ROM of approximately 19°–71°, the proximal interphalangeal (PIP) joints 23°–87°, and the distal interphalangeal (DIP) joints 10°–64° [6]. These values represent about 48%, 59%, and 60% of the active motion of the MCP, PIP, and DIP joints. Coordination among these joints is critical for effective gripping, with the PIP joints typically demonstrating greater flexion than the MCP and DIP joints during tasks such as cylinder gripping [7]. Wrist position also significantly influences grip strength, with maximal grip force typically achieved in a neutral wrist position [8].

Gripping Mechanisms

The human hand has approximately nine different types of grips (Fig. 8.1). Power grips and precision grips are the two primary types used by the human hand. Power grips involve the entire hand and are used in tasks that require significant

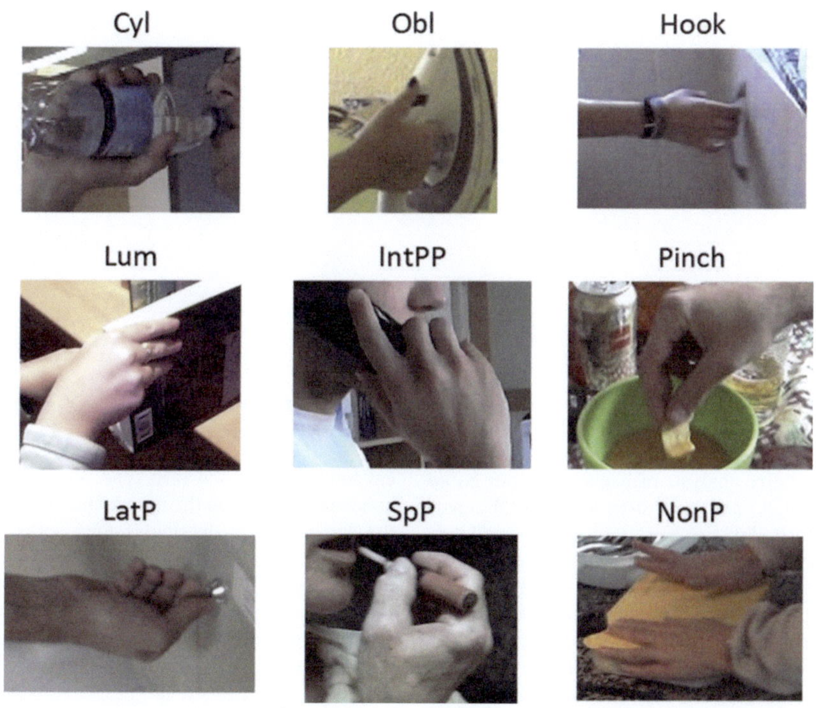

Fig. 8.1 Examples of functional hand grasp types used in daily activities. (*Cyl*) Cylindrical grasp: the palm is involved, with the thumb directly opposing the fingers (abducted or neutral). (*Obl*) Oblique palmar grasp: a variation of the cylindrical grasp where the palm is engaged but the thumb is adducted. (*Hook*) Hook grasp: the palm and thumb are not involved; the fingers support the object's weight. (*Lum*) Lumbrical grasp: the thumb and proximal phalanges, while excluding palm contact. (*IntPP*) Intermediate power-precision grasp: the palm is partially engaged, with grasp stabilization provided by the thumb and index finger. (*Pinch*) Pinch grasp: utilizes the thumb and one or more fingertips. (*LatP*) Lateral pinch: the lateral surface of one or more fingers is used, typically along with the thumb. (*SpP*) Special pinch: involves the thumb, the lateral surface of one or more fingers, and the fingertips of the other fingers. (*NonP*) Non-prehensile grasp: characterized by object manipulation without enclosing the object in the hand. ([15]; permission to reprint granted)

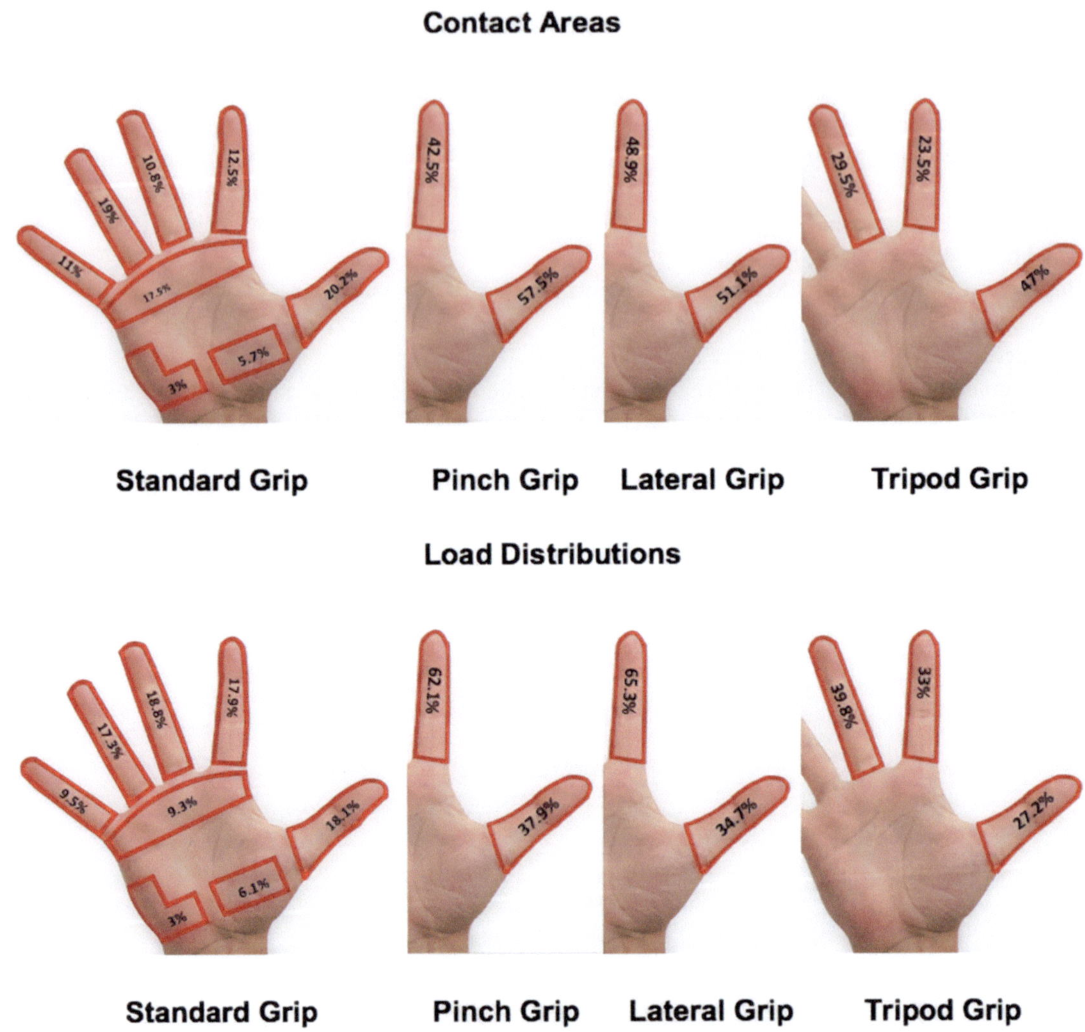

Fig. 8.2 Comparative analysis of hand grip types based on contact area (CA) and load distribution (LD) across anatomical regions. *Top row* (*contact areas*): visual representation of the percentage of surface area involved in standard, pinch, lateral, and tripod grips. Contact is predominantly distributed across the distal phalanges in the pinch and lateral grips. In contrast, the standard grip engages a broader area, including the palm—bottom row (load distributions), which is the distribution of mechanical load during each grip type. Pinch and lateral grips concentrate force on the thumb and index finger, whereas the standard grip distributes load across the palm and multiple digits. The tripod grip shows a more balanced distribution across the thumb and index-middle digits. Percentages represent relative contributions of each anatomical region to total contact or load [16]; permission to reprint granted)

force. The power grip primarily relies on the extrinsic muscles, particularly the flexor digitorum profundus (FDP) and flexor digitorum superficialis (FDS). These muscles generate substantial force, allowing for joint flexion (see Fig. 8.2). This grip is used when maximum force is required to hold or manipulate objects, such as lifting heavy items or using tools like hammers. The entire hand is engaged, and the fingers wrap tightly around the object.

In these grips, the fingers and thumb wrap around the object, and the extrinsic muscles—particularly the flexor digitorum profundus (FDP) and flexor digitorum superficialis (FDS)—generate the necessary force [3]. Neural control of power grips involves substantial cortical and subcortical contributions, with the reticular system playing a role in modulating corticospinal excitability [3]. Power grips rely less on satisfactory sensory feedback than precision grips as the primary goal is strength rather than precision. However, some sensory input is still necessary to ensure a secure hold.

In contrast, precision grips involve the thumb and one or more fingers (Fig. 8.3). The intrinsic muscles, including the thenar muscles (abductor pollicis brevis, flexor pollicis brevis, opponens pollicis), as well as the interossei and lumbricals, are crucial for precision grip function [18]. These grips rely heavily on direct corticospinal control, enabling independent finger movements [18]. This grip is essential for tasks that require accuracy and delicate manipulation, like holding a pen, threading a needle, or picking up small objects (see Fig. 8.3). It allows for control over the position and movement of the fingers. Fingers are positioned to create a more refined interaction with objects, often using the tips of the fingers and thumb. The opposing thumb and fingers make a "pinch" position, allowing for fine adjustments. Precision grips depend highly on sensory feedback for fine-tuning grip strength and coordination. Cutaneous receptors provide feedback regarding texture, weight, and the risk of slippage.

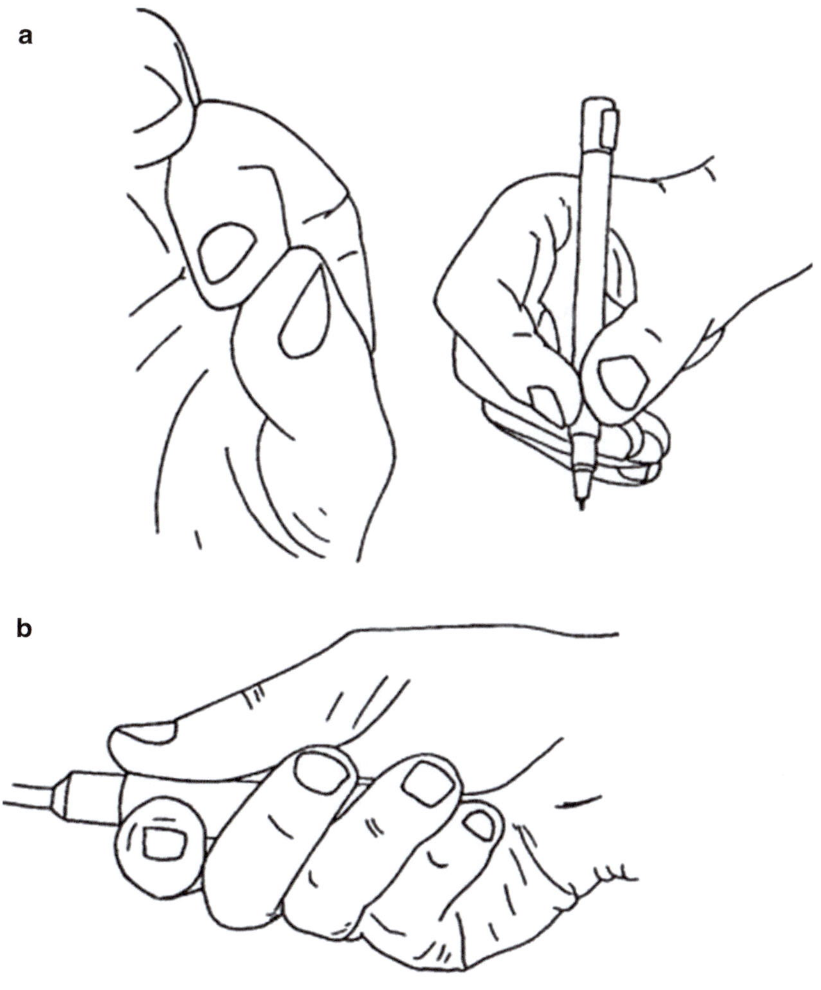

Fig. 8.3 Illustration contrasting two fundamental grasp types: (**a**) *Precision grip*, characterized by fine motor control involving the tips of the thumb and one or more fingers (e.g., writing or manipulating small objects), allowing precise positioning and delicate manipulation. (**b**) *Power grip*, involving wrapping the fingers and thumb around an object (e.g., holding a tool), stabilizes the object against the palm, enabling forceful and secure grasping. These grips reflect distinct neuromuscular control strategies and functional roles in hand–object interaction. ([17]; permission to reprint granted)

Another grip option is the soft grip, which engages a mix of intrinsic and extrinsic muscles and generally requires less force than a power grip. The muscles are involved in a way that allows for a gentler hold. This grip holds fragile or delicate items, such as glassware or small, easily breakable objects. The focus is on applying force in a controlled and gentle manner to prevent damage. The fingers are often spread slightly, and the grip may not involve full closure around the object. It allows for delicate handling without exerting excessive pressure. A firm reliance on sensory feedback is maintained to adjust grip pressure appropriately. This responsiveness helps prevent the object from being dropped or damaged [19].

Understanding the differences among these grips is essential for diagnosing hand pathologies and creating effective rehabilitation strategies. Each grip type serves distinct functional needs, utilizing the intricate biomechanics of the hand.

Neuromuscular coordination and sensory feedback are essential components for both grip types. Neuromuscular coordination involves the integration of signals from the motor cortex, corticospinal tract, and reticular system to regulate muscle activation patterns [3, 20].

Power grips typically exhibit a more significant reduction in intracortical inhibition, facilitating force generation [20]. Conversely, precision grips require more refined motor control, with stronger engagement of front-parietal networks to manage the increased demands of fine movement [3, 20].

Sensory feedback, particularly from cutaneous receptors, provides critical information about object properties such as texture, weight, and slip. This sensory input enables ongoing adjustments in grip force to maintain secure handling [21]. The feedback loop facilitates the dynamic adaptation of grip strength and coordination, enabling effective manipulation of objects under varying conditions [21].

Forces Involved in Gripping

Several biomechanical forces, including compression, tension, and shear forces, are involved in hand gripping. These forces are integral to hand movements and the generation of grip strength. They are commonly measured using dynamometers, instrumented handles, and force gloves. Hand size, age, gender, and surface friction significantly influence grip strength.

Compression forces act perpendicular to the surface of the object being gripped, compressing the object between the fingers and the palm. These forces are essential for maintaining a secure hold. Tension forces act along the length of muscles and tendons, pulling the fingers and thumb toward the palm. These are generated primarily by the flexor muscles during gripping actions. Shear forces, in contrast, act parallel to the object's surface and counteract any sliding motion between the hand and the object. They are significant for stabilizing the grip, especially when manipulating slippery items.

Grip forces are typically measured using dynamometers, which quantify the force exerted by the hand. Advanced methodologies include instrumented handles, force gloves, and multiaxis dynamometers. Multiaxis dynamometers assess the magnitude and direction of grip forces, providing comprehensive data on grip characteristics that offer a depth of information invaluable in our research and applications [22–24].

Several factors influence grip strength. Hand size and hand dominance play significant roles, with larger hands and the dominant hand typically demonstrating greater grip strength [25–27]. Grip strength generally declines with age and is typically higher in men than women [25–27]. Additionally, increased friction between the hand and the object can significantly enhance grip force generation, a practical implication we often encounter [25–27].

Modeling Gripping Forces

The modeling and estimation of gripping forces are paramount in biomechanical research and applications, such as ergonomic design, rehabilitation, and robotics. The anatomy of the hand, the properties of the objects being manipulated, and the specific task requirements all influence gripping forces. An accurate understanding of these forces can significantly enhance the functionality and safety of hand-related tasks.

The grasped object's weight, shape, texture, and size significantly influence the force required for a secure grip. For instance, a heavier or slippery object demands greater grip strength to prevent slipping.

Mathematical Modeling of Gripping Forces

Static Force Models These models estimate the grip force necessary to hold an object in a static position. A variety of models can be used, such as

Force Equilibrium Equations These account for the forces exerted by muscles and the object's weight. The sum of forces in the vertical direction must equal the weight of the gripped object (Fig. 8.4).

The setup shown in Fig. 8.4 provides a clear visual aid, helping readers understand the free-body diagrams and how various forces are modeled. These models are then combined with the first law of mechanics, which states that the sum of all forces acting along the X and Y axes in the case of 2D analysis must equal zero to satisfy the equilibrium conditions.

(a) Axial push force is modeled as the sum of frictional forces generated at multiple hand-handle contact points, where the friction force distributed along the grip is the product of the local normal force at each contact point multiplied by the coefficient of friction (assuming the friction is the same all around the hand grip). The diagram illustrates the grip force (Fg), the resultant normal force denoted by $\sum$ Fni and the directions of torque exertion about the handle.

(b) An experimental setup shows a cylindrical handle connected to a load cell via universal joints. An applied axial push force induces a reaction force directed proximally, generating an ulnar-directed moment about the wrist joint. Remember that torque is simply the cross product of a vector force and a distance (position vector).

(c) Application of a downward force by the hand to a fixed handle, such as when gripping a tool or object, produces an opposing radially directed moment. The resulting upward reaction force counters the ulnar-directed moment caused by the axial push force, thereby reducing the net wrist moment and potentially minimizing joint loading.

Friction Models Utilizing the coefficient of friction to calculate the minimum grip force necessary to prevent slipping:

$$F_g = \mu^* N$$

where F_g is the gripping force, μ is the coefficient of friction, which depends on the surface contact and sliding conditions, and N is the normal force (typically equal to the object's weight).

Dynamic Force Models

Dynamic models take into consideration the motion of the object and the hand. They can be modeled using various techniques, such as Lagrangian dynamics, Newton–Euler equations, or Kane's equations, which all stem from Newtonian mechanics. Applying these motion principles makes it possible to understand how gripping forces vary during dynamic tasks. A general differential equation takes the following form:

$$M(q).M(q)\cdot\ddot{q}+B\left(q,\ddot{q}\right)+C(q)\cdot\tau=0$$

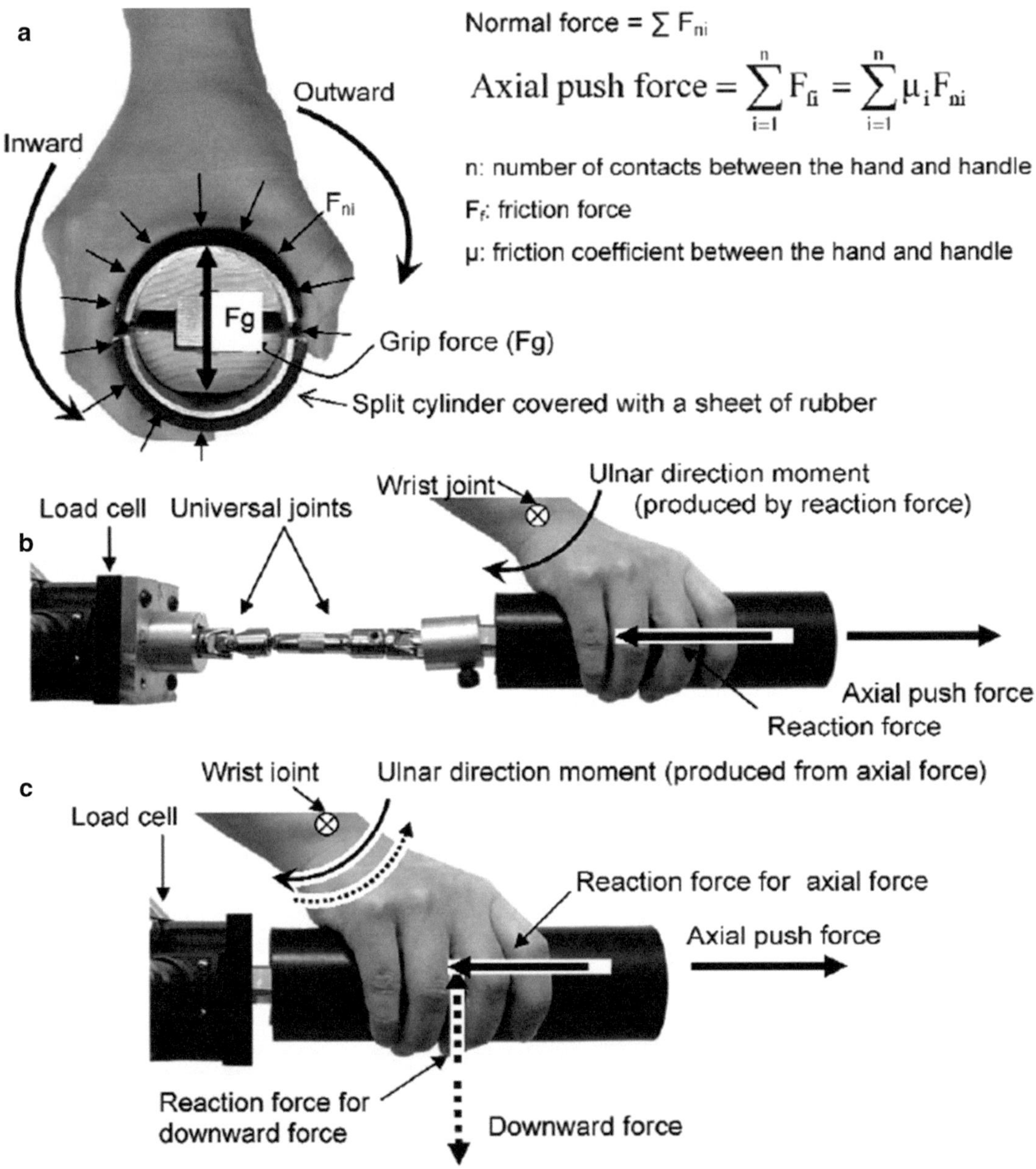

Fig. 8.4 Biomechanical model and experimental setup for analyzing axial push force. Wrist joint models (**a**, **b**, and **c**) depict different loading conditions. ([28]; permission to reprint requested)

M is the generalized mass matrix, and *B* is the nonlinear velocity contributions, such as Coriolis and other inertia terms. The last term denotes joint moments τ via the coefficient matrix *C*. τ is the summation of muscle moments and passive joint moments usually generated by the external forces, such as muscles and ligaments [29]. These equations can be used to represent the changing forces as the hand moves to grasp an object.

Muscle Force Models

The Hill-type muscle model is a classic model that describes the relationship between muscle force, muscle length, and muscle velocity. This model distinguishes between the contractile component, the series elastic component, and the parallel elastic component, allowing for a comprehensive understanding of muscle mechanics during contractions.

1. Contractile Component (CC): This represents the active force generated by the muscle fibers through contraction. The CC's force depends on the muscle's length and contraction velocity. According to the Hill equation, the force output is related to the muscle's activated cross-sectional area and the force–velocity characteristics of the muscle fibers.
2. Series Elastic Component (SEC): This component accounts for the elasticity of the tendons and any components in series with the muscle fibers. When the muscle contracts, it stretches the SEC, contributing to the overall force exerted by the muscle.
3. Parallel Elastic Component (PEC): This component represents the elastic properties of the muscle tissue itself when it is not actively generating force. The PEC can resist lengthening and contribute to the passive force when the muscle is elongated without activation.

The dynamics of the Hill-type muscle model are typically described by a set of equations that incorporate these components. For example, the total force provided by a muscle (F_total) can be described as

$$F\_\{total\} = F\_\{CC\} + F\_\{SEC\} - F\_\{PEC\}$$

where

F_{CC} is the force produced by the contractile component,
F_{SEC} is the elastic force from the series elastic component, and
F_{PEC} is the force contribution from the parallel elastic element.

The Hill-type model is beneficial in biomechanics as it enables simulations of muscle behavior during various activities. This aids in the analysis of injury prevention, rehabilitation protocols, and ergonomic assessments.

The Hill-type muscle model is a fundamental framework for understanding muscle mechanics in static and dynamic contexts.

The force produced by a muscle is intricately linked to both its length and the speed at which it contracts. These two essential characteristics are crucial in determining various important biomechanical properties, such as running speed, overall strength, and the distance one can jump. The relationship between muscle length and contraction velocity not only influences athletic performance but also affects the efficiency and effectiveness of movement in various physical activities.

Muscle Force–Length Relationship

Muscles possess an intrinsic elasticity primarily attributed to the presence of a remarkable protein known as titin. This unique characteristic allows skeletal muscles to stretch and recoil, embodying a natural resilience. These muscles are anchored to the bones through strong yet flexible tendons, crucial in maintaining the muscle's resting length—a state of gentle tension that enables optimal function (Fig. 8.5). However, envision a scenario where this vital connection is severed;

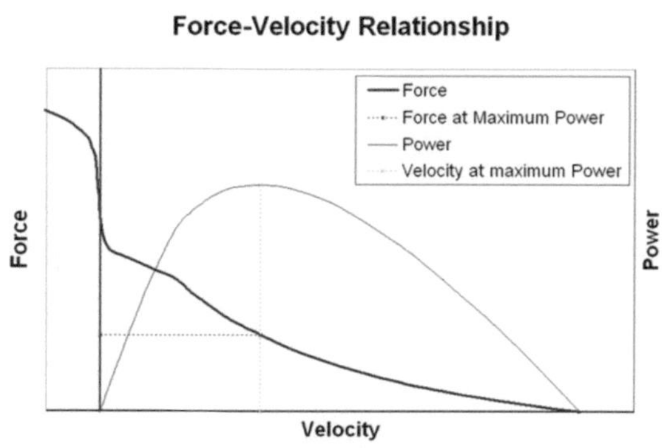

Fig. 8.5 *Force–velocity relationship*: the force and power produced decrease as velocity increases. Although force increases due to stretching with no velocity, zero power is made. Maximum power is generated at one-third of maximum shortening velocity

for instance, if the bicep were to detach from its points of attachment on the scapula or radius. In such a case, the muscle would react instinctively, contracting and ultimately shortening in length, losing its capacity for movement and efficiency.

Muscle fibers operate in a specific state to optimize the force generated during contraction, which is regulated by the interwoven myofilaments within the sarcomere. During sarcomere contraction, myosin heads bind to actin, forming cross-bridges. Subsequently, the thin filaments (actin) slide over the thick filaments (myosin) as the myosin heads exert a pulling force on the actin. This sliding mechanism leads to the shortening of the sarcomere, thereby generating the tension associated with muscle contraction. If a sarcomere is excessively stretched, inadequate overlap between the myofilaments occurs, resulting in diminished force production. Conversely, if a muscle is overly contracted, the capacity for further contraction is impaired, ultimately reducing the force that can be generated.

The tension produced in skeletal muscle fundamentally depends on the degree of overlap between actin and myosin myofilaments. In mammals, there is a considerable correlation between sarcomeres' optimal and actual resting lengths.

The force–velocity relationship in muscle relates the speed at which a muscle changes length to the force of this contraction and the resultant power output (force × velocity = power). The force generated by a muscle depends on the number of actin and myosin cross-bridges formed; a greater number of cross-bridges results in a greater amount of force. However, cross-bridge formation is not immediate, so if myofilaments slide over each other faster, the ability to form cross-bridges and the resultant force are reduced.

No cross-bridges can form at maximum velocity, so no force is generated, resulting in zero power (right edge of graph). The reverse is true for muscle stretching. Although the force of the muscle is increased, there is no velocity of contraction, and zero power is generated (left edge of graph). Maximum power is generated at approximately one-third of maximum shortening velocity.

Computational Techniques Used in Modeling Hands and Digits

Finite Element Analysis (FEA)

FEA allows for a detailed examination of the stress and strain distribution in the hand when exerting grip forces (Fig. 8.6). This modeling can help visualize how changes in grip strategy affect hand biomechanics.

Complex Multibody Models MBD simulates the interactions between hand segments and integrated muscles. It provides comprehensive insights into how gripping forces are distributed across joints and tissues.

In a study by Dimitra Blana et al., the authors created a simulation of object gripping and explored the model's use as a controller of grip force (Fig. 8.7). In this simulation, a virtual, smooth-sided cup was placed in the modeled hand and slowly filled with water. The loop included a controller to maintain a sufficient grip force, preventing the user from slipping. During a forward-dynamic simulation, the controller adjusted grip force by modulating the input muscle excitation level to the model.

Machine Learning Approaches

Recent trends involve using machine learning algorithms to predict grip forces based on sensor data. These models can be trained using data collected from various manipulation tasks to accurately estimate forces.

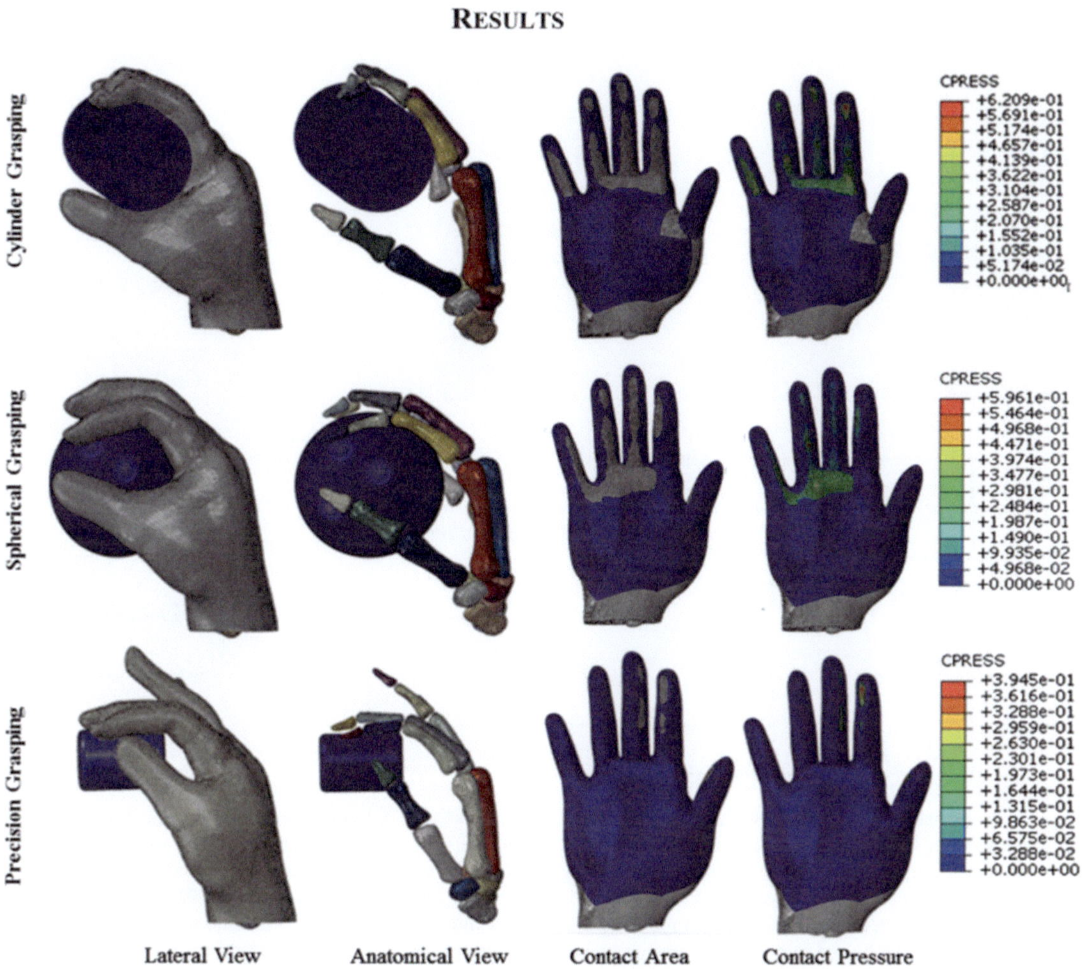

Fig. 8.6 Predicted contact area and pressure distribution during three types of grasps: cylindrical, spherical, and precision. Each row presents (from left to right) a lateral view of the grasp, anatomical visualization of musculoskeletal engagement, predicted palmar contact area, and contact pressure maps. Contact pressures (CPRESS) are displayed using a color gradient scale, with red indicating the highest pressures and blue indicating the lowest pressures. Cylindrical and spherical grasps exhibit broader contact areas with distributed pressure across the palm and fingers. In contrast, the precision grip exhibits more localized contact at the fingertips, characterized by a higher-pressure concentration. These simulations highlight biomechanical differences in load transmission and surface interaction across grasp types. ([30]; permission to reprint under CC BY 4.0)

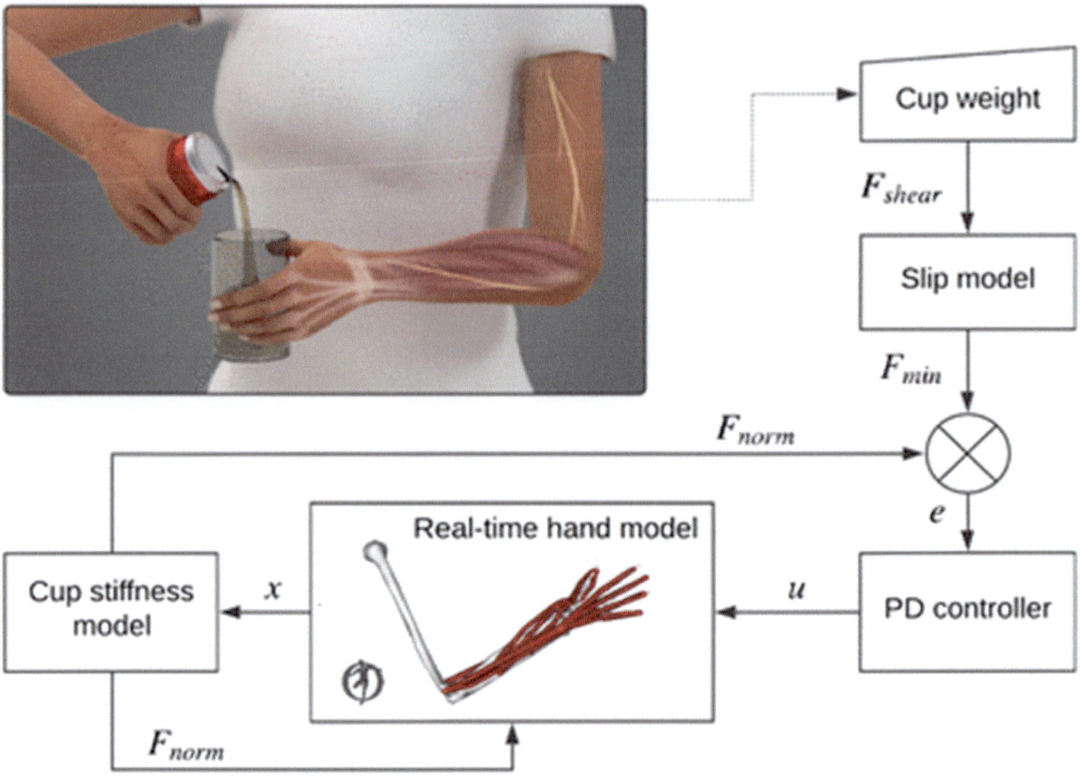

Fig. 8.7 Schematic representation of a simulated object gripping system. The model accounts for dynamic changes in cup weight to simulate fluid filling, generating a shear force (Fshear) on the fingers. This shear force is used within a slip model to estimate the minimum normal contact force (Fmin) required to prevent slippage. A real-time musculoskeletal hand model receives muscle excitation inputs (u) and produces finger displacement (x) as its output. This displacement is applied to a cup stiffness model to compute the resulting fingertip contact force (Fnorm). A proportional-derivative (PD) controller modulates muscle excitation in real time to maintain Fnorm >, ensuring a stable, slip-resistant grip [53]; permission requested)

Applications

Applications: Our understanding of gripping forces is not just theoretical. It has practical applications in developing effective rehabilitation programs to restore hand function (Fig. 8.8). By designing customized exercises to strengthen specific muscles or improve coordination, we can make a real difference in the lives of those with hand-related conditions.

Ergonomic Design Insights from grip force modeling can inform the design of tools, handles, and devices that enhance usability and reduce the risk of injury.

Robotics Accurate estimation of human-like gripping forces in robotic hand design enables the creation of more effective robotic systems that can interact safely and efficiently with a wide range of objects.

The modeling and estimation of gripping forces is not a solitary endeavor. It is a multidisciplinary effort integrating biomechanics, engi-

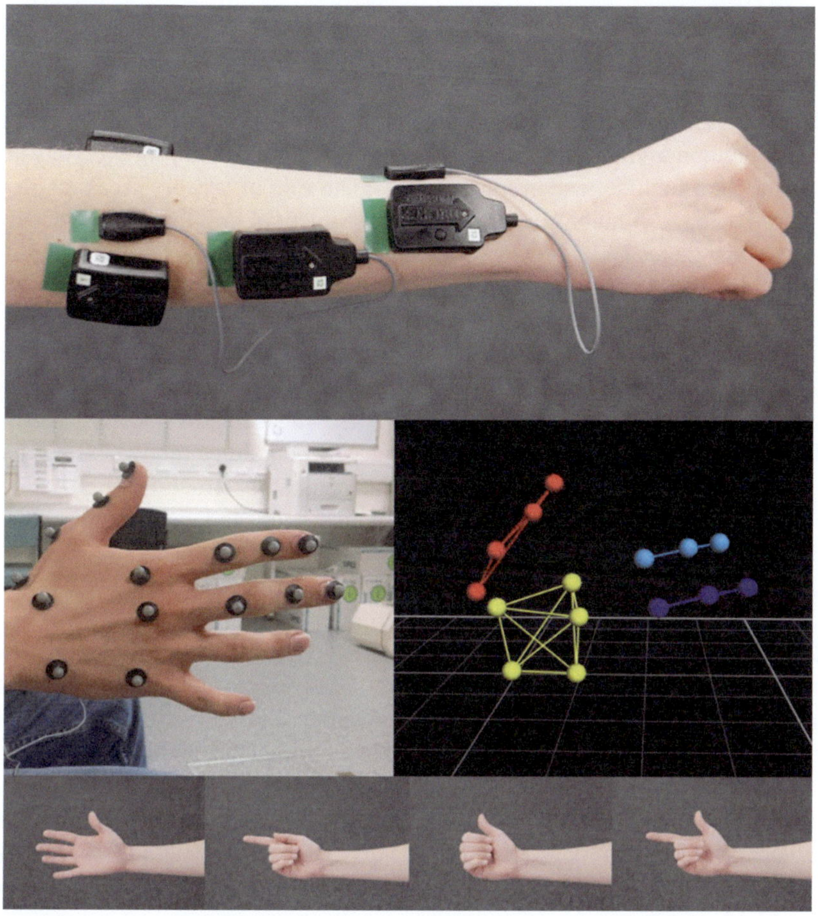

Fig. 8.8 Experimental setup for electromyographic (EMG) and kinematic analysis of hand postures. This setup is crucial for our research, enabling us to measure and analyze the forces and movements associated with different hand postures. *Top*: EMG electrodes were placed over four distinct forearm muscle regions to allow independent activation of five hand postures, including rest. The targeted muscles included the lateral and medial portions of the extensor digitorum communis (EDC), the flexor digitorum superficialis (FDS) located just distal to the superficial wrist flexors, and the extensor pollicis brevis (EPB). *Middle*: reflective markers were positioned on anatomical landmarks of the hand and fingers for motion capture-based kinematic tracking. *Bottom*: the four active hand postures used in the experiment: open hand, index point, fist, and thumbs-up, in addition to the resting posture (not shown). ([53]; permission requested)

neering, and computational methods. This collaborative approach, which includes researchers, professionals, and practitioners from various fields, is crucial in our collective effort to understand the dynamics of grip strength and develop more effective rehabilitation strategies, ergonomic tools, and advanced robotic systems. Ultimately, this collective effort improves functionality and safety in hand-related tasks, benefiting healthy individuals and those recovering from injury.

Pathologies Affecting Gripping Ability

Several hand pathologies can significantly impair gripping ability. Common injuries and conditions that affect grip function include tendon injuries, arthritis, and nerve damage. Tendon-related conditions such as trigger finger and De Quervain's tenosynovitis can diminish grip strength and functionality. The trigger finger is characterized by abnormal resistance during tasks. De Quervain's tenosynovitis, involving swelling of the extensor tendons at the wrist, affects thumb mobility and consequently reduces grip strength [32].

Arthritis, particularly osteoarthritis (OA) and rheumatoid arthritis (RA), is a major contributor to compromised grip strength. OA affecting the base of the thumb and interphalangeal joints reduces dexterity and grip due to pain and joint degeneration. RA, marked by synovitis of the metacarpophalangeal (MCP) joints, significantly impacts grip force, particularly when the thumb is involved [33–35].

Neurological conditions such as stroke, carpal tunnel syndrome (CTS), and hemiparesis are also associated with impaired grip strength. These conditions affect both power and precision grips, essential for daily activities. In CTS, nerve compression impairs sensory and motor functions, decreasing grip and pinch strength and increasing force variability during gripping tasks [33, 36]. Sensory impairment from CTS disrupts the feedback mechanisms necessary for fine motor control, leading to excessive grip forces and a reduced ability to balance torques across digits, ultimately compromising coordination in precision grips [33, 36].

Stroke survivors often experience deficits in neuromuscular coordination and sensory feedback, impairing grip strength and control. Studies have shown that stroke patients, even those with mild motor impairment, exhibit persistent difficulties in modulating finger forces and coordinating grip [37–39]. These challenges affect their ability to perform both power and precision grips. Similarly, individuals with hemiparesis exhibit impairments in anticipatory motor control and grip force modulation, whether congenital or acquired, which are critical for precision tasks. The capacity to plan and execute efficient grip strategies is often diminished, resulting in functional limitations during everyday activities [40].

The role of sensory feedback in the real-time adjustment of grip forces cannot be overstated. In conditions such as stroke and CTS, sensory deficits significantly impair the ability to adapt grip strength to object characteristics, thereby increasing the risk of inadequate or excessive grip and reducing handling efficiency [41, 42].

The mechanisms underlying grip impairment vary by pathology: tendon injuries cause pain and mechanical obstruction, arthritis leads to joint degeneration and pain, and nerve damage disrupts sensory feedback and motor control. Each of these factors contributes to diminished grip strength and dexterity.

Psychological factors are not to be overlooked in the context of hand function. Conditions such as depression, anxiety, and pain catastrophizing can significantly influence perceived disability and grip performance. While depression scores may show only minimal correlation with grip strength, these psychological factors can have a profound impact on patient-reported outcomes. Furthermore, traumatic hand injuries can lead to long-term psychological sequelae, including depression, anxiety, and post-traumatic stress disorder, which can further affect hand function and hinder rehabilitation outcomes. Understanding and addressing these factors is crucial for a holistic and effective rehabilitation process.

Rehabilitation Strategies

Rehabilitation strategies for improving grip strength in patients with hand injuries should be multifaceted, addressing physical and psychological components. Physical rehabilitation remains the cornerstone of treatment and is considered the most effective approach to recovery. Among the various rehabilitation protocols, exercise programs are particularly critical. Tailored exercise regimens are designed to address the specific needs of each condition. For tendon injuries, early and proactive grip strength training can promote faster recovery without compromising the integrity of the repair site. In arthritis, especially rheumatoid arthritis, hand exercises have demonstrated modest improvements in hand function and grip strength over the medium to long term. Task-specific training, as recommended by the American Physical Therapy Association, has been shown to improve strength and manual dexterity [43–45]. The most effective rehabilitation exercises for improving grip strength in patients with tendon injuries, such as trigger finger and De Quervain's tenosynovitis, include Intrinsic Plus Fist with Ball Exercise, Pulley Stretch Exercises, Rubber Band Resistance (RBR) Finger Abduction, and Early Motion Regimens.

Assessing grip strength and functional capabilities is critical for developing and tailoring rehabilitation strategies. Regular evaluations using dynamometers can objectively quantify grip strength. Functional assessments, such as the Jebson-Taylor Hand Function Test and the Action Research Arm Test, provide valuable insights into manual dexterity and overall hand function, facilitating the adjustment of rehabilitation plans as needed [45, 47]. Therapeutic exercises can be broadly categorized into strengthening and range of motion (ROM) improvement exercises. Strengthening exercises include the Intrinsic Plus Fist with Ball Exercise, which generates high electromyographic (EMG) activation of the first dorsal interosseous and opponens pollicis muscles, crucial for stabilizing the trapeziometacarpal joint [47]. Rubber Band Resistance (RBR) Finger Abduction effectively targets intrinsic hand muscles without over-fatiguing extrinsic muscles and involves repetitive finger abduction against a rubber band [46]. Early Motion Regimens, which include initiating controlled active and passive movements within 5 days post-repair, have been shown to enhance tendon healing and improve grip strength [48].

Range of motion exercises, including pulley stretch exercises and Passive Cyclical Stretching. Pulley Stretch Exercises, especially beneficial for conditions like trigger fingers, involve isometric resistance exercises that flex the finger against pressure to widen the space under the first annular pulley, thereby reducing symptoms and improving grip strength [45]. Passive Cyclical Stretching, combined with active-assisted, task-oriented training, can improve upper extremity motor control and task performance in post-stroke rehabilitation [49]. In addition to therapeutic exercises, assistive devices such as soft-robotic gloves can support grip strength and hand function during daily activities. These devices have significantly improved grip strength and functional performance over a 6-week intervention period [46]. Other supportive aids include orthoses and splints, which provide structural support, reduce pain, and enhance patient participation in rehabilitation exercises. These are particularly beneficial in conditions involving arthritis or nerve damage.

Occupational therapy plays a vital role in rehabilitating grip-related hand pathologies. Occupational therapists assist patients in making lifestyle modifications and preventing functional decline by using joint protection techniques, assistive technologies, and customized orthoses [43, 50]. Early implementation of occupational therapy interventions has significantly improved hand function and grip strength in conditions such as rheumatoid arthritis [43, 50].

Future Directions in Hand Biomechanics Research

Interdisciplinary collaboration is the cornerstone of future advancements in hand biomechanics research. By pooling our expertise and resources, we can develop more comprehensive and effective rehabilitation strategies, ultimately improving patient outcomes.

Wearable technologies, including soft-robotic and sensor gloves, have shown encouraging results in supporting daily activities and delivering therapeutic benefits. Soft-robotic gloves, such as the Carbon hand glove, enhance grip strength and hand function during routine tasks, with studies demonstrating substantial improvements over a 6-week intervention period [45].

Sensor gloves, which capture hand postures and provide real-time feedback, facilitate task-specific training and enhance motor control by enabling users to adjust movements during activity [45].

Robotic-assisted rehabilitation systems are gaining momentum due to their capacity to deliver high-intensity, high-frequency therapy. These systems, including soft wearable unilateral telerobotic platforms, support grasping tasks and promote cooperative interaction between patients and therapists, improving hand function and dexterity [51, 52]. In addition, bioinspired technologies such as smart triboelectric soft pneumatic actuators (TENG-SPA) have been developed to deliver sufficient stretching force for spastic fingers, showing considerable promise in digital rehabilitation medicine [51, 52].

When combined with assistive technologies such as virtual reality and motion-tracking systems, telerehabilitation enables remote monitoring and objective evaluation of hand function. The Virtual Glove System, for instance, offers real-time hand tracking and quantitative analysis of hand mobility, facilitating enhanced interaction between patients and therapists while supporting data sharing [53]. These digital platforms expand accessibility and enable continuous rehabilitation beyond traditional clinical settings [53].

Furthermore, advancements in motion and electromyography-based wearable technologies are creating new possibilities for remote therapy and monitoring. These devices provide real-time feedback and therapeutic guidance, making rehabilitation more accessible, personalized, and cost-effective.

Conclusion

This chapter has explored the intricate interplay between anatomy, neuromuscular coordination, sensory feedback, and external forces that underlie hand gripping. We examined the complementary roles of extrinsic and intrinsic muscles in both power and precision grips, alongside the kinematic and biomechanical forces—compression, tension, and shear—that contribute to grip strength. Various pathologies, from tendon injuries and arthritis to nerve damage, stroke, and carpal tunnel syndrome, were analyzed for their distinct impacts on grip ability while acknowledging the psychological factors that influence functional outcomes. The chapter presents comprehensive rehabilitation approaches, including targeted strengthening exercises, range-of-motion techniques, innovative assistive technologies, and specialized occupational therapy interventions. We also highlighted emerging innovations in telerehabilitation platforms.

The fundamental importance of hand biomechanics to effective rehabilitation cannot be overstated. For clinicians, therapists, and researchers to design targeted, evidence-based interventions, they must appreciate the complex neuromuscular and mechanical processes involved in gripping. Detailed insights into joint motion, muscle activation patterns, and force coordination directly inform the development of more precise diagnostic tools, surgical planning approaches, adaptive technologies, and personalized therapy programs. As hand function remains central to activities of daily living, occupational performance, and personal independence, improvements in grip strength and control can dramatically enhance the quality of life across diverse patient populations.

Looking forward, the future of hand biomechanics lies at the intersection of multiple disciplines, integrating engineering principles, neuroscientific discoveries, and clinical expertise to optimize patient outcomes. As rehabilitation technologies evolve and treatment approaches become increasingly personalized, the capacity to understand and modulate grip mechanics will remain essential for restoring functional independence in individuals with hand impairments. Continued research, clinical innovation, and interdisciplinary collaboration will significantly improve recovery and enhance hand function in real-world settings.

References

1. Valero-Cuevas FJ. An integrative approach to the biomechanical function and neuromuscular control of the fingers. J Biomech. 2005;38(4):673–84. https://doi.org/10.1016/j.jbiomech.2004.04.006. PMID: 15713287
2. O'Shea H, Redmond SJ. A review of the neurobiomechanical processes underlying secure gripping in object manipulation. Neurosci Biobehav Rev. 2021;123:286–300. https://doi.org/10.1016/j.neubiorev.2021.01.007. Epub 2021 Jan 23. PMID: 33497782
3. Tazoe T, Perez MA. Cortical and reticular contributions to human precision and power grip. J Physiol. 2017;595(8):2715–30. https://doi.org/10.1113/JP273679. Epub 2017 Feb 27. PMID: 27891607; PMCID: PMC5390869
4. Fox PM, Oliver JD, Nguyen V, Hentz VR, Curtin CM. Electromyographic analysis of grip. Orthopedics. 2019;42(6):e555–8. https://doi.org/10.3928/01477447-20190812-06. Epub 2019 Aug 14. PMID: 31408520
5. Arnet U, Muzykewicz DA, Fridén J, Lieber RL. Intrinsic hand muscle function, part 1: creating a functional grasp. J Hand Surg Am. 2013;38(11):2093–9. https://doi.org/10.1016/j.jhsa.2013.08.099. PMID: 24206973; PMCID: PMC3879019
6. Bain GI, Polites N, Higgs BG, Heptinstall RJ, McGrath AM. The functional range of motion of the finger joints. J Hand Surg Eur. 2015;40(4):406–11. https://doi.org/10.1177/1753193414533754. Epub 2014 May 23. PMID: 24859993
7. Gülke J, Wachter NJ, Geyer T, Schöll H, Apic G, Mentzel M. Motion coordination patterns during cylinder grip analyzed with a sensor glove. J Hand Surg Am. 2010;35(5):797–806. https://doi.org/10.1016/j.jhsa.2009.12.031. Epub 2010 Mar 1. PMID: 20189730
8. Caumes M, Goislard de Monsabert B, Hauraix H, Berton E, Vigouroux L. Complex couplings between joints, muscles, and performance: the role of the wrist in grasping. Sci Rep. 2019;9(1):19357. https://doi.org/10.1038/s41598-019-55443-w. PMID: 31852907; PMCID: PMC6920170
9. Li K, Wei N, Cheng M, Hou X, Song J. Dynamical coordination of hand intrinsic muscles for precision grip in diabetes mellitus. Sci Rep. 2018;8(1):4365. https://doi.org/10.1038/s41598-018-22588-z. PMID: 29531355; PMCID: PMC5847602
10. Dupan SSG, Stegeman DF, Maas H. Distinct neural control of intrinsic and extrinsic muscles of the hand during single finger pressing. Hum Mov Sci. 2018;59:223–33. https://doi.org/10.1016/j.humov.2018.04.012. Epub 2018 May 5. PMID: 29738941
11. Takei T, Seki K. Spinal interneurons facilitate coactivation of hand muscles during a precision grip task in monkeys. J Neurosci. 2010;30(50):17041–50. https://doi.org/10.1523/JNEUROSCI.4297-10.2010. PMID: 21159974; PMCID: PMC6634901
12. Adkinson JM, Johnson SP, Chung KC. The clinical implications of the oblique retinacular ligament. J Hand Surg Am. 2014;39(3):535–41. https://doi.org/10.1016/j.jhsa.2013.12.011. PMID: 24559632; PMCID: PMC3977782.
13. Hahn P, Heindl E. Does an ulnar nerve lesion influence the motion of the index finger? J Hand Surg Br. 1996;21(2):252–4. https://doi.org/10.1016/s0266-7681(96)80108-2. PMID: 8732411
14. Mentzel M, Apic G, Wachter NJ, Bauknecht S, Gulkin D, Gülke J. Erfassung der Kraftprofile verschiedener primärer Griffe durch dynamische Kraftmessung mit einem Sensorhandschuh [Assessment of force patterns of different primary grips through dynamic force measurement using a sensor glove]. Handchir Mikrochir Plast Chir. 2011;43(3):155–61. German. https://doi.org/10.1055/s-0031-1271805. Epub 2011 Mar 30. PMID: 21452111.
15. Richards WT, Vergara E, Dalaly DG, Coady-Fariborzian L, Mozingo DW. Acute surgical management of hand burns. J Hand Surg Am. 2014;39(10):2075-85.e2. https://doi.org/10.1016/j.jhsa.2014.07.032. PMID: 25257489
16. Ergen HI, Oksuz C. Evaluation of load distributions and contact areas in 4 common grip types used in daily living activities. J Hand Surg Am. 2020;45(3):251.e1–251.e8. https://doi.org/10.1016/j.jhsa.2019.06.006. Epub 2019 Sep 4. PMID: 31493940
17. Palastanga N, Field D, Soames R. Anatomy and Human Movement Structure and Function Book. Second Edition. 1994, https://doi.org/10.1016/C2013-0-04562-8
18. Dietz V. Neural coordination of bilateral power and precision finger movements. Eur J Neurosci. 2021;54(12):8249–55. https://doi.org/10.1111/ejn.14911. Epub 2020 Jul 28. PMID: 32682343
19. Watson J, Ring D. Influence of psychological factors on grip strength. J Hand Surg Am. 2008;33(10):1791–5. https://doi.org/10.1016/j.jhsa.2008.07.006. PMID: 19084180
20. Iturrate I, Chavarriaga R, Pereira M, Zhang H, Corbet T, Leeb R, Millán JDR. Human EEG reveals distinct neural correlates of power and precision grasping types. NeuroImage. 2018;181:635–44. https://doi.org/10.1016/j.neuroimage.2018.07.055. Epub 2018 Jul 27. PMID: 30056196
21. Witney AG, Wing A, Thonnard JL, Smith AM. The cutaneous contribution to adaptive precision grip. Trends Neurosci. 2004;27(10):637–43. https://doi.org/10.1016/j.tins.2004.08.006. PMID: 15374677
22. Enders LR, Seo NJ. Phalanx force magnitude and trajectory deviation increased during power grip with an increased coefficient of friction at the hand-object

interface. J Biomech. 2011;44(8):1447–53. https://doi.org/10.1016/j.jbiomech.2011.03.020. Epub 2011 Apr 14. PMID: 21496820

23. Austin EF Jr, Kearney CP, Chacon PJ, Winges SA, Acharya P, Choi JW. A fabricated force glove that measures hand forces during activities of daily living. Sensors (Basel). 2022;22(4):1330. https://doi.org/10.3390/s22041330. PMID: 35214233; PMCID: PMC8877267
24. Irwin CB, Towles JD, Radwin RG. Development and application of a multi-axis dynamometer for measuring grip force. Ergonomics. 2013;56(12):1841–9. https://doi.org/10.1080/00140139.2013.847212. Epub 2013 Oct 17. PMID: 24134054; PMCID: PMC3855544
25. Wichelhaus A, Harms C, Neumann J, Ziegler S, Kundt G, Prommersberger KJ, Mittlmeier T, Mühldorfer-Fodor M. Parameters influencing hand grip strength measured with the manugraphy system. BMC Musculoskelet Disord. 2018;19(1):54. https://doi.org/10.1186/s12891-018-1971-4. PMID: 29444676; PMCID: PMC5813357
26. Yu R, Ong S, Cheung O, Leung J, Woo J. Reference values of grip strength, prevalence of low grip strength, and factors affecting grip strength values in Chinese adults. J Am Med Dir Assoc. 2017;18(6):551.9–551.e16. https://doi.org/10.1016/j.jamda.2017.03.006. Epub 2017 Apr 29. PMID: 28465129
27. Martin-Brevet S, Jarrassé N, Burdet E, Roby-Brami A. Taxonomy-based analysis of force exchanges during object grasping and manipulation. PLoS One. 2017;12(5):e0178185. https://doi.org/10.1371/journal.pone.0178185. PMID: 28562617; PMCID: PMC5451044
28. Seo NJ, Armstrong TJ. Investigation of Grip Force, Normal Force, Contact Area, Hand Size, and Handle Size for Cylindrical Handles. Human Factors. 2008;50(5):734–44. https://doi.org/10.1518/001872008X354192 (Original work published 2008).
29. Amirouche F. Fundamentals of multibody dynamics: theory and applications. 1st ed. Springer Nature; 2007. https://doi.org/10.1007/b137682.
30. Cei G, Artoni A, Bianchi M. A review on finite element modelling of finger and hand mechanical behaviour in haptic interactions. Bio Model Mech. 2025;24(3):895–917. https://doi.org/10.1007/s10237-025-01943-w.
31. Blana D, Van Den Bogert AJ, Murray WM, Ganguly A, Krasoulis A, Nazarpour K, Chadwick EK. Model-Based Control of Individual Finger Movements for Prosthetic Hand Function. IEEE Trans Neural Syst Rehabil Eng. 2020;28(3):612–20. https://doi.org/10.1109/TNSRE.2020.2967901. Epub 2020 Jan 20. PMID: 31976900; PMCID: PMC7231850.
32. Currie KB, Tadisina KK, Mackinnon SE. Common hand conditions: a review. JAMA. 2022;327(24):2434–45. https://doi.org/10.1001/jama.2022.8481. Erratum in: JAMA. 2023;330(8):772. https://doi.org/10.1001/jama.2022.14783. PMID: 35762992.
33. Kim YH, Han EY, Kim J, Seo KB, Jeon YT, Im SH. Associations between hand function and electrophysiological measurements in hand osteoarthritis patients of different ages with or without carpal tunnel syndrome. Sci Rep. 2020;10(1):19278. https://doi.org/10.1038/s41598-020-74795-2. Erratum in: Sci Rep. 2021;11(1):3225. https://doi.org/10.1038/s41598-021-82090-x. PMID: 33159094; PMCID: PMC7648779.
34. Haugen IK, Aaserud J, Kvien TK. Get a grip on factors related to grip strength in persons with hand osteoarthritis: results from an observational cohort study. Arthritis Care Res. 2021;73(6):794–800. https://doi.org/10.1002/acr.24385. PMID: 32702166
35. Rydholm M, Sharma A, Jacobsson L, Turesson C. The relation between synovitis of individual finger joints and grip force over the first 5 years in early rheumatoid arthritis – a cohort study. Arthritis Res Ther. 2023;25(1):231. https://doi.org/10.1186/s13075-023-03212-6. PMID: 38037132; PMCID: PMC10688125
36. Chen PT, Jou IM, Lin CJ, Chieh HF, Kuo LC, Su FC. Is the control of applied digital forces during natural five-digit grasping affected by carpal tunnel syndrome? Clin Orthop Relat Res. 2015;473(7):2371–82. https://doi.org/10.1007/s11999-015-4189-x. Epub 2015 Feb 18. PMID: 25690168; PMCID: PMC4457761
37. Pennati GV, Plantin J, Carment L, Roca P, Baron JC, Pavlova E, Borg J, Lindberg PG. Recovery and prediction of dynamic precision grip force control after stroke. Stroke. 2020;51(3):944–51. https://doi.org/10.1161/STROKEAHA.119.026205. Epub 2020 Jan 7. PMID: 31906829
38. Ding Q, Patten C. External biomechanical constraints impair maximal voluntary grip force stability post-stroke. Clin Biomech (Bristol). 2018;57:26–34. https://doi.org/10.1016/j.clinbiomech.2018.06.001. Epub 2018 Jun 5. PMID: 29894857
39. Pollet AK, Patel P, Lodha N. Does the contribution of the paretic hand to bimanual tasks change with grip strength capacity following stroke? Neuropsychologia. 2022;168:108186. https://doi.org/10.1016/j.neuropsychologia.2022.108186. Epub 2022 Feb 18. PMID: 35189182; PMCID: PMC9007327
40. Bleyenheuft Y, Gordon AM. Precision grip in congenital and acquired hemiparesis: similarities in impairments and implications for neurorehabilitation. Front Hum Neurosci. 2014;30(8):459. https://doi.org/10.3389/fnhum.2014.00459. PMID: 25071502; PMCID: PMC4074995
41. Akaguchi R, Takamura Y, Okuno H, Morioka S, Kawashima N. Relative contribution of sensory and motor deficits on grip force control in patients with chronic stroke. Clin Neurophysiol. 2024;161:231–41. https://doi.org/10.1016/j.clinph.2024.02.033. Epub 2024 Mar 6. PMID: 38522269

42. Enders LR, Seo NJ. Effects of sensory deficit on Phalanx force deviation during power grip post stroke. J Mot Behav. 2017;49(1):55–66. https://doi.org/10.1080/00222895.2016.1191416. Epub 2016 Sep 3. PMID: 27592686; PMCID: PMC6101663
43. Kaji Y, Yamaguchi K, Nomura Y, Oka K, Nakamura O, Ishibashi Y, Yamamoto T. Postoperative early and proactive grip strength training program for distal radius fractures promotes earlier recovery of grip strength: a retrospective study. Medicine (Baltimore). 2022;101(19):e29262. https://doi.org/10.1097/MD.0000000000029262. PMID: 35583536; PMCID: PMC9276271
44. Williams MA, Srikesavan C, Heine PJ, Bruce J, Brosseau L, Hoxey-Thomas N, Lamb SE. Exercise for rheumatoid arthritis of the hand. Cochrane Database Syst Rev. 2018;7(7):CD003832. https://doi.org/10.1002/14651858.CD003832.pub3. PMID: 30063798; PMCID: PMC6513509
45. Osborne JA, Botkin R, Colon-Semenza C, DeAngelis TR, Gallardo OG, Kosakowski H, Martello J, Pradhan S, Rafferty M, Readinger JL, Whitt AL, Ellis TD. Physical therapist management of parkinson disease: a clinical practice guideline from the American physical therapy association. Phys Ther. 2022;102(4):pzab302. https://doi.org/10.1093/ptj/pzab302. Erratum in: Phys Ther. 2022 Aug 1;102(8): pzac098. https://doi.org/10.1093/ptj/pzac098. PMID: 34963139; PMCID: PMC9046970.
46. Kottink AIR, Nikamp CDM, Bos FP, Sluis CKV, Broek MVD, Onneweer B, Stolwijk-Swüste JM, Brink SM, Voet NBM, Rietman JS, Prange-Lasonder GB. Therapy effect on hand function after home use of a wearable assistive soft-robotic glove supporting grip strength. PLoS One. 2024;19(7):e0306713. https://doi.org/10.1371/journal.pone.0306713. PMID: 38990858; PMCID: PMC11239026
47. O'Connell C, Guo M, Soucy B, Calder M, Sparks J, Plamondon S. All hands-on deck: the multidisciplinary rehabilitation assessment and management of hand function in persons with neuromuscular disorders. Muscle Nerve. 2024; https://doi.org/10.1002/mus.28167. Epub ahead of print. PMID: 38845187.
48. Fischer HC, Triandafilou KM, Thielbar KO, Ochoa JM, Lazzaro ED, Pacholski KA, Kamper DG. Use of a portable assistive glove to facilitate rehabilitation in stroke survivors with severe hand impairment. IEEE Trans Neural Syst Rehabil Eng. 2016;24(3):344–51. https://doi.org/10.1109/TNSRE.2015.2513675. Epub 2015 Dec 31. PMID: 26731772
49. Dorf E, Blue C, Smith BP, Koman LA. Therapy after injury to the hand. J Am Acad Orthop Surg. 2010;18(8):464–73. https://doi.org/10.5435/00124635-201008000-00003. PMID: 20675639
50. Mathieux R, Marotte H, Battistini L, Sarrazin A, Berthier M, Miossec P. Early occupational therapy program increases hand grip strength at 3 months: results from a randomized, blind, controlled study in early rheumatoid arthritis. Ann Rheum Dis. 2009;68(3):400–3. https://doi.org/10.1136/ard.2008.094532. Epub 2008 Nov 17. PMID: 19015209
51. Kim DH, Lee Y, Park HS. Cooperative hand therapy via a soft, wearable, and unilateral telerobotic system. IEEE Trans Biomed Eng. 2023;70(1):366–77. https://doi.org/10.1109/TBME.2022.3191690. Epub 2022 Dec 26. PMID: 35849669
52. Li W, Luo F, Liu Y, Zou Y, Mo L, He Q, Lin PJ, Xu Q, Liu A, Zhang C, Cheng J, Cheng L, Ji L. Bioinspired smart triboelectric soft pneumatic actuator-enabled hand rehabilitation robot. Adv Mater. 2025;37(9):e2419059. https://doi.org/10.1002/adma.202419059. Epub 2025 Jan 10. PMID: 39797483
53. Placidi G, Di Matteo A, Lozzi D, Polsinelli M, Theodoridou E. Patient-therapist cooperative hand telerehabilitation through a novel framework involving the virtual glove system. Sensors (Basel). 2023;23(7):3463. https://doi.org/10.3390/s23073463. PMID: 37050523; PMCID: PMC10098681

Bone, Cartilage, and Biologics

9

Seth Roge, Mark Gonzalez, and Farid Amirouche

Introduction

Bone and cartilage are dynamic living tissues characterized by their ability to adapt structurally and functionally to various environmental stimuli, such as mechanical loading, biochemical factors, and physiological changes. This adaptability results in mechanical properties distinctly different from those of traditional homogeneous and isotropic materials like steel, which maintain uniformity in their physical characteristics throughout.

Bone, for instance, is a composite material primarily composed of collagen fibers and mineralized matrix, giving it both tensile strength and compressive stiffness. In contrast, cartilage is a highly hydrated tissue that relies on its unique composition of proteoglycans and collagen to provide elasticity and support under compressive forces.

By employing fundamental material science principles, researchers can gain a deeper insight into the biomechanical behavior of these tissues. A comprehensive understanding of the mechanical properties of bone and cartilage is crucial for advancing discussions on current research topics within hand biomechanics, such as injury prevention, rehabilitation strategies, and the design of biomimetic materials and prosthetics. This knowledge informs clinical practices and guides future directions for innovative interventions in treating musculoskeletal disorders.

Bone Biomechanics

Material Properties

The material properties of bone are the direct result of its structure and microarchitecture. Bone is a nonhomogeneous, anisotropic, viscoelastic solid. Regional variations in microarchitecture with differing organic-cellular, organic-noncellular, and inorganic components give rise to its nonhomogeneous behavior. Variations in microarchitecture, like the orientation of osteons, collagen fibers, and mineral distribution, give rise to its anisotropy. Similarly, this microarchitecture causes the bone to exhibit viscoelastic behavior instead of simple elastic behavior [1]. Further, the mechanical properties of bone may vary with the specific bone, the area of bone being tested, and

S. Roge · M. Gonzalez
Department of Orthopaedic Surgery, University of Illinois at Chicago, Chicago, IL, USA

F. Amirouche (✉)
Department of Orthopaedic Surgery, Orthopaedic and Spine Institute, Endeavor HealthSystem (Formerly Northshore University HealthSystem), an Affiliate of the University of Chicago Pritzker School of Medicine, Skokie, IL, USA

Department of Orthopaedic Surgery, University of Illinois at Chicago, Chicago, IL, USA
e-mail: famirouche@northshore.org

F. Amirouche et al. (eds.), *Biomechanics of the Hand*,
https://doi.org/10.1007/978-3-032-04688-8_9

the relative composition of cortical versus cancellous bone in a sample. Subject-specific factors contributing to variations in bone properties include a person's age, sex, and the presence of a disease that compromises bone density, like osteoporosis [2, 3].

Despite wide variation in the mechanical properties of any given bone sample, certain general principles regarding the biomechanics of bone are helpful to keep in mind. First, as a viscoelastic solid, the behavior of bone varies with the strain rate applied. Recent research using cadaveric human femur samples found that axially loaded bone undergoes more brittle failure above a physiologic strain rate of 1 s^{-1} [4, 5]. Elastic modulus and ultimate strength also generally increase with increasing strain rate [1, 6]. Moreover, cortical bone generally has a greater modulus of elasticity and ultimate strength when loaded in the longitudinal direction along the orientation of osteons compared to the transverse direction. Transverse loading results in more brittle failure than longitudinal loading [1, 2]. Bone is generally stronger in compression than in tension. Last, increasing bone density increases the elastic modulus and ultimate strength of a given sample, as shown in Fig. 9.1 [7].

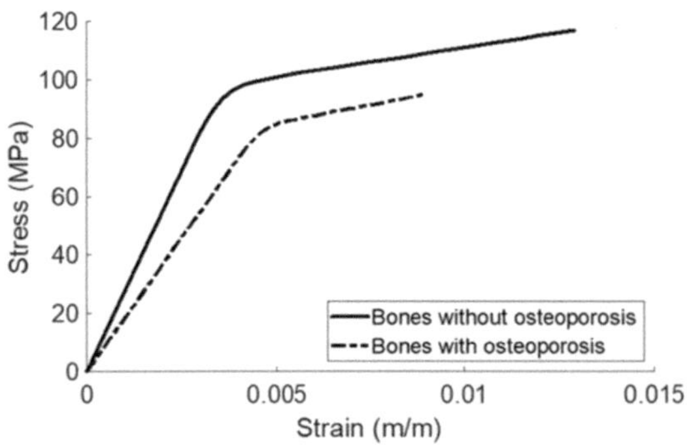

Fig. 9.1 A stress–strain curve comparing bone with and without osteoporosis. Osteoporotic bone exhibits a lower elastic modulus and ultimate strength than healthy bone

Bone Remodeling

In contrast to nonliving materials, bone is a dynamic tissue that alters its structure in response to the environment. Bone remodeling occurs when osteoclast cells resorb bone, and osteoblast cells form new bone to repair defects. These differing cell functions are the basis for the bone structure's ability to adapt [8] Relevant to biomechanics, Wolff's law states that bone will remodel to become stronger in response to repeated compressive stress. In contrast, in the absence of compressive loading or tension, bone will weaken due to the lack of this stress stimulus [9]. Research has demonstrated that osteocyte cells embedded within bone are responsible for transducing mechanical stress into electric signals that regulate remodeling [8]. For bone growing at the physis, the Hueter–Volkmann law dictates that compression inhibits bone growth, and tension accelerates bone growth [10].

Current Research on the Biomechanics of Healthy Hand Bones

Most recently, research on the biomechanical properties of bones of the hand has employed cadaver or finite-element analysis (FEA) to study the topic. One group performed Vickers hardness testing on various regions of cadaveric phalanges to characterize their hardness distribution relative to other long bones. Across phalanges, they found hardness was most significant in the shaft area and weakest in the head and base. These results were consistent with the properties of other long bones and reflective of the relative composition of cortical versus cancellous bone [11].

The focus of research on hand bones using finite-element methods is diverse. Similarly, the methods used to design FEA models vary between studies. When interpreting the results of the FEA-based studies below, one must consider several methodological considerations. First, the geometry of the bone model may be obtained from an idealized model, database, or patient imaging [12]. Bone is anisotropic but commonly modeled as a transversely isotropic or orthotropic material in FEA [13]. The bone may also be modeled as an elastic or viscoelastic solid, and the consequent model material properties, such as elastic modulus, bulk modulus, Poisson ratio, or stiffness parameters, must be based on prior work [3]. All of these factors may influence the results observed.

Varga et al. used an FEA model to analyze contact pressures on the scaphoid under normal in different hand positions, as shown in Fig. 9.2. In this study, they loaded cadaveric wrist flexors

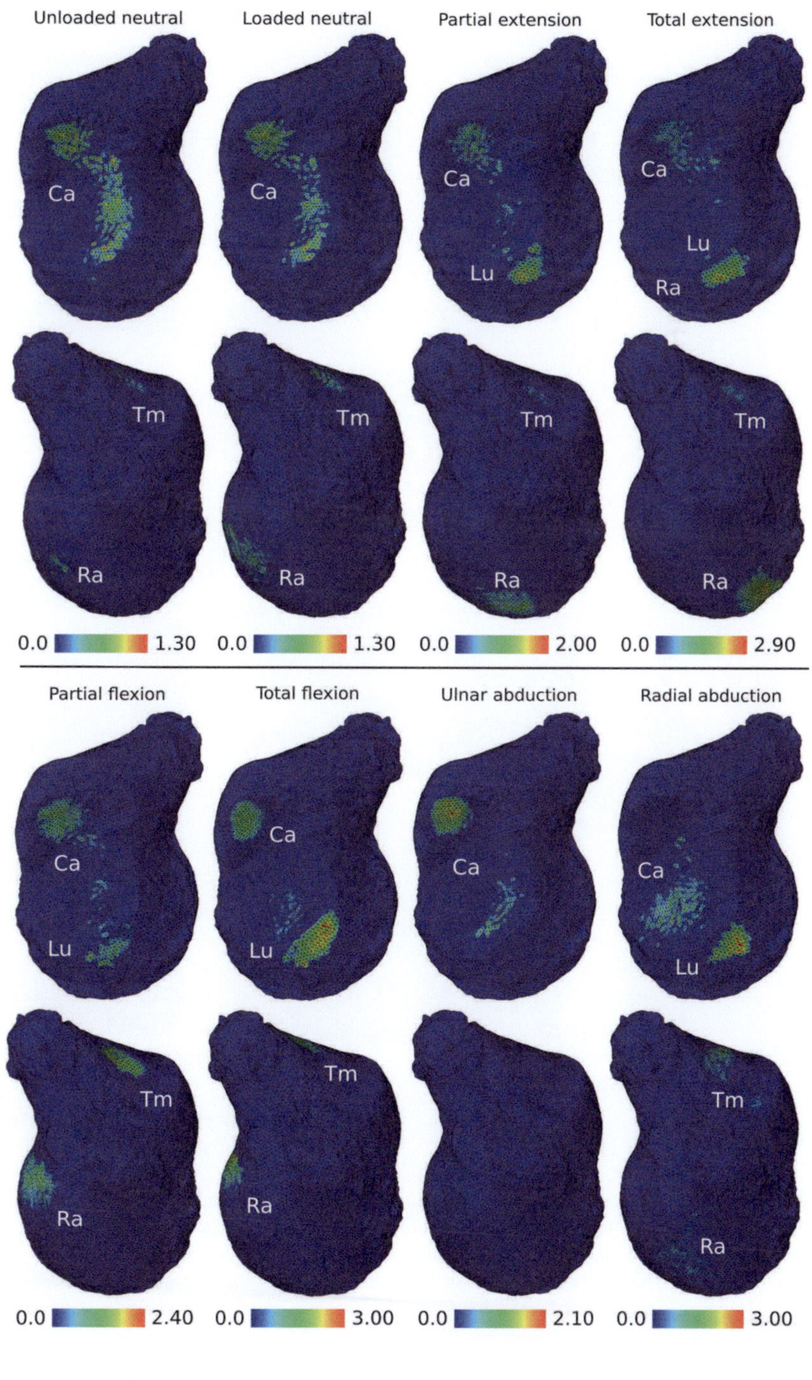

Fig. 9.2 Results from Varga et al. on FEA-based modeling of scaphoid contact forces. Contact forces are shown in MPa

and extensors to obtain different wrist positions (neutral, flexion, extension, radial deviation, ulnar deviation), and they then obtained CT images to model contact stresses caused by scaphoid contact with other carpal bones in each wrist position. They found peak pressures on the scaphoid consistent with other studies and high variation in stress according to the scaphoid region [14].

Other studies on healthy bones have used FEA to study the phalanges [15, 16]. Faudot et al. combined kinematic and medical imaging data to produce and validate a model to study contact pressures on the phalanges, as shown in Fig. 9.3. These authors recommend using this model for future biomechanical work and for modeling the effect of surgical procedures.

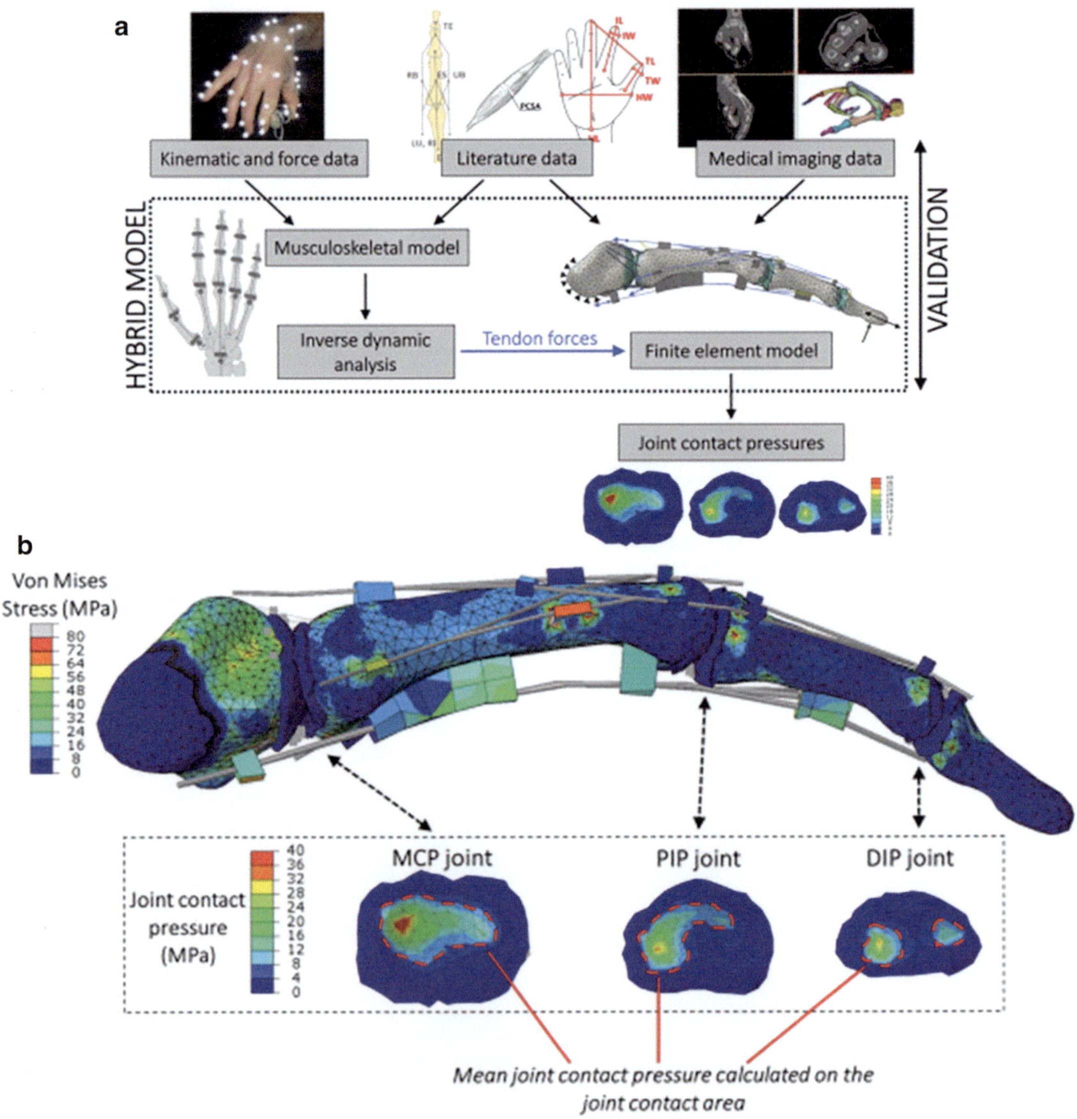

Fig. 9.3 (**a**) The design process used to model the joint contact pressures on the phalanges in a study by Faudot et al. The model was made using kinematic and force data in conjunction with biomechanical data published in the literature. Medical imaging data was used to obtain bone geometry and to validate the model. (**b**) An image of the final model produced by Faudot et al.

Various groups have also attempted to model the proximal carpal row, wrist, or entire hand. Modeling bone was a significant part of these studies [17–21]. Due to technological and methodological advances, these models have become more realistic. The most recent study models bone as a linear, isotropic material. However, these models still face challenges common to all FEA-based studies, including wide variation in reported bone properties [17, 22].

Current Research on the Biomechanics of Pathology and Treatment of Hand Bones

Cadaveric studies on biomechanics of bones of the hand cover topics, including the strength of scaphoid partial unions [23, 24], the strength of fixation constructs for metacarpal and phalanx fractures [25, 26], pullout strengths of screws used for distal radius fracture [27], and more. One group tested the hardness of different regions of the radius. They found a positive correlation between the hardness of the radius and the force required to pull a drilled screw out of the bone [27]. Brekke et al. simulated partial scaphoid union by performing osteotomies at 25, 50, and 75% of the thickness of the scaphoid waist. They subjected cadaveric samples to compressive cantilever loading, as shown in Fig. 9.4. Interestingly, the group found the distal pole rather than the osteotomy site to be the most common failure site. Samples with only 25% intact scaphoid waists could withstand physiologic loading [23].

Fig. 9.4 A sample scaphoid subjected to compressive cantilever loading. Osteotomy in this experiment was performed to simulate partial scaphoid union and was made through the dotted red line

FEA has been used to study the biomechanics of lunate fracture [28], osteotomy for Kienböck disease [29], scaphoid fracture [30, 31], metacarpal fracture [32], phalanx fracture [33], and more. As demonstrated in Fig. 9.5, Zhang et al. subjected a model of a metacarpal to cantilever bending to examine force distribution with different plate fixation constructs. These authors found maximal plate stress occurring over the fracture site in every case. The six-screw plate experienced the lowest plate stress. While extending to the eight-screw plate further minimized displacement over the six-screw plate, it exhibited increased metacarpal stress. This study discusses the trade-offs between different plate lengths, and the authors advocate using the six-screw design [32].

Camus et al. made a wrist model to simulate various osteotomies used to treat Kienböck disease, as shown in Fig. 9.6. The metacarpals were loaded in axial compression, and the resultant stresses were analyzed. They found the only osteotomy that completely unloaded the lunate was the Camembert osteotomy with Sennwald ulnar shortening. With the added benefit of being extra-articular, the authors recommend this method to prevent Kienböck disease-related lunate fracture [29].

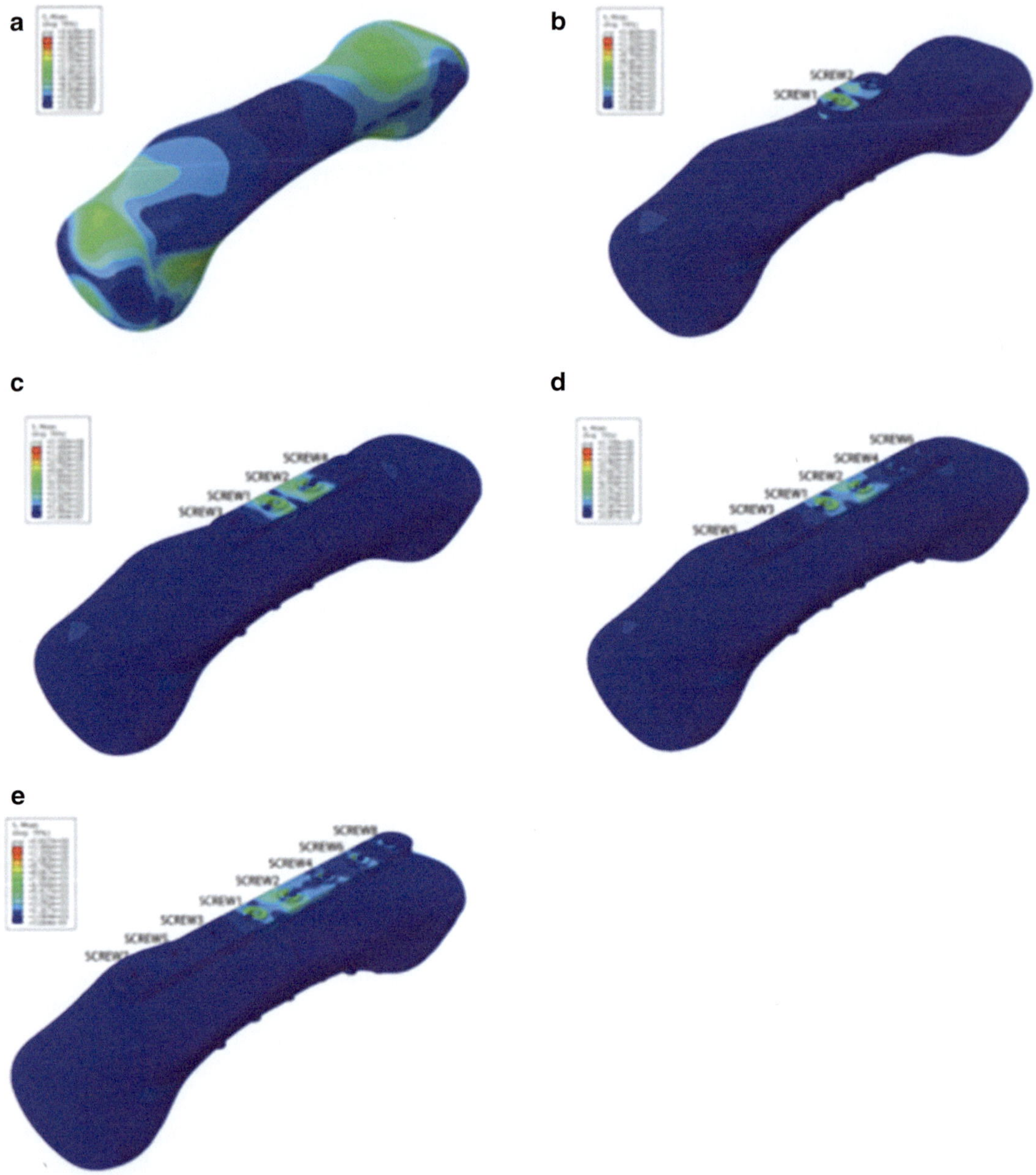

Fig. 9.5 The stress distribution observed in FEA models of a metacarpal subjected to cantilever bending with various plate fixation constructs

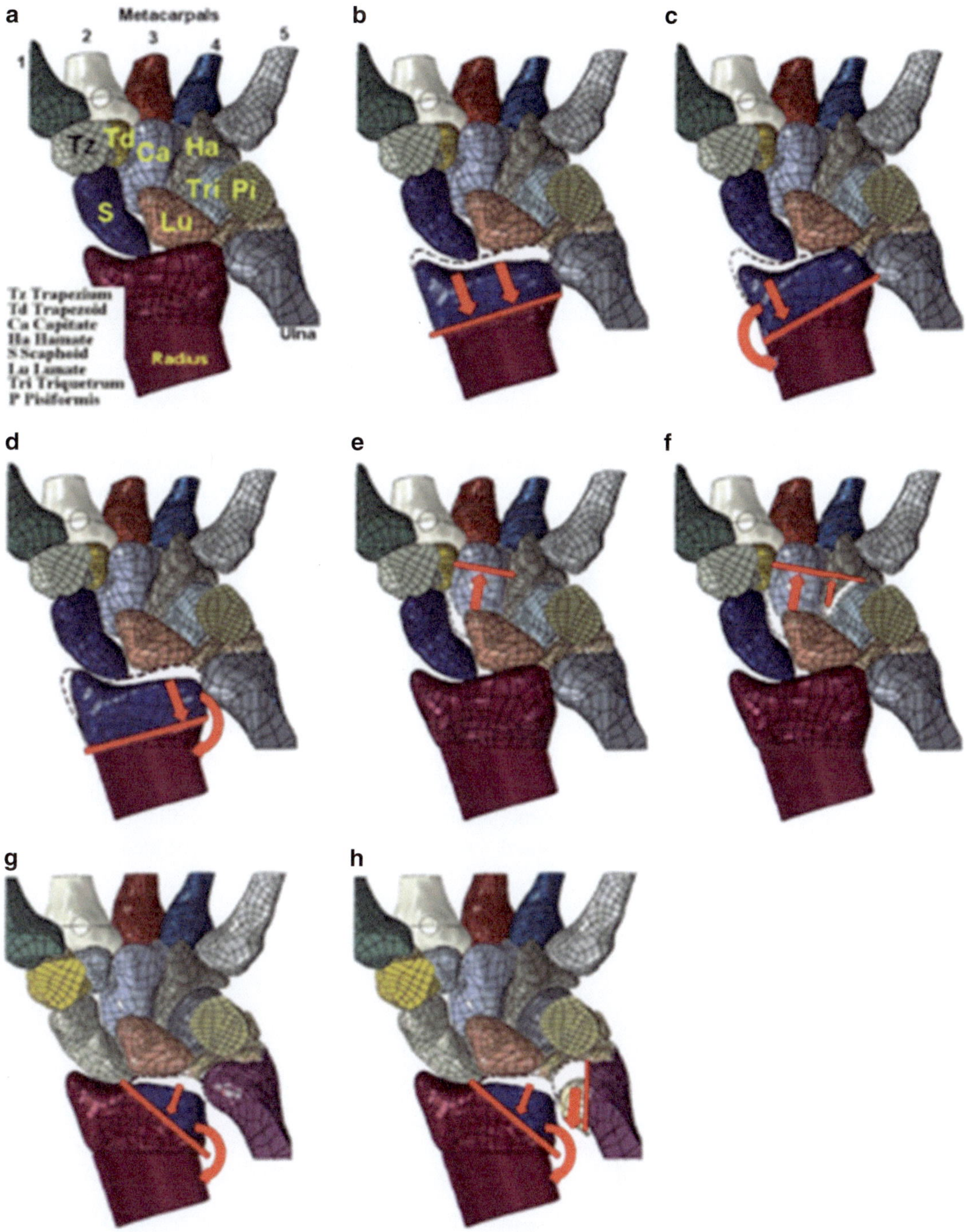

Fig. 9.6 The wrist model created by Camus et al. They simulated the following osteotomies for treatment of Kienböck disease: (**a**) Intact condition. (**b**) Radius transverse shortening osteotomy. (**c**) Lateral wedge osteotomy. (**d**) Medial wedge osteotomy. (**e**) Capitate osteotomy. (**f**) Capitate and hamate osteotomy. (**g**) Camembert osteotomy. (**h**) Camembert osteotomy and Sennwald ulnar shortening

Cartilage Biomechanics

The function of bone is to facilitate articulation, decrease stress on bone by increasing surface area, and lower bone wear by reducing friction. Cite the coefficient of friction as 0.01–0.04.

Material Properties

The material properties of articular cartilage, like bone, directly result from its composition, as demonstrated in Fig. 9.7. Articular cartilage generally contains 75% water-rich extracellular matrix (ECM) and 25% organic substances. This organic substance comprises type II collagen fibers, proteoglycans, and chondrocytes. The long collagen fibers' ability to uncoil before experiencing stress impart tensile strength, while proteoglycans and the ECM impart compressive strength.

Articular cartilage may be divided into three zones with differing structure and function. The superficial tangential zone contains transversely oriented collagen fibers that help resist shear stresses at the joint. The middle zone contains randomly oriented collagen fibers that facilitate the distribution of loads throughout the tissue, and the deep zone contains axially oriented collagen fibers that resist compressive forces. These regional variations in tissue composition and structure make articular cartilage a nonhomogeneous, anisotropic, viscoelastic solid. Further, articular cartilage is a dynamic tissue because compression will cause water to flow out of the tissue and into the joint space until an equilibrium is reached between the force acting on the cartilage and the cartilage's opposing force. Release of this compressive force will cause fluid flux back into the tissue as the hydrophilic proteoglycans attract water [1, 3, 34, 35]. Cartilage degeneration observed in diseases like osteoarthritis (OA) is characterized by an inability to control the water content of the tissue. Properly [36]. Like bone, long-term loading patterns or the lack thereof induce changes in cartilage composition mediated by chondrocytes [34].

Current Research on the Biomechanics of Healthy Hand Cartilage

Most modern research on the biomechanics of healthy hand cartilage utilizes finite-element methods. Cartilage geometry is particularly challenging to model, given its similarity to bone on CT scans. If not manually extruded from bone,

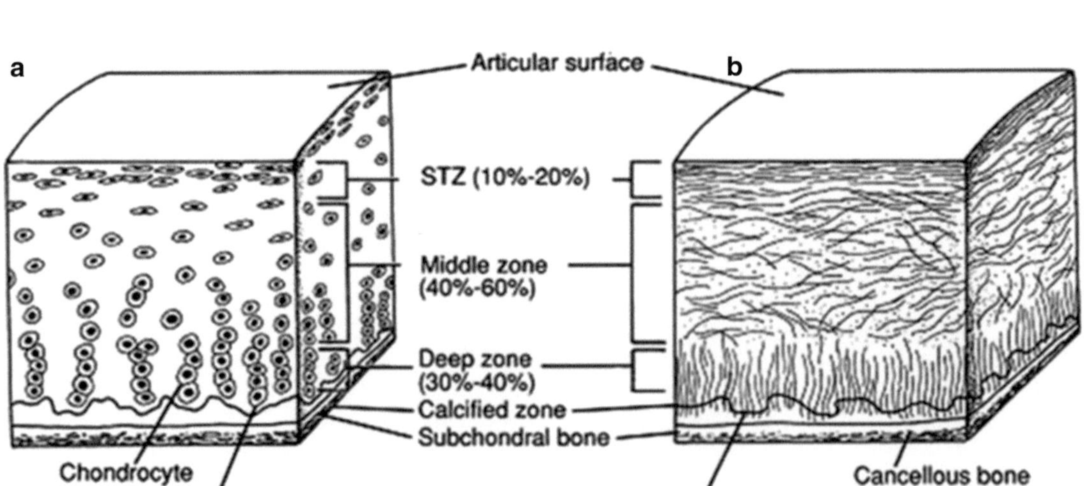

Fig. 9.7 The structure of articular cartilage. Articular cartilage can be divided into superficial, middle, and deep zones with varying compositions, functions, and properties. (**a**) cellular organization in the zones of articular cartilage; (**b**) collagen fiber architecture

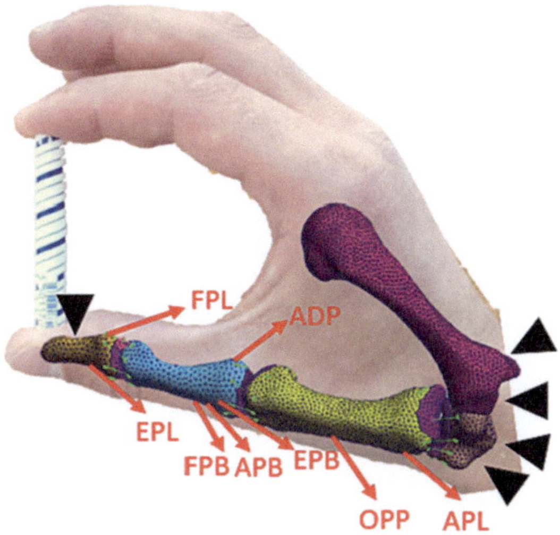

Fig. 9.8 The finite-element model used by Valerio et al. to study the effect of trapeziometacarpal stiffness on joint contact forces under pinch loading

the cartilage boundary must often be interpolated from the imaging [18]. One group used a series of CT images of the wrist in different positions to calculate kinematic data to aid the cartilage mapping [37]. Another issue in using finite-element methods to study cartilage is the selection of a material model. While linear isotropic models may be sufficient in some cases [14, 29]. Hyper-elastic models, such as the Mooney–Rivlin or Yeoh models, have been shown to better fit experimental data while minimizing computational complexity [38]. Regardless of the model chosen, the literature also presents a wide range of material parameters, such as elastic modulus, Poisson ratio, or hyper-elastic stiffness coefficients [39–41].

Valerio et al. used FEA to study the effect of passive trapeziometacarpal (TMC) joint stiffness on joint contact forces. They created the model shown in Fig. 9.8 and validated it against in vivo experimental data. TMC joint stiffness was modeled as springs. Increasing passive joint stiffness was negatively correlated with maximum joint contact forces. These findings were consistent with prior literature suggesting joint hypermobility as a risk factor for the development of TMC joint OA [42]. This same group found that joint morphology was more affected by joint passive stiffness or simulated loading than joint stiffness when modeling TMC joint contact forces [43].

Current Research on the Biomechanics of Pathology and Treatment of Hand Cartilage

Cadaveric research on pathological conditions of hand cartilage has yielded interesting results. Dourthe et al. sought an improved understanding of the in vivo properties of degenerating cartilage at the TMC joint. They measured cartilage thickness and material properties at this joint and then correlated them with computed tomography (CT) scans. The group found that CT-based cartilage attenuation indicative of OA was associated with increased elastic modulus of the tissue. The cartilage on the first metacarpal was also stiffer than overlying the trapezium [44]. Another group used a laser scanner to make cadaver-based models of the TMC joint bone and cartilage anatomy shown in Fig. 9.9. When the severity of OA was assessed using the Eaton grading system, the authors found cartilage wear spread evenly across the trapezium area. Still, they concentrated on the volar aspect of the first metacarpal. They correlate the lack of cartilage on the volar aspect of the first metacarpal and increased loading of this area with daily activities published in prior literature [45].

Bajuri et al. created a finite-element model of the entire wrist and simulated cartilage as a Mooney–Rivlin hyper-elastic material with parameters established previously. Among other changes, severe rheumatoid arthritis (RA) was modeled by obliterating all articular cartilage. Loading was applied to simulate static grip between the healthy and rheumatic models. Removal of articular cartilage from the model resulted in roughly ten times higher joint contact forces across many joints in the wrist. An example of the capitulate joint is shown in Fig. 9.10.

Dong et al. created a finite-element model of the entire hand to study the biomechanical effect of interventions for TMC joint OA. This group

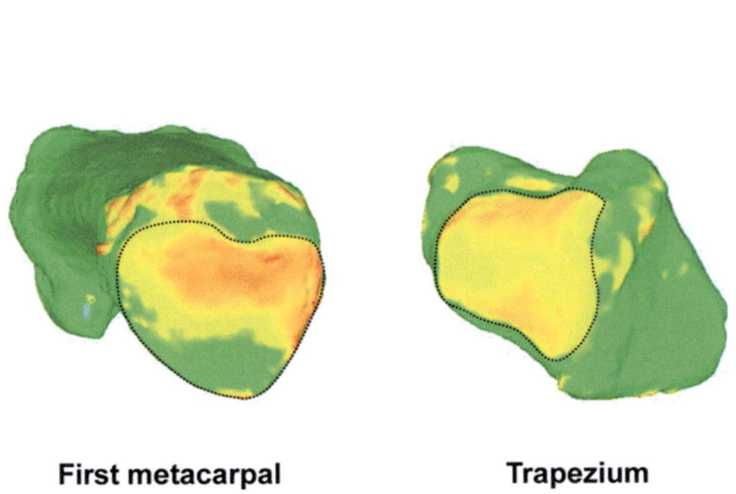

Fig. 9.9 Laser-scan models of the trapeziometacarpal joint demonstrating wear concentrated on the volar aspect of the first metacarpal

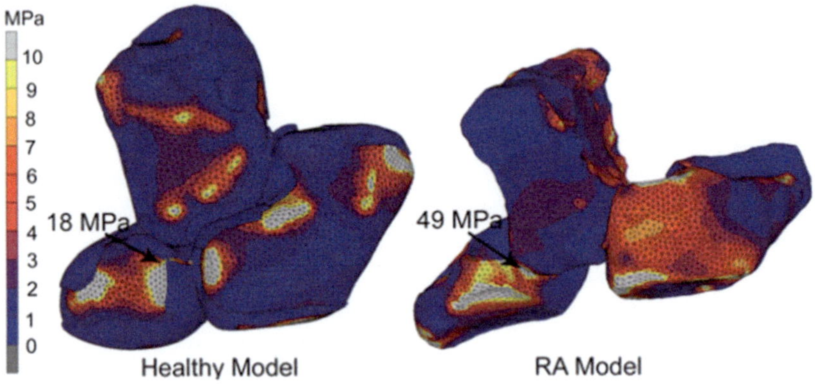

Fig. 9.10 An image showing contact forces in the model by Bajuri et al. Cartilage was removed from the model to simulate rheumatoid arthritis. After cartilage removal, joint contact forces increased in many joints. This image demonstrates the increased contact force at the capitulate joint from 18 to 49 MPa in the diseased model

simulated hand therapy by strengthening muscles and ligament imbrication by tightening ligaments, and they simulated four ligament reconstructive surgical interventions. These ligament reconstructions are shown in Fig. 9.11, and their results are shown in Fig. 9.12. TMC joint loading was analyzed using a lateral pinch task. Thenar eminence strengthening showed a 4% reduction in maximum joint contact force. Among imbrication procedures studied, dorsal ligament complex imbrication was the most effective. The Eaton–Littler volar reconstruction technique was the most effective surgical reconstructive procedure, with a maximum joint contact force reduction of 25%. These authors recommend further development of the model for individualized surgical planning [46].

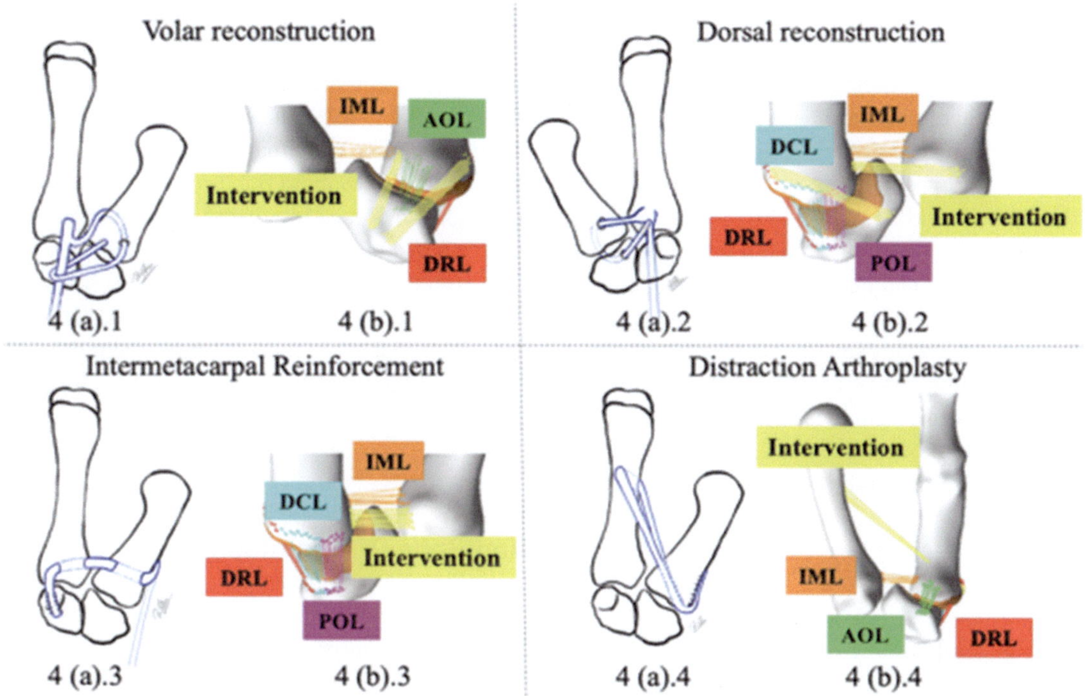

Fig. 9.11 The ligament reconstruction surgeries simulated in the study by Dong et al. These operations use various tendon grafts in vivo but were simulated by adding another ligament to the model

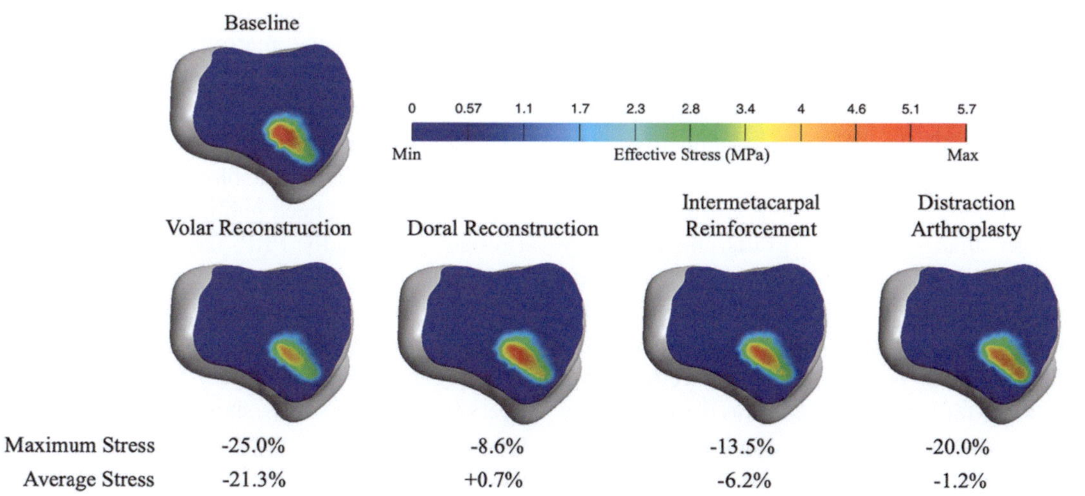

Fig. 9.12 Joint contact stresses on the trapezial surface of the trapeziometacarpal joint before and after various simulated ligament reconstructive surgeries. All operations resulted in a decrease in joint contact stresses

Biologics

Interest in using biological materials in various pathological conditions has increased in popularity. Although special focus has been given to the use of these products or interventions in areas like the hip and knee [47–49], recent research regarding their use in the hand holds promise. As with any biological intervention, the outcomes of orthobiologic treatment are heavily patient-dependent. Genetics and lifestyle factors involved in the healing process will result in different patient outcomes. This section will discuss using these interventions in the hand, emphasizing biomechanical considerations.

Bone Graft

A few general guiding principles must be considered when choosing a bone graft in vivo. Bone grafts may be osteoconductive, osteoinductive, or osteogenic. The more properties a graft possesses, the better it will be at remodeling and incorporating during in vivo use. Autograft, allograft, and xenograft bone grafts may be considered on a spectrum where autograft possesses all three properties, allograft lacks osteogenicity, and xenograft is only osteoconductive. The advantages of autograft are countered by donor site morbidity. While cortical grafts are generally stronger than cancellous grafts, their density slows the process of revascularization and incorporation into existing bone relative to cancellous grafts. Cortical allografts may be taken with a vascular pedicle to aid their remodeling. Cortical grafts can be used inlay to fill a bone defect or only in a fashion to reinforce existing bone. Demineralized bone matrix is another osteoconductive and osteoinductive option that may be used to augment bone growth, but it lacks biomechanical strength. Synthetic bone substitutes like calcium phosphate and calcium sulfate are osteoconductive substances that may be used to fill bone voids [50].

Bone graft or substitute may be used in various hand conditions like scaphoid nonunion [51], four-corner arthrodesis [52], Kienböck's disease [53], and more. The general mechanical properties discussed above hold for bone grafts in the hand, but research in this area is relatively limited. Vu et al. used cadavers to examine the biomechanics of a proximal hamate autograft for proximal pole scaphoid fractures. They measured joint contact forces in normal and reconstructed cadavers as they were taken through flexion-extension cycles. They found that radio scaphoid and radio lunate contact areas decreased after the procedure, but joint contact forces were equivocal. These authors suggested further exploration of the use of this autograft for scaphoid fracture [54].

Non-Bone Orthobiologics

Orthobiologics, including platelet-rich plasma (PRP), bone morphogenic protein (BMP), bone marrow aspirate, autologous adipose tissue, mesenchymal—or adipose-derived stem cells, and amniotic membrane, have been used during various orthopedic procedures to aid the healing of bone and cartilage. These biologics, derived from human plasma, bone marrow, or other tissue, have been demonstrated to accelerate healing in various operations [55]. Studies of their use in conditions of the hand have been relatively limited, and no studies on the biomechanics of these interventions have been published to date. These biologics, especially PRP, have been used for conditions including de Quervain's tenosynovitis, TMC joint OA, carpal tunnel syndrome, scaphoid nonunion, distal radius fracture, and tendon repair. While clinical studies on these interventions generally show positive results, they are limited to case reports or series, often lack control groups, and have limited follow-up [56, 57]. For this reason, the long-term outcomes of specific biological interventions compared to traditional therapies have yet to be described. One prospective study attempted to use an amniotic membrane graft to prevent adhesions after flexor tendon repair. Still, the study was terminated after operating on five of ten planned patients due to unfavorable results [58]. Although it is impossible to study these biologics in a cadaver model because their use depends on living tissue, future investigations may use kinematic or image-based modeling to study the biomechanical effects of these interventions.

The Role of Implants Alongside Biologics in Hand Surgery

Although not biologic, arthroplasty-related implants for the hand are a recent technology, and biologics may complement their use. A wide variety of devices are available for TMC joint arthroplasty, and they may replace only the first metacarpal aspect of the joint or both the metacarpal and trapezium. These devices have been shown to replicate the kinematics of a healthy thumb reliably [59]. Wrist arthroplasty is another hand-related implant that may benefit from the future development of biologics, such as implants for metacarpophalangeal (MCP) and proximal interphalangeal (PIP) joint arthroplasty. Implant options for MCP and PIP joint arthroplasty include the Neuflex and Swanson devices pictured in Fig. 9.13. Both options are made of silicone, and they differ in that the Neuflex device is bent to 30° of flexion, which decreases strain across the hinge by 35% [60]. Two clinical trials have demonstrated positive functional outcomes with either device but reported improved range of motion outcomes with the Neuflex over the Swanson [61, 62]. Future advances in the use of orthobiologics may accelerate healing during these procedures.

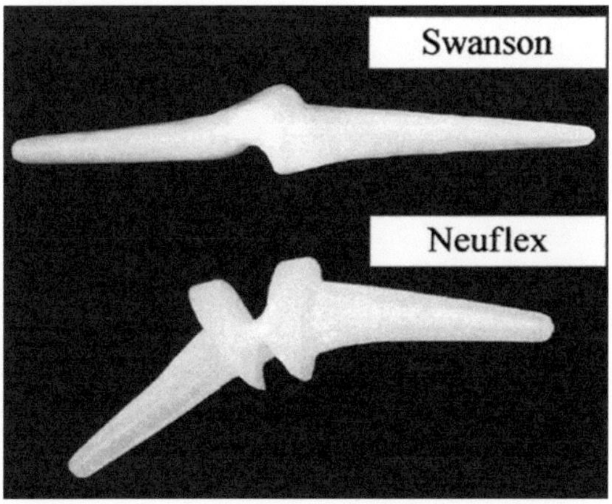

Fig. 9.13 Images of the Swanson and Neuflex silicone finger implants. The 30° of flexion present in the Neuflex allow for improved range of motion outcomes

Future Directions in Research

Research on the biomechanics of bone and cartilage utilizes sophisticated cadaveric and mathematical models. While the design of these models is often beyond reproach, future research may focus on improving the underlying assumptions and parameters in these models so that they may better approximate in vivo conditions. For example, most studies on the hand's bone and cartilage utilize material parameters originally derived from tissues outside the hand, like the femur or knee cartilage. These estimations may approximate the structures of the hand, but a more thorough understanding of the properties of structures specific to the hand would improve these models. Improved finite-element models of healthy and pathological conditions could be used to analyze the biomechanics of devices like Swanson and Neuflex finger implants and propose better-optimized designs. Future biomechanical studies should investigate how the properties of tissues vary with and without novel orthobiologic interventions.

Conclusion

This chapter began with a review of general biomechanical principles of bone and cartilage, and it built to a discussion of state-of-the-art research of these tissues as they apply to the hand. Advanced cadaveric and finite-element models have been used to investigate the biomechanics of healthy tissues, pathologic tissues, and various surgical interventions. This chapter also reviewed the emerging field of orthobiologics as they apply to hand conditions. General principles about the biomechanics of these interventions may be drawn from the literature, but more research is needed to understand how orthobiologics affect the properties of biological tissues.

References

1. Ozkaya N. Mechanical properties of biological tissues. In: Ozkaya N, Leger D, Goldsheyder D, Nordin M, editors. Fundamentals of biomechanics: equilibrium, motion, and deformation. 4th ed. Cham: Springer; 2017. p. 361–88.
2. Morgan EF, Unnikrisnan GU, Hussein AI. Bone mechanical properties in healthy and diseased states. Annu Rev Biomed Eng. 2018;20:119–43. https://doi.org/10.1146/annurev-bioeng-062117-121139.
3. Katz LJ. Mechanics of hard tissue. In: Schneck D, Bronzino J, editors. Biomechanics: principles and applications. Boca Raton: CRC Press; 2002.
4. Zioupos P, Hansen U, Currey JD. Microcracking damage and the fracture process in relation to strain rate in human cortical bone tensile failure. J Biomech. 2008;41(14):2932–9. https://doi.org/10.1016/j.jbiomech.2008.07.025.
5. Currey JD, Brear K, Zioupos P. Strain rate dependence of work of fracture tests on bone and similar tissues: reflections on testing methods and mineral content effects. Bone. 2019;128:115038. https://doi.org/10.1016/j.bone.2019.115038.
6. Lei J, Li Z, Su H, Li S, Liu Z. Investigating the mechanical properties of cortical bone under dynamic torsional loading. Theor Appl Mech Lett. 2024;14(4):100544. https://doi.org/10.1016/j.taml.2024.100544.
7. Hart NH, Nimphius S, Rantalainen T, Ireland A, Siafarikas A, Newton RU. Mechanical basis of bone strength: influence of bone material, bone structure and muscle action. J Musculoskelet Neuronal Interact. 2017;17(3):114–39. http://www.ismni.org
8. Rowe P, Koller A, Sharma S. Physiology, bone remodeling. Treasure Island (FL): StatPearls. Published online January 2025.
9. Wolff J. The law of bone remodeling, English Translation. Berlin: Springer; 1986.
10. Bartoníček J, Naňka O. The true history of the Hueter-Volkmann law. Int Orthop Published online October 1, 2024;48:2755–2762. https://doi.org/10.1007/s00264-024-06254-w.
11. Yin B, Guo JL, Wang JZ, Li S, Liu YK, Zhang YZ. Bone material properties of human phalanges using Vickers indentation. Orthop Surg. 2019;11(3):487–92. https://doi.org/10.1111/os.12455.
12. Taddei F, Pancanti A, Viceconti M. An improved method for the automatic mapping of computed tomography numbers onto finite element models. Med Eng Phys. 2004;26(1):61–9. https://doi.org/10.1016/S1350-4533(03)00138-3.
13. Cowin S, editor. Bone biomechanics handbook. 2nd ed. Boca Raton: CRC Press; 2001.
14. Varga P, Schefzig P, Unger E, Mayr W, Zysset PK, Erhart J. Finite element based estimation of contact areas and pressures of the human scaphoid in various functional positions of the hand. J Biomech. 2013;46(5):984–90. https://doi.org/10.1016/j.jbiomech.2012.11.053.
15. Tarniţă DN, Tarniţă D, Popa D, Tarniţă R. Analysis of stress and displacements of phalanx bone with the finite element method. Rom J Morphol Embryol. 2005;46(3):189–91.
16. Faudot B, Milan JL, Goislard de Monsabert B, Le Corroller T, Vigouroux L. Estimation of joint contact pressure in the index finger using a hybrid finite element musculoskeletal approach. Comput Methods Biomech Biomed Engin. Published online 2020;23(15):1225–1235. https://doi.org/10.1080/10255842.2020.1793965.
17. Marqués R, Melchor J, Sánchez-Montesinos I, Roda O, Rus G, Hernández-Cortés P. Biomechanical finite element method model of the proximal carpal row and experimental validation. Front Physiol. 2022;12:749372. https://doi.org/10.3389/fphys.2021.749372.
18. Kjartan Gíslason M, Nash DH. Finite element modelling of a multi-bone joint: the human wrist. London: InTech; 2012.
19. Gíslason MK, Stansfield B, Nash DH. Finite element model creation and stability considerations of complex biological articulation: the human wrist joint. Med Eng Phys. 2010;32(5):523–31. https://doi.org/10.1016/j.medengphy.2010.02.015.
20. Bajuri M, Abdul KM. Computational biomechanics of the wrist joint. New York: Springer; 2013.
21. Wei Y, Zou Z, Wei G, Ren L, Qian Z. Subject-specific finite element modelling of the human hand complex: muscle-driven simulations and experimental validation. Ann Biomed Eng. 2020;48(4):1181–95. https://doi.org/10.1007/s10439-019-02439-2.
22. Mena A, Wollstein R, Baus J, Yang J. Finite element modeling of the human wrist: a review. J Wrist Surg. 2023;12(06):478–87. https://doi.org/10.1055/s-0043-1768930.
23. Brekke A, Snoddy M, Lee D, Richard M, Desai M. Biomechanical strength of scaphoid partial unions. J Wrist Surg. 2018;07(05):399–403. https://doi.org/10.1055/s-0038-1661362.
24. Guss MS, Mitgang JT, Sapienza A. Scaphoid healing required for unrestricted activity: a biomechanical cadaver model. J Hand Surg. 2018;43(2):134–8. https://doi.org/10.1016/j.jhsa.2017.09.022.
25. Garcia-Lopez E, Siu JW, Kwong JW, Kamal RN, Schroeder N, Shapiro LM. A biomechanical analysis of oblique metacarpal metadiaphyseal fracture fixation in a cadaver model. J Hand Surg Published online 2024;50(8):1013.e1–1013.e7. https://doi.org/10.1016/j.jhsa.2024.08.009.
26. Sirota MA, Parks BG, Higgins JP, Means KR. Stability of fixation of proximal phalanx unicondylar fractures of the hand: a biomechanical cadaver study. J Hand Surg. 2013;38(1):77–81. https://doi.org/10.1016/j.jhsa.2012.09.026.
27. Wu WW, Zhu YB, Chen W, et al. Bone hardness of different anatomical regions of human radius and its

impact on the pullout strength of screws. Orthop Surg. 2019;11(2):270–6. https://doi.org/10.1111/os.12436.
28. Yan H, Sato K, Takahashi G, Mimata Y, Murakami K, Doita M. Load distribution in dorsally-angulated distal radius deformity using finite element analysis. J Hand Surg. 2023;48(10):1062.e1–6. https://doi.org/10.1016/j.jhsa.2022.02.022.
29. Camus EJ, Aimar A, Van Overstraeten L, Schuind F, Innocenti B. Lunate loads following different osteotomies used to treat Kienböck's disease: a 3D finite element analysis. Clin Biomech. 2020;78:105090. https://doi.org/10.1016/j.clinbiomech.2020.105090.
30. Martin J, Johnson NA, Shepherd J, Dias J, Martin J. Assessing the risk of re-fracture related to the percentage of partial union in scaphoid waist fractures. Bone Jt Open. 2023;4(8):612–20. https://doi.org/10.1302/2633-1462.48.BJO.
31. Varga P, Zysset PK, Schefzig P, Unger E, Mayr W, Erhart J. A finite element analysis of two novel screw designs for scaphoid waist fractures. Med Eng Phys. 2016;38(2):131–9. https://doi.org/10.1016/j.medengphy.2015.11.006.
32. Zhang JF, Li Y, Huo YX, et al. Biomechanical analysis of locking plates for fixation of metacarpal shaft fractures: a finite element analysis. Orthop Traumatol Surg Res. 2022;108(8):103340. https://doi.org/10.1016/j.otsr.2022.103340.
33. Pramudita JA, Kamiya S, Ujihashi S, et al. Estimation of conditions evoking fracture in finger bones under pinch loading based on finite element analysis. Comput Methods Biomech Biomed Engin. 2017;20(1):35–44. https://doi.org/10.1080/10255842.2016.1196197.
34. Zernicke R, Broglio S, Whiting W. Tissue biomechanics and adaptation. In: Biomechanics of Injury. 3rd ed. Champaign: Human Kinetics; 2008. p. 111–3.
35. Sophia Fox AJ, Bedi A, Rodeo SA. The basic science of articular cartilage: structure, composition, and function. Sports Health. 2009;1(6):461–8. https://doi.org/10.1177/1941738109350438.
36. Liu J, Xu S, Ma Z, Jiang Y, Zhao H, Ren L. Water loss and defects dependent strength and ductility of articular cartilage. J Mater Res Technol. 2022;21:1714–23. https://doi.org/10.1016/j.jmrt.2022.10.024.
37. Marai GE, Crisco JJ, Laidlaw DH. A kinematics-based method for generating cartilage maps and deformations in the multi-articulating wrist joint from CT images. Conf Proc IEEE Eng Med Biol Soc. 2006;2006:2079–82.
38. Brown C, Nguyen T, Moody H, Crawford R, Oloyede A. Assessment of common hyperelastic constitutive equations for describing normal and osteoarthritic articular cartilage. Inst Mech Eng. 2009;223(6):643.
39. Li Z, Kim JE, Davidson JS, Etheridge BS, Alonso JE, Eberhardt AW. Biomechanical response of the pubic symphysis in lateral pelvic impacts: a finite element study. J Biomech. 2007;40(12):2758–66. https://doi.org/10.1016/j.jbiomech.2007.01.023.
40. Little R, Wevers H, Siu D, Cooke T. A three-dimensional finite element analysis of the upper tibia. J Biomech Eng. 1986;108(2):111–9.
41. Vidal-Lesso A, Ledesma-Orozco E, Daza-Benítez L, Lesso-Arroyo R. Mechanical characterization of femoral cartilage under unicompartimental osteoarthritis. Ingeniería mecánica, tecnología y desarrollo. 2014;4(5):239–46.
42. Valerio T, Milan JL, Goislard de Monsabert B, Vigouroux L. The effect of trapeziometacarpal joint passive stiffness on mechanical loadings of cartilages. J Biomech. 2024;166:112042. https://doi.org/10.1016/j.jbiomech.2024.112042.
43. Valerio T, Milan JL, Goislard de Monsabert B, Vigouroux L. What is the most critical mechanical factor involved in trapeziometacarpal osteoarthritis development? A sensitivity analysis based on biomechanical modelling. Comput Methods Biomech Biomed Engin Published online 2024, 1.
44. Dourthe B, Nickmanesh R, Wilson DR, et al. Assessment of healthy trapeziometacarpal cartilage properties using indentation testing and contrast-enhanced computed tomography. Clin Biomech. 2019;61:181–9. https://doi.org/10.1016/j.clinbiomech.2018.12.015.
45. Miyamura S, Oka K, Sakai T, et al. Cartilage wear patterns in severe trapeziometacarpal joint osteoarthritis: a quantitative analysis. Osteoarthr Cartil. 2019;27(8):1152–62. https://doi.org/10.1016/j.joca.2019.03.006.
46. Dong M, Kerkhof F, Deleu GJ, Vereecke E, Ladd A. Using a finite element model of the thumb to study Trapeziometacarpal joint contact during lateral pinch. Clin Biomech. 2023;101:105852. https://doi.org/10.1016/j.clinbiomech.2022.105852.
47. Jones IA, Togashi RC, Thomas VC. The economics and regulation of PRP in the evolving field of orthopedic biologics. Curr Rev Musculoskelet Med. 2018;11(4):558–65. https://doi.org/10.1007/s12178-018-9514-z.
48. Hsu W, Mishra A, Rodeo S, et al. Platelet-rich plasma in orthopaedic applications: evidence-based recommendations for treatment. J Am Acad Orthop Surg. 2013;21(12):739–48.
49. Chen X, Jones IA, Park C, Vangsness CT. The efficacy of platelet-rich plasma on tendon and ligament healing: a systematic review and meta-analysis with bias assessment. Am J Sports Med. 2018;46(8):2020–32. https://doi.org/10.1177/0363546517743746.
50. Lee D, Poser J. Biomechanics of bone grafts and bone substitutes. In: Koh J, Zaffagnini S, Kuroda R, Longo UG, Amirouche F, editors. Orthopaeedic biomechanics in sports medicine. Springer; 2021. p. 37–57.
51. Dalcortivo RL, Kurland AM, Ignatiuk A, Kirschenbaum AE, Vosbikian MM, Ahmed IH. Scaphoid fractures requiring bone graft: does graft source matter? Eplasty. 24:e28 Published online 2024.

52. Mamede J, Castro Adeodato S, Aquino LR. Four-corner arthrodesis: description of surgical technique using headless retrograde crossed screws. Hand. 2018;13(2):156–63. https://doi.org/10.1177/1558944717702468.
53. Elhassan BT, Shin AY. Vascularized bone grafting for treatment of Kienböck's disease. J Hand Surg. 2009;34(1):146–54. https://doi.org/10.1016/j.jhsa.2008.10.014.
54. Vu CL, Telfer S, Huang JI. Biomechanics of proximal hamate autograft in scaphoid nonunion. J Wrist Surg. 2023;12(06):488–92. https://doi.org/10.1055/s-0043-1767671.
55. Toland GE, Makaram NS, Atzmon R, Donohue JP, Murray IR. Orthobiologics in the knee. Orthop Trauma. 2023;37(3):143–53.
56. Karim KE, Wu CM, Giladi AM, Murphy MS. Orthobiologics in hand surgery. J Hand Surg. 2021;46(5):409–15. https://doi.org/10.1016/j.jhsa.2021.01.006.
57. Steiner MM, Calandruccio JH. Biologic approaches to problems of the hand and wrist. Orthop Clin N Am. 2017;48(3):343–9. https://doi.org/10.1016/j.ocl.2017.03.010.
58. Leppänen OV, Karjalainen T, Göransson H, et al. Outcomes after flexor tendon repair combined with the application of human amniotic membrane allograft. J Hand Surg. 2017;42(6):474.e1–8. https://doi.org/10.1016/j.jhsa.2017.03.006.
59. Uchiyama S, Cooney WP, Niebur G, An KN, Linscheid RL. Biomechanical analysis of the trapeziometacarpal joint after surface replacement arthroplasty. J Hand Surg. 1999;24:483–90.
60. Erdogan A, Weiss APC. Metakarpophalangeale arthroplastik mit der neuflex-silastikprothese. Orthopade. 2003;32(9):789–93. https://doi.org/10.1007/s00132-003-0524-0.
61. Delaney R, Trail IA, Nuttall D. A comparative study of outcome between the Neuflex and Swanson metacarpophalangeal joint replacements. J Hand Surg. 2005;30(1):3–7. https://doi.org/10.1016/j.jhsb.2004.10.013.
62. Escott BG, Ronald K, Judd MGP, Bogoch ER. NeuFlex and Swanson metacarpophalangeal implants for rheumatoid arthritis: prospective randomized, controlled clinical trial. J Hand Surg. 2010;35(1):44–51. https://doi.org/10.1016/j.jhsa.2009.09.020.

10 Hand Tendon Biomechanics

Cadence Lee and Farid Amirouche

Composition and Structure

Macroscopic Structure

Tendons play an essential role in biomechanical function by transferring forces generated by muscle to bone to execute movement [1]. Tendons are fibrous connective tissue, primarily composed of water (55–70% of total mass), with the remaining dry mass being predominantly type I collagen (60–85%). Type I collagen is the major component responsible for the high tensile strength of tendons. Tendons are organized hierarchically, ascending from collagen molecules into fibrils, fibers, fascicles, and the tendon unit (Fig. 10.1) [2]. Each collagen molecule is roughly 1.5 nm in diameter and arranged along a longitudinal axis with 40 nm between other molecules [1]. Groups of five molecules crosslink to form microfibrils or pentafibrils. The microfibrils coalesce to form fibrils ranging from 10 to 500 nm in diameter. The fibrils further aggregate to form the collagen fiber, ranging from 1 to 20 um in diameter, with interspersed tenocytes. Fibers are then arranged to form the fascicle, ranging from 50 to 300 um in diameter. Finally, the fascicle, in combination with interspersed tenocytes, nerves, and vascular supply, is surrounded by connective tissue known as the interfascicular matrix (or endotendon) and comprises the tendon unit. The tendon unit's surface is then surrounded by epitenon, a connective tissue sheath, which can be surrounded by an additional connective tissue layer, the paratenon. When adjacent to a joint, this entire complex is encased by a synovial sheath.

Microscopic Structure

While most tendons are comprised of type I collagen, other key substances contribute to tendon material behavior. Type III collagen plays a role in tissue healing due to its ability to form rapid cross-links in areas of damage [2]. Type V collagen works with type I collagen to regulate fibril diameter. Type XII collagen has been shown to provide lubrication between fibers. Proteoglycans comprise 1–4% of tendon and are variably expressed depending on the mechanical forces applied during tension [3]. Proteoglycans are

C. Lee (✉)
Department of Orthopaedic Surgery, University of Illinois College of Medicine at Chicago, Chicago, IL, USA
e-mail: clee331@uic.edu

F. Amirouche
Department of Orthopaedic Surgery, Orthopaedic and Spine Institute, Endeavor HealthSystem (Formerly Northshore University HealthSystem), an Affiliate of the University of Chicago Pritzker School of Medicine, Skokie, IL, USA

Department of Orthopaedic Surgery, University of Illinois at Chicago, Chicago, IL, USA

F. Amirouche et al. (eds.), *Biomechanics of the Hand*,
https://doi.org/10.1007/978-3-032-04688-8_10

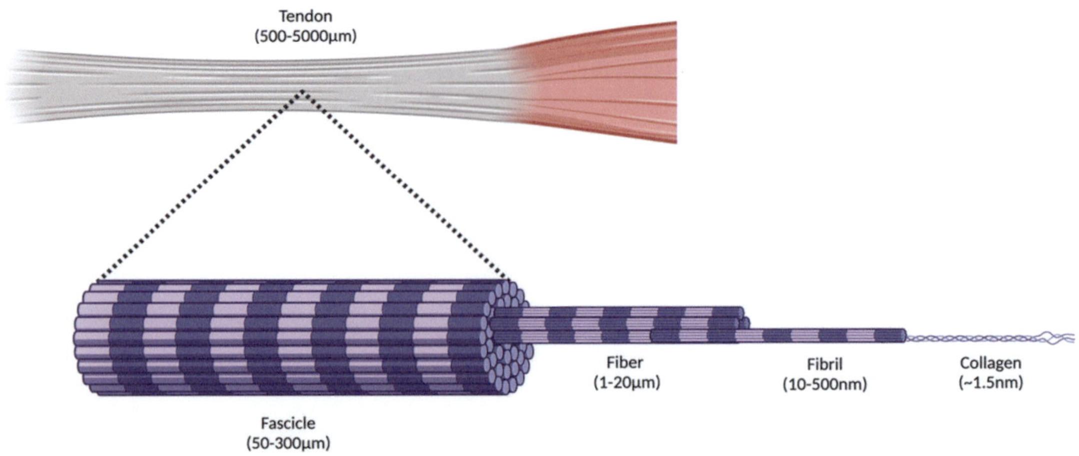

Fig. 10.1 Tendon hierarchical structure, depicting the breakdown of tendon unit, fascicle, fiber, fibril, and collagen, with associated sizes

responsible for the 3D matrix that interweaves collagen fibrils, including aggrecan, which retains water, thereby resisting compression, and decorin, which regulates fibril growth and adhesion. Elastin makes up 0.25–10% of the tendon's dry weight but is essential in providing recoil, resilience to stretch, and even healing [4, 5]. Tenascin-C and fibronectin further enhance mechanical stability and support elastin function. The predominant cell of tendons is the tenocyte, a fibroblast-like cell responsible for maintaining the components above and tendon homeostasis.

Enthesis and Myotendinous Junction

Significant differences exist in the transition between tendon and bone versus tendon and muscle [1]. When interfacing with the muscle, the tendon forms the myotendinous junction, where collagen fibers and the sarcolemma of muscle cells interdigitate along longitudinal projections to increase surface area. On the other hand, the tendon–bone interface, known as the enthesis, is joined by the tendon directly inserted onto the metaphysis or epiphysis for fibrous entheses or gradual zones of transition in fibrocartilaginous entheses. The gradual zones are split into four parts as the tendon conjoins with the bone: (1) tendon proper containing aligned collagen fibers, (2) unmineralized fibrocartilage containing type II collagen and proteoglycans, (3) mineralized fibrocartilage containing collagen type X and endochondral ossification, and (4) bone material. The multiscale transition between zones with increasing mineralization allows for a gradient of load distribution, thereby decreasing stress applied to the interface. At a microscale level, when the interdigitating fibers have high amplitude and frequency, there is increased toughness yet reduced strength due to increased stress concentration [6]. Therefore, fiber mineralization, length, amplitude variation at the interface, and interfacial variance play key roles in tendon–bone mechanics.

How Composition Enables Load-Bearing, Elasticity, and Gliding

The mechanical stability and strength of tendons are primarily credited to their predominant composition of type I collagen [7]. This strength is due to trivalent cross-linking between hydroxy lysyl pyridinoline and lysyl pyridinoline on adjacent collagen molecules, stabilizing the fibrils [8]. The orientation of molecule cross-linking also confers an anisotropic property, such that tendon material is stronger in the longitudinal direction of fibril assembly rather than perpendicularly [7]. As previously mentioned, proteoglycans also play an essential role in tendon composition. Versican and aggrecan are necessary for water retention and, thereby, resistance to compression, which explains their increased expression in areas surrounding joints [9]. Elastin and lubricin have been implicated in tendon elasticity, gliding, and recoil. These extracellular matrix proteins localize to the interfascicular matrix between fascicles and aid in sliding.

Biomechanical Properties

Stress–Strain Curve

The stress–strain curve of tendons provides insight into its functional utility [2]. The curve demonstrates four regions (Fig. 10.2): (1) the initial toe region, encompassing strain below 2% where the tendon fibrils are stretched during loading from a "crimped" flaccid position into straight alignment; (2) the linear region, encompassing strain below 4% where stress applied is proportional to deformation; (3) the yield region, where the tendon has surpassed its physiological limit and begins to experience micro-tears; and (4) the failure region, where tendon rupture occurs.

Young's Modulus

Young's modulus for tendons varies based on functional needs in the area of the body as well as age. The tibialis anterior tendon reportedly has a Young's modulus of ~1200 Mpa. In contrast, the patellar tendon is reported to have ~660 Mpa, further differentiated when evaluated by an aged patellar tendon specimen of ~504 Mpa [2]. Reported Young's modulus values for flexor tendons in the hand range between 1200–1800 Mpa [10], extensor tendons 1500–4000 Mpa, thumb tendons 900–1600 Mpa, and wrist tendons 1000–1700 Mpa [11]. Based on this data, tendons of the hand demonstrate a higher Young's modulus and, therefore, are more resistant to deformation and experience a more minor elastic strain for a given stress.

Viscoelastic Behavior

A key biomechanical property of tendons is their high viscoelasticity. This quality makes tendons more deformable at low strains and more resilient at higher ones. The stiffening at high strains

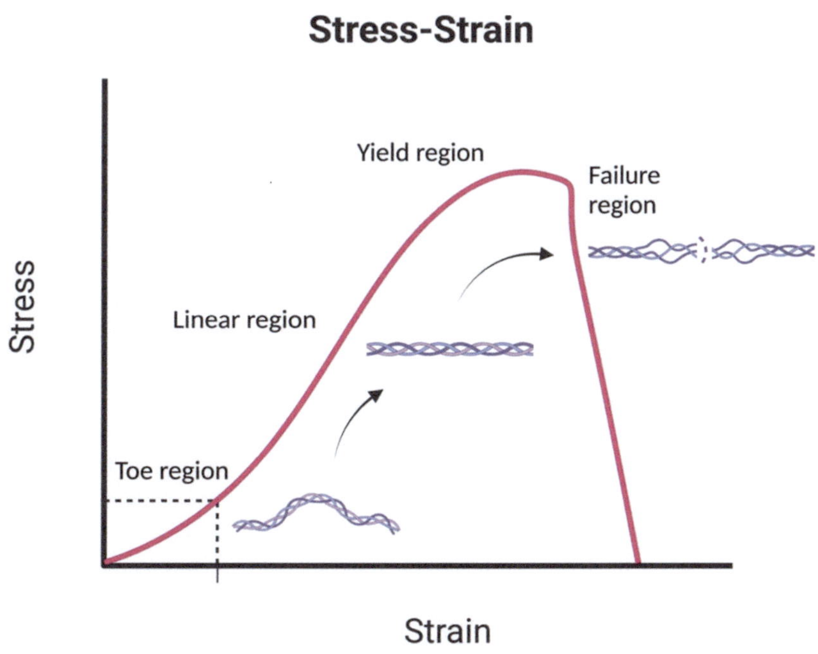

Fig. 10.2 The stress–strain curve of the tendon. The initial toe region comprises strain under 2%, the linear region representing proportional stress and deformation, the yield region comprising microtears, and the failure region, at which point the tendon ruptures

allows tendons to transfer greater muscular forces to bone. In contrast to purely elastic materials, viscoelastic materials demonstrate energy loss in heat (hysteresis) due to the discrepancy in loading and unloading curves on a stress–strain graph [12]. Viscoelasticity also determines tendon stress relaxation, by which a constant strain begins to exert less force over time due to collagen fiber rearrangement to reduce stress [13]. Viscoelasticity also impacts creep, whereby a continuous load applied to the tendon causes increased deformation over time.

Influencing Factors

Increased age is a well-known factor in altering tendon mechanical qualities. Not only is there an age-related decline in the number of tendon stem/progenitor cells, but also in the morphology, motility, and metabolism of cells that impact homeostasis of the extracellular matrix [14]. As extracellular matrix homeostasis is dysregulated, so is the synthesis and turnover of collagen. The accumulation of micro-injury, in combination with aging and cell senescence, leads to disorganized and inconsistently crosslinked collagen. In contrast, human and animal studies have noted that increased mechanical loads via physical exercise may positively impact tendon strength (perhaps explained by increased biomechanical collagen turnover), insertion site strength, and healing of acute tendon injury [2]. Furthermore, studies utilizing immunofluorescent staining, confocal and multiphoton imaging, nuclear shape analysis, and finite element analysis suggest that tendons exhibit multiscale properties in response to dynamic loading and during healing [15].

Anatomy of Hand Tendons

Flexor Tendons

The hand tendons can be grouped into two classifications: extensor tendons and flexor tendons. The majority of flexor tendons traverse the volar forearm under the transverse carpal ligament to adjoin the phalanges to produce finger flexion. Digits 2–4 are each associated with a superficial tendon, flexor digitorum superficialis (FDS), and an underlying deep tendon, flexor digitorum profundus (FDP). Digit 1 is associated with the flexor pollicis longus (FPL), which originates in the forearm, as well as flexor pollicis brevis (FPB) originating in the palm [16]. The FDS of the digits splits at the metacarpophalangeal (MCP) joint, termed the chiasma, and allows the FDP to continue towards the distal interphalangeal joint (DIP). At the wrist, the FDS and FDP of digits 3–5 are encased by a common synovial sheath known as the digital flexor sheath or ulnar bursa as they pass through the carpal tunnel [17]. The FPL is surrounded by the radial bursa for the first digit, which communicates with the FPL tendon sheath [18]. More distally, spanning MCP to DIP, digits 2–5 are associated with a flexor tendon sheath (also known as the digital synovial sheath), which aids in friction reduction to create a smooth plane for the tendons via a gliding mechanism [19]. Along the sheaths are fibrous condensations known as pulleys: annular (A1–A5), cruciate (C1–C3), and the palmar aponeurosis pulley [20]. These pulleys not only adhere the tendon to the bone to prevent "bowstringing" but also play a key role in converting translational muscle force into rotational movement at the joint. Among the flexor distribution associated with the volar surface of the hand, there are eight zones defined by anatomical landmarks (Fig. 10.3) [21]:

1. Zone I, distal to the insertion of FDS
2. Zone II, the proximal edge of the pulley system to the insertion of FDS
3. Zone III, the distal end of the carpal tunnel at the A1 pulley
4. Zone IV, the zone overlaying the carpal tunnel
5. Zone V, musculotendinous junction distal to the carpal tunnel
6. T1, distal to the interphalangeal (IP) joint on digit 1
7. TII, distal to the MP joint on digit 1
8. TIII, the distal edge of the carpal tunnel to the proximal edge of the A1 pulley

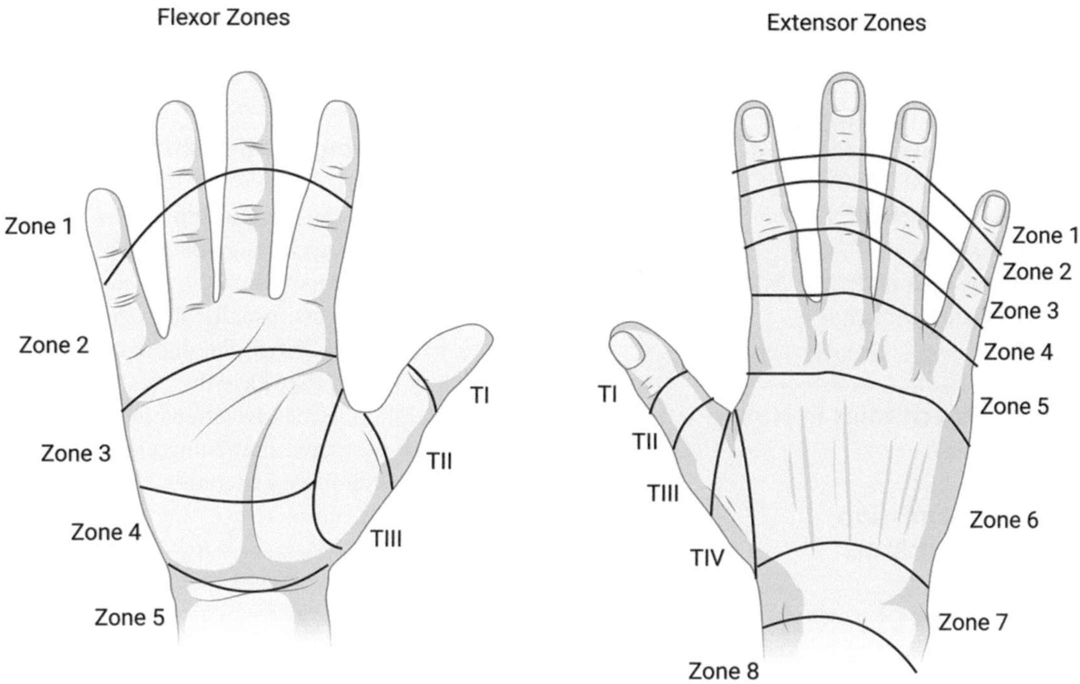

Fig. 10.3 Flexor and extensor zones of the hand, defined by anatomical landmarks. These zones guide diagnosis, treatment, and management

Extensor Tendons

The extensor tendons of the hand are divided into intrinsic and extrinsic extensors, defined by association with a muscle originating from the hand versus the forearm, respectively [22]. Extrinsic extensors enter the hand through six compartments segregated by the extensor retinaculum at the wrist, with each compartment being associated with a synovial sheath for friction reduction. The first compartment contains the abductor pollicis longus (APL) and extensor pollicis brevis (EPB), the second compartment contains the extensor carpi radialis longus and brevis (ECRL and ECRB), the third compartment contains the extensor pollicis longus (EPL), the fourth compartment contains the extensor indicis (EI) and extensor digitorum communis (EDC), the fifth compartment contains the extensor digiti minimi (EDM), and the sixth compartment contains the extensor carpi ulnaris (ECU). Extrinsic extensor tendons include the extensor digitorum (ED) tendons, which correspond to digits 2–5 and originate from the EDC. The EI tendon is associated with digit 2 and EDM is associated with digit 5. An intertendinous connection connects the ED tendons, whereas EI and EDM operate from independent muscle groups. When ED tendons reach the metacarpal head, they segregate to form the dorsal hood, lateral bands, and central band, which fan out across the distal phalanges and aid in stabilization during flexion and extension [23].

Intrinsic tendons include four dorsal interossei, whose function is the abduction of digits 2–4, three volar interossei, whose function is to adduct, and four lumbrical muscles that attach to the FDP [22]. The dorsal hand associated with extensor tendons is categorized into 13 zones, defined by anatomical landmarks (Fig. 10.3) [21]:

1. Zone I, distal to the IP joint
2. Zone II, the middle phalanx
3. Zone III, proximal to the IP joint
4. Zone IV, the proximal phalanx
5. Zone V, the MP joint
6. Zone VI, the dorsum of the hand, including metacarpals and carpals

7. Zone VII, zone overlying the extensor retinaculum
8. Zone VIII, dorsal forearm
9. T1, distal to the IP joint on digit 1
10. TII, proximal phalanx on digit 1
11. TIII, the MP joint on digit 1
12. TIV, the metacarpal bone associated with digit 1
13. TV, the area under the extensor retinaculum associated with digit 1

Tendon Mechanics in Hand Function

Force Transmission and Moment Arms

The moment arm is mechanically defined as the perpendicular distance between the line of action and the axis of rotation. In biomechanics, the moment arm indicates a muscle's ability to produce mechanical advantage or leverage in a given direction around a joint [24]. In the context of the hand, moment arms are essential in creating substantial grip and detailed movements around the MCP, PIP, and DIP. Larger moment arms around the MCP correspond with greater torque, whereas smaller moment arms around the PIP and even smaller for the DIP correspond with lesser amounts of flexion leverage. The FDS of the third finger has the largest flexor moment arm in the hand, whereas the EDC of the middle finger has the largest extensor moment arm. However, there are nuances between flexor and extensor moment arms. Flexors around the MCP joint demonstrate a linear moment arm until late flexion when the relationship between the moment arm and joint angle is nonlinear. Extensors, on the other hand, demonstrate a constant moment arm. Constant moment arms are attributed to the pulley system that prevents bowstringing of the tendons and, therefore, connects the tendon's line of action to the bone. When a muscle contracts, the tendon moves a given distance around the rotating joint, known as tendon excursion. Tendon excursion is directly proportional to the amount of rotation of the joint [25].

Length–Tension Relationship

The length–tension relationship is the force a muscle can generate for a given length. Optimal force generation depends on the overlap of actin and myosin filaments, usually at its peak during resting length [26]. When the digits of the hand are slightly flexed, they have the greatest optimal force generation. Excessive stretching via hyperextension or excessive shortening via full flexion reduces potential force generation [27].

Mechanical Response to Loads and Injuries

Mechanoadaptation

Although we have thus far discussed the impressive mechanical qualities of tendons, they are not simply collagen bundles. They possess bio-adaptability and a complex homeostatic regimen. Exercise studies have shown that microinjury and mechanical loading can activate myofibroblasts within tendons and lead to greater concentrations of procollagen type I C-terminal propeptide and that increased collagen turnover can ultimately result in net increased collagen composition after just 4 weeks of physical training [2]. Mechanical loading has also been associated with increased tendon stem cells (TSCs) concentration; interestingly, small mechanical stretching promotes TSC differentiation into tenocytes. In contrast, sizeable mechanical stretching promotes differentiation into adipocytes, osteocytes, and chondrocytes. Overuse of tendons has been associated with increased inflammatory mediators such as prostaglandin E2 and leukotriene B4 production, which can stimulate differentiation of TSCs into non-tenocyte aggregations (such as osteocytes and adipocytes) within the tissue. In contrast, underuse, sleep, and stress induce a higher catabolic rate and cause degeneration.

Common Injuries

Tendon injuries are divided into acute and chronic types. Acute injury includes rupture and laceration, usually due to trauma, whereas chronic injury includes overuse and degeneration, usually due to repeated stress. Common acute tendon injuries of the hand include Mallet finger (ED tendon rupture at the DIP), Jersey Finger (avulsion of FDP at DIP), Boutonniere deformity (central slip extensor tendon rupture at the dorsal middle phalanx), and Gamekeeper's/Skier's thumb (hyperabduction of ulnar collateral ligament at the MCP joint) [28]. Chronic injuries include trigger finger, aka stenosing tenosynovitis (a tendon is too large for its associated tendon sheath, causing catching of the joint), de Quervain tenosynovitis (thickening of the synovial sheath of the APL and EPB), and tendonitis associated with aging or excessive use [29].

Impact of Injury on Biomechanics

In addition to producing significant pain and discomfort, tendon injuries can produce functional disability. Pathophysiology of conditions such as trigger finger and de Quervain's tenosynovitis involve the increasing breakdown of the inner gliding layer of the A1 pulley [30]. In chronic cases, the gliding interface of the sheath can be colonized by hyperplastic fibrotic tissue. One theory suggests that years of friction between a tendon and pulleys or sheaths, despite the gliding mechanism, can result in tissue fraying, degeneration, and impingement. There remains uncertainty as to whether tendon inflammation is a primary process or secondary to mechanical stress. In either case, tendon injury causes significant functional limitations, either through direct rupture of the tendon itself or through stenosis of a tendon sheath, leading to impairment of the inherent gliding mechanism.

Healing and Repair

Early studies set a precedent for the belief that tendons heal via intrinsic mechanisms rather than by extrinsic or external migration of cells into the tissue [31]. More recent theories accept that a mix of intrinsic and extrinsic healing is likely in vivo. The overall goal of tendon repair is to restore the gliding function of tendons by bridging new tendon fibers across the injury site. This involves typical wound healing processes such as releasing chemotactic agents, growth factors, and cellular proliferation to produce a scar [31]. Inappropriate mechanical loading of the healing tissue results in excessive scar proliferation and adhesion development, resulting in a mechanically disadvantageous repair site. Conversely, excessive mechanical loading can result in rerupture or injury of the tendon. Appropriate repair of an injured tendon requires a balance of mechanical stimulation to ensure proper growth factor release, cellular proliferation, and collagen turnover.

Emerging Studies in Tendon Biomechanics

Physics-Based Models

Finite element analysis (FEA) mathematically simulates a material's behavior or structure given material constraints and forces. A recent FEA study utilized an adapted hand model of the well-established neuromuscular human body models (NHBMs) to simulate the tendon strain during Jersey finger injury computationally [33]. The mathematical gripping force, flexor activation, and injury assessment results were consistent with clinical data from medical studies. Another recent FEA study evaluated FDS, FDP, and ED tendon contraction behavior during synergistic flexion [32]. The authors found that FDS and FDP contraction provided alternating power during flexion, with FDS providing the primary driving force and the ED tendon acting in opposition. FEA has additionally been used to simulate tendon excursion in the carpal tunnel, median nerve compression during wrist flexion and extension, post-stroke hand grip, and behavior of surgical anchoring devices during tendon repair [33–36]. FEA models can be used to further optimize machine-learning studies for improving simulated tendon behavior (Fig. 10.4). Previous studies utilize rigid links with joint torques or line-of-force muscles to evaluate hand tendon biomechanics [37]. Newer research has assessed the tendon network in the context of a Eulerian-on-Lagrangian model to capture better the sheath and pulley system of the native hand. Investigators have modeled the tendon network to provide coupled digit motions, even simulating Boutonniere and Swan Neck deformities.

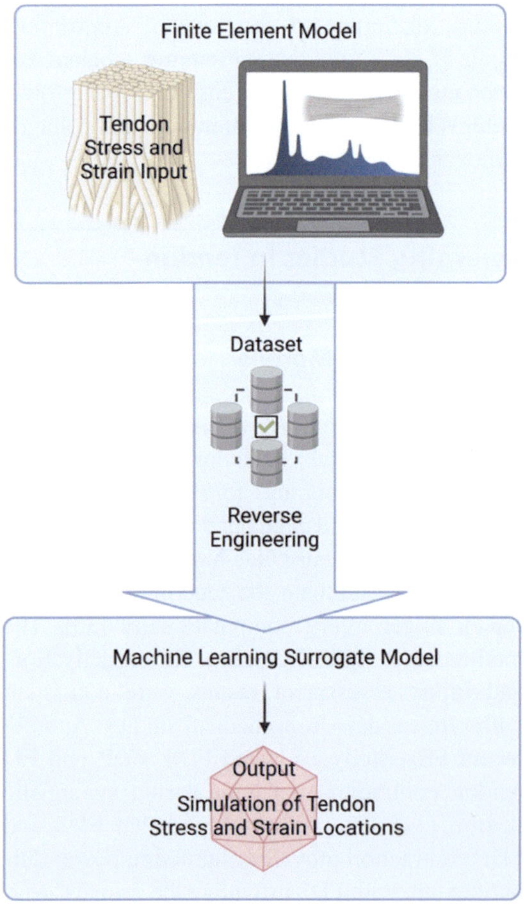

Fig. 10.4 A schematic representing the flow process of FEA model input to machine learning surrogate model output. (Figure adapted from Khayam et al. [41])

Emerging Techniques (Imaging and Motion)

A recent study has found that ultrasound has 100% sensitivity and 95% specificity in diagnosing partial-thickness tears of hand flexor tendons [38]. However, given the small diameter of the hand tendons, clinical ultrasound elastography cannot accurately determine tendon stiffness. Emerging studies have identified novel imaging modalities for mapping hand tendons. High-frequency ultrasound shear elastography (HFUSE) can determine the shear wave velocity of the flexor tendons [39]. The HFUSE transducer is mounted with a handheld external vibration system, allowing for ease of use. This method provided 2D shear wave velocity images that depict healthy and injured tendons in patients with tendon injuries. More detailed information has been imaged using HFUSE combined with dual-direction shear wave imaging (DDSWI) to obtain both transverse and longitudinal shear waves [40]. The HFUSE-DDSWI imaging technique could create a 3D pixel map of the flexor carpi radialis tendon in simulated models and human volunteers.

Clinical Applications of Biomechanical Studies

Recent studies have noted significant findings concerning surgical innovation of hand tendon repair. One study evaluating tendon transfer of FDS to FPL used a new DC suture with a salt-infused silicone core, which preserves tissue approximation and minimizes laxity [42]. The investigators found that biomechanically, the DC suture allowed for a shorter interweaving zone and extrapolated that this suture use would allow for a more aggressive postoperative rehabilitation course.

Biomechanical analysis has compared the utility in cadaveric models, of either complete excision of A2 and A4 pulleys, repair of the A2 with one ring of tendon graft, with two rings of tendon graft with and without repair of the A4 with one ring [43]. This study found that the three-ring configuration, with two-ring A2 and one-ring A4 repair, provided the most remarkable recovery of PIP flexion compared to the other groups. This is an essential consideration in evaluating surgical tendon repair.

Clinical Implications

Rehabilitation

Surgeries to repair acute tendon injuries of the hand are associated with good outcomes, with 75% of patients experiencing a return to total active motion [44]. Re-rupture is described in 4–10% of cases for the FDS and FDP and 3–17% of cases for the FPL. Postoperative rehabilitation includes early limb mobilization to reduce adhesion formation, improve tendon healing, and improve functional outcomes. This includes passive and active flexion and extension up to 6 weeks when light use resumes. The patient should utilize an orthosis or splint during the first 8 weeks to protect the tendon. At 8 weeks, the patient may begin resisted flexion, and at 12 weeks, the patient should have a full return to baseline activity. Wound management can include compression therapy, elevation, cryotherapy, and topical creams. Evidence shows that neuromuscular electrical stimulation can increase blood flow, wound healing, and cellular proliferation [45].

Tissue Engineering

Recent research has identified numerous potential methods for supporting tendon repair. Investigators have proposed growth factors influencing adipose-derived stromal cells to differentiate into advantageous cell types within the native tissue [31]. Clinical trials have shown positive outcomes for patients receiving platelet-rich plasma injections to treat Achilles and rotator cuff tendinopathies, though specific treatment for hand pathologies remains understudied [46, 47]. Alternatively, biomimetic tools can provide structural support to healing tendons. Electrospinning is a process that pumps biocompatible polymer through a high-voltage electric field to create delicate, nanometric porous scaffolding, which is collected and reinforced with carbon nanotubes (Fig. 10.5). Soft lithography produces a similar polymer scaffold but utilizes cast molding and micro-engineered hydrogels to create a 3D-printed structure. However, despite exciting innovations in tissue engineering, in vivo studies primarily utilize animal models [48].

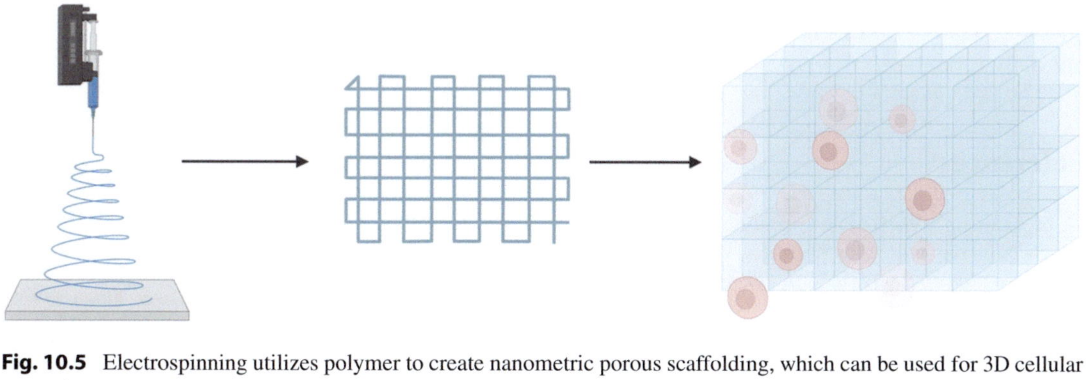

Fig. 10.5 Electrospinning utilizes polymer to create nanometric porous scaffolding, which can be used for 3D cellular support

Further clinical study is needed to elucidate the full potential of scaffolding and biomimetics in human tendon repair.

Conclusion

Tendons are vital connective tissues that serve as bridges between muscles and bones, enabling the effective transmission of forces necessary for movement. Their unique composition and intricate structure are specifically designed to handle the demands placed on them during various physical activities. With high tensile strength and viscoelastic properties, tendons can withstand significant loads while absorbing and dissipating energy.

In addition to these mechanical characteristics, the adaptability and responsiveness of tendon cells—known as tenocytes—play a crucial role in their function. These cells can sense changes in mechanical stress and signal appropriate biological responses, ensuring that tendons can adjust to different physical demands.

The hands, in particular, require tendons that can support a wide range of activities, from generating substantial grip strength for lifting heavy objects to performing delicate and precise movements needed for tasks like writing or playing a musical instrument. Consequently, the anatomy of tendons in the hands is robust and versatile, enabling them to fulfill various roles essential for our daily functions.

References

1. Ackermann PW, Hart DA, editors. Metabolic influences on risk for tendon disorders, vol. 920. Springer; 2016. https://doi.org/10.1007/978-3-319-33943-6.
2. Wang JHC, Guo Q, Li B. Tendon biomechanics and mechanobiology--a minireview of basic concepts and recent advancements. J Hand Ther. 2012;25(2):133–40.; quiz 141. https://doi.org/10.1016/j.jht.2011.07.004.
3. Canata GL, d'Hooghe P, Hunt KJ, editors. Muscle and tendon injuries: evaluation and management. Berlin Heidelberg: Springer; 2017. https://doi.org/10.1007/978-3-662-54184-5.
4. Svärd A, Hammerman M, Eliasson P. Elastin levels are higher in healing tendons than intact tendons and influence tissue compliance. FASEB J. 2020;34(10):13409–18. https://doi.org/10.1096/fj.202001255R.
5. Hill JR, Eekhoff JD, Brophy RH, Lake SP. Elastic fibers in orthopedics: form and function in tendons and ligaments, clinical implications, and future directions. J Orthop Res. 2020;38(11):2305–17. https://doi.org/10.1002/jor.24695.
6. Deymier-Black AC, Pasteris JD, Genin GM, Thomopoulos S. Allometry of the tendon enthesis: mechanisms of load transfer between tendon and bone. J Biomech Eng. 2015;137(11):111005. https://doi.org/10.1115/1.4031571.
7. Fratzl P. Collagen: structure and mechanics. Springer; 2008.
8. Couppé C, Hansen P, Kongsgaard M, et al. Mechanical properties and collagen cross-linking of the patellar tendon in old and young men. J Appl Physiol. 2009;107(3):880–6. https://doi.org/10.1152/japplphysiol.00291.2009.
9. Screen HRC, Berk DE, Kadler KE, Ramirez F, Young MF. Tendon functional extracellular matrix. J Orthop Res. 2015;33(6):793–9. https://doi.org/10.1002/jor.22818.
10. Singh R, Rymer B, Theobald P, Thomas PBM. A review of current concepts in flexor tendon repair: physiology, biomechanics, surgical technique and rehabilitation. Orthop Rev. 2015;7(4):6125. https://doi.org/10.4081/or.2015.6125.
11. Weber JF, Agur AMR, Fattah AY, Gordon KD, Oliver ML. Tensile mechanical properties of human forearm tendons. J Hand Surg Eur Vol. 2015;40(7):711–9. https://doi.org/10.1177/1753193415584715.
12. Lake SP, Snedeker JG, Wang VM, Awad H, Screen HRC, Thomopoulos S. Guidelines for ex vivo mechanical testing of tendon. J Orthop Res. 2023;41(10):2105–13. https://doi.org/10.1002/jor.25647.
13. Baar K. Stress relaxation and targeted nutrition to treat patellar tendinopathy. Int J Sport Nutr Exerc Metab. 2019;29(4):453–7. https://doi.org/10.1123/ijsnem.2018-0231.
14. Kwan KYC, Ng KWK, Rao Y, et al. Effect of aging on tendon biology, biomechanics and implications for treatment approaches. Int J Mol Sci. 2023;24(20):15183. https://doi.org/10.3390/ijms242015183.
15. Freedman BR, Rodriguez AB, Leiphart RJ, et al. Dynamic loading and tendon healing affect multiscale tendon properties and ECM stress transmission. Sci Rep. 2018;8(1):10854. https://doi.org/10.1038/s41598-018-29060-y.
16. Gray H. In: Standring S, Anand N, Tunstall R, editors. Gray's anatomy: the anatomical basis of clinical practice. 42nd ed. Elsevier; 2021.
17. Esmaeilnejad-Ganji SM, Baghianimoghadam B. Flexor Digitorum Superficialis and Flexor Digitorum Profundus with separated sheaths: a new

normal variation in human. Colomb Med (Cali). 2015;46(4):199–201.

18. Lui TH. Flexor Pollicis longus Tendoscopy. Arthrosc Tech. 2017;6(1):e249–54. https://doi.org/10.1016/j.eats.2016.09.030.
19. Titan AL, Foster DS, Chang J, Longaker MT. Flexor tendon: development, healing, adhesion formation, and contributing growth factors. Plast Reconstr Surg. 2019;144(4):639e–47e. https://doi.org/10.1097/PRS.0000000000006048.
20. Doyle JR. Anatomy of the flexor tendon sheath and pulley system: a current review. J Hand Surg. 1989;14(2 Pt 2):349–51. https://doi.org/10.1016/0363-5023(89)90110-x.
21. Germann G, Levin LS, Sherman R. Reconstructive surgery of the hand and upper extremity. 1. Auflage. Thieme Medical Publishers; 2017.
22. Rockwell WB, Butler PN, Byrne BA. Extensor tendon: anatomy, injury, and reconstruction. Plast Reconstr Surg. 2000;106(7):1592–603.; quiz 1604, 1673. https://doi.org/10.1097/00006534-200012000-00024.
23. Kichouh M, Vanhoenacker F, Jager T, et al. Functional anatomy of the dorsal hood or the hand: correlation of ultrasound and MR findings with cadaveric dissection. Eur Radiol. 2009;19(8):1849–56. https://doi.org/10.1007/s00330-009-1383-9.
24. Francis-Pester FW, Thomas R, Sforzin D, Ackland DC. The moment arms and leverage of the human finger muscles. J Biomech. 2021;116:110180. https://doi.org/10.1016/j.jbiomech.2020.110180.
25. An KN. Tendon excursion and gliding: clinical impacts from humble concepts. J Biomech. 2007;40(4):713–8. https://doi.org/10.1016/j.jbiomech.2006.10.008.
26. Robbins D. Muscle biomechanics. In: Human Orthopaedic Biomechanics. Elsevier; 2022. p. 121–35. https://doi.org/10.1016/B978-0-12-824481-4.00009-3.
27. Hauraix H, Goislard De Monsabert B, Herbaut A, Berton E, Vigouroux L. Force–length relationship modeling of wrist and finger flexor muscles. Med Sci Sports Exerc. 2018;50(11):2311–21. https://doi.org/10.1249/MSS.0000000000001690.
28. Maxwell S, Olson D. Common tendon injuries in the hand. Curr Sports Med Rep. 2015;14(6):428–9. https://doi.org/10.1249/JSR.0000000000000210.
29. Currie KB, Tadisina KK, Mackinnon SE. Common hand conditions: a review. JAMA. 2022;327(24):2434. https://doi.org/10.1001/jama.2022.8481.
30. McAuliffe JA. Tendon disorders of the hand and wrist. J Hand Surg. 2010;35(5):846–53. https://doi.org/10.1016/j.jhsa.2010.03.001.
31. James R, Kesturu G, Balian G, Chhabra AB. Tendon: biology, biomechanics, repair, growth factors, and evolving treatment options. J Hand Surg. 2008;33(1):102–12. https://doi.org/10.1016/j.jhsa.2007.09.007.
32. Lv Y, Zheng Q, Chen X, Hou C, An M. Analysis on synergistic cocontraction of extrinsic finger flexors and extensors during flexion movements: a finite element digital human hand model. Zhang X, ed. PLoS One. 2022;17(5):e0268137. https://doi.org/10.1371/journal.pone.0268137.
33. Henderson J, Thoreson A, Yoshii Y, Zhao KD, Amadio PC, An KN. Finite element model of subsynovial connective tissue deformation due to tendon excursion in the human carpal tunnel. J Biomech. 2011;44(1):150–5. https://doi.org/10.1016/j.jbiomech.2010.09.001.
34. Peshin S, Karakulova Y, Kuchumov AG. Finite element modeling of the fingers and wrist flexion/extension effect on median nerve compression. Appl Sci. 2023;13(2):1219. https://doi.org/10.3390/app13021219.
35. Rodriguez Reinoso M, Civera M, Burgio V, et al. 3D printing and testing of rose thorns or limpet teeth inspired anchor device for tendon tissue repair. Acta Bioeng Biomech. 2021;23(4):63–74. https://doi.org/10.37190/ABB-01899-2021-01.
36. Ramlee MH, Beng GK, Bajuri N, Abdul Kadir MR. Finite element analysis of the wrist in stroke patients: the effects of hand grip. Med Biol Eng Comput. 2018;56(7):1161–71. https://doi.org/10.1007/s11517-017-1762-3.
37. Sachdeva P, Sueda S, Bradley S, Fain M, Pai DK. Biomechanical simulation and control of hands and tendinous systems. ACM Trans Graph. 2015;34(4):1–10. https://doi.org/10.1145/2766987.
38. Bekhet CNH, Ghaffar MKA, Nassef MA, Khattab RT. Role of ultrasound in flexor tendon injuries of the hand: a new insight. Ultrasound Med Biol. 2021;47(8):2157–66. https://doi.org/10.1016/j.ultrasmedbio.2021.02.023.
39. Chen PY, Yang TH, Kuo LC, Hsu HY, Su FC, Huang CC. Evaluation of hand tendon elastic properties during rehabilitation through high-frequency ultrasound shear elastography. IEEE Trans Ultrason Ferroelectr Freq Control. 2021;68(8):2716–26. https://doi.org/10.1109/TUFFC.2021.3077891.
40. Xu GX, Chen PY, Huang CC. Visualization of human hand tendon mechanical anisotropy in 3-D using high- frequency dual-direction shear wave imaging. IEEE Trans Ultrason Ferroelectr Freq Control. 2023;70(11):1457–69. https://doi.org/10.1109/TUFFC.2023.3312273.
41. Khayam SU, Ajmal A, Park J, Kim IH, Park JW. Tendon stress estimation from strain data of a bridge girder using machine learning-based surrogate model. Sensors. 2023;23(11):5040. https://doi.org/10.3390/s23115040.
42. Pastor T, Zderic I, Dhillon M, et al. New dynamic suture material for tendon transfer surgeries in the upper extremity—a biomechanical comparative analysis. Arch Orthop Trauma Surg. 2024;144(6):2905–14. https://doi.org/10.1007/s00402-024-05322-5.
43. Soulii L, Amirouche F, Solitro G, et al. Evaluation of A2 and A4 hand pulley repair using tendon graft rings. J Hand Surg. 2021;46(7):626.e1–6. https://doi.org/10.1016/j.jhsa.2020.11.025.

44. WilsonJones N, Laing H. Acute injuries to the flexor and extensor tendons of the hand. Surg Oxf. 2006;24(12):441–5. https://doi.org/10.1053/j.mpsur.2006.10.009.
45. Peters SE, Jha B, Ross M. Rehabilitation following surgery for flexor tendon injuries of the hand. Cochrane Bone, Joint and Muscle Trauma Group, ed. Cochrane Database Syst Rev. Published online January 7, 2017;1(1):CD012479. https://doi.org/10.1002/14651858.CD012479.
46. Mi B, Liu G, Zhou W, et al. Platelet rich plasma versus steroid on lateral epicondylitis: meta-analysis of randomized clinical trials. Phys Sportsmed. Published online February 21, 2017;45(2):97–104. https://doi.org/10.1080/00913847.2017.1297670.
47. Yang J, Sun Y, Xu P, Cheng B. Can patients get better clinical outcomes by using PRP in rotator cuff repair: a meta-analysis of randomized controlled trials. J Sports Med Phys Fitness. 2016;56(11):1359–67.
48. Mao Z, Fan B, Wang X, et al. A systematic review of tissue engineering scaffold in tendon bone healing in vivo. Front Bioeng Biotechnol. 2021;9:621483. https://doi.org/10.3389/fbioe.2021.621483.

11 Nerve Compression Neuropathies of the Hand

John Alvarez, Mark Gonzalez, and Farid Amirouche

Anatomy

Anatomy and Biomechanical Properties of Peripheral Nerves

Peripheral nerves are composed of axons (both myelinated and unmyelinated), three distinct connective tissue layers, and a microvascular supply, as seen in Fig. 11.1. The outermost layer, the epineurium, is formed by loose connective tissue that encases the entire nerve, providing cushioning against external compressive forces [33]. It also allows nerve gliding within its surrounding tissues, reducing friction during movement [22]. The perineurium surrounds individual nerve fascicles and is composed of dense connective tissue that resists tensile forces [32]. This layer is critical for maintaining the internal environment of the nerve, acting as a selective barrier to protect axons from external mechanical and chemical insults [26]. The innermost layer, the endoneurium, forms a barrier around axons within a collagenous matrix and contributes to tensile strength, though to a lesser degree than the perineurium [6]. An undulating configuration of axons and connective tissue layers is present within peripheral nerves, which enhances resilience against mild tension stress [27].

Peripheral nerves have been described as viscoelastic tissues, allowing adaptation to mechanical stressors within physiological limits [37]. The tension stress-strain relationship of these nerves is nonlinear, as shown in Fig. 11.2. Initial application of tension stress results in minimal strain due to the straightening of the undulating nerve fibers. This nonlinear behavior protects nerves during normal limb movement by distributing forces evenly across their structural components. However, irreversible injury can occur beyond specific thresholds.

The mechanical resilience of peripheral nerves is derived from their extracellular matrix (ECM) composition and cellular architecture. Collagen type I, predominant in the epineurium [18], provides tensile strength and disperses compressive forces, while collagen type IV, localized to the perineurium and endoneurium [26], stabilizes the basement membrane surrounding Schwann cells

J. Alvarez · M. Gonzalez
Department of Orthopaedic Surgery, University of Illinois at Chicago, Chicago, IL, USA

F. Amirouche (✉)
Department of Orthopaedic Surgery, Orthopaedic and Spine Institute, Endeavor HealthSystem (Formerly Northshore University HealthSystem), an Affiliate of the University of Chicago Pritzker School of Medicine, Skokie, IL, USA

Department of Orthopaedic Surgery, University of Illinois at Chicago, Chicago, IL, USA
e-mail: famirouche@northshore.org; amirouch@uic.edu

F. Amirouche et al. (eds.), *Biomechanics of the Hand*,
https://doi.org/10.1007/978-3-032-04688-8_11

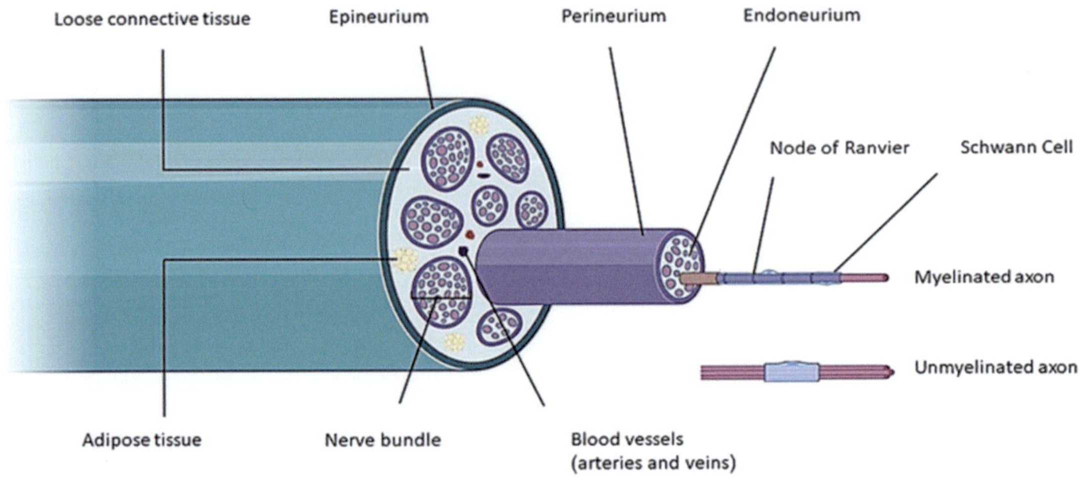

Fig. 11.1 The structure of the peripheral nerve is composed of three layers, with its blood supply within the epineural compartment. (Adapted from Klimovich et al. [16])

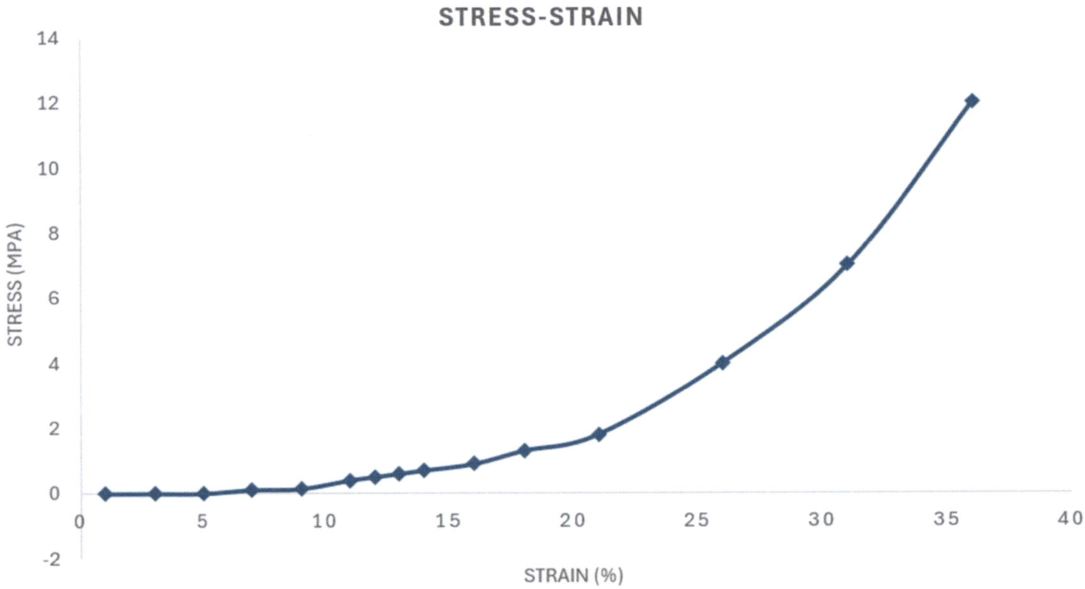

Fig. 11.2 The stress–strain curve of peripheral nerves under tension load demonstrates a nonlinear relationship

and axons. Atomic force microscopy (AFM) studies on developing mouse sciatic nerves revealed that collagen digestion resulted in a drop of approximately 50%, 37%, and 54% in the apparent elastic modulus of young, juvenile, and adult sciatic nerves [29].

Forces Acting on Peripheral Nerves

Compression

Compression is perhaps the force clinicians are most intuitively familiar with due to the prevalence of carpal tunnel in the general population [1]. This force occurs when external pressure is applied to the nerve continuously by adjacent thick connective tissue or episodically by repetitive wrist flexion/extension. The vasa nervorum, a microvascular network embedded within the nerve, exhibits bidirectional blood flow to mitigate ischemic injury [35]. Still, these vessels are susceptible to compression; pressures as low as 20 mmHg collapse capillaries, reducing endoneurial blood flow by 50% within 30 min [31]. In diabetic nerves or those with pre-existing vascular compromise, this vulnerability can be exacerbated by reduced vascular lumen diameter under transverse compression [24]. One study showed that epineurial vessels were more sensitive to compression than endoneurial vessels as light compression (50 mmHg for 2 h) induced epineurial edema, with higher pressures or increased duration required to cause injury to endoneurial vessels [30].

Insights into the effects of controlled crush of nerve trunks can be useful when studying nerve compression. A recent study characterized the neuromuscular response during and after crush injury using a standardized method with forceps, as shown in Fig. 11.3 [11]. With a standardized compression model, there is potential for testing interventions to enhance subsequent regeneration and behavioral recovery, offering hope for future advancements in nerve compression treatment.

Degrees of Nerve Injury

The degrees of nerve injury have been studied extensively and are useful to consider under the context of nerve compression. The first degree of nerve injury, also known as neurapraxia, is defined by a decrease in conduction velocity due to demyelination of the nerve, with sparing of axonal components. With first-degree nerve injuries, the symptoms of pain, numbness, and proprioceptive issues are commonly present, but there is an essential absence of muscle wasting. The second degree of nerve injury, also known as axonotmesis, involves injury to both nerve myelin and axonal components, but the endoneurium, perineurium, and epineurium remain intact. Second-degree nerve injury involves pain and numbness as well, but this is when muscle wasting begins to arise, such as with thenar muscle wasting in severe cases of carpal tunnel syndrome.

Another important feature of second-degree nerve injury is that Wallerian degeneration can occur from axonal disruption. The third degree of nerve injury, also known as neurotmesis, involves injury to nerve myelin, axons, and the endoneurium, sparing the perineurium and epineurium. Muscle wasting continues at this stage, but an important feature is the absence of pain, highlighting the severity of nerve injury. The symptoms of third-degree nerve injury are consistent with fourth-degree injury, but the perineurium is damaged. The same is true with fifth-degree nerve injury in terms of symptoms, but it is defined by complete disruption of the nerve trunk. These different degrees of nerve injury have varying implications on hand function, from mild sensory disturbances to severe motor deficits, and understanding these implications is crucial for effective diagnosis and treatment.

Tension

Peripheral nerves regularly experience tension during joint movement, a force that might be overlooked compared to compression but is particularly significant during activities involving repetitive motions or more extensive ranges of motion [15]. The ulnar nerve undergoes approximately 4.7–8 mm of strain during elbow flexion in healthy individuals as it courses around the medial epicondyle of the elbow [5].

The amount of stretch a nerve can endure without mechanical failure is an area of great debate [20]. One study concluded that a 6% stretch of nerve length could be tolerated before

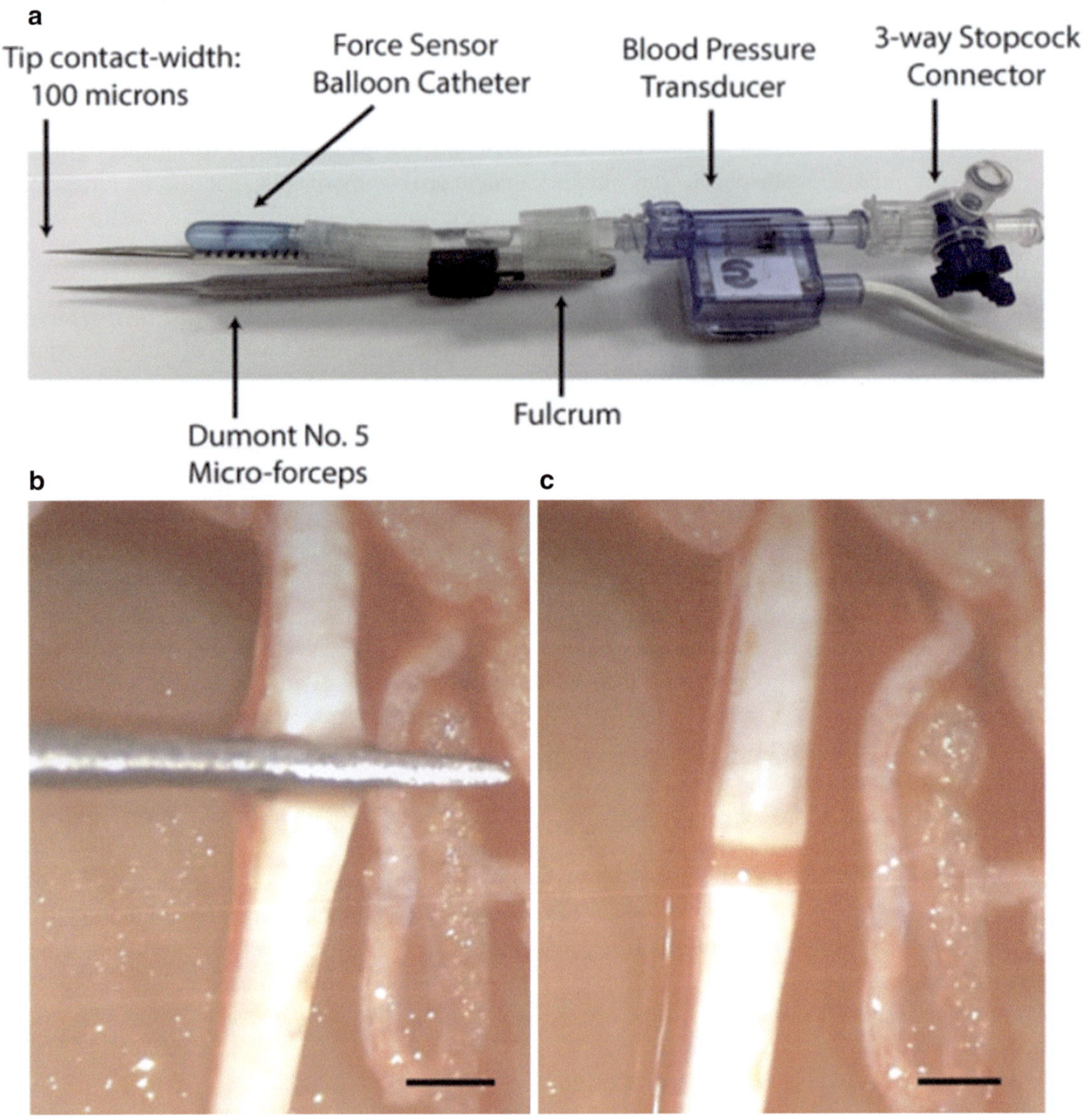

Fig. 11.3 Controlled nerve crush was accomplished with the use of instrumented micro-forceps. (Adapted from Hamad et al. [11])

[12], while another reported a tolerated stretch of 25–75% [13]. While it is not clear what the exact threshold is, it is well-documented that peripheral nerves can withstand a degree of tension stress. With peripheral nerves sliding in tight proximity to adjacent tissues and anatomical structures such as bones or fibrous bands, shear forces could potentially be a source of stress on the nerve as they undergo excursion during normal movement. However, the outermost epineurial layer is loosely associated with the perineurium, which minimizes frictional forces [23].

As with nerve compression, insights into the effects of controlled stretching of nerve trunks can be helpful when studying nerve tension. A recent study provided a proper perspective on nerve stretch's histological and mechanical impacts using Krackow sutures on a peripheral

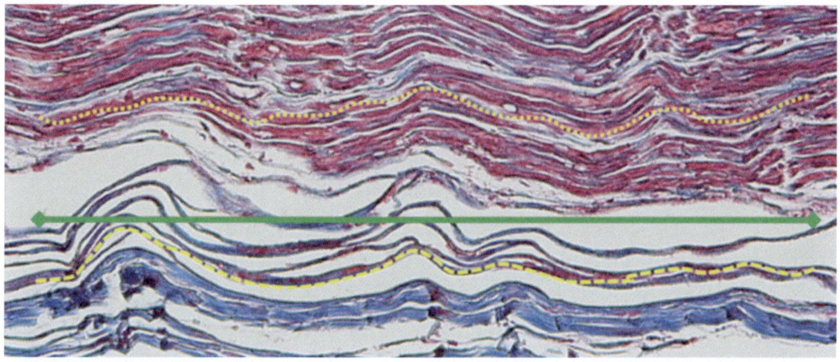

Fig. 11.4 Without tensile stress, peripheral nerves can be histologically observed in an undulating pattern of conformation. (Adapted from Kerns et al. [14])

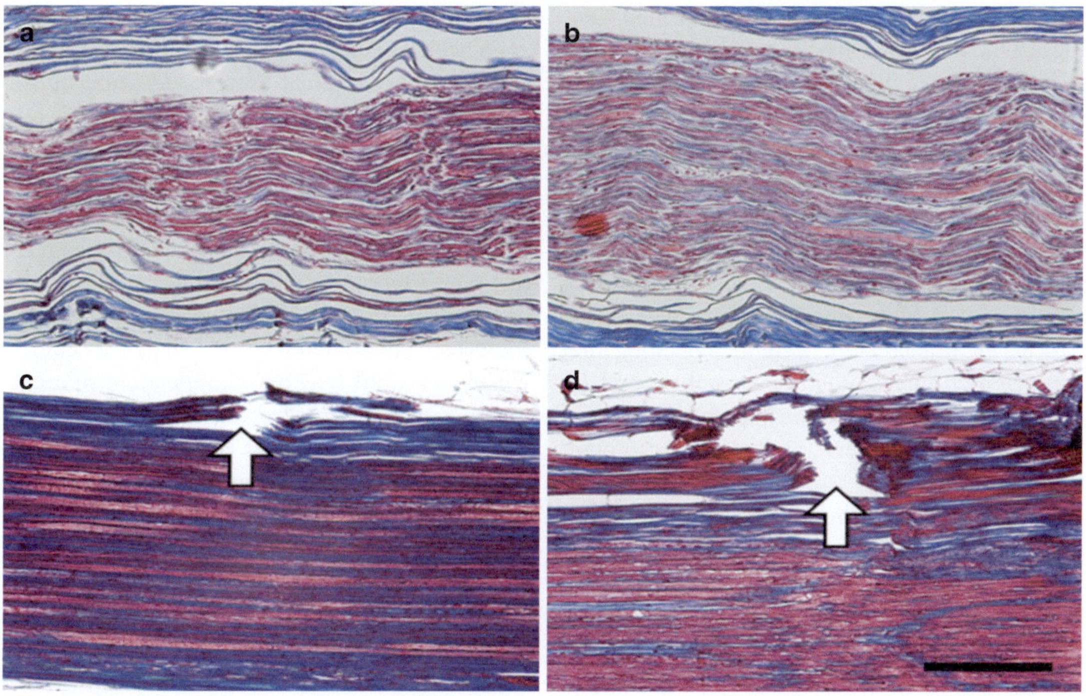

Fig. 11.5 Tensile stressing results in mechanical failure of the perineurial layer, highlighting its role as a structural component against stretch injury. (Adapted from Kerns et al. [14])

nerve [14]. First, the relaxed nerve segment was examined, and a histological view of the undulating pattern nerve tissue was appreciated, as seen in Fig. 11.4. After being stretched to failure, the nerve tissue was once again examined histologically and revealed a high prevalence of perineurial tears (as seen in Fig. 11.5) with increased tensile force, which supports the role of the perineurium in peripheral nerve structural integrity.

Biomechanical Studies of Compression Syndromes

Median Nerve

The most well-known compressive syndrome, carpal tunnel syndrome (CTS), arises from median nerve compression against the transverse carpal ligament. Although many cases of CTS are idiopathic, one study utilized a wick catheter to show that pressures within the carpal tunnel in idiopathic CTS were higher than in healthy subjects [2]. Normal pressure in the carpal tunnel has been reported to be around 2–10 mmHg [40]. With wrist extension and flexion, pressures can increase by 10x and 8x, respectively [40]. A study utilizing ultrasound to measure image characteristics in patients while relaxed vs typing on a keyboard found that cross-sectional area and swelling ratio increased significantly after only 30 min of typing, and the authors highlighted the role that ulnar deviation had on ultrasound findings of the median nerve [36].

The impact of ulnar deviation at the wrist was also reported in an ultrasound study examining the image characteristics of individuals handling a smartphone [41]. Another study of the median nerve involved wheelchair users on a dynamometer who had nerve conduction measured at the wrist while self-propelling. Median nerve latency was measured more significantly in wheelchair patients of higher body weight, indicating that the increased forces required for self-propulsion may have a positive relationship with the forces encountered by the nerve as it traverses through the carpal tunnel [3].

Median nerve sliding in response to joint movements was quantitatively studied using ultrasound, and it was found that shoulder adduction and elbow flexion resulted in decreased nerve loading [9]. When applied in a radioulnar direction across the wrist, the effect of compressive force on ultrasound changes in the carpal arch and median nerve was studied in healthy volunteers. It was found that with 10–20 N of force, the carpal arch height and area increased significantly, and the median nerve shape became less flattened, suggesting that noninvasive application of radioulnar force may potentially relieve median nerve compression in some patients [21]. Median nerve compression has also been demonstrated to occur at the pronator syndrome proximal to the carpal tunnel. However, quantitative biomechanical data for these regions remain sparse.

Ulnar Nerve

Ulnar nerve compression syndromes occur most commonly at the cubital tunnel and distally at that wrist. Cubital tunnel syndrome is the second most common compressive syndrome and occurs due to compression of the ulnar nerve as it courses through the arcuate ligament of the elbow [5]. A study found that the pressure within the cubital tunnel is at its lowest at full elbow extension, around 7 mmHg. At 130° of flexion, the pressure within the cubital tunnel reached 40 mmHg [10].

Wheelchair users were examined in another study concerning ulnar nerve conduction during self-propulsion. The authors looked specifically at the variation in the range of motion utilized by different subjects and found decreased conduction velocities in patients who demonstrated shorter ranges of motion. The authors suggest that this difference was due to fewer strokes made possible by more extensive ranges of motion. Increased strokes in shorter range of motion wheelchair users may have implications of repetitive use pathology [4]. Another common site of compression of the ulnar nerve is in Guyon's canal. Although many cases of ulnar nerve compression in Guyon's canal are attributed as idiopathic, the most known cause is compression from an adjacent ganglion cyst.

Radial Nerve

Of the peripheral nerves that innervate the hand, the least commonly implicated with compressive pathology is the radial nerve [25]. Posterior interosseous nerve (PIN) compression syndrome most commonly occurs with compression of this radial nerve branch at the arcade of Frohse, a fibrous band within the supinator muscle. One study found pressure in the arcade of Frohse to be significantly increased during maximal pronation, whether in full elbow flexion or extension [17].

Pathophysiological Consequences of Mechanical Forces

A cascade of pathological changes may result when peripheral nerves are subjected to mechanical force insults through compression or tension. Normal nerve function and long-term damage can occur from vascular compromise, axonal injury, demyelination, and fibrosis.

Microvascular Compromise

Compression exceeding certain thresholds collapses the vasa nervorum, leading to significant reductions in microvascular supply to the nerve. Studies in animal models have demonstrated that pressures of 20 mmHg can result in decreases in intraneural blood flow, with 80 mmHg causing a complete block of perfusion [31]. If compression is prolonged and with sufficient pressure, the resulting ischemia can disrupt energy availability for ATP-dependent processes of nerve cells, such as axonal transport, which is crucial for maintaining nerve function [38]. Prolonged ischemia can also lead to the breakdown of the blood–nerve barrier, resulting in edema, which may cause even higher increases in pressure within a nerve fascicle [19].

In cases where vascular perfusion is already compromised, such as nerves of poorly controlled diabetic patients, compression thresholds for ischemic injury are lower, making these nerves particularly vulnerable to the etiologies of compression syndromes [34]. The final stage of prolonged ischemia and microvascular compromise of peripheral nerve is Wallerian degeneration, a process in which axons and myelin sheaths disintegrate distally from the injury location.

Axonal Transport

In a healthy neuron, both fast and slow axonal transport of cellular products occurs from the body to the terminal end. Fast axonal transport is energy-dependent and often involves the transport of neurotransmitters. Slow axonal transport mainly consists of the movement of cytoskeletal components such as actin and tubulin.

In experimental studies, compression was demonstrated to disrupt both fast and slow axonal transport from pressure levels of 20–30 mmHg [8]. With decreased transport of critical cellular products distally of the compression site, it has been hypothesized that this would increase susceptibility to compression at a distal site of the same nerve [39].

Fibrosis

When mechanical stressors on the nerve are chronic, fibrotic changes to nerve tissue can occur. A histological study of nerves at different time points after undergoing compression found the first pathological change to be nerve edema and correlated with the severity of axonal injury [28]. Histological evaluation within a few days demonstrated proliferation of fibrous tissue, and at 28 days, there was fibrosis of nerve tissue with infiltration of macrophages.

Intraneural fibrosis reduces nerve compliance and increases the risk of injury from compressive and tensile forces. Over time, this fibrotic remodeling creates a self-perpetuating cycle of entrapment neuropathy that can become resistant to conservative management strategies such as splinting or activity modification.

Demyelination

Sustained compression leads to demyelination, mainly affecting the paranodal architecture of myelinated fibers. Experimental studies on rat sciatic nerves subjected to graded compression have demonstrated that pressures of 30 mmHg for 2 h predominantly result in demyelination [28]. Electron microscopy evaluation observations showed that demyelination was correlated with Schwann cell necrosis and was most pronounced in fibers nearest to the perineurium when compared to those at the core of the fascicle, which are especially vulnerable due to their proximity to perineurial vessels that they are dependent on. The localized ischemia caused by compression likely disrupts the Schwann cell's ability to maintain myelin integrity.

Electromyographic Correlates

Electromyography (EMG) is a valuable tool that evaluates conduction velocity and amplitude changes to provide insights into the functional consequences of nerve compression neuropathies. Demyelination from mechanical insults can be observed from conduction velocity reductions of 5–10 m/s across the site of compression. When forces lead to the pathophysiological consequence of axonal loss, amplitude reductions in EMG may be observed, especially in more advanced stages of compression neuropathy. A reduction in compound muscle action potential (CMAP) amplitude by more than 50% correlates with significant axonal degeneration.

Despite its utility, EMG has limitations. False negatives are particularly common in cases with intermittent or positional symptoms, as these may not manifest during static testing conditions. For example, patients with ulnar neuropathy at the cubital tunnel may exhibit typical conduction velocities with the elbow between 0° and 90° of flexion despite symptomatic nerve compression after prolonged flexion at 130°.

Operator dependency is another limitation; improper electrode placement can result in significant variability in recorded amplitudes and latencies. Inadequate skin preparation is another variable that may result in inaccurate findings [7]. Furthermore, EMG cannot reliably distinguish between compression neuropathies caused by external forces and those arising from intrinsic structural abnormalities such as cysts, necessitating complementary imaging studies for definitive diagnosis.

Conclusion

Peripheral nerve compression neuropathies result from mechanical forces exerted on the nerves that exceed the tissue's capacity for adaptation. The epineurium facilitates the smooth gliding of nerve tissue relative to adjacent structures while providing limited protection against external compressive forces. In contrast, the perineurium serves to resist tensile deformation. The viscoelastic properties of nerves enable them to withstand mild mechanical stressors encountered during normal limb movement.

However, critical thresholds, such as specific compression pressures or elongation percentages, delineate the transition from physiological adaptation to pathological injury. The pathophysiological mechanisms associated with these neuropathies encompass ischemia, disruption of axonal transport, demyelination of nerves, and fibrosis induced by chronic inflammation. Electromyography remains a pivotal diagnostic tool for quantifying functional impairment; nevertheless, its sensitivity is often diminished in the early stages of neuropathies or instances involving positional symptoms.

References

1. Atroshi I, Gummesson C, Johnsson R, Ornstein E, Ranstam J, Rosén I. Prevalence of carpal tunnel syndrome in a general population. JAMA. 1999;282(2):153–8. https://doi.org/10.1001/jama.282.2.153.
2. Bauman TD, Gelberman RH, Mubarak SJ, Garfin SR. The acute carpal tunnel syndrome. Clin Orthop Relat Res. 1981;156:151–6.
3. Boninger ML, Cooper RA, Baldwin MA, Shimada SD, Koontz A. Wheelchair pushrim kinetics: body weight and median nerve function. Arch Phys Med Rehabil. 1999;80(8):910–5. https://doi.org/10.1016/S0003-9993(99)90082-5.
4. Boninger ML, Impink BG, Cooper RA, Koontz AM. Relation between median and ulnar nerve function and wrist kinematics during wheelchair propulsion1. Arch Phys Med Rehabil. 2004;85(7):1141–5. https://doi.org/10.1016/j.apmr.2003.11.016.

5. Bozentka DJ. Cubital tunnel syndrome pathophysiology. Clin Orthop Relat Res. 1998;351:90–4.
6. Carp S. Peripheral nerve injury an anatomical and physiological approach for physical therapy intervention. Philadelphia: F.A. Davis; 2015.
7. Cram JR, Rommen D. Effects of skin preparation on data collected using an EMG muscle-scanning procedure. Biofeedback Self Regul. 1989;14(1):75–82. https://doi.org/10.1007/BF00999342.
8. Dahlin LB, McLean WG. Effects of graded experimental compression on slow and fast axonal transport in rabbit vagus nerve. J Neurol Sci. 1986;72(1):19–30. https://doi.org/10.1016/0022-510X(86)90032-8.
9. Dilley A, Lynn B, Greening J, DeLeon N. Quantitative in vivo studies of median nerve sliding in response to wrist, elbow, shoulder and neck movements. Clin Biomech. 2003;18(10):899–907. https://doi.org/10.1016/S0268-0033(03)00176-1.
10. Gelberman RH, Yamaguchi K, Hollstien SB, Winn SS, Heidenreich FPJ, Bindra RR, Hsieh P, Silva MJ. Changes in interstitial pressure and cross-sectional area of the cubital tunnel and of the ulnar nerve with flexion of the elbow. An experimental study in human cadavera. J Bone Joint Surg Am. 1998;80(4):492.
11. Hamad MN, Boroda N, Echenique DB, Dieter RA, Amirouche FML, Gonzalez MH, Kerns JM. Compound motor action potentials during a modest nerve crush. Front Cell Neurosci. 2022;16:798203. https://doi.org/10.3389/fncel.2022.798203.
12. Highet WB, Sanders FK. The effects of stretching nerves after suture. Br J Surg. 1943;30(120):355–69. https://doi.org/10.1002/bjs.18003012012.
13. Hoen TI, Brackett CE. Peripheral nerve lengthening. I. Experimental. J Neurosurg. 1956;13(1):43–62. https://doi.org/10.3171/jns.1956.13.1.0043.
14. Kerns J, Piponov H, Helder C, Amirouche F, Solitro G, Gonzalez M. Mechanical properties of the human tibial and peroneal nerves following stretch with histological correlations. Anat Rec. 2019;302(11):2030–9. https://doi.org/10.1002/ar.24250.
15. Kleinrensink GJ, Stoeckart R, Vleeming A, Snijders CJ, Mulder PGH. Mechanical tension in the median nerve. The effects of joint positions. Clin Biomech (Bristol). 1995;10(5):240–4. https://doi.org/10.1016/0268-0033(95)99801-8.
16. Klimovich P, Rubina K, Sysoeva V, Semina E. New frontiers in peripheral nerve regeneration: concerns and remedies. Int J Mol Sci. 2021;22(24):13380. https://doi.org/10.3390/ijms222413380.
17. Links AC, Graunke KS, Wahl C, Green JR, Matsen FA. Pronation can increase the pressure on the posterior interosseous nerve under the arcade of Frohse: a possible mechanism of palsy after two-incision repair for distal biceps rupture--clinical experience and a cadaveric investigation. J Shoulder Elb Surg. 2009;18(1):64–8. https://doi.org/10.1016/j.jse.2008.07.001.
18. Luque EH, Angulo E, Montes GS. A histochemical and electron microscopic study on the collagen of nerves in the domestic fowl. J Anat. 1983;137(Pt 1):171–6.
19. Mackinnon SE, Dellon AL, Hudson AR, Hunter DA. Chronic human nerve compression--a histological assessment. Neuropathol Appl Neurobiol. 1986;12(6):547–65. https://doi.org/10.1111/j.1365-2990.1986.tb00159.x.
20. Mahan MA. Nerve stretching: a history of tension. J Neurosurg. 2020;132(1):252–9. https://doi.org/10.3171/2018.8.JNS173181.
21. Marquardt TL, Gabra JN, Li Z-M. Morphological and positional changes of the carpal arch and median nerve during wrist compression. Clin Biomech. 2015;30(3):248–53. https://doi.org/10.1016/j.clinbiomech.2015.01.007.
22. Millesi H, Zöch G, Rath T. The gliding apparatus of peripheral nerve and its clinical significance. Ann Chir Main Memb Super. 1990;9(2):87–97. https://doi.org/10.1016/s0753-9053(05)80485-5.
23. Millesi H, Zöch G, Reihsner R. Mechanical properties of peripheral nerves. Clin Orthop Relat Res. 1995;314:76–83.
24. Mojaddidi MA, Ahmed MS, Ali R, Jeziorska M, Al-Sunni A, Thomsen NOB, Dahlin LB, Malik RA. Molecular and pathological studies in the posterior interosseous nerve of diabetic and non-diabetic patients with carpal tunnel syndrome. Diabetologia. 2014;57(8):1711–9. https://doi.org/10.1007/s00125-014-3271-3.
25. Nakano KK. Nerve entrapment syndromes. Curr Opin Rheumatol. 1997;9(2):165–73. https://doi.org/10.1097/00002281-199703000-00015.
26. Peltonen S, Alanne M, Peltonen J. Barriers of the peripheral nerve. Tissue Barriers. 2013;1(3):e24956. https://doi.org/10.4161/tisb.24956.
27. Pourmand R, Ochs S, Jersild RA. The relation of the beading of myelinated nerve fibers to the bands of Fontana. Neuroscience. 1994;61(2):373–80. https://doi.org/10.1016/0306-4522(94)90238-0.
28. Powell HC, Myers RR. Pathology of experimental nerve compression. Lab Investig. 1986;55(1):91–100.
29. Rosso G, Guck J. Mechanical changes of peripheral nerve tissue microenvironment and their structural basis during development. APL Bioeng. 2019;3(3):036107. https://doi.org/10.1063/1.5108867.
30. Rydevik B, Lundborg G. Permeability of intraneural microvessels and perineurium following acute, graded experimental nerve compression. Scand J Plast Reconstr Surg. 1977;11(3):179–87. https://doi.org/10.3109/02844317709025516.
31. Rydevik B, Lundborg G, Bagge U. Effects of graded compression on intraneural blood blow. An in vivo study on rabbit tibial nerve. J Hand Surg Am. 1981;6(1):3–12. https://doi.org/10.1016/s0363-5023(81)80003-2.

32. Rydevik BL, Kwan MK, Myers RR, Brown RA, Triggs KJ, Woo SL, Garfin SR. An in vitro mechanical and histological study of acute stretching on rabbit tibial nerve. J Orthop Res. 1990;8(5):694–701. https://doi.org/10.1002/jor.1100080511.
33. Sunderland S. The connective tissues of peripheral nerves. Brain. 1965;88(4):841–54. https://doi.org/10.1093/brain/88.4.841.
34. Tang C-W, Ju M-S, Lin C-CK. Morphological and hemodynamic changes of sciatic nerves and their vasa nervorum during circular compression and relaxation. J Biomech. 2020;110:109974. https://doi.org/10.1016/j.jbiomech.2020.109974.
35. Teixeira MJ, Almeida DB, Yeng LT. Concept of acute neuropathic pain. The role of *nervi nervorum* in the distinction between acute nociceptive and neuropathic pain. Rev dor. 2016;17:5–10. https://doi.org/10.5935/1806-0013.20160038.
36. Toosi KK, Hogaboom NS, Oyster ML, Boninger ML. Computer keyboarding biomechanics and acute changes in median nerve indicative of carpal tunnel syndrome. Clin Biomech (Bristol). 2015;30(6):546–50. https://doi.org/10.1016/j.clinbiomech.2015.04.008.
37. Topp KS, Boyd BS. Structure and biomechanics of peripheral nerves: nerve responses to physical stresses and implications for physical therapist practice. Phys Ther. 2006;86(1):92–109. https://doi.org/10.1093/ptj/86.1.92.
38. Tuček S, Hanzlíková V, Stráníková D. Effect of ischemia on axonal transport of choline acetyltransferase and acetylcholinesterase and on ultrastructural changes of isolated segments of rabbit nerves in situ. J Neurol Sci. 1978;36(2):237–46. https://doi.org/10.1016/0022-510X(78)90084-9.
39. Upton AR, McComas AJ. The double crush in nerve entrapment syndromes. Lancet. 1973;2(7825):359–62. https://doi.org/10.1016/s0140-6736(73)93196-6.
40. Werner RA, Andary M. Carpal tunnel syndrome: pathophysiology and clinical neurophysiology. Clin Neurophysiol. 2002;113(9):1373–81. https://doi.org/10.1016/s1388-2457(02)00169-4.
41. Woo H-C, White P, Ng H-K, Lai CWK. Development of kinematic graphs of median nerve during active finger motion: implications of smartphone use. PLoS One. 2016;11(7):e0158455. https://doi.org/10.1371/journal.pone.0158455.

Biomechanics of Carpal Tunnel Syndrome

12

Ye Lin, Farid Amirouche, and Gautam Malhotra

Anatomy and Pathoanatomy of the Carpal Tunnel

Overview

The carpal tunnel is an anatomic space at the wrist formed by the carpal bones and ligaments [25]. The floor of the carpal tunnel is formed by both the proximal and distal carpal rows. The roof or most volar aspect of the carpal tunnel is defined by the transverse carpal ligament (TCL), which attaches to the scaphoid and trapezium on the lateral side of the wrist and the hamate and pisiform on the medial side [30] (Fig. 12.1). The carpal tunnel is an elaborate biomechanical structure whose pathomechanics play an essential role in developing carpal tunnel syndrome. This article aims to review the carpal tunnel's movement-related biomechanics and anatomical and morphological features and describe the pathomechanics and pathophysiology associated with carpal tunnel syndrome. Topics of discussion include biomechanics of the median nerve, flexor tendons, subsynovial tissue, transverse carpal ligament, carpal tunnel pressure, morphological properties, and biomechanical improvement and physiological restoration mechanisms. We hope that the biomechanical knowledge of carpal tunnel will improve the understanding and management of carpal tunnel syndrome [25, 30]. Cadaveric studies have shown that the narrowest portion of the carpal tunnel occurs at the level of the hook of the hamate. The carpal tunnel width at the hook of the hamate (20 mm) was significantly smaller than its proximal (24 mm) or distal (25 mm) extent. Traversing this space are the nine tendons of the extrinsic flexors of the thumb and fingers, along with the median nerve. The borders of the carpal tunnel are rigid, and subtle changes leading to crowding within the carpal tunnel can contribute to the pathophysiology of CTS.

Y. Lin (✉) · G. Malhotra
Department of Orthopaedic Surgery, University of Illinois at Chicago, Chicago, IL, USA
e-mail: yelin@uic.edu

F. Amirouche
Department of Orthopaedic Surgery, Orthopaedic and Spine Institute, Endeavor HealthSystem (Formerly Northshore University HealthSystem), an Affiliate of the University of Chicago Pritzker School of Medicine, Skokie, IL, USA

Department of Orthopaedic Surgery, University of Illinois at Chicago, Chicago, IL, USA

The Transverse Carpal Ligament

The transverse carpal ligament is the distal continuation of the antebrachial fascia of the forearm. The TCL is roughly 10 times thicker than the antebrachial fascia [12]. The transverse carpal ligament's thickness varies from 1.5–2.5 mm proximally to 2.5–6.0 mm distally. Imaging and

F. Amirouche et al. (eds.), *Biomechanics of the Hand*,
https://doi.org/10.1007/978-3-032-04688-8_12

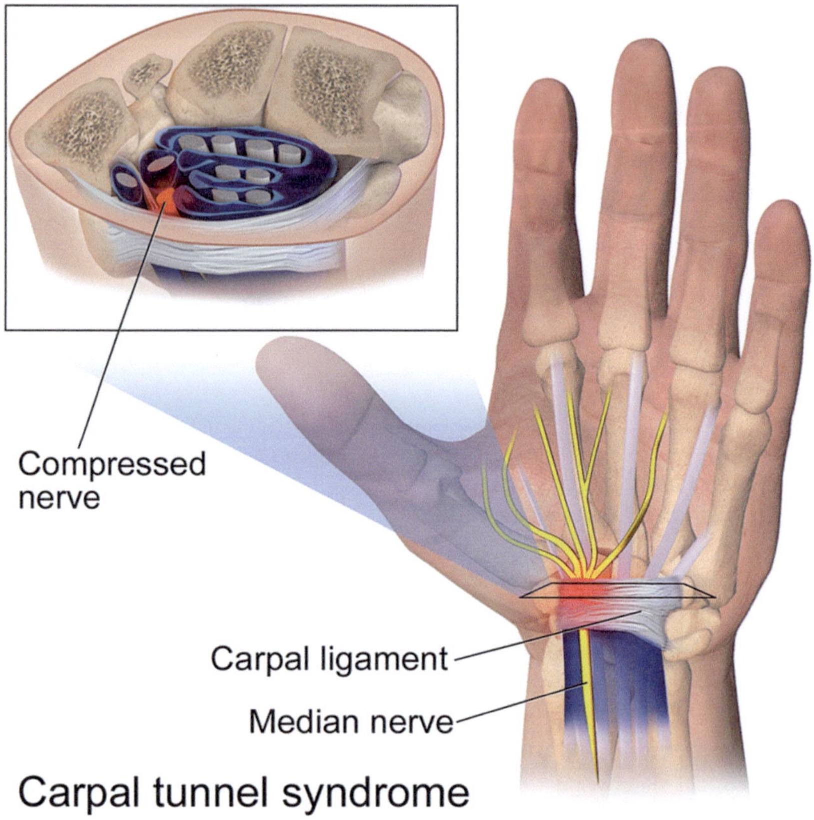

Fig. 12.1 Anatomy of the carpal tunnel and pathophysiology of carpal tunnel syndrome illustrating the median nerve compression. This image is available under the Creative Commons CC-BY-NC license and permits noncommercial use, distribution, and reproduction in any medium, provided the original work is properly cited [6]

histologic studies have demonstrated that hypertrophy of the transverse carpal ligament contributes to the development of CTS. In an ultrasonography study, the transverse carpal ligament was thickened in 88% of patients with diagnosed CTS [9]. Histologically, the transverse carpal ligament in patients with CTS has mucoid change, inflammation, increased and fibrocartilaginous metaplasia, and faster growth of contractile cells [2, 33]. The ligament's altered composition may change the flexor retinaculum's elasticity and contractility, and the thickening leads to diminished space within the carpal tunnel. Both changes are likely common contributing factors in the pathological process.

Complete transection of the transverse carpal ligament is commonly employed to treat carpal tunnel syndrome. One must consider the functions of the transverse carpal ligament to evaluate the consequences of dividing this ligament critically. These functions include providing transverse carpal stability through its bony attachments, serving as an anchor for the thenar and hypothenar muscles, and acting as a pulley for the flexor tendons to prevent bowstringing [18, 25].

Flexor Tendons and the Subsynovial Connective Tissue

The extrinsic finger tendons have significant excursion during finger motion, between 3 and 5 cm [47]. A radial and ulnar-sided bursa help limit the friction between the tendons and osseous structures. The tendons themselves are surrounded by a loose connective tissue described as the subsynovial connective tissue (SSCT). This tissue is rich in lymphatics and blood vessels. This connective tissue is arranged in alternating layers of horizontally and vertically oriented fibrous tissue. Each layer is loosely connected to the adjacent layer, binding all the tendons and the median nerve together. As a tendon moves, the

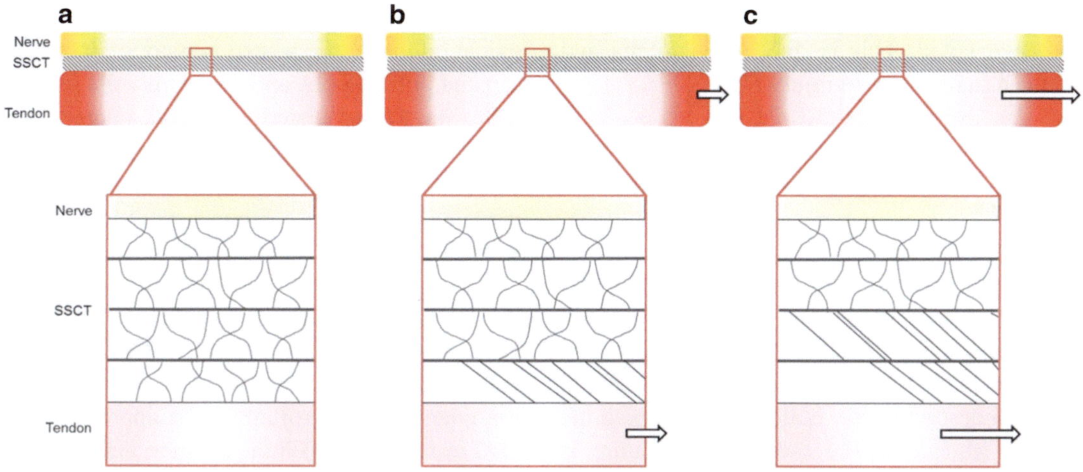

Fig. 12.2 Schematic representation of sequential recruitment of SSCT layers during tendon excursion. On top, the median nerve, SSCT, and a flexor tendon are depicted. Below are the interconnecting vertical thinner fibers. (**a**) Without tendon excursion, the vertical SSCT fibers hang loosely. (**b**) During the first part of tendon excursion, in the middle pictograph, the row of SSCT fibers closest to the tendon is loaded first, stretching out the fibers. (**c**) With increasing excursion, the second layer of SSCT fibers is recruited [38]

closest fibrous layer moves, subsequent layers are recruited, and so on. This pattern creates a flexible scaffold for the blood vessels and lymphatics. Synovial fluid separates the layers and serves as a lubricant, allowing a measure of gliding [38]. This structure enables differential movement of the tendons, although it also limits this differential movement. It has been suggested that exceeding this will lead to a shearing injury of the fibrous layers, which can result in scarring and altered motion (Fig. 12.2). Ettema et al. demonstrated that SSCT layers closest to the tendons were more injured than those lying further away [14]. Tissue damage creates progressive fibrosis, increased SSCT thickness, and increased SSCT stiffness, leading the SSCT to move en bloc with the tendon rather than in differential layers, creating abnormal tendon gliding [15, 43]. This also impairs nerve gliding and motion.

The Median Nerve

The median nerve is the volarmost structure within the carpal tunnel. Like the flexor tendons, it is surrounded by and bound to the subsynovial connective tissue. During physiological motion, the median nerve glides proximally and distally and exhibits a transverse motion within the carpal tunnel [44, 49].

Mechanical compression of peripheral nerves can lead to nerve edema, ischemia, and perineural fibrosis. This compression can result in demyelination and axonal degeneration that leads to progressive neurological symptoms [32]. All these changes to the median nerve have been reported in carpal tunnel syndrome. Restricted motion in both the longitudinal and transverse planes has been demonstrated. This restriction is likely related to the limitation of the space available to the nerve and scarring of the subsynovial connective tissue, thereby contributing to increased mechanical strain on the nerve with finger and wrist motion. Increased flattening of the nerve has been noted as a result of mechanical compression. The median nerve is found to be flattened within and distal to the carpal tunnel in nearly 97% of individuals with CTS [35]. The flattening ratio increases as the nerve travels further distally in the carpal tunnel [29]. Flattening of the median nerve is consistent with the expected increase in carpal tunnel pressures with CTS. This finding quickly normalizes after carpal tunnel surgery [3, 13]. Finally, an enlargement or increase in the cross-sectional area of the nerve has been noted. This is due to hyperemia and edema as a response to injury [16].

Volume and Pressure of the Carpal Tunnel

In a neutral wrist at rest, the carpal tunnel pressure is <15 mmHg [21, 34, 48]. Wrist position can, however, significantly alter pressure within the carpal tunnel. The transverse carpal ligament has been described as a pulley around which the tendons glide [5]. The radius of curvature is influenced by wrist position. In a flexed position, the tendons essentially move closer to the transverse carpal ligament, and the radius of curvature is reduced. As force is exerted with the wrist in a flexed position (small radius of curvature), it is hypothesized that more pressure is placed on the adjacent wrist structures, with median nerve compression clinically relevant to carpal tunnel syndrome. This increase in carpal tunnel pressure in wrist flexion has been measured, where flexion of the wrist to ninety degrees nearly tripled the pressure within the carpal tunnel [21]. Positioning the wrist in flexion has also been associated with increased gliding resistance of the tendons, as the tendons move differentially compared to neighboring tendons rather than in parallel [51].

Increased carpal tunnel pressures have been seen across studies in individuals with CTS [21, 34]. This pressure elevation has several etiologies, including increased volume of the carpal tunnel contents (whether the native contents or abnormal masses), general factors for tissue swelling, and wrist positioning outside of neutral alignment. The increase in pressure contributes to the development of neurological symptoms. In animal studies, 20–30 mmHg of pressure on tibial nerves was sufficient to constrain venous epineural blood flow and cause edema and hyperemia [37]. Interestingly, as CTS becomes more advanced, long-term compression of the carpal tunnel contents results in tissue atrophy and counterintuitive normal carpal tunnel pressure [42].

Biomechanical studies support the epidemiological finding that repeated vibratory tool usage, grip and pinch forces, and awkward wrist posture in flexion are associated with work-related CTS [41]. They may also support using wrist splinting during activity and sleep as a conservative treatment modality for carpal tunnel syndrome, which decreases symptom severity [20].

Carpal tunnel volume has been estimated with multiple imaging modalities, including ultrasound, magnetic resonance imaging (MRI), and computed tomography (CT). MRI evaluation quantified the volume to be approximately 5000 mm^3 and 2500 mm^3 for men and women, respectively [7]. Studies have not consistently shown that patients with CTS have decreased carpal tunnel volume compared to their asymptomatic counterparts. However, the relative volume of the contents of the carpal tunnel compared to the total volume of the carpal tunnel was shown to be increased in patients with CTS compared to normal controls, 0.57 versus 0.50, respectively [12].

Role of Ultrasound in Diagnosis

Traditionally, carpal tunnel syndrome was considered a clinical diagnosis based on symptoms and provocative testing aided by electromyography. Advances in image quality and affordability of ultrasound have increased interest in understanding its role in the diagnostic pathway for carpal tunnel syndrome [50]. Static ultrasound can characterize expected mechanical changes in the carpal tunnel in patients with CTS. The median nerve is seen to be compressed at the level of the pisiform and paradoxically hyperemic and swollen proximally and distally, forming an hourglass shape. Technically, the cross-sectional area of the nerve can be visualized as a honeycomb within a hyperechoic epineurium. A cross-sectional area of >9 mm^2 at the pisiform level, the level of greatest compression, or a swelling ratio >1.3 is considered to have a diagnostic value for CTS [4, 9, 10] (Fig. 12.3).

Static ultrasound is complemented with dynamic cross-sectional analyses. Ultrasound

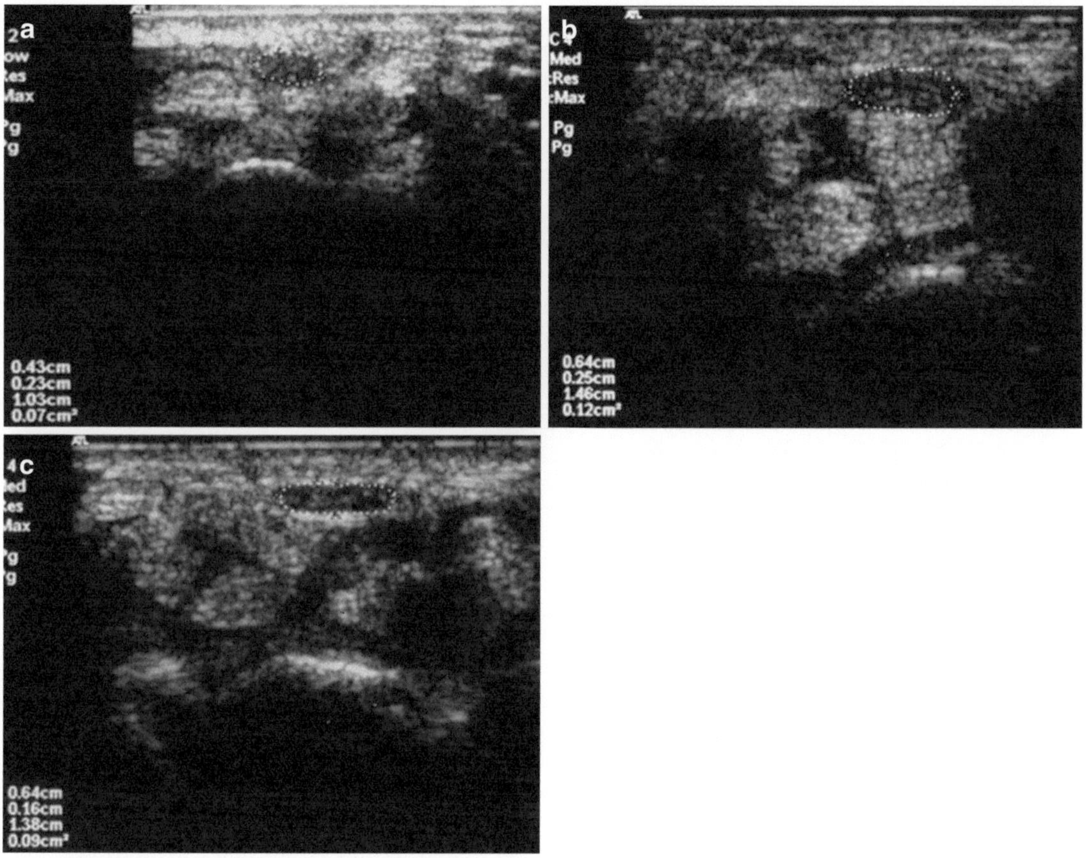

Fig. 12.3 Sonograms of a symptomatic CTS patient. (**a**) The median nerve is of normal size at the proximal level (7 mm^2). (**b**) At the pisiform level, the size of the median nerve was more significant (12 mm^2) compared with the proximal level. The swelling ratio was 1.7. (**c**) Increased flattening of the median nerve at the distal level. [4]

recordings of the tendons, median nerve, and subsynovial connective tissue are made while the patient moves their fingers from extension into flexion. At the point of maximum displacement in extension and flexion, the displacement and the median nerve morphology are measured. In flexion and extension, the median nerve deformation, circularity (complexity of shape), and the change in circumference were significantly more significant in CTS patients than in controls [45].

Finally, elastography can quantify the mechanical properties of tissue in vivo. Specifically strain and shear experienced. Strain imaging allows tissue displacement measurement when a compressive force is applied, whereas shear wave imaging measures physical tissue displacement parallel to the applied stress [40]. Median nerve stiffness is higher in patients with moderate to severe CTS than in mild CTS or controls [11, 26, 31]. As with all ultrasound-based techniques, user variability is seen, and an experienced technician is needed to perform and interpret such studies.

Biomechanical Changes After Carpal Tunnel Release

Whether open or endoscopic, carpal tunnel release (CTR) involves transecting the transverse carpal ligament and the distal aspect of the antebrachial fascia, although patients have significant clinical benefits with CTR, the biomechanical consequences postoperatively are still being further investigated.

Carpal Arch Shape

As one role of the transverse carpal ligament is to provide carpal arch stability, CTR flattens the transverse arch of the hand. Radiographic measurements postoperatively demonstrate that patients had an average of 13.6% (~3 mm) of carpal arch widening 19.7 months after open surgery [19]. There appears to be less carpal arch widening after endoscopic surgery, approximately 7% in the first 6 weeks after surgery [46]. Cadaveric studies have shown that CTR decreased the transverse carpal arch stiffness by 7.5% [18]. When finite element modeling was used to study simulated grasp, there were increased contact stresses on the midcarpal joints [22]. The biomechanical consequences of this change in carpal tunnel arch shape are unclear. Increased pisotriquetral pressures may result in pillar pain at the hypothenar eminence [39]. Clinical studies have shown that a more than 20% increase in carpal arch width was associated with decreased grip strength [19].

The Volume of the Carpal Tunnel

CTR increases carpal tunnel volume, thereby decreasing pressure within the carpal tunnel. After carpal tunnel release, a computed tomography (CT) scan of the carpus demonstrated increased carpal tunnel canal caliber. Using MRI to evaluate the volar cross-sectional area, authors found that after endoscopic CTR, the cross-sectional area of the carpal tunnel increased 3.6 times, whereas the dorsal cross-sectional surface area only increased minimally (approximately 3%) [27, 36]. They also found that the transverse carpal ligament bowed convexly after release. In that study, they did not see that increased carpal arch persisted at 8 months follow-up, suggesting recoil of the carpal arch postoperatively. The authors thus concluded that the diastasis of the transverse carpal ligament allowed for the increase in carpal tunnel volume post-surgical rather than a flattening in the carpal arch [8].

Median Nerve

The flattening in the median nerve observed in carpal tunnel syndrome is seen to improve after carpal tunnel release, as soon as two weeks after surgery on ultrasound [13]. The nerve is also seen to become more volarly displaced as the volume of the carpal tunnel itself is increased [1]. Using dynamic ultrasound, the median nerve was observed to have increased overall displacement during finger flexion after carpal tunnel release [38]. These changes suggest decreased compression on the median nerve.

Tendon Function

The extrinsic wrist tendons are also volarly displaced after carpal tunnel release [8]. As the transverse carpal ligament serves as a pulley for the flexor tendons to pass over, carpal tunnel release has been shown to increase tendon excursion after 20 to 30 degrees of wrist flexion at the flexor digitorum profundus and flexor digitorum superficialis [28]. Some postoperative wrist flexion weakness may be attributed to this bowstringing, which consumes tendon motion.

Grip Strength

Finally, carpal tunnel release may be associated with grip weakness [17, 39]. Weakness in grip is attributed to the volar displacement of the flexor tendons and the relaxed anchor for the thenar and hypothenar muscles [17, 23, 28]. More recently, the position of the scaphoid during wrist movement after carpal tunnel release has been investigated, demonstrating more significant scaphoid extension, which is hypothesized to contribute to postoperative weakness [24].

Future Directions

Most patients who undergo carpal tunnel surgery experience significant relief from their symptoms, with studies demonstrating high success rates in alleviating discomfort and restoring hand function. However, some individuals may still encounter complications postoperatively, such as a persistent pillar pain, diminished grip strength, and incomplete resolution of symptoms. These unresolved issues may make revision surgery necessary to achieve satisfactory outcomes.

Despite the general effectiveness of the procedure, current biomechanical research has not yet succeeded in reliably predicting or stratifying the risk for these postoperative complications before surgery. This gap in predictive capability underscores the need for a deeper understanding of the intricacies of carpal tunnel mechanics. By expanding our knowledge in this area, healthcare professionals and researchers may be better positioned to develop targeted therapeutic interventions and foster advancements in surgical techniques, ultimately enhancing patient outcomes and minimizing the risk of complications.

References

1. Ablove RH, Peimer CA, Diao E, Oliverio R, Kuhn JP. Morphologic changes following endoscopic and two-portal subcutaneous carpal tunnel release. J Hand Surg Am. 1994;19(5):821–6. https://doi.org/10.1016/0363-5023(94)90194-5.
2. Allampallam K, Chakraborty J, Bose KK, Robinson J. Explant culture, immunofluorescence and electron-microscopic study of flexor retinaculum in carpal tunnel syndrome. J Occup Environ Med. 1996;38(3):264–71. https://doi.org/10.1097/00043764-199603000-00011.
3. Allmann KH, Horch R, Uhl M, Gufler H, Altehoefer C, Stark GB, Langer M. MR imaging of the carpal tunnel. Eur J Radiol. 1997;25(2):141–5. https://doi.org/10.1016/s0720-048x(96)01038-8.
4. Altinok T, Baysal O, Karakas HM, Sigirci A, Alkan A, Kayhan A, Yologlu S. Ultrasonographic assessment of mild and moderate idiopathic carpal tunnel syndrome. Clin Radiol. 2004;59(10):916–25. https://doi.org/10.1016/j.crad.2004.03.019.
5. Armstrong TJ, Chaffin DB. Some biomechanical aspects of the carpal tunnel. J Biomech. 1979;12(7):567–70. https://doi.org/10.1016/0021-9290(79)90045-9.
6. Blausen.com staff: Medical gallery of Blausen Medical 2014. Wiki Journal of Med. 2014, 1:https://doi.org/10.15347/wjm/2014.010
7. Bower JA, Stanisz GJ, Keir PJ. An MRI evaluation of carpal tunnel dimensions in healthy wrists: implications for carpal tunnel syndrome. Clin Biomech (Bristol, Avon). 2006;21(8):816–25. https://doi.org/10.1016/j.clinbiomech.2006.04.008.
8. Brooks JJ, Schiller JR, Allen SD, Akelman E. Biomechanical and anatomical consequences of carpal tunnel release. Clin Biomech. 2003;18(8):685–93. https://doi.org/10.1016/S0268-0033(03)00052-4.
9. Buchberger W, Schön G, Strasser K, Jungwirth W. High-resolution ultrasonography of the carpal tunnel. J Ultrasound Med. 1991;10(10):531–7. https://doi.org/10.7863/jum.1991.10.10.531.
10. Cartwright MS, Hobson-Webb LD, Boon AJ, Alter KE, Hunt CH, Flores VH, Werner RA, Shook SJ, Thomas TD, Primack SJ, Walker FO, American Association of Neuromuscular and Electrodiagnostic Medicine. Evidence-based guideline: neuromuscular ultrasound for the diagnosis of carpal tunnel syndrome. Muscle Nerve. 2012;46(2):287–93. https://doi.org/10.1002/mus.23389.
11. Cingoz M, Kandemirli SG, Alis DC, Samanci C, Kandemirli GC, Adatepe NU. Evaluation of median nerve by shear wave elastography and diffusion tensor imaging in carpal tunnel syndrome. Eur J Radiol. 2018;101:59–64. https://doi.org/10.1016/j.ejrad.2018.02.005.
12. Cobb TK, Dalley BK, Posteraro RH, Lewis RC. Anatomy of the flexor retinaculum. J Hand Surg Am. 1993;18(1):91–9. https://doi.org/10.1016/0363-5023(93)90251-W.
13. El-Karabaty H, Hetzel A, Galla TJ, Horch RE, Lücking CH, Glocker FX. The effect of carpal tunnel release on median nerve flattening and nerve conduction. Electromyogr Clin Neurophysiol. 2005;45(4):223–7.
14. Ettema AM, Amadio PC, Zhao C, Wold LE, O'Byrne MM, Moran SL, An K-N. Changes in the functional structure of the tenosynovium in idiopathic carpal tunnel syndrome: a scanning electron microscope study. Plast Reconstr Surg. 2006;118(6):1413–22. https://doi.org/10.1097/01.prs.0000239593.55293.c7.
15. Ettema AM, Zhao C, Amadio PC, O'Byrne MM, An K-N. Gliding characteristics of flexor tendon and tenosynovium in carpal tunnel syndrome: a pilot study. Clin Anat. 2007;20(3):292–9. https://doi.org/10.1002/ca.20379.
16. Filius A, Scheltens M, Bosch HG, Doorn PA van, Stam HJ, Hovius SER, Amadio PC, Selles RW (2015) Multidimensional ultrasound imaging of the wrist: changes of shape and displacement of the median nerve and tendons in carpal tunnel syndrome. J Orthop Res 33(9):1332. doi:https://doi.org/10.1002/jor.22909.
17. Fuss FK, Wagner TF. Biomechanical alterations in the carpal arch and hand muscles after carpal tunnel release: a further approach toward understanding the function of the flexor retinaculum and the cause of postoperative grip weakness. Clin Anat. 1996;9(2):100–8. https://doi.org/10.1002/(SICI)1098-2353(1996)9:2<100::AID-CA2>3.0.CO;2-L.
18. Garcia-Elias M, An KN, Cooney WP, Linscheid RL, Chao EY. Transverse stability of the carpus. An analytical study. J Orthop Res. 1989;7(5):738–43. https://doi.org/10.1002/jor.1100070516.

19. Gartsman GM, Kovach JC, Crouch CC, Noble PC, Bennett JB. Carpal arch alteration after carpal tunnel release. J Hand Surg Am. 1986;11(3):372–4. https://doi.org/10.1016/s0363-5023(86)80144-7.
20. Gatheridge MA, Sholty EA, Inman A, Pattillo M, Mindrup F, Sanderson DL. Splinting in carpal tunnel syndrome: the optimal duration. Mil Med. 2020;185(11–12):e2049–54. https://doi.org/10.1093/milmed/usaa222.
21. Gelberman RH, Hergenroeder PT, Hargens AR, Lundborg GN, Akeson WH. The carpal tunnel syndrome. A study of carpal canal pressures. J Bone Joint Surg Am. 1981;63(3):380–3.
22. Guo X, Fan Y, Li Z-M. Effects of dividing the transverse carpal ligament on the mechanical behavior of the carpal bones under axial compressive load: a finite element study. Med Eng Phys. 2009;31(2):188–94. https://doi.org/10.1016/j.medengphy.2008.08.001.
23. Hunter JM. Recurrent carpal tunnel syndrome, epineural fibrous fixation, and traction neuropathy. Hand Clin. 1991;7(3):491–504.
24. Ishiko T, Puttlitz CM, Lotz JC, Diao E. Scaphoid kinematic behavior after division of the transverse carpal ligament. J Hand Surg Am. 2003;28(2):267–71. https://doi.org/10.1053/jhsu.2003.50055.
25. Joshi A, Patel K, Mohamed A, Oak S, Zhang MH, Hsiung H, Zhang A, Patel UK. Carpal tunnel syndrome: pathophysiology and comprehensive guidelines for clinical evaluation and treatment. Cureus. 2022;14(7):e27053. https://doi.org/10.7759/cureus.27053.
26. Kantarci F, Ustabasioglu FE, Delil S, Olgun DC, Korkmazer B, Dikici AS, Tutar O, Nalbantoglu M, Uzun N, Mihmanli I. Median nerve stiffness measurement by shear wave elastography: a potential sonographic method in the diagnosis of carpal tunnel syndrome. Eur Radiol. 2014;24(2):434–40. https://doi.org/10.1007/s00330-013-3023-7.
27. Kato T, Kuroshima N, Okutsu I, Ninomiya S. Effects of endoscopic release of the transverse carpal ligament on carpal canal volume. J Hand Surg Am. 1994;19(3):416–9. https://doi.org/10.1016/0363-5023(94)90055-8.
28. Kiritsis PG, Kline SC. Biomechanical changes after carpal tunnel release: a cadaveric model for comparing open, endoscopic, and step-cut lengthening techniques. J Hand Surg Am. 1995;20(2):173–80. https://doi.org/10.1016/S0363-5023(05)80003-6.
29. Kotevoglu N, Gülbahce-Saglam S. Ultrasound imaging in diagnosing carpal tunnel syndrome and its relevance to clinical evaluation. Joint Bone Spine. 2005;72(2):142–5. https://doi.org/10.1016/j.jbspin.2004.03.012.
30. Li Z-M, Jordan DB. Carpal tunnel mechanics and its relevance to carpal tunnel syndrome. Hum Mov Sci. 2023;87:103044. https://doi.org/10.1016/j.humov.2022.103044.
31. Lin C-P, Chen I-J, Chang K-V, Wu W-T, Özçakar L. Utility of ultrasound elastography in evaluation of carpal tunnel syndrome: a systematic review and meta-analysis. Ultrasound Med Biol. 2019;45(11):2855–65. https://doi.org/10.1016/j.ultrasmedbio.2019.07.409.
32. Mackinnon SE. Pathophysiology of nerve compression. Hand Clin. 2002;18(2):231–41. https://doi.org/10.1016/s0749-0712(01)00012-9.
33. Nakamichi K, Tachibana S. Distance between the median nerve and ulnar neurovascular bundle: clinical significance with ultrasonographically assisted carpal tunnel release. J Hand Surg Am. 1998;23(5):870–4. https://doi.org/10.1016/s0363-5023(98)80165-2.
34. Okutsu I, Ninomiya S, Hamanaka I, Kuroshima N, Inanami H. Measure pressure in the carpal canal before and after endoscopic management of carpal tunnel syndrome. J Bone Joint Surg Am. 1989;71(5):679–83.
35. Radack DM, Schweitzer ME, Taras J. Carpal tunnel syndrome: are the MR findings a result of population selection bias? AJR Am J Roentgenol. 1997;169(6):1649–53. https://doi.org/10.2214/ajr.169.6.9393185.
36. Richman JA, Gelberman RH, Rydevik BL, Hajek PC, Braun RM, Gylys-Morin VM, Berthoty D. Carpal tunnel syndrome: morphologic changes after release of the transverse carpal ligament. J Hand Surg Am. 1989;14(5):852–7. https://doi.org/10.1016/s0363-5023(89)80089-9.
37. Rydevik B, Lundborg G, Bagge U. Effects of graded compression on intraneural blood blow. An in vivo study on rabbit tibial nerve. J Hand Surg Am. 1981;6(1):3–12. https://doi.org/10.1016/s0363-5023(81)80003-2.
38. Schrier V, Amadio P. The biomechanics of subsynovial connective tissue in health and its role in carpal tunnel syndrome. J Electromyogr Kinesiol. 2018;38:232–9. https://doi.org/10.1016/j.jelekin.2017.10.007.
39. Seradge H, Seradge E. Piso-triquetral pain syndrome after carpal tunnel release. J Hand Surg Am. 1989;14(5):858–62. https://doi.org/10.1016/s0363-5023(89)80090-5.
40. Sigrist RMS, Liau J, Kaffas AE, Chammas MC, Willmann JK. Ultrasound elastography: review of techniques and clinical applications. Theranostics. 2017;7(5):1303–29. https://doi.org/10.7150/thno.18650.
41. Silverstein BA, Fine LJ, Armstrong TJ. Occupational factors and carpal tunnel syndrome. Am J Ind Med. 1987;11(3):343–58. https://doi.org/10.1002/ajim.4700110310.
42. Szabo RM, Chidgey LK. Stress carpal tunnel pressures in patients with carpal tunnel syndrome and normal patients. J Hand Surg Am. 1989;14(4):624–7. https://doi.org/10.1016/0363-5023(89)90178-0.
43. Tat J, Wilson KE, Keir PJ. Pathological changes in the subsynovial connective tissue increase with self-reported carpal tunnel syndrome symptoms. Clin Biomech (Bristol, Avon). 2015;30(4):360–5. https://doi.org/10.1016/j.clinbiomech.2015.02.015.
44. Ugbolue UC, Hsu W-H, Goitz RJ, Li Z-M. Tendon and nerve displacement at the wrist during finger movements. Clin Biomech (Bristol, Avon).

2005;20(1):50–6. https://doi.org/10.1016/j.clinbiomech.2004.08.006.
45. Van Doesburg MHM, Henderson J, Yoshii Y, Mink van der Molen AB, Cha SS, An K-N, Amadio PC. Median nerve deformation in differential finger motions: ultrasonographic comparison of carpal tunnel syndrome patients and healthy controls. J Orthop Res. 2012;30(4):643–8. https://doi.org/10.1002/jor.21562.
46. Viegas SF, Pollard A, Kaminski K. Carpal arch alteration and related clinical status after endoscopic carpal tunnel release. J Hand Surg Am. 1992;17(6):1012–6. https://doi.org/10.1016/s0363-5023(09)91048-6.
47. Wehbé MA, Hunter JM. Flexor tendon gliding in the hand. Part I. In vivo excursions. J Hand Surg Am. 1985;10(4):570–4. https://doi.org/10.1016/s0363-5023(85)80085-x.
48. Weiss ND, Gordon L, Bloom T, So Y, Rempel DM. Position of the wrist associated with the lowest carpal-tunnel pressure: implications for splint design. J Bone Joint Surg Am. 1995;77(11):1695–9. https://doi.org/10.2106/00004623-199511000-00008.
49. Wright TW, Glowczewskie F, Wheeler D, Miller G, Cowin D. Excursion and strain of the median nerve. J Bone Joint Surg Am. 1996;78(12):1897–903. https://doi.org/10.2106/00004623-199612000-00013.
50. Yoshii Y, Zhao C, Amadio PC. Recent advances in ultrasound diagnosis of carpal tunnel syndrome. Diagnostics (Basel). 2020;10(8):596. https://doi.org/10.3390/diagnostics10080596.
51. Zhao C, Ettema AM, Osamura N, Berglund LJ, An K-N, Amadio PC. Gliding characteristics between flexor tendons and surrounding tissues in the carpal tunnel: a biomechanical cadaver study. J Orthop Res. 2007;25(2):185–90. https://doi.org/10.1002/jor.20321.

Nerve Repair and Reconstruction

13

Farid Amirouche and John Alvarez

Introduction

Nerve Injuries in the Hand and Upper Extremity

Peripheral nerve injuries can result from various mechanisms, each with distinct biomechanical implications for repair and reconstruction. First, we examine lacerations, which represent one of the most common mechanisms of nerve injury, typically resulting from sharp objects such as knives, glass, or other cutting instruments. These injuries produce clean transections of nerve tissue that, while damaging, often allow for direct repair when identified early. The biomechanical characteristics of laceration injuries generally preserve the architecture of the nerve ends, facilitating tension-free coaptation when the gap is minimal [1]. In cases of lacerations with nerve transection, the complete discontinuity of the nerve requires surgical intervention for any chance of recovery. The biomechanical challenge in these cases consists of maintaining appropriate tension across the repair site [2, 3].

Crush injuries typically occur from high-energy compression forces, such as those encountered in motor vehicle accidents or accidents with heavy machinery. From a biomechanical perspective, crush injuries present unique challenges as they may sometimes cause axonotmesis, where axons are damaged, but the peripheral nerve connective tissue architecture may remain partially intact. Concerning the classification systems developed by Seddon and expanded by Sunderland [2], crush injuries can range from focal demyelination to more severe damage involving the endoneurium, perineurium, and, in some cases, full-thickness nerve destruction.

Ballistic nerve injuries present some more complex biomechanical challenges in nerve repair. Approximately 30–60% of GSWs to the upper extremity result in nerve injuries requiring repair or reconstruction [1, 3, 4]. The biomechanical mechanism of ballistic injuries is particularly devastating due to the combination of direct tissue destruction along the bullet path and shockwave forces causing damage to surrounding tissues. This combination simultaneously results in transection and crush mechanisms, creating a large zone of injury. One study reported that surgical exploration after ballistic trauma revealed

F. Amirouche (✉)
Department of Orthopaedic Surgery, Orthopaedic and Spine Institute, Endeavor HealthSystem (Formerly Northshore University HealthSystem), an Affiliate of the University of Chicago Pritzker School of Medicine, Skokie, IL, USA

Department of Orthopaedic Surgery, University of Illinois at Chicago, Chicago, IL, USA
e-mail: famirouche@northshore.org; amirouch@uic.edu

J. Alvarez
Department of Orthopaedic Surgery, University of Illinois at Chicago, Chicago, IL, USA

F. Amirouche et al. (eds.), *Biomechanics of the Hand*,
https://doi.org/10.1007/978-3-032-04688-8_13

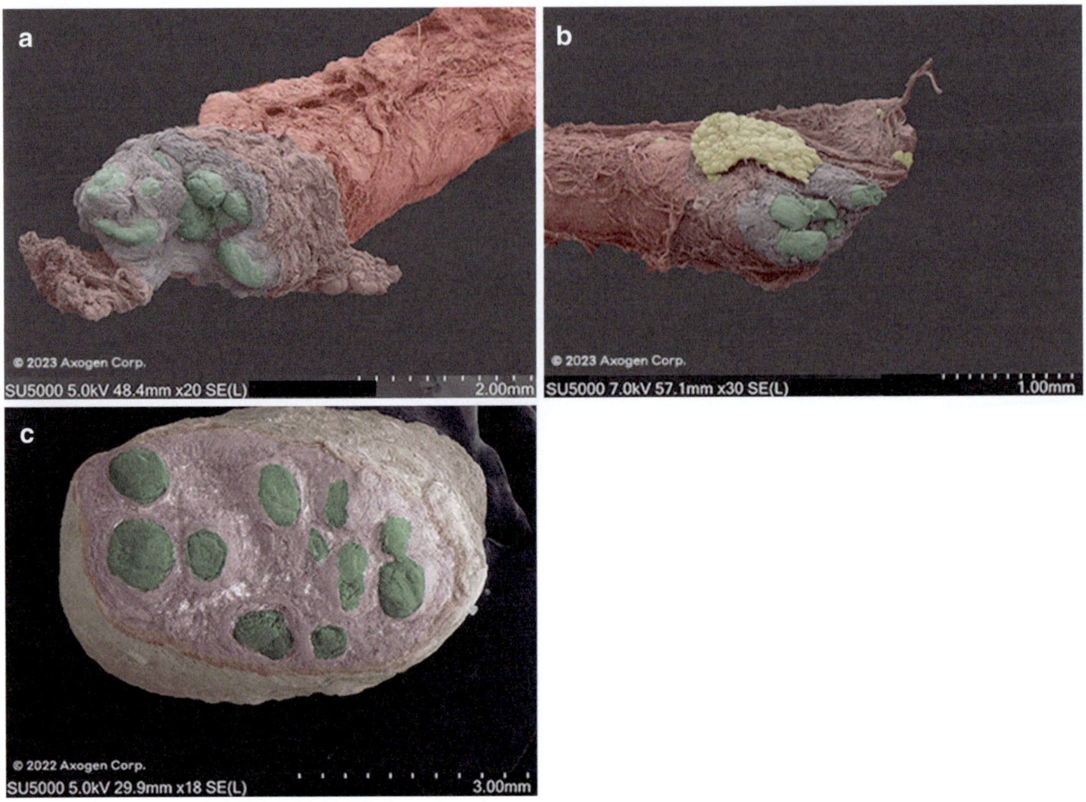

Fig. 13.1 These images show scanning electron microscope (SEM) images of nerves cut by a sharp piece of glass (**a**), and a fresh kitchen knife (**b**), compared to nerves trimmed with a fresh surgical scalpel (**c**), demonstrating the impact that even the sharpest injury mechanisms can have on surrounding nerve tissue. (Reproduced with permission from Axogen, "for small-gap repair, new evidence challenges primary repair as the universal gold standard?", https://www.axogeninc.com/hcp/direct-nerve-repair-alternatives/. No changes were made)

20–55% of nerves classified as requiring repair or grafting [5–7].

Extensive research indicates that proper nerve stabilization followed by transection using a thin, sharp blade is paramount when it comes to nerve procedures. Studies have shown that employing sharp scissors or a no. 11 scalpel blade results in considerably less surface area roughness, indicative of tissue damage, than using a traditional razor blade. Interestingly, while some mechanisms of injury, such as knife or sharp glass injuries, seem to produce a "clean" cut, the evidence reveals that they often fall short in achieving the flat, even nerve ends necessary for optimal nerve regeneration once repaired (Fig. 13.1) [8, 9].

Brief History of Modern Nerve Repair Advancements

Nerve repair has a rich history that dates back centuries, but significant advancements have been made primarily in the twentieth and twenty-first centuries. The evolution of nerve repair techniques could be described as follows:

The foundation for modern nerve repair began in the nineteenth century with recognizing peripheral nerve anatomy and the understanding that nerves could regenerate under certain conditions. Early surgical techniques were rudimentary, primarily focused on suturing damaged nerves without fully understanding their healing processes [10–13]. In the 1930s–1950s, we saw the introduction of the nerve grafting technique, which marked a pivotal moment in nerve repair where surgeons began exploring autologous nerve grafts (using the patient's nerves) to bridge severed nerve gaps, significantly improving surgical outcomes [14]. Microsurgical techniques revolutionized nerve repair, and the development of microscopes and delicate instrumentation allowed for more precise suturing methods, enhancing the ability to repair smaller nerves. It is important to note that the work of surgeons like McKinnon [15–17] and others laid the groundwork for meticulous nerve repair. Progress has been made between the 1980s and the 1990s, when the introduction of biodegradable nerve conduits provided an alternative to autologous nerve grafts. Researchers began experimenting with synthetic materials and tubes to facilitate nerve regeneration in damaged areas where direct suturing was impossible. In the past 20 years, we have seen a surge in studying growth factors, such as nerve growth factor (NGF) and neurotrophic factors, which have the potential to enhance nerve regeneration. Researchers began incorporating these biological factors into nerve repair strategies to improve outcomes [18, 19].

Today, nerve repair is a multidisciplinary arena that integrates biology, materials science, and engineering. Key advancements include biomaterials and modern synthetic conduits made from biocompatible materials that serve as structural supports and promote cell migration and axonal growth. Innovations include conduits impregnated with growth factors or stem cells to enhance regeneration. These scaffolds are made using bioprinting techniques, which allow for transplanting potential solutions to nerve repairs for nerve regeneration and regaining full functionality [20–26].

The decellularization of nerve grafts involves using donor nerve tissue with its cellular components removed. This approach provides a natural scaffold for regeneration while minimizing the risk of immune rejection [27]. Research into the use of stem cells for nerve repair is ongoing. Stem cells can potentially differentiate into Schwann cells (essential for nerve regeneration), providing a biological approach to enhance nerve healing. Techniques using Schwann cell transplants have shown promise in enhancing nerve regeneration and functional recovery [28, 29]. Furthermore, we see the potential of nerve repair depicted by some researchers working to optimize the delivery of neurotrophic factors to injury sites. This involves techniques such as local injections or engineered delivery systems that use nanotechnology to release these factors over a controlled period [30, 31].

The engineering of the nerve also uses pulsed electrical stimulation, which is integrated into some repair strategies to enhance regeneration, as electricity can promote cell signaling and support nerve regrowth [32]. Tissue engineering and the advent of 3D bioprinting open new horizons for creating customized nerve scaffolds that can accurately mimic nerve anatomy. Robotics and technology integration will be crucial in computer-assisted surgical techniques, allowing for even greater precision in nerve repair operations, improving nerve alignment and suturing accuracy [33, 34].

Principles of Nerve Biomechanics

Repair Distances and Tension

The tension-free nerve repair principle has long been considered fundamental for optimal outcomes. When injuries result in gapping of the proximal and distal ends, nerve grafting has been utilized to avoid repairs with tension present. Early studies demonstrated that nerve tissue elongation compromised blood flow within the vasa nervorum. As a result, local ischemia and fibrotic changes may occur at the repair site, greatly hindering regenerative probabilities [10]. Gapping may occur in cases where crush injuries have caused direct peripheral nerve tissue damage along its length, but it may also result from transection. For example, nerve ends retract from each other after transection due to the low level of constant longitudinal tension [35, 36] (Fig. 13.2).

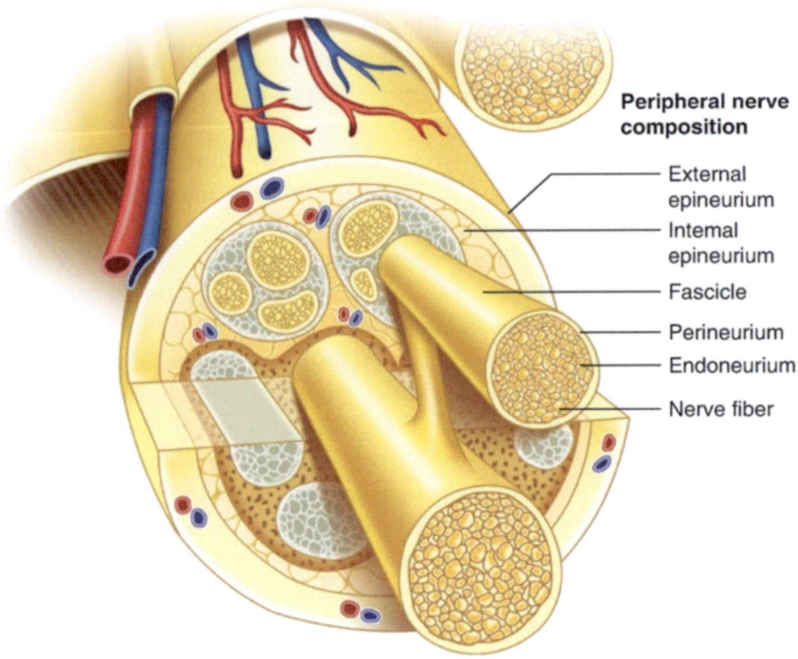

Fig. 13.2 Peripheral nerve composition. (Reproduced from Lanier et al [1] with permission)

Strategies for mitigating tension along the repair site are numerous and perhaps speak to the importance of this nerve repair principle. Tension redistribution devices, such as nerve tubes, have been developed to enable end-to-end repair of more significant nerve gaps and work by distributing the tension toward more structurally sound tissue away from the injury site. Neurolysis and mobilization, a technique in which nerve tissue is released from adjacent tissue, is another approach to mitigate tension. This is done by releasing the proximal and distal stumps and mobilizing them to enable increased freedom of movement. Nerve transposition, another technique fundamentally concerned with mobilization, can also be employed when strategizing tension reduction and can be particularly helpful with more significant gap defects. Given that nerves experience increased tensile forces at varying positions of joint flexion and extension, another method of tension reduction at the repair site has been fixing the joint in flexion for a given period to facilitate regeneration. Then, the joint is progressively splinted in decreasing flexion. This approach takes advantage of the viscoelastic properties of peripheral nerves, with stress relaxation occurring under sustained loading.

The comprehensive repair of a peripheral nerve transection injury is the gold standard in neurological recovery. Timely intervention is crucial, as delayed repair, ineffective nerve debridement, and premature functional mobilization can exacerbate tension at the repair site, hindering the vital process of axon regeneration. Prompt diagnosis and referral to a specialist are essential to optimize outcomes, allowing for meticulous exploration that can significantly mitigate nerve retraction, minimize debridement, and reduce the overall gap size. An organized, standardized approach to management can yield substantial benefits for society as a whole.

Nonetheless, early exploration presents challenges, particularly in accurately defining the extent of the injury zone and assessing the adequacy of nerve debridement. Innovative techniques such as "sutureless" nerve approximation utilizing conduits, the interposition of autologous grafts, or the application of processed nerve allografts can effectively alleviate repair site tension. In some instances, avoiding sutures through interposed de-tensioning grafts and applying tissue adhesives may enhance the repair process.

However, it is important to note that a significant gap within a conduit may fail to support effective regeneration. Grafts are often characterized by two neurorrhaphy sites, presenting additional hurdles for axons in their journey toward reestablishing functional connections.

The phenomenon of nerve gap arises following transection, primarily due to the disruption of biotensegrity—an essential principle that posits all biological tissues contain longitudinal fibers that are naturally pre-tensioned. When these crucial structures are severed, the corresponding longitudinal fibers lose their inherent pre-tension, leading to the retraction of the cut ends. Should there be a delay in the repair process, the nerve experiences an increase in its modulus of elasticity, resulting in heightened tension forces acting on the closure of the gap. This is partly because the pre-tensioning within the nerve has been restored during this time. Furthermore, delays can give rise to scar formation at the proximal and distal ends of the nerve, creating a significant barrier to effective neural regeneration.

To facilitate healing, the nerve ends must undergo debridement, which unfortunately enlarges the injury gap, requiring reconstruction. When direct suturing is employed, it often generates tension at the repair site, with stress concentrations occurring at the interface between the suture and the nerve. This localized strain can lead to ischemic conditions, inflammation, and subsequent fibrosis. Moreover, if the suture fails under load, it can extend the potential zone of injury and scar tissue formation, potentially culminating in a neuroma-in-continuity at the repair site. Even in the absence of significant delays or loss of nerve tissue, there remains an intrinsic stress concentration at the repair site due to the compromised integrity of the nerve itself. Thus, creating a conducive environment for axon regeneration is paramount in alleviating intraneural scarring.

While direct end-to-end tension-free repair of an injured nerve is widely considered the gold standard in surgical practice, inconsistent clinical outcomes often necessitate alternative approaches. The process of axon regeneration is inherently slow, and regenerating axons tend to face considerable challenges in traversing the disrupted and disorganized intraneural architecture and any scar tissue present within the trauma and repair zone. Optimizing the conditions for regeneration involves meticulous debridement of damaged nerve segments, ensuring the precise alignment of fascicles, and taking measures to minimize scar formation at the repair location. When the nerve gap is too extensive for direct repair without introducing tension, the interposition of a nerve graft can serve as a viable option to alleviate that tension. However, it is crucial to note that the regenerating axons must successfully navigate not one but two coaptation sites, which adds complexity to the repair process.

Extrinsic mechanical forces cause functional changes in a peripheral nerve. A peripheral nerve's stress–strain relationship may be nonlinear, meaning that after a threshold value, the strain on a nerve causes the intrafascicular environment to change rapidly [37]. Various peripheral nerves may demonstrate a characteristic modulus of elasticity based on several factors, including their local environment, size, and mechanical properties (Fig. 13.3). However, after an initial loading (toe) region, nerves demonstrate an inverse relationship between stress and strain that has implications on a cellular level during periods of stretch, tension, and injury [38].

Materials for Optimal Nerve Gliding

The biomechanical microenvironment must support proper nerve gliding. As nerves regenerate, they must have a certain freedom of movement to accommodate joint movements. For this reason, selecting materials utilized in nerve repair represents a vital area for proper nerve reconstruction efforts. Assessing adequate nerve gliding after reconstruction ensures optimal recovery and functionality. Two effective methods for this assessment are ultrasound technology and clini-

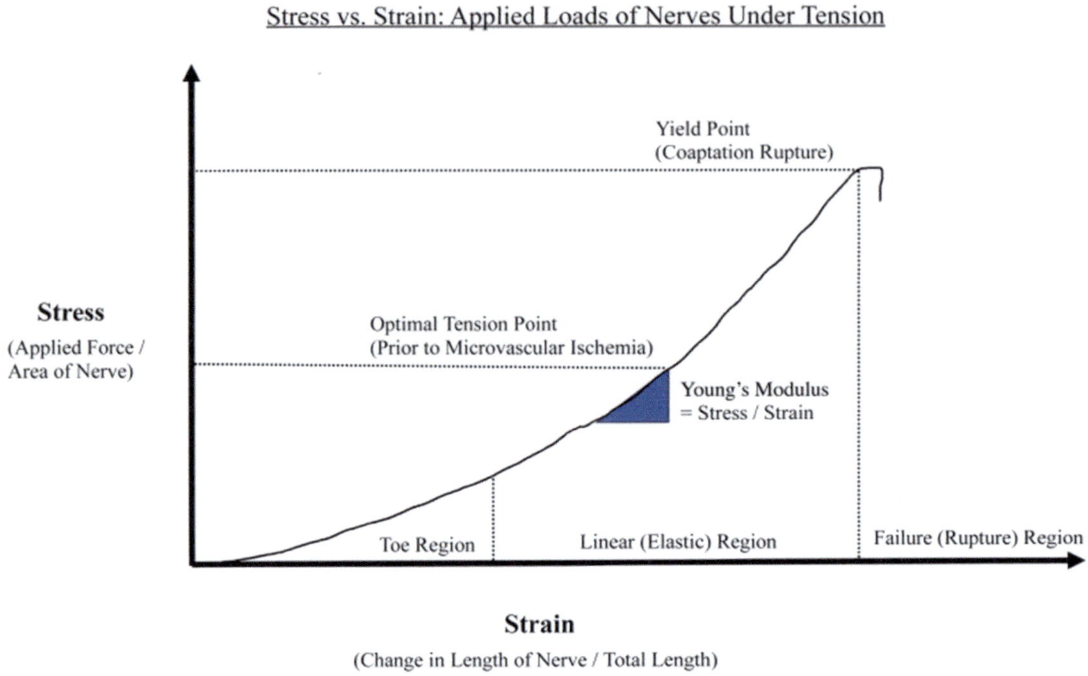

Fig. 13.3 Stress versus strain: applied loads of nerves under tension. (Reproduced from McQuillan et al. [4] under the Creative Commons Attribution 4.0 International License (CC BY 4.0). No changes have been made)

cal provocation tests. Ultrasound imaging allows clinicians to visualize nerve structures in real time, providing information about their mobility and surrounding soft tissues.

It can be applied as follows:

1. Dynamic Ultrasound Assessment: By performing dynamic ultrasound examinations, clinicians can observe the movement of the nerve during specific motions. For example, moving the wrist or elbow can help visualize the glide of the median or ulnar nerve. An absence of standard glide might indicate adhesions or inefficiencies in movement post-reconstruction.
2. Nerve Cross-sectional Area Measurement: Measuring the cross-sectional area of the nerve before and after reconstruction can indicate changes in swelling or fibrosis. A reduced cross-sectional area post-reconstruction, ideally returning to at least baseline levels, can suggest improved nerve gliding. This technique is beneficial in evaluating outcomes after surgical interventions.
3. Comparative Studies: Using ultrasound to compare nerve gliding in reconstructed versus nonreconstructed limbs can provide insight into the effectiveness of the surgical repair. For instance, a study examining the gliding of the radial nerve before and after a surgical intervention can help understand the reconstruction's success.

Clinical Provocation Tests:

Provocation tests assess how well a nerve can glide through its anatomical pathways under stress. Some examples include

1. Tinel's Sign: This test involves tapping over the nerve to elicit tingling sensations in the distribution area. Post-reconstruction, the absence or reduction of symptoms may indicate improved nerve gliding.
2. Phalen's Test: Involves holding the wrist in maximum flexion for 60 s to see if symptoms of paresthesia appear. A decrease in symptoms post-reconstruction would indicate improved nerve mobility.

3. Neurodynamic Testing: Specific tests, such as the straight leg raise or slump test, can assess nerve tension and glide. If a patient demonstrates improved range and reduced discomfort in these tests post-surgery, it can indicate adequate nerve gliding.
4. Functional Tests: Assessing the patient's ability to perform tasks that require nerve function (like fine motor skills or grip strength tests) can also provide practical insights into nerve gliding efficiency.

By utilizing these methods—ultrasound for detailed visualization and clinical tests for functional assessment—clinicians can effectively gauge the status of nerve gliding post-reconstruction. This comprehensive approach helps inform rehabilitation strategies and optimize patient outcomes.

Adhesions and tethering can significantly impair nerve regeneration by creating physical barriers that inhibit the growth and repair of nerve tissues. When nerves are injured, the natural healing process involves the regeneration of axons, the long projections of nerve cells that transmit signals. Adhesions can form from inflammation or surgery, leading to scar tissue development. This scar tissue can bind nerves to surrounding tissues or organs, restricting their movement and leading to tension that may impede regeneration.

Tethering can also affect blood flow and nutrient delivery to the injured area, which are crucial for healing. The presence of adhesions can cause pain and limit the range of motion, further complicating rehabilitation efforts. Both adhesions and tethering can create a hostile environment for nerve regeneration, leading to incomplete recovery and chronic complications. Addressing these issues early in treatment is essential for promoting optimal healing and functional recovery.

Stiffness Balance

The stiffness of a repair construct is another important biomechanical consideration. In an ideal scenario, materials used in nerve repair should match the stiffness of the native nerve. Too stiff materials increase the risk of compressing regenerated tissue at the repair site, and those that are too compliant increase the risk of insufficient support to the delicate, injured nerve tissue [39].

While balancing stiffness can improve the chances of successful repair, the variation of mechanical properties among peripheral nerves depending on age, species, and anatomical location can present a considerable challenge in this pursuit.

Differences in nerve mechanics among species can significantly impact the applicability of research developments using animal models to human medicine. Each species has evolved adaptations that influence its nervous system's structure, function, and regeneration capabilities. These variations can include differences in the myelination of nerve fibers, the composition of the extracellular matrix, and neuronal connectivity.

For instance, rodents are commonly used in neurological research due to their rapid breeding and genetic similarity to humans. However, the faster nerve conduction velocity and different synaptic structures in rodents compared to humans can lead to disparities in how diseases or treatments manifest. Additionally, the intrinsic properties of nerve cells, such as ion channel expression, can differ, which may result in different pharmacological responses to treatments. Hence, while initial findings in animal models may show promise, translating these results to human applications can be challenging due to these fundamental differences.

Anatomical locations also play a crucial role in how nerves function and respond to injury or disease. For example, the sciatic nerve, the largest nerve in the human body, runs from the lower back down the leg and is responsible for many functions, including motor control and sensation of the lower extremities. Its structure, which includes a thick layer of myelin and a relatively large diameter, allows for rapid transmission of nerve impulses. In contrast, the ulnar nerve, which runs along the inner side of the arm to the hand, has a different anatomical course and may be more susceptible to compression injuries, such as cubital tunnel syndrome.

The mechanical properties associated with different nerve locations can also vary, with some nerves being more resilient to injury than others. For example, nerves in the upper limbs, like the ulnar nerve, may be more prone to damage due to the proximity to bony structures and less surrounding protective tissue than deeper nerves like the sciatic. Additionally, regeneration capabilities can be quite different; At the same time, the peripheral nerves (like the sciatic) have a greater capacity for regeneration due to their supportive cellular environment; other nerves, such as those in the central nervous system, demonstrate much more limited recovery after injury.

These differences highlight the importance of context when interpreting research from animal models. Understanding human nerves' unique mechanical and anatomical characteristics compared to animal models is essential for developing effective treatments and therapies that can be successfully translated into clinical application for human patients.

The Different Nerve Repair Constructs

Autografts and Allografts

Biomechanical Strengths of Autografts vs. Allografts

Choosing between autografts and allografts in nerve repair surgeries is critical as it influences the biomechanical outcomes and the broader clinical implications for patients. This discussion will explore key factors, including regeneration quality, structural quality, stiffness, immune reactions, and donor site morbidity associated with each graft type.

Regeneration Quality

Autografts represent the established standard in nerve repair, attributed to their superior regenerative properties. These grafts utilize the patient's nerve tissue, maintaining essential structural elements, including the extracellular matrix (ECM) and Schwann cells. The presence of Schwann cells is particularly significant, as these cells facilitate axonal regeneration and contribute to the remyelination process, which is vital for achieving functional recovery. Clinical studies demonstrate that autografts are associated with higher rates of successful nerve regeneration, with restored sensory and motor functions recorded more frequently compared to allografts. In contrast, while allografts serve as a viable alternative, their regenerative capacity may be compromised due to the processing necessary for tissue preservation. This processing can lead to a reduction in the availability of critical cellular components and ECM constituents that are integral to effective nerve repair.

Structural Quality

The structural integrity of a graft is vital for maintaining its function after implantation. Autografts retain their natural ECM and structural architecture, providing a robust scaffold for nerve regeneration. Their biomechanical properties closely mimic those of the native nerve, facilitating better integration and load-bearing capabilities.

In contrast, the structural quality of allografts can be compromised by the decellularization process used to remove donor cells. This treatment can alter the graft's physical properties, potentially leading to weaker scaffolds that may not support successful nerve growth or withstand mechanical stresses as effectively as autografts.

Stiffness of Grafts

Stiffness is critical in a graft's ability to function biomechanically when implanted. Autografts usually possess stiffness characteristics similar to those of the original nerve, which helps maintain mechanical stability and reduces the risk of graft failure during the healing process. This compatibility in mechanical properties aids in minimizing stress concentration at the repair site, thus supporting a smoother healing process.

On the other hand, allografts, with their potential variability in stiffness due to processing techniques, can introduce a layer of complexity. This inconsistency can lead to mismatches in mechanical properties compared to the surrounding tissue, potentially complicating the healing process and affecting the overall functional recovery. This underscores the need for careful consideration and planning when using allografts in nerve repair.

Immune Reaction to Foreign Nerve Tissue

The immune response plays a crucial role when considering the use of allografts. Since these grafts introduce foreign materials into the recipient's body, they can trigger an immune reaction. This may lead to inflammation and, in some cases, graft rejection. Such reactions not only hinder the healing process but can also complicate the patient's postoperative care.

Autografts derived from the patient's body are inherently biocompatible, eliminating the risk of an undesirable immune response. This aspect is critical in enhancing clinical outcomes and reducing graft failure or prolonged recovery complications.

Donor Site Morbidity and Pain with Autografts

While autografts offer numerous advantages, they are not without drawbacks. Harvesting donor tissue can lead to significant morbidity, including pain, sensory loss, and potential complications at the donor site. Patients may experience long-term discomfort or functional impairment in the area from which the graft was taken, raising essential considerations for overall treatment strategies.

Allografts, on the other hand, do not involve the same level of morbidity associated with donor site issues. They offer a compelling alternative for patients, especially in cases where preserving the surrounding nerve tissue is vital for post-surgical function.

Both autografts and allografts present unique biomechanical strengths and considerations in nerve repair. Autografts generally excel in regeneration potential, structural quality, and biocompatibility, whereas allografts offer advantages in avoiding donor site complications. Clinicians must weigh these factors carefully, considering the individual patient's circumstances, to decide on the most appropriate graft type for optimal healing and functional recovery.

Nerve Guidance Conduits

Nerve guidance conduits provide certain advantages. For example, longitudinal strength is improved because the suture–conduit interface is stronger than the suture–nerve interface. This can be useful when completely tensionless repairs are not feasible.

Nerve guidance conduits are innovative devices used to repair peripheral nerve injuries, particularly when traditional direct repair is not feasible due to significant gaps between the nerve ends. These conduits offer several advantages that enhance the overall outcomes of nerve repair, particularly in structural integrity and biomechanical performance.

One of the main advantages of nerve guidance conduits is their ability to enhance the longitudinal strength at the repair site. When a nerve is sutured back together after an injury, the strength of the connection heavily relies on the interface between the suture material and the nerve tissue. However, this suture–nerve interface can often be weak, resulting in inadequate stability, especially when tension is applied. In contrast, the interface between the suture and the conduit—known as the suture–conduit interface—generally provides a stronger connection. This increased strength is primarily due to the material properties of the conduits, which are designed to distribute forces better and accommodate the placement of sutures.

In cases where achieving a completely tensionless repair is not possible due to the extent of the injury or the retraction of nerve ends, nerve guidance conduits become particularly valuable. By utilizing these conduits, surgeons can create a favorable environment that encourages nerve growth and regeneration while minimizing tension at the repair site. The enhanced strength of the suture–conduit interface means that even if some tension is present, the risk of failure at the sutured connection is reduced.

Nerve guidance conduits are typically designed to encourage cellular infiltration and support the regrowth of axons along the conduit. This design helps bridge the gap created by an injury. Additionally, the enhanced longitudinal strength of the conduits allows them to maintain structural integrity while supporting the healing process. This provides a scaffold that withstands mechanical stresses and promotes biological healing mechanisms.

Using nerve guidance conduits enhances the biomechanics of nerve repair by strengthening the connection between sutures and the repair construct. This is especially beneficial when a completely tension-free repair cannot be achieved. This structural advantage can lead to better functional recovery and increased prospects for successful nerve regeneration.

Furthermore, we have nerve guidance conduits (NGCs), which are medical devices used to support the regeneration of damaged or injured nerves. The effectiveness of these conduits heavily depends on the materials used in their construction. Various materials can be employed, each offering distinct biomechanical properties influencing the conduit's performance. A breakdown of the key properties includes tensile strength, stiffness, and elasticity, and why they are essential.

1. Tensile Strength: This property measures how much force a material can withstand when stretched or pulled before breaking. For NGCs, a high tensile strength is crucial because the conduit needs to remain intact when subjected to the mechanical forces that occur during the body's natural movements. If the conduit fails under tension, it can disrupt the healing process by allowing nerve fibers to retract or misalign. Materials such as polylactic acid (PLA) and certain types of polyurethanes are often chosen for their favorable tensile strength, ensuring that the conduit can endure physical stresses without compromising the regeneration of the nerve.
2. Stiffness refers to a material's resistance to deformation under an applied load. In the context of NGCs, the material's stiffness should balance support and flexibility. A too stiff conduit may not allow the nerve fibers to grow correctly, as it could restrict their natural movement and adaptability to surrounding tissues. On the other hand, a material that is too flexible might not provide sufficient support. Polycaprolactone (PCL), for example, has an ideal stiffness level that promotes nerve healing while still permitting some movement, which is essential for dynamic biological environments.
3. Elasticity: Elasticity is the ability of a material to return to its original shape after being stretched or compressed. This property is significant for NGCs because nerves undergo various mechanical forces as the body moves. If a conduit can stretch and then revert to its initial shape, it can accommodate the movements of surrounding tissues without hindering the nerve regeneration process. Silicone and specific elastomers are often utilized in NGCs for their excellent elastic properties, allowing them to flex and adapt while providing a supportive environment for nerve growth.

Nanotechnology-based therapies that previously seemed unreachable are now being considered to repair and reconstruct PNIs. They have the power to deliver bioactive molecules in a controlled manner, tune cellular behavior, and ultimately guide tissue regeneration effectively. Nanotechnology also offers opportunities in the imaging field, with a precision never achieved before, which is helpful for diagnosis, surgery, and patient follow-up (Fig. 13.4).

The choice of materials for nerve guidance conduits is critical to their function. Each material possesses unique biomechanical advantages that can enhance or hinder nerve regeneration. Researchers and clinicians are often challenged to design effective NGCs that promote successful nerve healing in various clinical scenarios by optimizing attributes like tensile strength, stiffness, and elasticity.

Autologous nerve transplantation is widely regarded as the most dependable clinical intervention for peripheral nerve injury (PNI) [40,

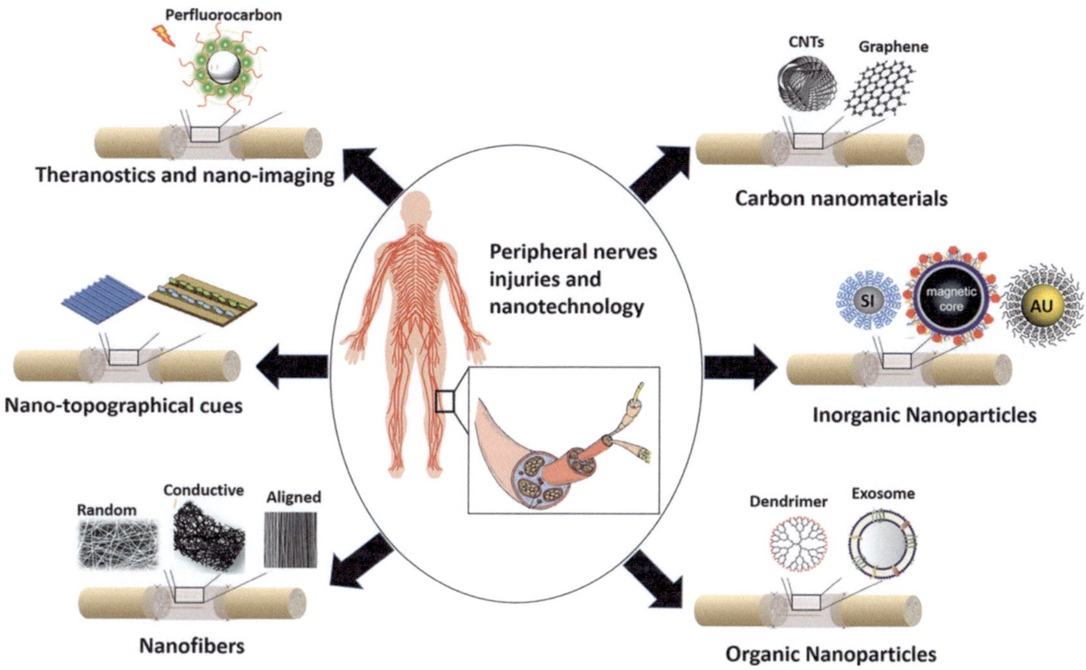

Fig. 13.4 Nanotechnology-based therapies. (Reproduced from Carvalho et al. [31] with permission)

41]. One of the notable advantages of this approach is its inherent capacity for revascularization, a critical process that supports the regeneration of damaged tissues [42]. Revascularization plays a pivotal role by alleviating oxygen deprivation at the injury site, thereby enhancing the supply of essential nutrients and cellular components that facilitate nerve repair and regeneration [43].

Despite its numerous benefits, the application of autologous nerve transplantation is not without significant challenges and unavoidable drawbacks. Among these are the potential damage inflicted on the donor site, the often limited availability of suitable donor nerves, and the risk of incompatibility between the donor and recipient nerves [44, 45]. These factors underscore the complexity of this treatment method and highlight the need for continued research and innovation in the field of nerve repair.

Tissue-engineered grafts have recently been a hot research topic for treating PNI. NGCs synthesizing various cues have received increasing attention and extensive studies (Fig. 13.5).

In recent years, the remarkable capabilities of nerve guidance conduits (NGCs) in repairing peripheral nerve injuries (PNI) have garnered increasing attention, positioning them as the potential gold standard to supplant traditional autologous nerve grafts. This review critically examines the repair effects of various NGCs, each designed with distinct strategies, through the lenses of mechanical cues, biochemical cues, and the role of supporting cells.

The NGCs analyzed incorporate a range of mechanical configurations, including hollow conduits, multi-channel designs, grooved structures, and fiber-filled variants. Each of these designs, while innovative, presents specific limitations. For instance, hollow NGCs often lead to the regeneration of nerve axons that are scattered and poorly aligned, which significantly diminishes their ability to bridge the gap to target tissues effectively.

The production methods for multi-channel and grooved NGCs are notably intricate, involving complex manufacturing processes that may pose challenges in scalability and application. On

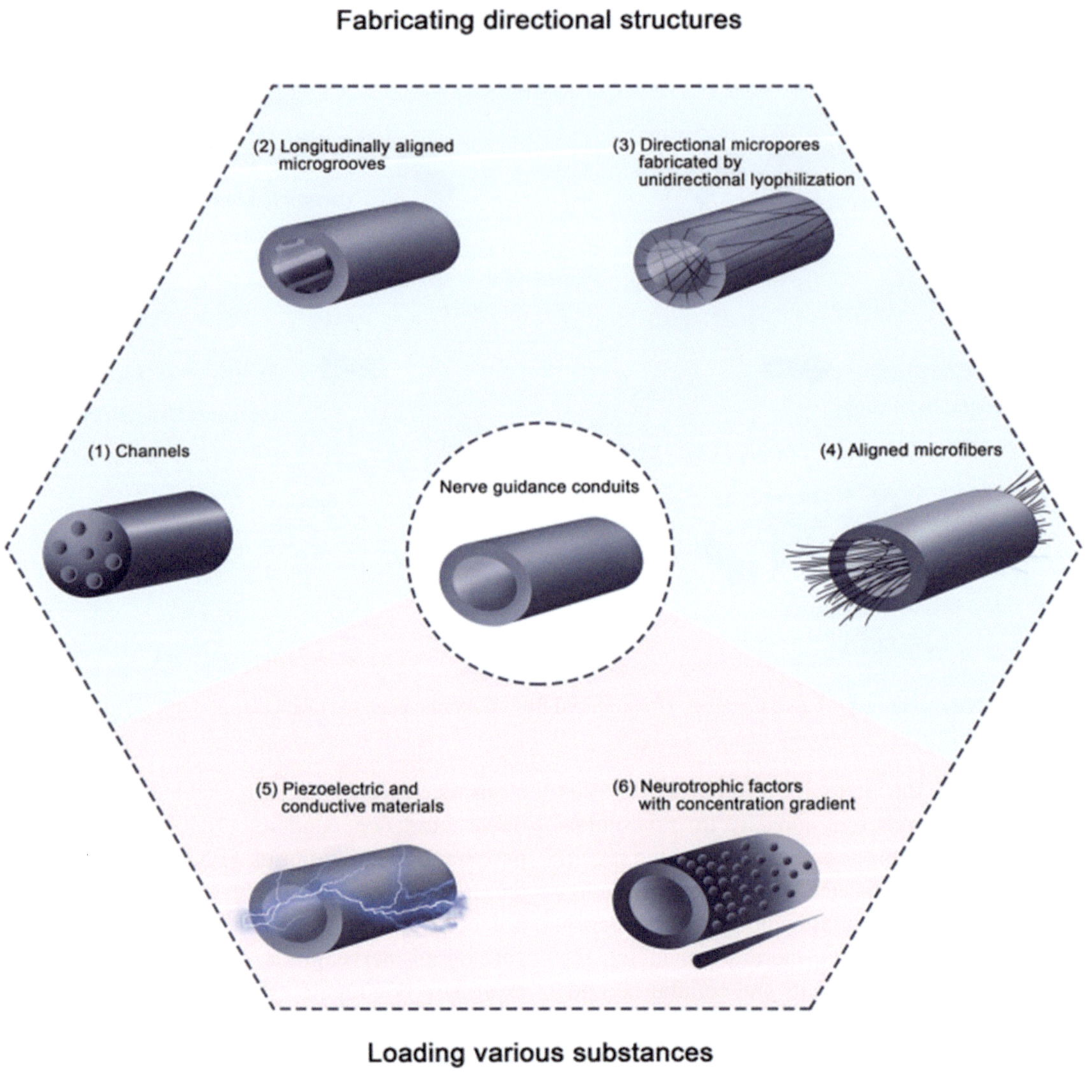

Fig. 13.5 Schematic representation of main strategies to enhance the ability of NGCs to promote the directional growth of nerves. Depicted strategies include fabricating directional structures and loading various substances within NGCs. (Reproduced from Zhang et al. [46] under the Creative Commons Attribution 4.0 International License (CC BY 4.0). No changes have been made)

the other hand, fiber-filling NGCs highlight the need for advancements in materials science, as they require the creation of biodegradable fibers that can dissolve appropriately, thus eliminating the need for cumbersome secondary surgeries to remove the conduit after nerve regeneration.

As we look to the future, exploring and innovating new materials or guiding fillers that can more effectively facilitate nerve growth and ensure optimal healing outcomes is essential. There is a pressing demand for research that delves into the design of NGCs tailored to meet these critical needs, potentially revolutionizing how we approach nerve repair.

Future Directions in Nerve Repair and Reconstruction: Topological and Piezoelectric Scaffolds

Topological scaffolds are part of neural tissue engineering efforts to improve nerve regeneration after injury by recreating the natural extracellular matrix structure.

Topological scaffolds are crucial in neural tissue engineering, particularly for improving nerve regeneration after injury. By recreating the structural and functional characteristics of the natural extracellular matrix (ECM), these scaffolds aim to provide an optimal environment for nerve repair and regrowth.

Concept of Topological Scaffolds

Topological scaffolds are engineered materials designed to mimic the natural architecture of nerves and the surrounding ECM. This architecture includes specific topographical features, such as gradients, channels, and orientations that facilitate cellular growth and communication. The design often considers the anisotropic nature of nerves, which grow in a specific direction and require guidance for effective regeneration.

Importance of Mimicking the Extracellular Matrix

The natural ECM provides several critical functions that support cell survival, proliferation, and differentiation. By replicating the ECM's physical and biochemical properties, topological scaffolds enhance these biological processes:

1. Cell Adhesion and Migration: The scaffold's surface characteristics are designed to promote the adhesion and migration of Schwann cells (the glial cells responsible for supporting nerve regeneration) along the scaffold. This is crucial for bridging gaps in damaged nerves.
2. Guidance for Axonal Growth: The scaffold's orientation and structure can direct axons' growth, helping them reach their target tissues. Patterns that mimic the alignment of natural nerve fibers can provide the necessary guidance for directional growth.
3. Nutrient Support: Scaffolds can be designed with porosity and channeling, allowing for the diffusion of nutrients and oxygen, essential for cellular health and nerve regeneration.
4. Biochemical Cues: Topological scaffolds can be functionalized with growth factors or other bioactive molecules that promote cellular activities necessary for regeneration, such as neurotrophic factors that stimulate nerve growth.

Types of Topological Designs

1. Aligned Fibers: Scaffolds made from aligned nanofibers or microfibers can closely resemble the natural organization of nerves, promoting directional nerve growth.
2. Porous Structures: Porous scaffolds with controlled pore sizes can facilitate cell infiltration, provide a space for new tissue formation, and allow for nutrient and waste exchange.
3. Gradient Scaffolds: These scaffolds possess varying properties (e.g., stiffness, biochemical composition) along their length, which can further influence cell behavior and the regeneration process.
4. 3D Printed Scaffolds: Advancements in 3D printing technology allow for the creation of highly customized scaffolds that precisely match the topological features of the nerves they aim to repair.

While topological scaffolds show promise, several challenges remain. These include ensuring biocompatibility, long-term stability, and integration with host tissues. Ongoing research is focused on optimizing scaffold designs, exploring the incorporation of bioactive molecules, and understanding the complex interactions between the scaffold and the cellular environment.

Topological scaffolds are a promising advancement in neural tissue engineering. By closely mimicking the natural extracellular matrix, they aim to enhance nerve regeneration after injury, providing new avenues for effective treatment strategies in peripheral nerve injuries.

Piezoelectric materials are gaining attention as a potential innovative approach to nerve repair. Recent research underscores the significant role that electrical currents play in facilitating tissue repair processes. By harnessing the unique properties of piezoelectric materials, which can convert mechanical stress into electrical energy, we can enhance nerve regeneration when these materials are integrated into specially designed nerve tubes placed around injury sites.

Integrating piezoelectric components into nerve repair methods could enhance recovery and foster a supportive environment for nerve growth and functional restoration. This strategy offers potential for improving the efficacy of nerve repair technologies, similar to the way topological scaffolds enhance structural stability and guide cellular behavior. As we further investigate these materials, their possible use in clinical applications might signify a significant advancement in tackling nerve injuries and conditions.

Electrostimulation, magnetic stimulation, laser therapy, and the application of growth factors are the main strategies used to enhance nerve repair and regeneration. These innovative nonsurgical techniques offer promising possibilities for promoting limited regeneration in cases of short-gap injuries in peripheral nerves, providing hope for better recovery outcomes. However, critically evaluating the potential drawbacks associated with these methods is essential. For instance, selecting the appropriate frequency ranges and durations for electrostimulation is crucial; improper use, characterized by high frequencies or prolonged stimulation, may inadvertently worsen nerve injuries [47–58].

Numerous studies have highlighted the various applications of magnetic stimulation in promoting nerve regeneration [59–61], demonstrating its ability to support axonal growth and aid functional recovery. However, some research raises questions about the effectiveness of variable magnetic fields in nerve repair [62]. A notable study by Walker et al. involved administering three different modes of magnetic stimulation to rats with sciatic nerve injuries. Unfortunately, the results showed no significant improvement in functional recovery among the different treatment methods [63].

Current research continues to push the boundaries, aiming to enhance functional recovery through innovative materials, biological therapies, and advanced surgical tools, including AI. These advancements offer hope for patients with nerve injuries and have the potential to revolutionize the standards of care in reconstructive nerve surgery. As research progresses, integrating technology and biology promises even more exciting developments in the near future.

Conclusion

The intricacies of peripheral nerve injuries affecting the hand and upper limbs pose considerable difficulties for clinicians and patients. This chapter has emphasized the diverse characteristics of nerve damage, ranging from lacerations and crush injuries to ballistic trauma. Each has distinct biomechanical considerations that should be considered when devising repair approaches. A thorough comprehension of these mechanisms guides surgical methods and improves our capacity to forecast results and customize treatments to meet individual patient requirements.

The historical perspective on nerve repair advancements reveals a trajectory of innovation driven by scientific inquiry and clinical necessity. Significant progress has been made in the field of nerve repair, moving from rudimentary techniques to sophisticated methodologies that leverage biological principles and advanced materials. The exploration of nerve grafts, conduits, and cutting-edge technologies such as piezoelectric scaffolds underscore the dynamic nature of this field. These innovations open avenues for improved healing by facilitating the regeneration of nerve tissue and promoting reconnection across injury sites.

The biomechanical principles discussed, particularly those relating to repair distances and tension management, emphasize the importance of achieving ideal conditions for nerve regeneration. The evidence indicates that tension-free repair is paramount to successful outcomes; thus, strategies that mitigate tension at the repair site must remain a priority in clinical practice. As we continue to refine our approaches to nerve repair, it is essential to integrate these principles with advancements in surgical technology and biomaterials.

Future research should further elucidate the underlying nerve repair and regeneration mechanisms, especially in complex injuries resulting from ballistic trauma. Investigating the roles of cellular responses, biomechanical forces, and innovative engineering designs will pave the way for novel therapies that enhance functional recovery.

The quest for improved nerve repair outcomes continues. The evolving landscape of nerve repair reflects the potential for enhanced clinical practices.

References

1. Lanier ST, Brogan DM. Nerve compression, nerve injury, and nerve regeneration: an overview. In: Dy CJ, Brogan DM, Wagner ER, eds. Peripheral nerve issues after orthopedic surgery. Springer International Publishing; 2022:3–26. https://doi.org/10.1007/978-3-030-84428-8_1
2. Davis GA. Reflections on the history of nerve repair—Sir Sydney Sunderland's final presentation to the Neurosurgical Society of Australasia. Neurosurgery. 2020;87(3):E373–82. https://doi.org/10.1093/neuros/nyaa059.
3. Peltonen S, Alanne M, Peltonen J. Barriers of the peripheral nerve. Tissue Barriers. 2013;1(3):e24956. https://doi.org/10.4161/tisb.24956.
4. McQuillan TJ, Zelenski NA, Ghareeb PA. The role of nerve tension on nerve repair success. J Hand Surg Glob Online. 2024;6(5):691–3. https://doi.org/10.1016/j.jhsg.2024.01.018.
5. Stepan JG. Ballistic nerve injuries: state of the evidence and approach to the patient based on experience. J Hand Surg Glob Online. 2024;6(5):743–8. https://doi.org/10.1016/j.jhsg.2024.01.021.
6. Straszewski AJ, Schultz K, Dickherber JL, Dahm JS, Wolf JM, Strelzow JA. Gunshot-related upper extremity nerve injuries at a level 1 trauma center. J Hand Surg. 2022;47(1):88.e1–6. https://doi.org/10.1016/j.jhsa.2021.03.020.
7. Pannell WC, Heckmann N, Alluri RK, Sivasundaram L, Stevanovic M, Ghiassi A. Predictors of nerve injury after gunshot wounds to the upper extremity. Hand. 2017;12(5):501–6. https://doi.org/10.1177/1558944716675294.
8. Tang JB, Giddins G, Omokawa S, Boeckstyns MEH, Tay SC, Giesen T. Common hand problems with different treatments in countries in Asia and Europe. Hand Clin. 2017;33(3):561–9. https://doi.org/10.1016/j.hcl.2017.04.010.
9. Deutinger M, Girsch W, Burggasser G, et al. Peripheral nerve repair in the hand with and without motor sensory differentiation. J Hand Surg. 1993;18(3):426–32. https://doi.org/10.1016/0363-5023(93)90085-H.
10. Sunderland S. A classification of peripheral nerve injuries producing loss of function. Brain. 1951;74(4):491–516. https://doi.org/10.1093/brain/74.4.491.
11. Seddon HJ. Three types of nerve injury. Brain. 1943;66(4):237–88. https://doi.org/10.1093/brain/66.4.237.
12. Naff NJ, Ecklund JM. History of peripheral nerve surgery techniques. Neurosurg Clin N Am. 2001;12(1):197–209. x
13. Friedman AH, Elias WJ, Midha R. Introduction: peripheral nerve surgery--biology, entrapment, and injuries. Neurosurg Focus. 2009;26(2):E1. https://doi.org/10.3171/FOC.2009.26.2.E1.
14. Kim DH, Kline DG, editors. Kline & Hudson's nerve injuries: operative results for major nerve injuries, entrapments and tumors. 2nd ed. Saunders Elsevier; 2008.
15. Mackinnon SE. Pathophysiology of nerve compression. Hand Clin. 2002;18(2):231–41. https://doi.org/10.1016/s0749-0712(01)00012-9.
16. Arriaga MA. Nerve surgery: Susan E. Mackinnon, ed.; New York: Thieme Medical Publishers Inc., 2016. Otol Neurotol. 2017;38(5):625. https://doi.org/10.1097/MAO.0000000000001369.
17. Wood MD, Mackinnon SE. Pathways regulating modality-specific axonal regeneration in peripheral nerve. Exp Neurol. 2015;265:171–5. https://doi.org/10.1016/j.expneurol.2015.02.001.
18. Sachanandani NF, Pothula A, Tung TH. Nerve gaps. Plast Reconstr Surg. 2014;133(2):313–9. https://doi.org/10.1097/01.prs.0000436856.55398.0f.
19. Riccio M, Marchesini A, Pugliese P, De Francesco F. Nerve repair and regeneration: biological tubulization limits and future perspectives. J Cell Physiol. 2019;234(4):3362–75. https://doi.org/10.1002/jcp.27299.
20. Berry D, Ene J, Nathani A, Singh M, Li Y, Zeng C. Effects of physical cues on stem cell-derived extracellular vesicles toward neuropathy applications. Biomedicine. 2024;12(3):489. https://doi.org/10.3390/biomedicines12030489.
21. Hopkins TM, Little KJ, Vennemeyer JJ, et al. Short and long gap peripheral nerve repair with magnesium metal filaments. J Biomed Mater Res A. 2017;105(11):3148–58. https://doi.org/10.1002/jbm.a.36176.
22. Mottaghitalab F, Farokhi M, Zaminy A, et al. A biosynthetic nerve guide conduit based on silk/SWNT/fibronectin nanocomposite for peripheral nerve regeneration. PLoS One. 2013;8(9):e74417. https://doi.org/10.1371/journal.pone.0074417.
23. Chang CJ, Hsu SH. The effect of high outflow permeability in asymmetric poly(dl-lactic acid-co-glycolic acid) conduits for peripheral nerve regeneration. Biomaterials. 2006;27(7):1035–42. https://doi.org/10.1016/j.biomaterials.2005.07.003.
24. Tao J, Hu Y, Wang S, et al. A 3D-engineered porous conduit for peripheral nerve repair. Sci Rep. 2017;7:46038. https://doi.org/10.1038/srep46038.
25. Du J, Liu J, Yao S, et al. Prompt peripheral nerve regeneration induced by a hierarchically aligned fibrin nanofiber hydrogel. Acta Biomater. 2017;55:296–309. https://doi.org/10.1016/j.actbio.2017.04.010.

26. Zhao YH, Niu CM, Shi JQ, Wang YY, Yang YM, Wang HB. Novel conductive polypyrrole/silk fibroin scaffold for neural tissue repair. Neural Regen Res. 2018;13(8):1455–64. https://doi.org/10.4103/1673-5374.235303.
27. Lackington WA, Ryan AJ, O'Brien FJ. Advances in nerve guidance conduit-based therapeutics for peripheral nerve repair. ACS Biomater Sci Eng. 2017;3(7):1221–35. https://doi.org/10.1021/acsbiomaterials.6b00500.
28. Jessen KR, Arthur-Farraj P. Repair Schwann cell update: adaptive reprogramming, EMT, and stemness in regenerating nerves. Glia. 2019;67(3):421–37. https://doi.org/10.1002/glia.23532.
29. Yi C, Dahlin LB. Impaired nerve regeneration and Schwann cell activation after repair with tension. Neuroreport. 2010;21(14):958–62. https://doi.org/10.1097/WNR.0b013e32833e787f.
30. Sedaghati T, Seifalian AM. Nanotechnology and biofunctionalisation for peripheral nerve regeneration. Neural Regen Res. 2015;10(8):1191–4. https://doi.org/10.4103/1673-5374.162678.
31. Carvalho CR, Silva-Correia J, Oliveira JM, Reis RL. Nanotechnology in peripheral nerve repair and reconstruction. Adv Drug Deliv Rev. 2019;148:308–43. https://doi.org/10.1016/j.addr.2019.01.006.
32. Chu XL, Song XZ, Li Q, et al. Basic mechanisms of peripheral nerve injury and treatment via electrical stimulation. Neural Regen Res. 2022;17(10):2185–93. https://doi.org/10.4103/1673-5374.335823.
33. Mashoudy KD, Brooks SG, Andrade LF, Wagner JD, Yosipovitch G. From compression to itch: exploring the link between nerve compression and neuropathic pruritus. Am J Clin Dermatol. 2025;26(1):23–33. https://doi.org/10.1007/s40257-024-00898-5.
34. Han S, Zhao X, Cheng L, Fan J. Recent progresses in neural tissue engineering using topographic scaffolds. Am J Stem Cells. 2024;13(1):1–26. https://doi.org/10.62347/WMDZ8890.
35. Terzis J, Faibisoff B, Williams B. The nerve gap: suture under tension vs. graft. Plast Reconstr Surg. 1975;56(2):166–70.
36. Weber M, Marshall A, Timircan R, et al. Touch localization after nerve repair in the hand: insights from a new measurement tool. J Neurophysiol. 2023;130(5):1126–41. https://doi.org/10.1152/jn.00271.2023.
37. Kwan MK, Wall EJ, Massie J, Garfin SR. Strain, stress and stretch of peripheral nerve rabbit experiments in vitro and in vivo. Acta Orthop Scand. 1992;63(3):267–72. https://doi.org/10.3109/17453679209154780.
38. Sunderland IRP, Brenner MJ, Singham J, Rickman SR, Hunter DA, Mackinnon SE. Effect of tension on nerve regeneration in rat sciatic nerve transection model. Ann Plast Surg. 2004;53(4):382–7. https://doi.org/10.1097/01.sap.0000125502.63302.47.
39. Millesi H, Zöch G, Reihsner R. Mechanical properties of peripheral nerves. Clin Orthop. 1995;314:76–83.
40. Kornfeld T, Borger A, Radtke C. Reconstruction of critical nerve defects using allogenic nerve tissue: a review of current approaches. Int J Mol Sci. 2021;22(7):3515. https://doi.org/10.3390/ijms22073515.
41. Mitsuzawa S, Ikeguchi R, Aoyama T, et al. The efficacy of a scaffold-free bio 3D conduit developed from autologous dermal fibroblasts on peripheral nerve regeneration in a canine ulnar nerve injury model: a preclinical proof-of-concept study. Cell Transplant. 2019;28(9–10):1231–41. https://doi.org/10.1177/0963689719855346.
42. Saffari TM, Mathot F, Friedrich PF, Bishop AT, Shin AY. Revascularization patterns of nerve allografts in a rat sciatic nerve defect model. J Plast Reconstr Aesthet Surg. 2020;73(3):460–8. https://doi.org/10.1016/j.bjps.2019.11.048.
43. Mathot F, Rbia N, Bishop AT, Hovius SER, Shin AY. Adipose derived mesenchymal stem cells seeded onto a decellularized nerve allograft enhances angiogenesis in a rat sciatic nerve defect model. Microsurgery. 2020;40(5):585–92. https://doi.org/10.1002/micr.30579.
44. Zhang H, Zhang H, Wang H, Zhao Y, Chai R. Natural proteins-derived asymmetric porous conduit for peripheral nerve regeneration. Appl Mater Today. 2022;27:101431. https://doi.org/10.1016/j.apmt.2022.101431.
45. Wang ZZ, Sakiyama-Elbert SE. Matrices, scaffolds & carriers for cell delivery in nerve regeneration. Exp Neurol. 2019;319:112837. https://doi.org/10.1016/j.expneurol.2018.09.020.
46. Zhang Z, Ma M. Strategies to enhance the ability of nerve guidance conduits to promote directional nerve growth. Biomed Eng Online. 2024;23(1):40. https://doi.org/10.1186/s12938-024-01233-z.
47. Zuo KJ, Gordon T, Chan KM, Borschel GH. Electrical stimulation to enhance peripheral nerve regeneration: update in molecular investigations and clinical translation. Exp Neurol. 2020;332:113397. https://doi.org/10.1016/j.expneurol.2020.113397.
48. Guo R, Liao M, Ma X, et al. Cochlear implant-based electric-acoustic stimulation modulates neural stem cell-derived neural regeneration. J Mater Chem B. 2021;9(37):7793–804. https://doi.org/10.1039/D1TB01029H.
49. Guo R, Xiao M, Zhao W, et al. 2D Ti3C2TxMXene couples electrical stimulation to promote proliferation and neural differentiation of neural stem cells. Acta Biomater. 2022;139:105–17. https://doi.org/10.1016/j.actbio.2020.12.035.
50. Liao M, Cui Q, Hu Y, et al. Recent advances in the application of MXenes for neural tissue engineering and regeneration. Neural Regen Res. 2024;19(2):258–63. https://doi.org/10.4103/1673-5374.379037.
51. Liao M, Hu Y, Zhang Y, et al. 3D $Ti_3C_2T_x$ MXene–Matrigel with electroacoustic stimulation to promote the growth of spiral ganglion neurons. ACS Nano. 2022;16(10):16744–56. https://doi.org/10.1021/acsnano.2c06306.

52. Hu J, Hu W, Tang L, Wang Y. Fundamental neurocircuit of anti-inflammatory effect by electroacupuncture stimulation identified. Neurosci Bull. 2022;38(7):837–9. https://doi.org/10.1007/s12264-022-00849-2.
53. Qian Y, Cheng Y, Cai J, et al. Advances in electrical and magnetic stimulation on nerve regeneration. Regen Med. 2019;14(10):969–79. https://doi.org/10.2217/rme-2018-0079.
54. Hu Y, Li D, Wei H, et al. Neurite extension and orientation of spiral ganglion neurons can be directed by superparamagnetic iron oxide nanoparticles in a magnetic field. Int J Nanomedicine. 2021;16:4515–26. https://doi.org/10.2147/IJN.S313673.
55. Andreo L, Ribeiro BG, Alves AN, et al. Effects of photobiomodulation with low-level laser therapy on muscle repair following a peripheral nerve injury in Wistar rats. Photochem Photobiol. 2020;96(5):1124–32. https://doi.org/10.1111/php.13255.
56. Li R, Hui LD, Yu ZH, Wang J, Kun LX, Xiao J. Growth factors-based therapeutic strategies and their underlying signaling mechanisms for peripheral nerve regeneration. Acta Pharmacol Sin. 2020;41(10):1289–300. https://doi.org/10.1038/s41401-019-0338-1.
57. Chen X, Ren L, Zhang H, et al. Basic fibroblast growth factor-loaded methacrylate gelatin hydrogel microspheres for spinal nerve regeneration. Smart Med. 2023;2(2):e20220038. https://doi.org/10.1002/SMMD.20220038.
58. Xia L, Zhao X, Ma X, et al. Controllable growth of spiral ganglion neurons by magnetic colloidal nanochains. Nano Today. 2022;44:101507. https://doi.org/10.1016/j.nantod.2022.101507.
59. Keyan Z, Liqian Z, Xinzhong X, Juehua J, Chungui X. Pulsed electromagnetic fields improved peripheral nerve regeneration after delayed repair of one month. Bioelectromagnetics. 2023;44(7–8):133–43. https://doi.org/10.1002/bem.22443. Epub 2023 Jun 5. PMID: 37277911.
60. Su DB, Zhao ZX, Yin DC, Ye YJ. Promising application of pulsed electromagnetic fields on tissue repair and regeneration. Prog Biophys Mol Biol. 2024;187:36–50. https://doi.org/10.1016/j.pbiomolbio.2024.01.003. Epub 2024 Jan 26. PMID: 38280492.
61. Kerns JM, Fakhouri AJ, Weinrib HP, Freeman JA. Electrical stimulation of nerve regeneration in the rat: the early effects evaluated by a vibrating probe and electron microscopy. Neuroscience. 1991;40(1):93–107. https://doi.org/10.1016/0306-4522(91)90177-p. PMID: 2052157.
62. Mayer Y, Shibli JA, Saada HA, Melo M, Gabay E, Barak S, Ginesin O. Pulsed electromagnetic therapy: literature review and current update. Braz Dent J. 2024;35:e246109. https://doi.org/10.1590/0103-6440202406109. PMID: 39476109; PMCID: PMC11506130.
63. Walker JL, Evans JM, Resig P, Guarnieri S, Meade P, Sisken BS. Enhancement of functional recovery following a crush lesion to the rat sciatic nerve by exposure to pulsed electromagnetic fields. Exp Neurol. 1994;125(2):302–5. https://doi.org/10.1006/exnr.1994.1033. PMID: 8313945.

14 Pulley Function and Repair

Nirav Mungalpara and Farid Amirouche

Overview of Flexor Tendon Sheath and Pulley System

Pulleys are thickened regions in flexor tendon sheaths, acting as fulcrums for digit flexion and extension. The flexor tendon sheath comprises a retinacular portion, including five annular pulleys, three cruciform pulleys, and the palmar aponeurosis pulley [1–3]. It overlays a synovial portion, which spans from the metacarpal neck to the distal interphalangeal (DIP) joint, functioning as a lubricating and nutritional system for tendons and pulleys [1, 2]. This sheath's vascularized synovium produces synovial fluid and avoids friction surfaces, ensuring tendon movement without impingement or buckling. Figure 14.1 represents the pulley anatomy published in the *Journal of Hand Surgery* in 1988 [4].

N. Mungalpara
Department of Orthopaedics, University of Illinois at Chicago College of Medicine, Chicago, IL, USA
e-mail: Nirav@uic.edu

F. Amirouche (✉)
Department of Orthopaedics Surgery, Orthopaedic and Spine Institute, Endeavor HealthSystem (Formerly Northshore University HealthSystem), an Affiliate of the University of Chicago Pritzker School of Medicine, Skokie, IL, USA

Department of Orthopaedic Surgery, University of Illinois at Chicago, Chicago, IL, USA
e-mail: Amirouch@uic.edu

Variations in Finger Pulley Anatomy

A1 Pulley and Its Variations

Many times, the A1 pulley is continuous with the palmar aponeurosis sheath. A recent study by Hevesi et al. [5] suggested that this confluence varies on a digit-to-digit basis, with no observed confluence in the thumb (0%) and the most common confluence observed in the middle finger (67%). The length is also highest in the middle finger at 5.7 ± 1.1 mm and shortest in the thumb at 4.6 ± 1.8 mm.

The study of 510 cadaveric fingers and thumbs done by Jongjirasiri et al. [6] demonstrated that the A1 pulley length and its distance from the perpendicular line drawn from the posterior superior prominence of the metacarpal head to the volar aspect could reliably locate the proximal edge of the A1 pulley. This correlation is significant for minimally invasive or percutaneous trigger finger and thumb release. The A1 pulley lengths were as follows: thumbs (5.30 ± 0.53 mm), index finger (6.32 ± 0.17 mm), middle finger (6.58 ± 0.19 mm), ring finger (6.32 ± 0.20 mm), and small finger (5.30 ± 0.49 mm), with an overall average of 6.13 ± 0.17 mm. The proximal margin aligned with the perpendicular line in 327 cases (64.1%) and was within ≤1 mm in 464 cases (91.0%) and within ≤2 mm in 509 cases (99.8%).

F. Amirouche et al. (eds.), *Biomechanics of the Hand*,
https://doi.org/10.1007/978-3-032-04688-8_14

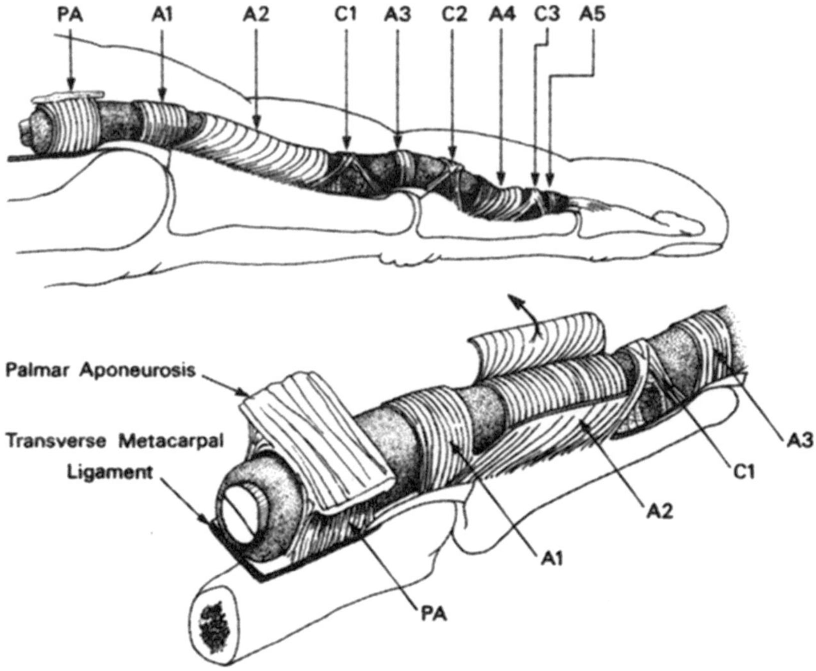

Fig. 14.1 Pulley anatomy (Reprinted with permission from Doyle et al. [4], License No: 5961630516581)

Other Finger Pulleys (A2–A5) and Anatomical Variability

A2 and A4 pulleys are fibroosseous pulleys, while A1, A3, and A5 are volar plate pulleys. There are substantial differences in the pulleys' incidence, shape, and size, depending on the type of pulley, the finger, and the side of the hand [7]. There is controversy regarding the incidence, shape, and length of the pulleys [4, 8–11]. The A1, A2, and A4 pulleys are usually present in all fingers, while the incidence of other pulleys varies. Thirty-two different patterns of pulley combinations were found. The structure of the pulleys also varied. Most important was the variation of A1, which consisted of one to four separate rings. In 18%, there was no gap between the A1 and A2 pulleys. A greater understanding of the anatomical variation of the pulley system is beneficial for hand surgeons performing open or percutaneous trigger finger releases and radiologists examining the area sonographically.

Katzman et al. [12] studied the A5 pulley in 32 cadaveric fingers, reporting an average proximal-to-distal length of 3.8 mm and an average width of 8.9 mm. The distal edge of the A5 pulley was located 7.7 mm proximal to the profundus tendon insertion, just proximal to the distal interphalangeal joint. Histological analysis revealed three distinct layers: an outer fibrofatty tissue layer; a middle layer of connective tissue containing collagen bundles, fibrocytes, and elastin fibers; and an innermost layer of hyaluronic acid-secreting cells. These findings establish the A5 pulley as a discrete structure, anatomically and histologically, consistent with the composition and characteristics described for other annular pulleys.

Thumb Pulley Anatomy

The human thumb's unique complexity enables fine movements and forceful grasps. Normal flexion relies on the integrity of the flexor pollicis longus (FPL) tendon and its retinacular pulley system [13]. These pulleys are thickened regions of the tendon sheath that ensure smooth tendon gliding and maintain bone-tendon apposition.

They act as an efficient fulcrum for thumb flexion and extension without bowstringing [13].

There is controversy regarding the number and morphology of thumb pulleys. Doyle and Blythe [14] described three pulleys: A1, A2, and an oblique pulley. Schmidt and Fischer [15] identified a Y-shaped fiber complex with proximal annular and distal oblique parts between A1 and A2. Bayat et al. [13] reported a distinct variable (Av) pulley in 90% of cases, while Schubert et al. [16] confirmed its presence in 93% of specimens, classifying pulley morphology into four types. Hirschmann et al. [17] identified all four pulleys (A1, Av, oblique, and A2) via MRI in 100% of cases. Four types of pulley arrangement can be seen in the thumb [16, 18].

- *Type I:* The Av pulley was absent (least common) (Fig. 14.2a).
- *Type II:* The Av pulley was present along with the other three pulleys, with its fibers running transversely to the long axis of the FPL tendon (Fig. 14.2b).
- *Type III:* Similar to type II, but the fibers of the Av pulley were oblique, running proximodistally from the ulnar to the radial side relative to the FPL tendon (most common) (Fig. 14.2c).
- *Type IV:* All pulleys were present, and the Av pulley fibers were transverse and fused with the A1 pulley (Fig. 14.2d).

The mean widths (± SD) of the A1, Av, oblique, and A2 pulleys were 5.06 ± 0.87 mm, 5.38 ± 1.22 mm, 4.68 ± 1.13 mm, and 6.04 ± 1.41 mm [18]. Figure 14.2 illustrates the same.

Figure 14.2a–d illustrates the progressive refinement of the thumb pulley system anatomy as described by Bayat et al. [13] and Schubert et al. [16]

Bayat et al. [13] (2002) first described the traditional three-pulley model by identifying annular (Av) pulley between the A1 and oblique pulleys, which they categorized into three anatomical configurations: a transverse Av (type I, Fig. 14.2a), a fused A1–Av complex (type II, Fig. 14.2b), and an oblique or triangular Av pulley (type III, Fig. 14.2c). Their study established the Av as an independent and clinically relevant structure. A decade later, Schubert et al. [16] (2012) built upon these findings with a larger cadaveric series, introducing a four-type classification system that formally acknowledged both the presence and absence of the Av pulley. Notably, they defined a type I configuration lacking the Av pulley entirely, reverting to the classical tripulley anatomy (A1, oblique, A2), as shown in Fig. 14.2d. To maintain clarity, they reclassified Bayat's original types I–III as types II, III, and IV, respectively, aligning their nomenclature with the newly introduced type I.

Recent Advances in Thumb Pulley Anatomy

Recently, Sullivan et al. (2024) [19] identified a new pulley. Three distinct transverse elements were identified proximal to the A1 pulley in the thumb, including (1) the named ligament of Grapow, (2) the proximal commissural ligament and associated transverse fascial tissues, and (3) a definable A0, identified in 10/12 (83.3%) cadavers and 15/24 (62.5%) thumbs. Ten A0 pulleys occurred bilaterally, while five occurred unilaterally. An A0 pulley in greater than 80% of the studied thumbs is an important anatomic finding that should be considered in surgical treatment. An additional finding of importance is that the pulley system anatomy demonstrated variability in laterality.

The Biomechanical Role of the Palmar Aponeurosis (A0 Pulley)

A prospective, randomized clinical trial [20] investigated the role of palmar aponeurosis (A0 pulley) in trigger finger pathology. Thirty fingers from 24 patients underwent randomized initial release of either the A0 or A1 pulley, followed by sequential release of the remaining pulley if

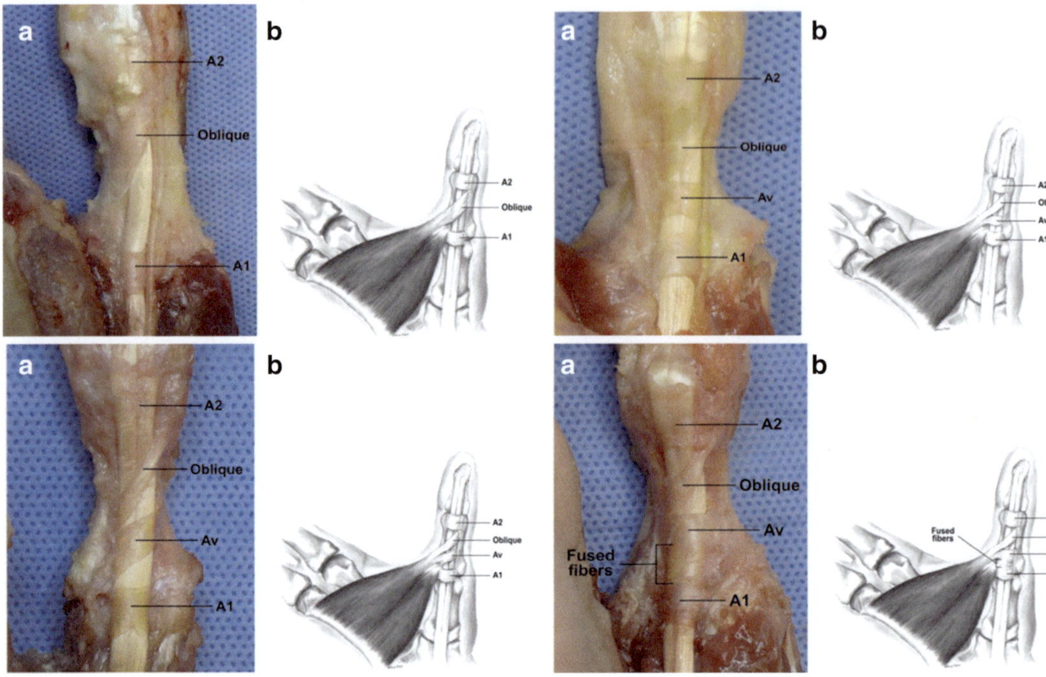

Fig. 14.2 Thumb pulley in a cadaveric hand with the illustrated image beside. (Reprinted with permission from Schubert et al. [16], License No: 6004981183508 (**a**), Reprinted with permission from Bayat et al. [13], License No: 5962140374613 (**b–d**))

symptoms persisted. After the initial A0 release, 47% of fingers had complete symptom resolution, 24% showed improvement but incomplete resolution, and 29% showed no improvement. Similarly, after the initial A1 release, 46% had complete resolution, 23% improved partially, and 31% showed no change. Complete resolution was achieved in all cases after releasing both pulleys. The study suggests that the A0 pulley contributes to 31–47% of trigger finger cases and that releasing both A0 and A1 pulleys may result in more excellent symptom resolution than releasing the A1 pulley alone. However, more extensive trials are needed for confirmation.

Venting of the A2 pulley leads to an incremental increase in tendon bowstringing and tendon slack [21]. However, differences in metacarpophalangeal flexion angle were not observed until full A2 pulley release and only in the index finger, and no differences were observed in proximal interphalangeal flexion angles. Therefore, the benefit of releasing the A2 pulley when clinically necessary will likely outweigh the risks of loss of motion or strength. Another study [22] identified two key factors contributing to the increased exposure of the ring and middle fingers. First, these fingers exhibited higher fingertip force intensities than the index and little fingers. Second, the pulley forces of the ring and middle fingers were near their rupture thresholds, unlike the other fingers. This heightened pulley force was attributed to the specific anatomical positioning of the ring and middle finger pulleys, which increases their mechanical load.

The palmar aponeurosis pulley plays a significant biomechanical role, comparable to annular and cruciate flexor tendon pulleys. Efficiency parameters, including load, excursion, and work efficiencies, were analyzed by M C Philis et al. [23] in an intact pulley system and after sequential sectioning of the palmar aponeurosis and proximal annular pulleys (A1, A2). Sectioning the palmar aponeurosis pulley alone did not significantly alter efficiency, but combining it with proximal annular pulley sectioning significantly decreased excursion efficiency. Complete sectioning of all three pulleys increased load efficiency while decreasing work efficiency due to reduced friction and counteractive forces and increased tendon angulation during flexion. These findings highlight the role of the palmar aponeurosis pulley in preventing bowstringing at the metacarpophalangeal joint (MCP) and its contribution to overall hand function. While the palmar aponeurosis pulley is often regarded as an accessory pulley, its preservation is essential, as it can serve as a compensatory structure in cases where the annular one or annular two pulleys are absent. This pulley is particularly vulnerable during fasciectomy for Dupuytren's disease [24]. Surgeons should remember that the transverse fibers of the palmar fascia, which form the roof of the palmar aponeurosis pulley, are typically unaffected by Dupuytren's disease. Preserving these transverse elements while excising the diseased longitudinal and superficial components of the palmar fascia is crucial.

Biomechanics of the Pulley System

The finger and thumb pulley system enables the flexor tendons to function efficiently and guarantee unrestricted and forceful flexion [25]. Injuries to these anatomical structures can be detrimental to hand function [26, 27]. Lesions of the pulley system render complete and forceful grip and precise finger motion difficult to perform [28–31]. The biomechanics of the pulley system have been examined thoroughly [4, 9, 11, 30, 32]. Breaking and tensile strength vary between the different annular and cruciate pulleys [33, 34]. Bowstringing is not always clinically evident and renders this examination not very precise and subtle [35].

Strength of Pulleys

G T Lin et al. [30] investigated the biomechanical strength of the pulleys. To analyze the mechanical properties of the fibroosseous pulleys in the hand, different pulley types—diaphyseal annular pulleys, volar plate annular pulleys, and cruciate or condensable portions—were evaluated. Among the annular pulleys, the A2 pulley emerged as the strongest, with the A1 and A4 pulleys being the next strongest. The A3 pulley, although comparable in mean breaking strength to other annular pulleys, had a lower absolute breaking load due to its shorter length. The A4 pulley was noted to be the least compliant of all tested pulleys, highlighting its stiffness. The fibroosseous A2 and A4 pulleys demonstrated superior mechanical strength and stiffness compared to the A1, A3, and A5 (volar plate) pulleys.

Angular Displacement and Moment Arm

A radian represents the arc length on a circle's circumference equal to its radius. When a radius moves through 57.29° (one radian), it covers an arc equivalent to its length (Fig. 14.3). This relationship is critical in understanding tendon excursion and moment arm mechanics: for every 57° of joint motion, tendon excursion matches the moment arm. The MCP (*metacarpophalangeal*) joint, with a standard range of 85°, corresponds to 1.48 radians ($85 \div 57.29$). The concept of the radian provides a valuable tool [24], as dividing the required joint ROM by one radian and multiplying by the moment arm yields the necessary tendon excursion.

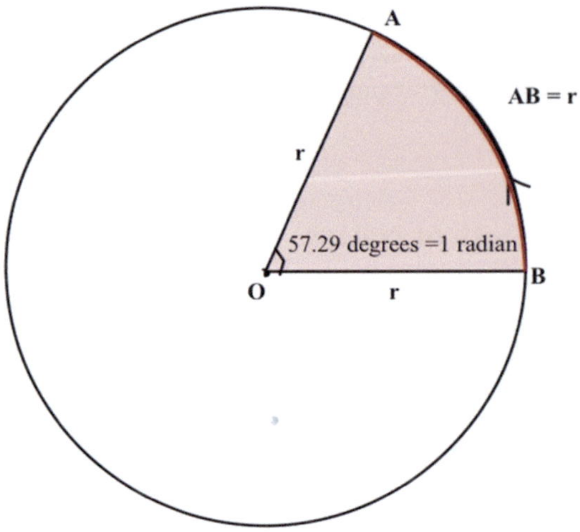

Fig. 14.3 Illustration of one radian in a circle. The diagram demonstrates the definition of a radian as the angle subtended at the center of a circle (O) when the arc length (AB) is equal to the radius (r). The corresponding angle measures approximately 57.29 degrees, emphasizing the fundamental relationship between arc length, radius, and angular measurement in radians. [Courtesy: Nirav Mungalpara]

The moment arm is the perpendicular distance between the line of force application and the joint's center of rotation (axis), defined as the distance between the joint's axis of motion and the central longitudinal axis of the adjacent flexor tendon. Typical measurements for joint moment arms are 10 mm for the metacarpophalangeal joint, 7.5 mm for the proximal interphalangeal (PIP) joint, and 5 mm for the distal interphalangeal joint. Evidence from a biomechanical study by Brand et al. [36] using fresh cadavers showed that the moment arm remains relatively constant during motion, with a straight-line relationship between joint angle changes and tendon excursion. While some joints, like the wrist and metacarpophalangeal joint, have two axes of rotation, the axis of the metacarpophalangeal joint is primarily considered in the coronal plane. Although moment arms in intact pulley systems likely remain constant, disruption of the pulley mechanism, if not adequately reconstructed, can significantly alter the moment arm. As most of the available excursion is utilized for full finger flexion, any increase in the moment arm could result in a considerable loss of function (Figs. 14.4 and 14.5).

Excursion

Excursion refers to the sum of a muscle's ability to stretch beyond its resting length and contract below it. Muscle fibers can actively contract and passively lengthen by approximately 40% each [37]. Excursion, proportional to muscle fiber length, is limited and plays a key role in the relationship between excursion and moment arm. Brand et al. [36] categorized excursions into potential, required, and available types. Potential excursion represents the resting fiber length without connective tissue constraints. The excursion necessary refers to the maximum length change needed for a muscle to move its joints through its full range of motion. As described by Freehafer et al. [38], available excursion measures the maximum excursion after detaching the muscle from its insertion. Typically aligning with required excursion, the available excursion can adapt to use patterns, increase through stretching or exercise, or decrease due to postinjury adhesions, connective tissue restrictions, or scarring. Active muscle contraction accounts for approximately half of the total excursion and is responsible for tendon and joint movement. In their study on differential tendon gliding in the hand, Wehbe and Hunter [39] observed a mean active flexor profundus tendon excursion of 32 mm (range: 15–43 mm) during finger movement from full extension to full flexion with the wrist in neutral. This range for flexor digitorum superficialis tendon is 24 mm (range: 14–37 mm). Significant loss of finger motion and function becomes evident when muscle excursion is reduced due to intrinsic or extrinsic factors like adhesions or scarring of the sheath or joint capsule. Table 14.1 [24] shows the relationship of the joint ROM, moment arms, and excursion.

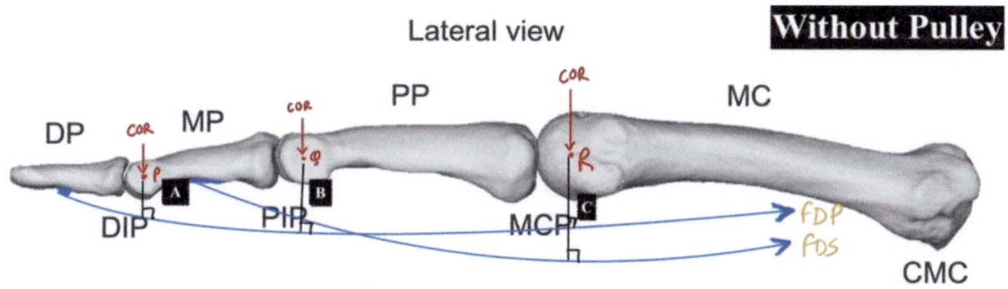

Fig. 14.4 Lateral view of the finger without the pulley system. The absence of a functional pulley leads to increased bowstringing of the flexor tendons (FDP, FDS), altering their moment arms (A, B, C) at the DIP, PIP, and MCP joints. This results in decreased mechanical efficiency and reduced force transmission, significantly affecting finger flexion control. The altered moment arms impact joint torques and may lead to compensatory mechanisms, increasing strain on the flexor tendons and surrounding structures. The lack of a pulley reduces flexor tendon force transmission efficiency, underscoring the vital role of the pulley system in maintaining finger function. The COR stands for Center of Rotations in this figure. [Courtesy: Nirav Mungalpara]

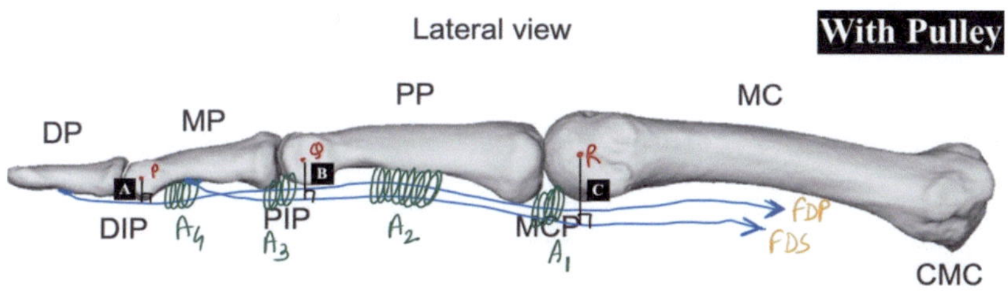

Fig. 14.5 Lateral view of the finger with the pulley system intact. The functional pulleys (A1, A2, A3, A4) are vital in keeping the flexor tendons (FDP, FDS) close to the bone, thereby reducing bowstringing and optimizing force transmission. The intact pulley system ensures efficient moment arms (A, B, C) at the DIP, PIP, and MCP joints, enhancing mechanical advantage and reducing strain on the flexor tendons. This enhancement in finger flexion control and prevention of excessive tendon excursion underscores the significant role of the pulley system in promoting stable and efficient finger movement. The COR stands for Center of Rotations in this figure. [Courtesy: Nirav Mungalpara]

Table 14.1 Relationship of the joint ROM, moment arms, and excursion

Joint (ROM)	Moment arm (millimeters)	Excursion (millimeters)+
Metacarpophalangeal (85)	10[a]	14.8[a]
	12	17.8
Proximal interphalangeal (110)	7.5[a]	14.4[a]
	9	17.3
Distal interphalangeal (65)	5[a]	5.7[a]
	34.9[a] versus 41.9	
+joint ROM × MA = excursion		
57.29°		

The relationship between required range of motion, normal and increased moment arm, and the resultant excursion. Joint range of motion divided by one radian multiplied by the moment arm yields excursion (see text).
MA moment arm
[a]Normal values

Work of Flexion

To better understand the dynamics of the flexor tendon pulley system, it is essential to consider the concept of work of flexion, first introduced by Lane et al. [40, 41] and later applied to flexor pulley function by Peterson et al. [25]. While tendon excursion is a valid measure of pulley efficiency, work of flexion offers a more comprehensive evaluation by representing the resistance to gliding that the tendon encounters during flexion. This parameter is particularly valuable in assessing factors beyond the mechanical integrity of the pulleys, including viscoelastic forces of the skin and subcutaneous tissues, tendon adhesions, joint stiffness, and tendon-sheath incongruities. Peterson et al. [25] demonstrated that pulley loss impacts flexion work more significantly than tendon excursion; for example, loss of the A2 pulley

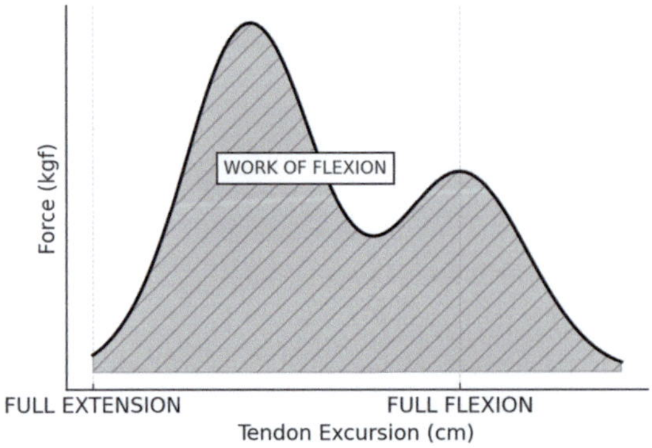

Fig. 14.6 Graph illustrating the work of flexion as a function of tendon excursion. (Adapted from Peterson et al. [25])

Table 14.2 Comparative changes in tendon excursion and work of flexion with pulley excision

Pulley(s) excised	Intact	Tendon excursion	Work of flexion
A1	A2, A4	−0.64 ± 2.04	10.00 ± 8.16
A2	A1, A4	8.50 ± 1.62	44.10 ± 7.15
A4	A1, A2	9.93 ± 3.40	19.95 ± 10.21
A1, A2	A4	20.56 ± 3.67	62.36 ± 13.91
A2, A4	A1	33.66 ± 4.60	107.04 ± 22.77
A1, A4	A2	9.69 ± 1.91	39.83 ± 8.70
A1, A2, A4	None	65.09 ± 6.21	172.35 ± 15.30

Adapted from Peterson et al. [25]
A annular pulleys

resulted in only an 8.5% increase in tendon excursion but a substantial 44% increase in flexion work. Therefore, flexion work is a more sensitive indicator of overall tendon function, integrating multiple resistive factors. Figure 14.6 illustrates the work of flexion in the graph. Comparative changes in tendon excursion and flexion work with pulley excision are detailed in Table 14.2.

Lubrication Factor

The lubrication mechanism between the human flexor digitorum profundus tendon and the A2 pulley was investigated in vitro [42], revealing that gliding resistance at the tendon-pulley interface significantly increased following hyaluronidase treatment. This finding, supported by Alcian blue staining, suggests the presence of hyaluronate or proteoglycan complexes in the synovial membrane and tendon matrix, acting as boundary lubricants to facilitate smooth gliding and reduce resistance. These insights have critical clinical implications: intrasynovial tendon donor grafts, which retain this natural boundary lubricant, may provide superior postoperative tendon gliding compared to traditional extrasynovial grafts, such as the palmaris longus tendon. This underscores the potential for improving tendon repair and reconstruction surgical outcomes. Ethnographic studies were conducted by S Mester et al. [43] on eight isolated fingers and two intact cadaveric hands. Observations revealed bulging and overlapping synovial pockets between ligamentous structures during flexion and their flattening during extension. The proximal cul-de-sac's dynamic expansion and contraction suggest a potential synovial fluid mixing mechanism. During flexion and extension, a natural joint-like bowstringing of the flexor tendons was also noted at the PIP (proximal interphalangeal) and DIP joints. This supports the hypothesis that the flexor tendon moment arm increases at these joints.

Schweizer et al. [44] developed a method to indirectly measure friction between the flexor tendons and pulleys of the middle and ring fingers in vivo. An isokinetic movement device was constructed to evaluate maximum force during wrist flexion, interphalangeal joint flexion (rolling in and out), and isolated proximal interphalangeal (PIP) joint flexion. These movements, involving varying degrees of flexor tendon sheath gliding (minimal in wrist flexion), were assessed in 51 hands of 26 male subjects. The most significant difference between eccentric and concentric maximum force (29.9%) was observed in PIP joint flexion, compared to more minor differences in rolling in and out movements (26.8%) and wrist flexion (14.5%). These results suggest that friction between flexor tendons and pulleys, accounting for 9% of the eccentric flexion force in the PIP joint, contributes significantly to the overall force, challenging the low-friction hypothesis under high loads. This study highlights the biomechanical role of friction in tendon-pulley interactions, particularly during high-load activities.

Static and Dynamic Torque

Schweizer et al. [45] studied how preceding flexion or extension movements influence the static interaction between finger flexor tendons and pulleys in generating flexion torque. Six fresh-frozen cadaveric-long fingers were tested using an isokinetic movement device for the PIP joint. Both flexor tendons were loaded with 40 N during flexion and extension, and torque was recorded dynamically and statically at various PIP joint positions (0–105°). Static torque was consistently higher following extension than flexion, with the most significant difference (11%) occurring at approximately 83° of flexion. Static torque was lower than dynamic torque during extension but higher during flexion. The influence of preceding movements on torque generation was also replicated using a mechanical finger model.

Biomechanics of the Pulley Injury

Impact of Pulley Injury on Work and Excursion Efficiency

In vitro, the biomechanical effect of the pulley injury has been studied before [46]. Cutting an A2 or A4 pulley significantly impacted work and excursion efficiency. Notably, the loss of A4 resulted in an 85% reduction in both work and excursion efficiency, while the loss of A2 caused a 94% reduction in excursion efficiency [46]. In this model by Rispler et al. [46], excluding the influence of the skin, A4 absence led to the most remarkable biomechanical efficiency changes. Preserving work and excursion efficiency required combined A2, A3, and A4 pulleys. The flexion work is more significantly impacted by pulley loss than tendon excursion, making it a more sensitive measure of tendon function [47]. The A2 pulley was identified as the most critical for flexor tendon function, followed by the A4 pulley. Near-normal hand function required the presence of both A2 and A4 while sacrificing the A1 pulley did not lead to a significant loss of flexion. The study by Mitsionis [47] also highlighted the supplemental role of skin and soft tissue in reducing tendon bow-stringing, complementing the fibroosseous pulleys. While this study focused on tendon excursion and work of flexion to assess the effects of pulley loss, these parameters are equally valuable for evaluating other flexor tendon studies, including pulley reconstruction.

Bowstringing and Tendon Excursion Changes

Gau-Tyan Lin et al. [30] concluded that serial pulley sectioning showed two types of bowstringing, affecting the relationship between tendon excursion and joint motion. The most constant, resulting in a 15% loss of motion for a fixed tendon excursion, occurred over the concave surfaces of the proximal and middle phalanges. Bowstringing at the proximal interphalangeal joint was present only after 30 degrees of flexion because of the convexity of the phalangeal condyles.

Low et al. [32] concluded that A1 pulley injury will decrease the force required to flex the digit fully. The force decreased with more extensive A1 pulley release, which became significant when more than 50% of the pulley was released. These changes were probably due to loss of friction force rather than bowstringing of the tendon. The pulley's role is more on the eccentric loading of the flexor muscles, hence the static crimp grip in rock climbers; using the eccentric load on flexor tendons will significantly affect when a pulley is injured because of the increased moment arms and bowstringing of the tendon at MCP and PIP joints.

Influence of Partial Pulley Injury on Gliding Resistance, Stiffness, and Strength

After A2 pulley excision [48], the size and location of the remaining segment influence gliding resistance, stiffness, and failure strength. At the extreme excision level tested, the distal 25% segment exhibited higher gliding resistance, strength, and stiffness than the proximal 25% segment. However, limiting excision to 50% proximally or distally results in minimal increases in gliding resistance while retaining substantial pulley strength. These findings support the clinical feasibility of partial A2 pulley excision, up to 50%, for improved exposure during tendon repair.

A study [32] investigating factors contributing to tendon bowstringing at the proximal phalanx revealed that even a partial rupture of the A2 pulley causes significant bowstringing. Using ultrasonography to quantify the bowstringing area in cadaveric fingers under a 100 N FDP load, the researchers found that a 30% excision of the A2 pulley led to noticeable bowstringing. Partial rupture at the distal rim of the A2 pulley resulted in greater bowstringing than at the proximal rim. Interestingly, additional A3 pulley rupture or excision of the proximal tendon sheath did not exacerbate bowstringing, but removing the overlying skin significantly increased it. These findings highlight the critical role of the A2 pulley and its distal segment in preventing bowstringing and maintaining tendon alignment.

Effects of Proximal Versus Distal Pulley Excision on Finger Kinematics

The flexor tendon pulley system, mainly the A2 and A4 pulleys, is crucial in maintaining finger kinematics. Understanding the biomechanical consequences of partial pulley excisions is essential for surgical decision-making. Chow et al. [33] investigated the effects of proximal versus distal 50% of the A2 and A4 pulley incisions on finger motion and joint dynamics. They hypothesized that proximal excisions would have a more pronounced impact on kinematics due to their anatomic proximity to key joints. The research, which utilized 16 cadaveric hands, focused on the index and ring fingers. These were instrumented with micropotentiometers at the metacarpophalangeal (MCP), proximal interphalangeal (PIP), and distal interphalangeal (DIP) joints to measure angular displacements during tendon excursions. The flexor digitorum profundus (FDP) tendon was isolated and connected to a computer-controlled servomotor to simulate tendon motion, while joint inertial torques were calculated using kinematic data.

The results demonstrated that proximal incisions of the A2 pulley caused the most significant reduction in total finger range of motion (ROM), particularly at the PIP joint, which exhibited an earlier initiation of motion and decreased maximum flexion (Fig. 14.7).

Similarly, proximal A4 pulley incisions also reduced PIP joint ROM, though the effect was less pronounced than A2 excisions (Fig. 14.8).

Distal incisions of both pulleys had minimal impact on kinematics. Statistical analysis revealed significant differences in PIP ROM and total finger ROM for proximal excisions ($p < 0.001$), while inertial torques remained unchanged across all conditions. These findings were attributed to bowstringing, where tendon displacement away from the joint axis reduces mechanical efficiency. Proximally located pulley incisions likely exacerbated this effect due to their closer anatomic relationship to the PIP joint. Clinically, these findings support the preservation of proximal pulley segments during surgical interventions, such as pulley venting, to maintain finger function.

Effects of Pulley Repair on Finger Joint Kinematics and Dynamics

The finger pulley system is essential for maintaining proper tendon function and joint dynamics during finger motion. Amirouche et al. [49] explore the biomechanics of the finger pulley system, focusing on the effects of various pulley excisions and repairs on finger joint kinematics and dynamics. That study hypothesizes that the

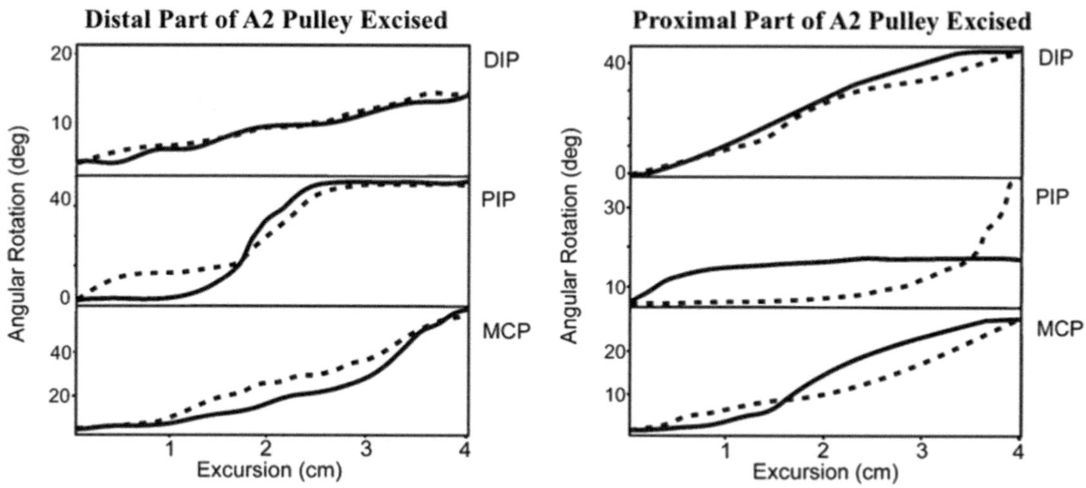

Fig. 14.7 Comparison of angular rotation at the metacarpophalangeal (MCP), proximal interphalangeal (PIP), and distal interphalangeal (DIP) joints following distal and proximal A2 pulley excision. The graphs illustrate the effects of distal (left) and proximal (right) A2 pulley excision on joint kinematics relative to tendon excursion. Distal A2 excision shows moderate changes in angular rotation across all joints, whereas proximal A2 excision results in a marked increase in DIP joint rotation and a reduction in PIP joint motion. These findings highlight the differential biomechanical consequences of pulley excision location on finger flexion dynamics. (From Chow et al. [33])

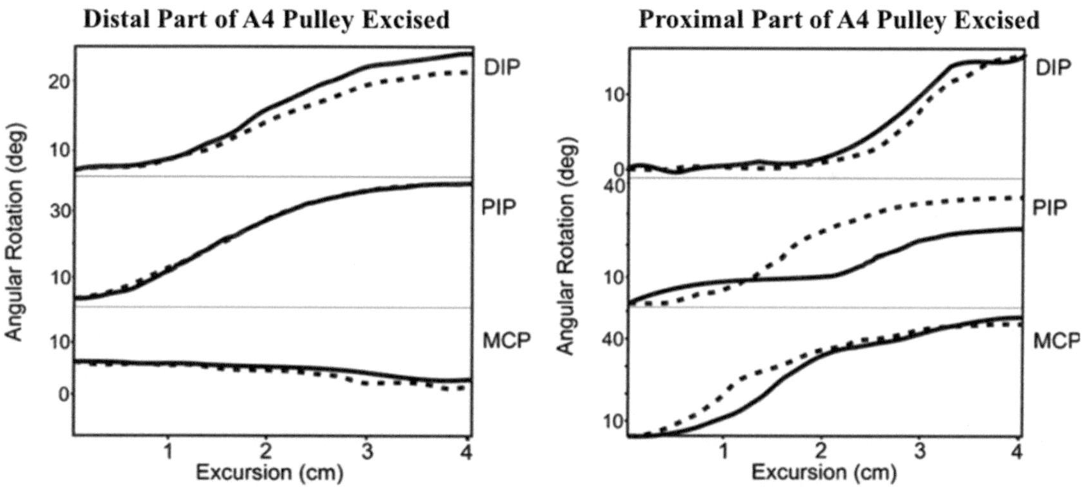

Fig. 14.8 Comparison of angular rotation at the metacarpophalangeal (MCP), proximal interphalangeal (PIP), and distal interphalangeal (DIP) joints following distal and proximal A4 pulley excision. The graphs depict the effects of distal (left) and proximal (right) A4 pulley excision on joint kinematics relative to tendon excursion. Distal A4 excision results in moderate changes in DIP and PIP joint rotation, while proximal A4 excision leads to a more pronounced increase in DIP flexion and a reduction in PIP motion. The MCP joint remains relatively unaffected in both conditions. These findings underscore the role of the A4 pulley in maintaining balanced flexor tendon function and joint stability. (From Chow et al. [33])

excision of specific pulleys (A1, A2, and A4) will significantly alter the range of motion and joint torques. At the same time, surgical reconstruction can restore normal finger function, particularly with the A2 and A4 pulleys playing a critical role.

Experimental Setup and Measurement

The research utilized six fresh-frozen cadaver hands, with the index and middle fingers as the primary focus. The experimental setup involved micropotentiometers to measure joint angles and a pulley system to simulate tendon excursion. The hands were mounted in a neutral position, and the flexor digitorum profundus (FDP) tendon was isolated and connected to a drive pulley. Data acquisition was performed using a DT VEE system, and simultaneous joint angles and tendon excursions were recorded. A mathematical model based on a planar three-link rigid body system was developed to analyze the finger's dynamics, incorporating each finger segment's mass and inertia properties to compute the joint torques required for motion.

Effects of Pulley Excision on Joint Kinematics

The experimental results revealed significant changes in finger joint kinematics and dynamics following pulley excisions and repairs. Excision of the A2 pulley alone resulted in minimal changes in the overall range of motion; still, the maximum angle at the metacarpophalangeal (MCP) joint decreased slightly, as shown in Fig. 14.9, which illustrates the range of motion for the PIP, DIP, and MCP joints under various excision and repair conditions.

Excision of the A1 and A2 pulleys led to a marked decrease in the proximal interphalangeal (PIP) joint's range of motion, with bowstringing effects observed at both the MCP and PIP joints. Complete excision of all pulleys resulted in bowstringing across all three joints, with the distal interphalangeal (DIP) joint experiencing increased motion, as depicted in Fig. 14.10.

Pulley Repair and Restoration of Joint Kinematics

Repair of the A1 pulley alone significantly increased the PIP joint's range of motion but caused a total loss of motion at the DIP joint and a decrease in MCP joint motion. Repair of the A2 and A4 pulleys was most effective in restoring the finger's kinematics to that of the intact hand, with the distal A2 pulley repair proving particularly efficient, as demonstrated in Fig. 14.11.

Torque Analysis Under Different Pulley Conditions

The torque analysis, illustrated in Fig. 14.12, revealed that the most significant differences in joint torques occurred with the excision of the A1 and A4 pulleys. Repairing these pulleys significantly increased the required torque, with the A2 and A4 pulley repairs restoring the most incredible torque difference.

The study's findings highlight the critical role of the A2 and A4 pulleys in maintaining proper finger joint kinematics and dynamics. The loss in joint flexion observed is more significant than previously reported, likely due to the more reliable data acquisition and filtering methods employed. The comprehensive nature of the study, including the development of a mathematical model, provides a valuable tool for further research, particularly in understanding the dynamic behavior of the finger joints under different pulley conditions. It offers new insights into the energy requirements for finger motion and the effects of pulley excisions and repairs on finger joint dynamics. This study provides a comprehensive analysis of the finger pulley system's biomechanics, offering new insights into the impact of pulley excisions and repairs on finger joint dynamics. It underscores the importance

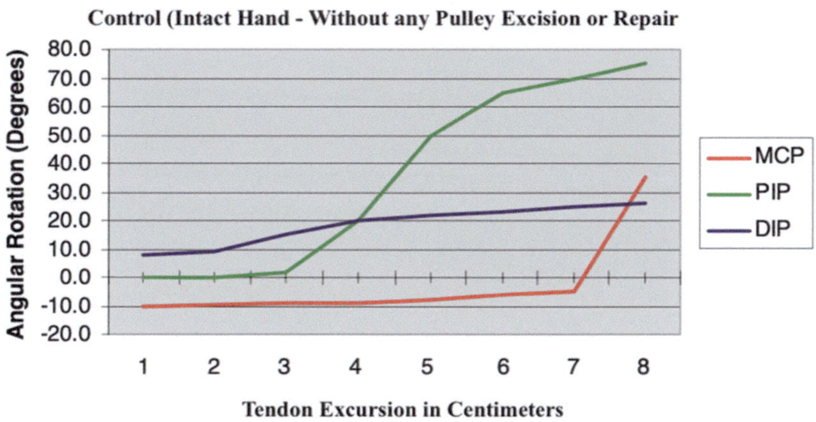

Fig. 14.9 Angular rotation of the metacarpophalangeal (MCP), proximal interphalangeal (PIP), and distal interphalangeal (DIP) joints relative to tendon excursion in an intact hand. The graph illustrates the relationship between tendon excursion (cm) and angular rotation (degrees) at the MCP, PIP, and DIP joints. The PIP joint exhibits the highest angular rotation, followed by the DIP joint, while the MCP joint demonstrates minimal rotation until a sharp increase near maximum excursion. (From Amirouche et al. [49])

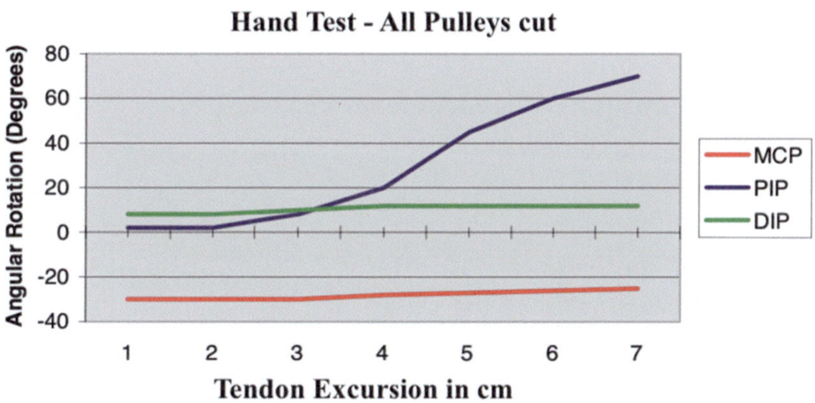

Fig. 14.10 Angular rotation of the metacarpophalangeal (MCP), proximal interphalangeal (PIP), and distal interphalangeal (DIP) joints relative to tendon excursion in a hand with all pulleys cut. The graph demonstrates the altered joint kinematics following complete pulley excision. Compared to the intact hand, the DIP joint exhibits significantly increased angular rotation, while the PIP joint shows reduced movement, and the MCP joint remains relatively unchanged. These findings highlight the critical role of the pulley system in maintaining balanced tendon excursion and joint motion efficiency. (From Amirouche et al. [49])

of the A2 and A4 pulleys in maintaining proper finger function and highlights the potential of surgical reconstruction to restore normal kinematics. The mathematical model serves as a foundation for future research, enabling a deeper understanding of the finger pulley system's mechanics and the development of improved surgical techniques.

Biomechanical Outcomes of Pulley Repair Configurations

Soulii et al. [50] evaluated the mechanical characteristics of A2 and A4 pulley repairs in cadaveric fingers to simulate clinical pulley damage and repair scenarios. The primary objective was to assess how varying repair configurations influ-

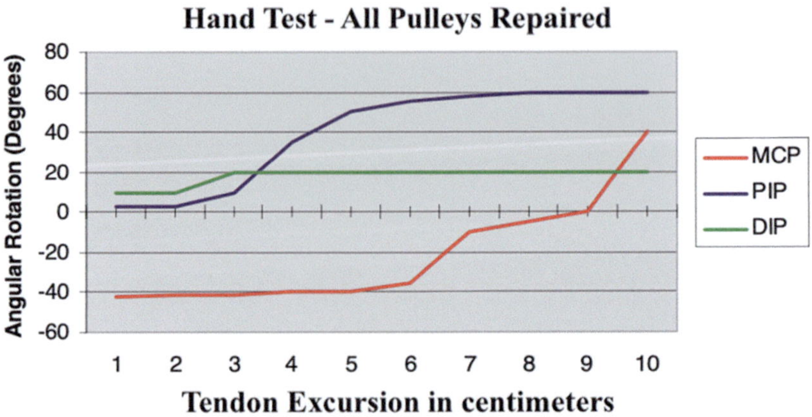

Fig. 14.11 Angular rotation of the metacarpophalangeal (MCP), proximal interphalangeal (PIP), and distal interphalangeal (DIP) joints relative to tendon excursion in a hand with all pulleys repaired. The graph illustrates the restoration of joint motion following pulley reconstruction. Compared to the condition with all pulleys cut, the repaired system shows improved balance in angular rotation across all joints. The PIP joint regains near-normal function, while the MCP joint demonstrates increased flexion at higher excursions. These findings highlight the effectiveness of pulley reconstruction in restoring normal tendon excursion and joint biomechanics. (From Amirouche et al. [49])

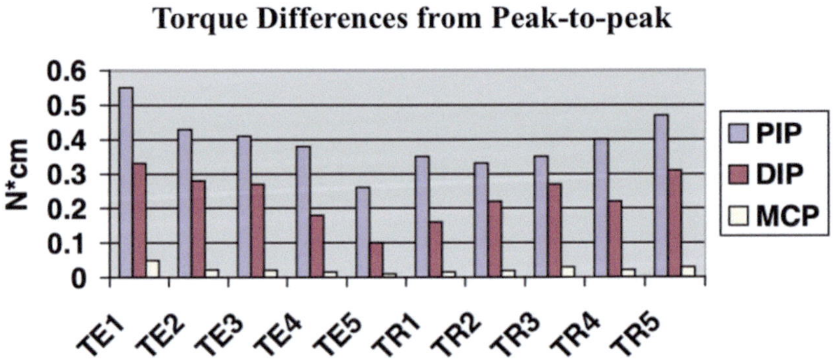

Where:

TE1 - A1, A2, A4 uncut
TE2 - A2, A4 uncut
TE3 - Proximal A2, A4 uncut
TE4 - A4 uncut
TE5 - All cut

TR1 - A1, A4 repair
TR2 - Proximal A2, A4 repair
TR3 - Distal A2, A4 repair
TR4 - Distal A2 repair
TR5 - A2, A4 repair

Fig. 14.12 The difference in torque (peak-to-peak) across metacarpophalangeal (MCP), proximal interphalangeal (PIP), and distal interphalangeal (DIP) joints under different pulley conditions. The bar graph compares torque variations in various experimental conditions, including intact, excised, and repaired pulley configurations. TE1–TE5 represent varying degrees of pulley excision, while TR1–TR5 denotes different pulley repair strategies. The highest torque values are observed in intact conditions (TE1), while complete pulley excision (TE5) leads to the most significant torque reduction. Pulley repairs (TR1–TR5) partially restore torque, with distal A2 and A4 repairs (TR4, TR5) showing the most effective recovery. These findings highlight the biomechanical significance of pulley integrity in maintaining optimal torque transmission during finger flexion. (From Amirouche et al. [49])

ence finger kinematics and range of motion. Researchers simulated pulley damage using 11 frozen human fingers, followed by different repair strategies using tendon graft rings. Configurations included excision of both pulleys, single-ring and two-ring repairs at A2, and combined repairs at A2 and A4. Measurements of tendon excursion and joint rotation were recorded using a servomotor and potentiometers. The results revealed that complete excision of A2 and A4 pulleys led to a significant decrease in maximum rotational angles—30% at the PIP joint and 17% at the MCP joint. Repairs improved joint motion proportionally to the number and placement of graft rings. A combined three-ring configuration (two at A2 and one at A4) yielded the best restoration of kinematics, reducing the deficit at the PIP joint to 17% and at the MCP joint to 4%. However, increasing the number of graft rings at A2 alone did not produce significant additional benefits and occasionally impeded motion due to increased tension.

The findings emphasize that while reconstruction of both A2 and A4 pulleys optimizes finger biomechanics, repairing A2 alone may suffice in many clinical situations, particularly in reducing surgical complexity and tissue dissection. The study also highlights that intrasynovial tendon grafts, such as flexor digitorum profundus (FDP) tendons, offer superior gliding properties compared to extrasynovial options. These insights are relevant not only for traumatic injuries, such as those seen in rock climbing, but also for planning flexor tendon repairs and staged reconstructions.

Conclusion

In conclusion, the intricate interplay of the finger pulley system, mainly the A2 and A0 pulleys, plays a pivotal role in maintaining tendon alignment and joint dynamics during finger movement. The studies discussed highlight the significance of the A2 pulley in preventing bowstringing, revealing that even minor damage can lead to substantial functional consequences. The findings on the palmar aponeurosis underscore its biomechanical importance, countering the common perception of it as merely an accessory structure. Specifically, its preservation during surgical interventions, such as fasciectomy for Dupuytren's disease, is essential for maintaining overall hand function.

Moreover, the evidence suggests that surgical interventions targeting the A0 and A1 pulleys can provide considerable relief from trigger finger symptoms, with a recommendation for sequential release to optimize outcomes. The biomechanics of the finger pulley system show that the excision of pulleys significantly impacts range of motion, indicating that careful consideration during surgery is crucial to restore and preserve finger function.

Overall, this chapter underscores the complex relationships within the finger pulley system, reinforcing the need for further research to fully understand these dynamics and improve clinical practices in hand surgery. Understanding these nuances is essential for enhancing therapeutic strategies and ensuring better patient outcomes in hand-related pathologies.

References

1. Yehia SR, Duncan H. Synovial fluid analysis. Clin Orthop Relat Res. 1975;107:11–24. https://doi.org/10.1097/00003086-197503000-00003.
2. Lundborg G, Myrhage R, Rydevik B. The vascularization of human flexor tendons within the digital synovial sheath region—structural and functional aspects. J Hand Surg Am. 1977;2(6):417–27. https://doi.org/10.1016/s0363-5023(77)80022-1.
3. Manske PR, Lesker PA. Palmar aponeurosis pulley. J Hand Surg Am. 1983;8(3):259–63. https://doi.org/10.1016/s0363-5023(83)80154-3.
4. Doyle JR. Anatomy of the finger flexor tendon sheath and pulley system. J Hand Surg Am. 1988;13(4):473–84. https://doi.org/10.1016/s0363-5023(88)80082-0.
5. Hevesi M, Logli AL, Ramazanian T, Kakar S. Defining the digit-specific confluence of the A1 pulley. J Hand Surg Am. 2023;48(8):835.e1–4. https://doi.org/10.1016/j.jhsa.2022.02.011.

6. Jongjirasiri Y. Length and landmark of A1 pulley in hand: an anatomical study. J Med Assoc Thai. 2009;92(1):41–6.
7. De las Heras J, Sanudo JR, Simón de Blas C, Krompas I, SalaBerri GJ, Aragonés P. The incidence and shape of the digital pulleys: a study of 192 fingers in 48 cadaveric hands. J Hand Surg Eur Vol. 2022;47(8):818–24. https://doi.org/10.1177/17531934221101716.
8. Strauch B, de Moura W. Digital flexor tendon sheath: an anatomic study. J Hand Surg Am. 1985;10(6 Pt 1):785–9. https://doi.org/10.1016/s0363-5023(85)80152-0.
9. Barton NJ. Experimental study of optimal location of flexor tendon pulleys. Plast Reconstr Surg. 1969;43(2):125–9. https://doi.org/10.1097/00006534-196902000-00002.
10. de la Caffinière JY. Functional anatomy of the proximal pulley of the fingers. Arch Anat Pathol (Paris). 1971;19(4):357–66.
11. Doyle JR. Anatomy of the flexor tendon sheath and pulley system: a current review. J Hand Surg Am. 1989;14(2 Pt 2):349–51. https://doi.org/10.1016/0363-5023(89)90110-x.
12. Katzman BM, Klein DM, Garven TC, Caligiuri DA, Kung J, Collins ED. Anatomy and histology of the A5 pulley. J Hand Surg Am. 1998;23(4):653–7. https://doi.org/10.1016/S0363-5023(98)80051-8.
13. Bayat A, Shaaban H, Giakas G, Lees VC. The pulley system of the thumb: anatomic and biomechanical study. J Hand Surg Am. 2002;27(4):628–35. https://doi.org/10.1053/jhsu.2002.34008.
14. Doyle JR, Blythe WF. Anatomy of the flexor tendon sheath and pulleys of the thumb. J Hand Surg Am. 1977;2(2):149–51. https://doi.org/10.1016/s0363-5023(77)80101-9.
15. Schmidt HM, Fischer G. Pulleys of the tendon sheath of the flexor pollicis longus muscle. Handchir Mikrochir Plast Chir. 1999;31(6):362–6. https://doi.org/10.1055/s-1999-13553.
16. Schubert MF, Shah VS, Craig CL, Zeller JL. Varied anatomy of the thumb pulley system: implications for successful trigger thumb release. J Hand Surg Am. 2012;37(11):2278–85. https://doi.org/10.1016/j.jhsa.2012.08.005.
17. Hirschmann A, Sutter R, Schweizer A, Pfirrmann CWA. MRI of the thumb: anatomy and spectrum of findings in asymptomatic volunteers. AJR Am J Roentgenol. 2014;202(4):819–27. https://doi.org/10.2214/AJR.13.11397.
18. Gnanasekaran D, Veeramani R, Karuppusamy A. Morphometric study of pulleys of the thumb. Anat Cell Biol. 2018;51(2):71. https://doi.org/10.5115/acb.2018.51.2.71.
19. Sullivan JM, Edwards S, Waetjen E, Dogar S, Walker ME. Back to the anatomy lab to redefine the pulley system of the thumb: is there an A0 pulley? Hand (New York). Published online December 11; 2024. p. 15589447241302356. https://doi.org/10.1177/15589447241302356.
20. Differential Pulley Release in Trigger Finger: A Prospective, Randomized Clinical Trial—PubMed. Accessed 28 Jan 2025. https://pubmed.ncbi.nlm.nih.gov/33648377/
21. Shipley T, Vanhoof-Villalba SL, Lee S, et al. Effects of A2 pulley venting on bowstringing and tendon slack: a biomechanical investigation. Hand (N Y). Published online June 21; 2024. p. 15589447241259804. https://doi.org/10.1177/15589447241259804.
22. Vigouroux L, Quaine F, Paclet F, Colloud F, Moutet F. Middle and ring fingers are more exposed to pulley rupture than index and little during sport-climbing: a biomechanical explanation. Clin Biomech (Bristol). 2008;23(5):562–70. https://doi.org/10.1016/j.clinbiomech.2007.12.009.
23. Phillips C, Mass D. Mechanical analysis of the palmar aponeurosis pulley in human cadavers. J Hand Surg Am. 1996;21(2):240–4. https://doi.org/10.1016/S0363-5023(96)80107-9.
24. Doyle JR. Palmar and digital flexor tendon pulleys. Clin Orthop Relat Res (1976–2007). 2001;383:84–96. Accessed 29 Dec 2024. https://journals.lww.com/corr/abstract/2001/02000/palmar_and_digital_flexor_tendon_pulleys_.11.aspx
25. Peterson WW, Manske PR, Bollinger BA, Lesker PA, McCarthy JA. Effect of pulley excision on flexor tendon biomechanics. J Orthop Res. 1986;4(1):96–101. https://doi.org/10.1002/jor.1100040112.
26. Gabl M, Rangger C, Lutz M, Fink C, Rudisch A, Pechlaner S. Disruption of the finger flexor pulley system in elite rock climbers. Am J Sports Med. 1998;26(5):651–5. https://doi.org/10.1177/03635465980260050901.
27. Schöffl V, Einwag F, Strecker W, Schöffl I. Strength measurement and clinical outcome after pulley ruptures in climbers. Med Sci Sports Exerc. 2006;38:637–43. https://doi.org/10.1249/01.mss.0000210199.87328.6a.
28. Vandeputte G, Dubert T. Closed traumatic rupture of the flexor pulleys of a long finger associated with avulsion of the flexor digitorum superficialis. J Hand Surg Br. 2001;26(3):266–8. https://doi.org/10.1054/jhsb.2001.0559.
29. Van Demark RE, Kimber KA, Jorgensen BD. A4 annular flexor pulley injury in a baseball pitcher. J Hand Surg Eur. 2015;40(6):649–51. https://doi.org/10.1177/1753193413518700.
30. Lin GT, Amadio PC, An KN, Cooney WP. Functional anatomy of the human digital flexor pulley system. J Hand Surg Am. 1989;14(6):949–56. https://doi.org/10.1016/S0363-5023(89)80043-7.
31. Arora R, Lutz M, Haug L, Struve P, Deml C, Gabl M. Secondary reconstruction of the digital A2-flexor pulley. Oper Orthop Traumatol. 2013;25(5):499–504. https://doi.org/10.1007/s00064-011-0053-7.
32. Leeflang S, Coert JH. The role of proximal pulleys in preventing tendon bowstringing: pulley rupture and tendon bowstringing. J Plast Reconstr Aesthet Surg. 2014;67(6):822–7. https://doi.org/10.1016/j.bjps.2014.01.041.

33. Chow JC, Sensinger J, McNeal D, Chow B, Amirouche F, Gonzalez M. Importance of proximal A2 and A4 pulleys to maintaining kinematics in the hand: a biomechanical study. Hand (New York). 2014;9(1):105–11. https://doi.org/10.1007/s11552-013-9547-0.
34. Walbeehm ET, McGrouther DA. An anatomical study of the mechanical interactions of flexor digitorum superficialis and profundus and the flexor tendon sheath in zone 2. J Hand Surg Br. 1995;20(3):269–80. https://doi.org/10.1016/s0266-7681(05)80077-4.
35. Marco RA, Sharkey NA, Smith TS, Zissimos AG. Pathomechanics of closed rupture of the flexor tendon pulleys in rock climbers. JBJS. 1998;80(7):1012. Accessed 29 Dec 2024. https://journals.lww.com/jbjsjournal/fulltext/1998/07000/Pathomechanics_of_Closed_Rupture_of_the_Flexor.10.aspx
36. Brand PW, Cranor KC, Ellis JC. Tendon and pulleys at the metacarpophalangeal joint of a finger. J Bone Joint Surg Am. 1975;57(6):779–84.
37. Elftman H. Biomechanics of muscle with particular application to studies of gait. J Bone Joint Surg Am. 1966;48(2):363–77.
38. Freehafer AA, Peckham PH, Keith MW. Determination of muscle-tendon unit properties during tendon transfer. J Hand Surg Am. 1979;4(4):331–9. https://doi.org/10.1016/s0363-5023(79)80069-6.
39. Wehbé MA, Hunter JM. Flexor tendon gliding in the hand. Part I. In vivo excursions. J Hand Surg Am. 1985;10(4):570–4. https://doi.org/10.1016/s0363-5023(85)80085-x.
40. Lane JM, Bora FW, Black J. cis-hydroxyproline limits work necessary to flex a digit after tendon injury. Clin Orthop Relat Res. 1975;109:193–200. https://doi.org/10.1097/00003086-197506000-00028.
41. Lane JM, Black J, Bora FW. Gliding function following flexor-tendon injury. A biomechanical study of rat tendon function. J Bone Joint Surg Am. 1976;58(7):985–90.
42. Uchiyama S, Amadio PC, Ishikawa JI, An KN. Boundary lubrication between the tendon and the pulley in the finger. JBJS. 1997;79(2):213. Accessed 29 Dec 2024. https://journals.lww.com/jbjsjournal/fulltext/1997/02000/Boundary_Lubrication_between_the_Tendon_and_the.8.aspx
43. Mester S, Schmidt B, Derczy K, Nyarady J, Biro V. Biomechanics of the human flexor tendon sheath investigated by tenography. J Hand Surg. 1995;20(4):500–4. https://doi.org/10.1016/S0266-7681(05)80163-9.
44. Schweizer A, Frank O, Ochsner PE, Jacob HAC. Friction between human finger flexor tendons and pulleys at high loads. J Biomech. 2003;36(1):63–71. https://doi.org/10.1016/S0021-9290(02)00242-7.
45. Schweizer A, Moor BK, Nagy L, Snedecker JG. Static and dynamic human flexor tendon–pulley interaction. J Biomech. 2009;42(12):1856–61. https://doi.org/10.1016/j.jbiomech.2009.05.025.
46. Rispler D, Greenwald D, Shumway S, Allan C, Mass D. Efficiency of the flexor tendon pulley system in human cadaver hands. J Hand Surg Am. 1996;21(3):444–50. https://doi.org/10.1016/S0363-5023(96)80361-3.
47. Feasibility of partial A2 and A4 pulley excision: effect on finger flexor tendon biomechanics—PubMed. Accessed 28 Jan 2025. https://pubmed.ncbi.nlm.nih.gov/10194015/
48. Tanaka T, Amadio PC, Zhao C, Zobitz ME, An KN. The effect of partial A2 pulley excision on gliding resistance and pulley strength in vitro. J Hand Surg Am. 2004;29(5):877–83. https://doi.org/10.1016/j.jhsa.2004.04.014.
49. Amirouche F, Gonzalez M, Koldoff J, Tioco J, Ham K. A biomechanical study of the finger pulley system during repair. Technol Health Care. 2002;10(1):23–31. https://doi.org/10.3233/THC-2002-10103.
50. Soulii L, Amirouche F, Solitro G, et al. Evaluation of A2 and A4 hand pulley repair using tendon graft rings. J Hand Surg Am. 2021;46(7):626.e1–6. https://doi.org/10.1016/j.jhsa.2020.11.025.

15 FEA in Pulley Repair

Nirav Mungalpara and Farid Amirouche

What Is Finite Element Analysis (FEA)?

Finite element analysis (FEA) is a computational method that simulates and predicts how complex structures behave under various loading conditions by dividing or meshing the geometry representing the structure at hand into smaller recognizable shapes or discrete elements connected by so-called nodes. Tracking the displacements of the nodes under different boundary conditions is essentially what makes FEA so powerful. Several other characteristics, such as strain, stress, deformation, energy, etc., can be deduced mathematically from the analyzed structure. FEA was initially developed for structural and aerospace engineering applications, but it is now widely used in biomechanics and orthopedics to investigate the mechanical properties of biological tissues and implants. The FEA method formulates a set of algebraic and differential equations based on the law of physics (first and second law of mechanics). It solves for stress, strain, and deformation, considering the applied loads and boundary conditions that mimic the conditions under which the current body is being tested and the corresponding material properties assigned to each part of the structure such as the bone, cartilage, ligaments, or fluids. In biomechanics, FEA is a valuable tool for evaluating bone-implant interactions, fracture mechanics, and soft tissue behavior with high precision. One of FEA's key advantages is its ability to simulate multiple loading conditions, identify potential failure points, and optimize implant designs before clinical application.

FEA has gained prominence in orthopedic biomechanics due to advances in computational power and engineering modeling software. It allows researchers to simulate a wide range of scenarios, such as physiological processes, disease mechanisms, injury patterns, and the effectiveness of surgical interventions [1]. FEA enhances clinical research by offering a platform to compare surgical treatments that may not be feasible in human or animal studies while linking simulation outcomes to clinical results. Traditional ex vivo experiments are limited by the variability of biological tissues, complexity in setup, cost, and challenges in repeatability [1]. On the other hand, FEA provides the opportunity

N. Mungalpara
Department of Orthopaedics, University of Illinois at Chicago College of Medicine, Chicago, IL, USA
e-mail: Nirav@uic.edu

F. Amirouche (✉)
Department of Orthopaedics Surgery, Orthopaedic and Spine Institute, Endeavor HealthSystem (Formerly Northshore University HealthSystem), an Affiliate of the University of Chicago Pritzker School of Medicine, Skokie, IL, USA

Department of Orthopaedic Surgery, University of Illinois at Chicago, Chicago, IL, USA
e-mail: Amirouch@uic.edu; famirouche@northshore.org

F. Amirouche et al. (eds.), *Biomechanics of the Hand*,
https://doi.org/10.1007/978-3-032-04688-8_15

to measure clinically relevant biomechanical parameters consistently. Generic and patient-specific anatomical models can simulate various clinical and surgical conditions under physiological loading [2].

Despite its many capabilities, the full clinical applicability and translational potential of FEA are still not fully understood. For spine surgeons, a deeper understanding of the strengths and limitations of FEA models could significantly improve clinical decision-making and personalized treatment. Similarly, for hand surgeons, comprehensive knowledge of FEA models focusing on the hand, digits, and pulley systems could provide insights into personalized treatment options and surgical planning. The key to unlocking the full potential of FEA in the clinical setting lies in strengthening collaborations between engineers and clinicians, which may further expand the practical application of these models.

Brief History and Background

The earliest FEA models were developed and validated using ex vivo cadaveric studies, which utilized pulleys and weights to replicate biomechanical forces on the hand during physiological activities. Although ex vivo testing has been a staple for studying hand-related disorders and surgical interventions, it has significant limitations. These include the variable quality of tissue samples, often sourced from older donors, the complexity of test setups, limited repeatability, restricted biomechanical parameters, and high costs [3]. Advances in computational technology have allowed FEA models to overcome these challenges, offering greater flexibility in simulating real-life conditions at a fraction of the cost. FEA models enable the consistent measurement of clinically relevant parameters under repeatable conditions. With current technology, it is possible to create personalized FEA models that incorporate individual patient characteristics such as bone density and anatomy. However, the fundamental assumptions of these basic models still influence simulation outcomes [2, 4, 5].

For example, spine FEA models are designed to replicate the intrinsic biomechanical response of the human spine under various loading conditions. Cervical spine flexion-extension produces diverse segmental ranges of motion (ROMs), while the weight of the head and torso contributes to cervical and dorsal spine movement. Numerous FEA models exist, ranging from isolated tissue models (e.g., ligaments) to thoroughly segmented spine models that simulate the intact spine [6–8]. These models have been tested under multiple loading paradigms, including axial forces, combined axial loads and bending moments, vibration, and impact. Such studies are essential for understanding the biomechanics of the spine in healthy, diseased, and surgically altered conditions.

Similarly, FEA has been used to develop detailed three-dimensional human hand models. Several studies have focused on creating biomechanical models for artificial limb design and ergonomic research [9, 10]. Advanced imaging techniques such as computed tomography (CT) and magnetic resonance imaging (MRI) are often employed to generate accurate three-dimensional models. The images are imported into computer-aided design (CAD) programs to construct three-dimensional geometries [11].

D N Rocha et al [12] developed an FEA model of the human hand, analyzing its behavior during simulated index finger flexion. The creation of any FEA model involves several critical steps:

1. Building a three-dimensional geometric model
2. Using specialized software to generate an FEA model from the three-dimensional model
3. Applying boundary conditions and performing simulations
4. Validating the simulation results against experimental data
5. Using the validated model to assess biomechanical properties under various surgical and pathological conditions

For instance, the human hand consists of 27 bones and numerous muscles, tendons, and ligaments, resulting in 23 degrees of freedom. Due to the hand's complexity, D N Rocha et al. [12] proposed a simplified model consisting of 16 bones—4 metacarpals and 12 phalanges. To ensure biomechanical accuracy, this geometric model represented the soft tissue, bones, joint capsules, and cartilage.

In the model described by D N Rocha et al. [12], the base geometries were constructed from medical images and imported into SolidWorks for further refinement. Anatomical structures such as bones, cartilage, and capsules were modeled using reference points and splines to generate new solid geometries. Joints were defined at the metacarpophalangeal (MCP), proximal interphalangeal (PIP), and distal interphalangeal (DIP) levels, incorporating articular cartilage and capsules at the bone ends for biomechanical accuracy.

Building of Geometrical Model

There are various methods to build a three-dimensional geometrical model described by D N Rocha et al. [12].

A base geometric model derived from medical images was used as a reference for creating a new geometrical model because the base model was illustrative. Bone, cartilage, and capsule geometries were imported into SolidWorks [13], where reference points were made and interconnected with smoothing curves (splines) to generate new solid geometries. Anatomical images from an atlas [14] (Fig. 15.1a) guided the assembly of a new finger model using these geometries (Fig. 15.1b). Joints were constructed to connect the bones at the metacarpophalangeal (MCP), proximal interphalangeal (PIP), and distal inter-

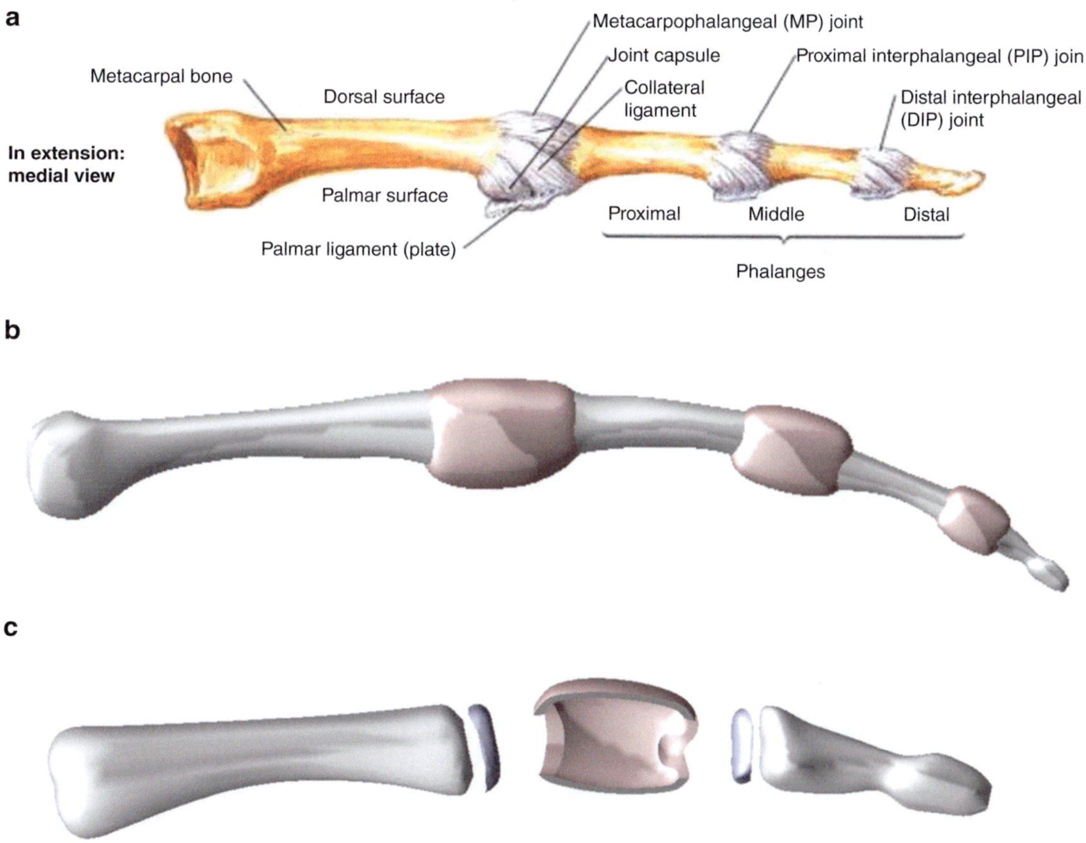

Fig. 15.1 (**a**) Representative image, (**b**) assembly of the index finger geometry, and (**c**) the proximal interphalangeal joint. (From Rocha et al. [12])

phalangeal (DIP) levels, incorporating articular capsules and cartilage at bone ends (Fig. 15.1c). Skin tissue was modeled by linking several cross-sectional sketches along the finger assembly with reference points, which were displaced outward to match soft tissue thickness values from the literature [15].

Another prevalent method is to derive from the CT or MR images. Wei et al. [16] have described the geometrical information for a subject-specific FE human hand obtained from CT and MR images of a 23-year-old healthy male. The two-dimensional data were processed using Mimics (Materialise, Leuven, Belgium), with CT images manually segmented for bones and skin, while MR images were used to reconstruct subcutaneous tissues and tendons. The geometry was exported as an STL mesh and converted into solid models in Creo (PTC, Creo Parametric, US). The anatomical positions of bones and soft tissues were aligned based on MR images. All structures, including bones (14 phalangeal, 5 metacarpal, 8 carpal), subcutaneous tissues, and skin, were imported into Abaqus (Simulia, Providence, US). Solid elements (C3D4H) were used to mesh the model (Fig. 15.2), while ligaments were modeled with spring elements to represent supporting tissues around the joints, with anatomical locations determined from MR images.

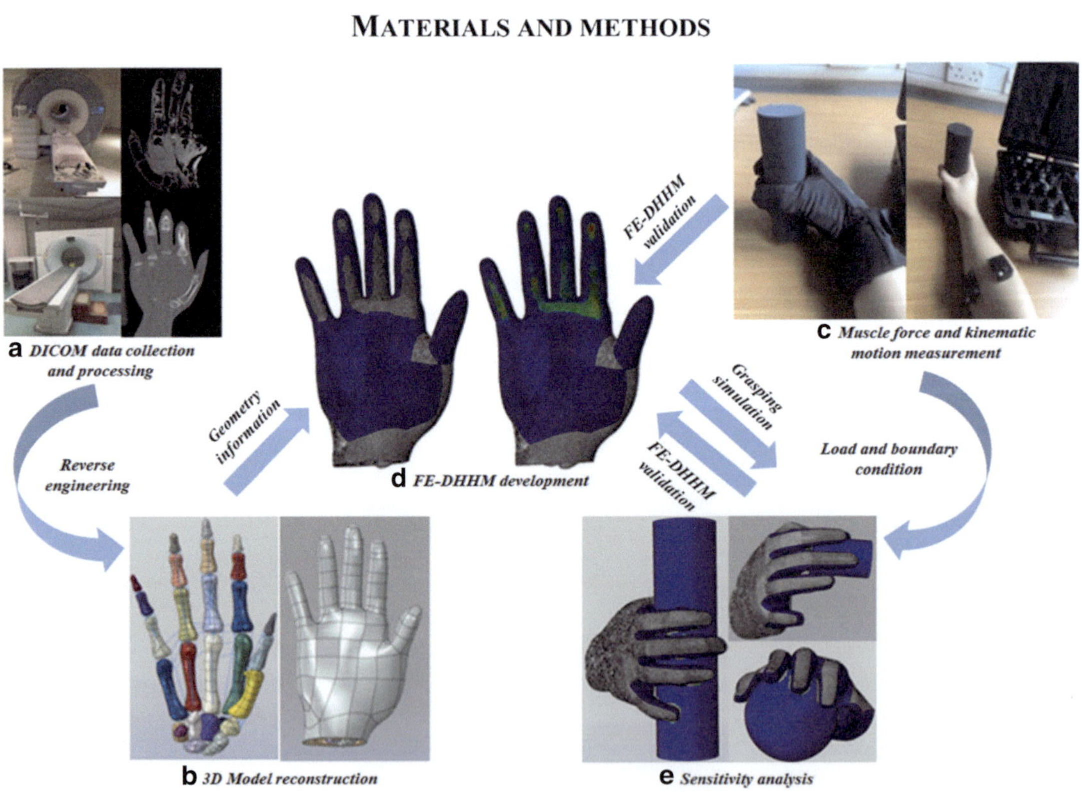

Fig. 15.2 (**a**) DICOM data collected from a specific subject. (**b**) Three-dimensional model reconstruction. (**c**) The kinematic motion data and muscle forces were collected from the in vivo experiments. (**d**) Predicted contact area and contact pressure. (**e**) The sensitivity analysis. (From Wei et al. [16])

FEM Model and Simulation

R N Rocha et al. [12] developed a new hand geometry model through atlas images which were later imported into Abaqus for finite element method (FEM) analysis. The capsules were characterized using the medial collateral ligament properties of the knee, modeled as Neo-Hookean hyperelastic material with constant $C_1 = 6.43$ MPa [17]. Two simulations were conducted with different skin tissue properties: the first used a Mooney-Rivlin hyperelastic material with $C_{10} = 9.4$ kPa = 9.4 and $C_{11} = 82$ kPa [18] and the second with $C_{10} = 1.7$ kPa and $C_{11} = 1$ kPa.

The FEM process involved meshing the geometric model. Bones and cartilages, assumed to be rigid bodies due to minimal deformation during finger motion, were modeled with triangular shell elements (R3D3), while capsules and skin tissue, considered deformable bodies, used tetrahedral hybrid elements (C3D4H). Contact pairs and tie constraints were applied to ensure accurate biomechanical representation. Contact pairs allowed frictionless sliding between cartilages and ensured capsules enveloped the bones. Tie constraints were used between capsules, cartilages, and bones to maintain structural unity by aligning displacements between connected nodes.

Element size was optimized based on geometry detail, contact pairs, and regions of interest. To reduce penetration, the slave surface elements were set at half the size of the master surface elements. Regions of interest, such as the fingers, had finer elements for more accurate analysis than the wrist (Fig. 15.3a, b).

In another method deployed by Wei et al. [16], the skin was modeled as a hyperelastic material using the Ogden model, which has been shown best to represent the complex, anisotropic behavior of human skin [19]. Material parameters were extracted from previous studies on uniaxial tensile tests of the human skin and fat tissue. These parameters were fitted into the Ogden model using the "material evaluate" module in Abaqus to simulate the nonlinear properties of the skin. Bones were modeled as isotropic linear elastic with Young's modulus of 17 GPa and Poisson's ratio of 0.3 [15, 17]. Ligament stiffness values were derived from the literature and used to model spring elements representing joint-supporting structures. Frictional surface contact between the skin and external objects was defined with a coefficient of 0.74, ensuring sliding interaction without penetration [17]. The subject used for the FE hand model also participated in in vivo grasping tests, which included three common grasping postures: cylindrical, spherical, and precision grasping [18]. For these tests, motion capture data were recorded using a VMG30 data glove (Virtual Motion Lab, USA), which accurately measured joint angles at the interphalangeal and metacarpophalangeal joints (Fig. 15.4). Fingertip contact pressures were simultaneously recorded with sensors embedded in the finger pads, enabling correlation of kinematic data with pressure distribution. EMG signals were captured using a Delsys wireless EMG system (Delsys

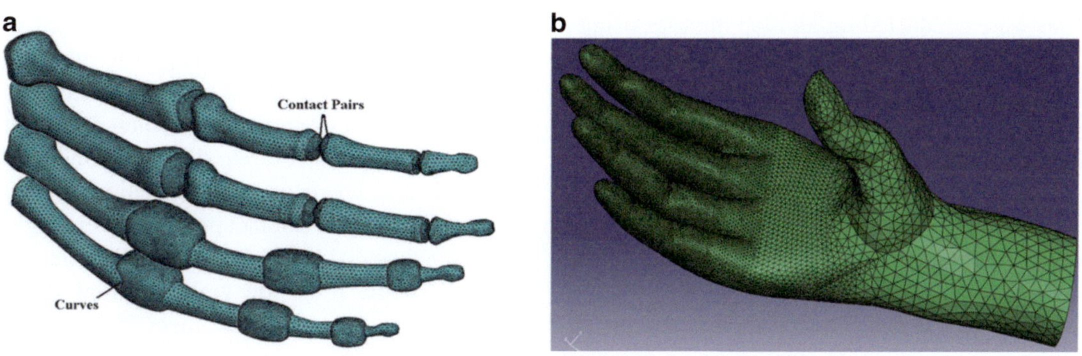

Fig. 15.3 (**a**) Mesh of bones, cartilage, and capsules; (**b**) skin tissue mesh. (From Rocha et al. [12])

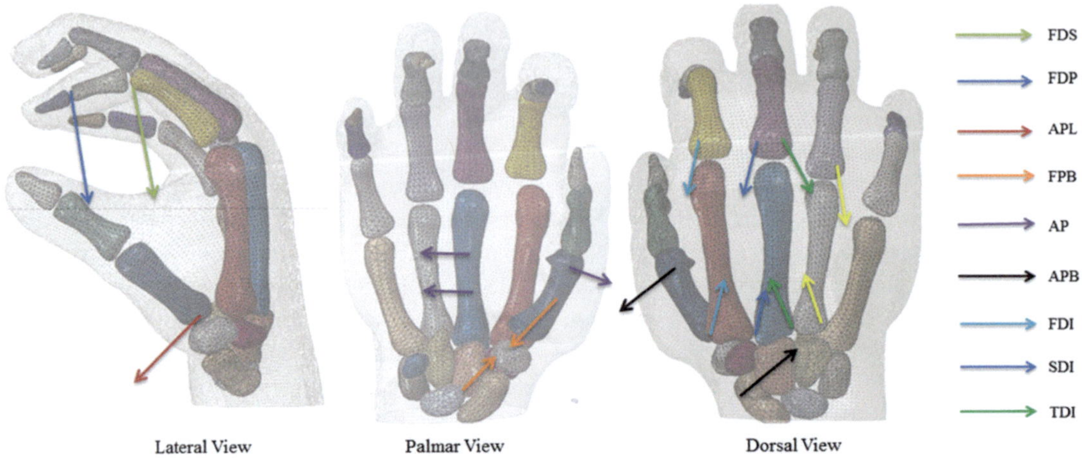

Fig. 15.4 Muscle force definition in the FE human hand. The arrows represent directions and anatomical positions of the applied muscle forces for all three defined grasping. (From Wei et al. [16])

Table 15.1 Muscle forces during grasping one of the six trials

Muscle	Cylindrical grasping (N)	Spherical grasping (N)	Precision grasping (N)
FDS	138.78	240.03	111.72
FDP	89.37	155.84	162.20
APL	32.01	14.87	51.07
APB	6.74	6.04	6.11
AP	26.68	11.10	115.36
FPB	105.23	57.00	102.60
FDI	41.40	46.01	42.17
SDI	68.11	64.50	61.70
TDI	41.20	39.49	32.62

Source: Wei et al. [16]

Inc., USA) to estimate muscle forces involved in grasping [19, 20]. Following the guidelines for surface electromyography, Trigno sensors were positioned along muscle fibers to record EMG signals. Maximum voluntary contraction (MVC) tests were conducted for all nine muscles using a Jamar dynamometer (Table 15.1). EMG data were bandpass filtered (20–400 Hz) with a Butterworth filter, rectified, and normalized to MVC values for each muscle [21]. The muscle forces were then estimated from the filtered EMG signals, assuming a linear relationship between EMG activity and muscle force during isometric contraction [22].

Nine muscle forces were applied as concentrated loads at the anatomical insertion points of tendons and ligaments identified from MR images. The forces were used to flexor digitorum superficialis (FDS), flexor digitorum profundus (FDP), abductor pollicis longus (APL), flexor pollicis brevis (FPB), adductor pollicis (AP), abductor pollicis brevis (APB), and three dorsal interossei muscles (FDI, SDI, TDI). The kinematics of the hand were defined using reference points and rigid wire connectors at the joints to simulate flexion, extension, and MCP joint adduction/abduction. Two reference points were created at each interphalangeal joint: one at the proximal head and another at the rotational center. A rigid wire connected these points to restrict motion to a single axis. This approach was repeated for all joints, ensuring a realistic biomechanical representation of hand movements. The FE hand model was assigned 27 degrees of freedom across the 15 interphalangeal and wrist joints, with angular displacements specified at each joint based on measured angles to maintain biomechanical accuracy and numerical stability.

The FE model accurately predicted contact pressure and contact area during cylindrical, spherical, and precision grasping, aligning well with in vivo results. Contact pressures were highest on the thumb during cylindrical and spherical grasping, while precision grasping showed the highest pressure on the index finger. Relative differences for contact pressure remained below 20% for cylindrical and spherical grasps and under 16% for precision grasping, with contact areas differing by less than 15%. Simulated pressures were generally higher than in vivo results, while experimental contact areas exceeded simulated values by 7.9–14.1%.

Material property sensitivity analysis demonstrated that contact pressure increased with hardened subcutaneous tissue, while softened skin caused more considerable pressure variations. Skin properties influenced contact pressure more, but the contact area was more sensitive to subcutaneous tissue changes. Muscle force variations also significantly affected contact pressure. A 20% increase in FDP force raised middle fingertip pressure by 23.7% and affected other fingertips by up to 22.8%. In contrast, FDS force had minimal impact, causing only a 0.3–9.1% change in fingertip pressure. AP force increased thumb contact pressure by 15.3% and expanded contact area by 13%. Nonlinear responses reflected the complex biomechanical behavior of skin, subcutaneous tissue, and muscle forces during grasping.

Use of FEA in Pulley Repair

Amirouche et al. [23] previously published a biomechanical study analyzing the kinematic impact of looped autograft reconstruction following A2 pulley rupture in cadaveric models. The findings demonstrated significant reductions in finger joint flexion with double- and triple-loop repairs compared to single-loop repairs. To build on this work, we developed and applied a validated FEA model to further explore the mechanical behavior of these reconstructions under simulated conditions, mainly focusing on the crimp grip posture, known to place substantial stress on the flexor tendon pulley system.

The FEA model was constructed to assess von Mises stress distribution at the tendon-pulley interface and analyze the effects of pulley length and height reductions on grip force and pulley stress. Cadaveric kinematic data validated the computational model and ensured the simulated results accurately reflected real-world observations.

Figure 15.5 illustrates the surgical excision and repair of the A2 pulley using single-, double-, and triple-ring grafts. These configurations were replicated in the FEA model to simulate their biomechanical effects. The FEA analysis confirmed that single-loop repairs restored better kinematics but resulted in significantly higher stresses at the tendon-pulley interface than multiple-loop repairs. Figure 15.6 depicts the four simulated pulley configurations in the FEA model: intact, complete excision, and two variations of reduced pulley thickness and length. These alterations allowed for a comprehensive analysis of how pulley modifications impact stress distribution and grip force during the crimp grip posture.

Figure 15.7 shows the crimp grip's finite element simulations, illustrating the digit's dynamic musculoskeletal motion in this posture [24]. The simulation revealed that pulley length reductions increased grip force significantly compared to height reductions. Figure 15.8 highlights the grip force achieved with 30 mm of tendon excursion for the various simulated pulley alterations, dem-

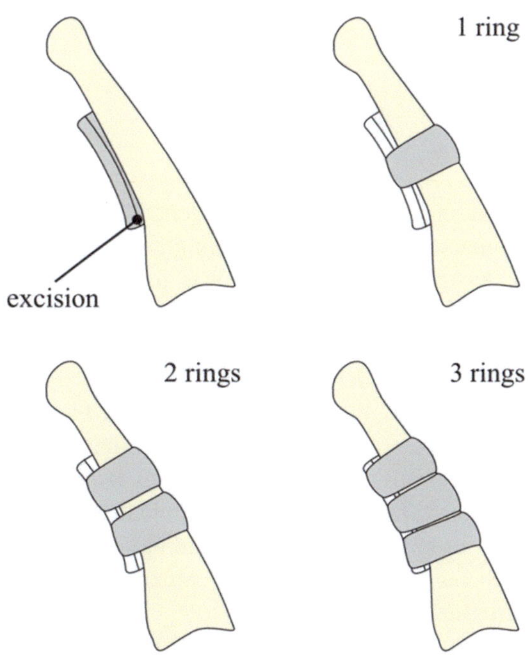

Fig. 15.5 Excision and repair of the A2 pulley using one-, two-, and three-ring FDP tendon grafts via loop reconstruction method. (From Amirouche et al. [24])

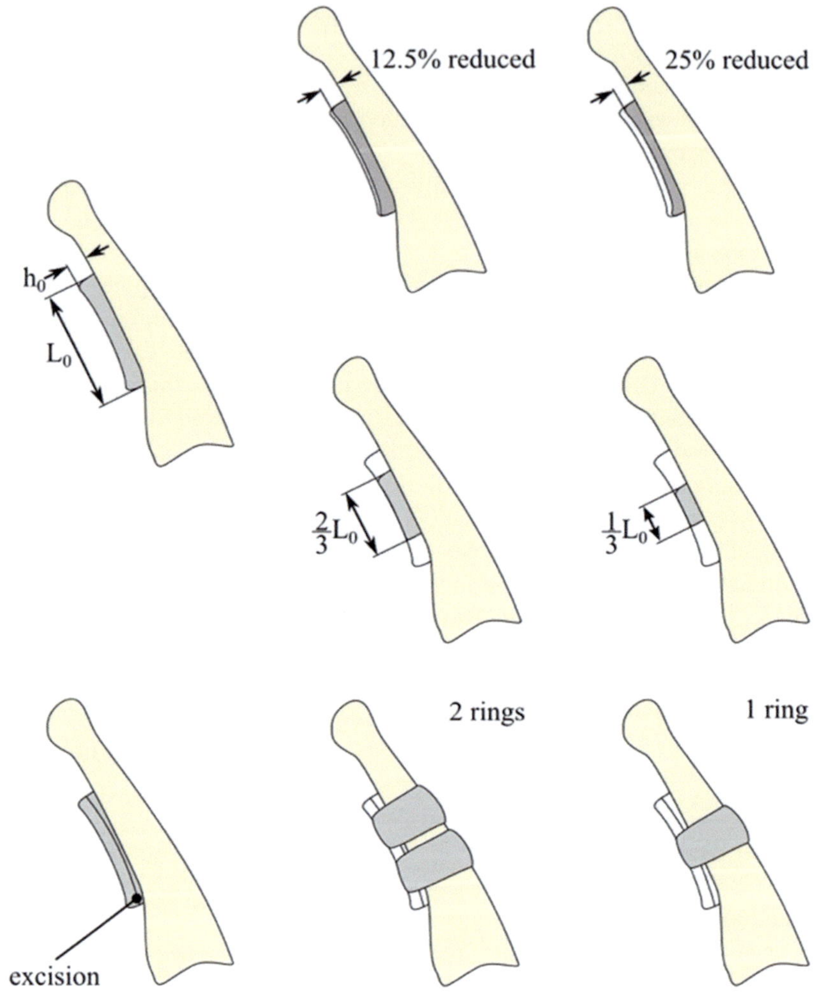

Fig. 15.6 Four alterations of the A2 pulley were simulated with the FEA model. (From Amirouche et al. [24])

onstrating that reducing pulley length had a more pronounced impact on increasing grip force.

A significant difference was found between the one- and two-loop repairs ($p < 0.05$) and only for the PIP flexion angle (Table 15.2).

Figure 15.9 depicts the von Mises stress distribution across the A2 pulley. It shows that the highest stress concentrations were found at the distal portion of the pulley when its length was shortened. This elevated stress could increase the risk of failure under high-loading conditions, such as rock climbing or other high-stress activities.

The simulated von Mises stress at different tendon excursions and grip forces for the four simulated pulley configurations was described. The stress values were significantly higher in the shortened pulley configurations, especially at more extensive tendon excursions. These findings emphasize that while restoring kinematics is crucial, reducing pulley length can create detrimental stress patterns that compromise repair integrity.

The results of this study suggest that optimizing pulley repair requires balancing, restoring kinematic function, and minimizing mechanical

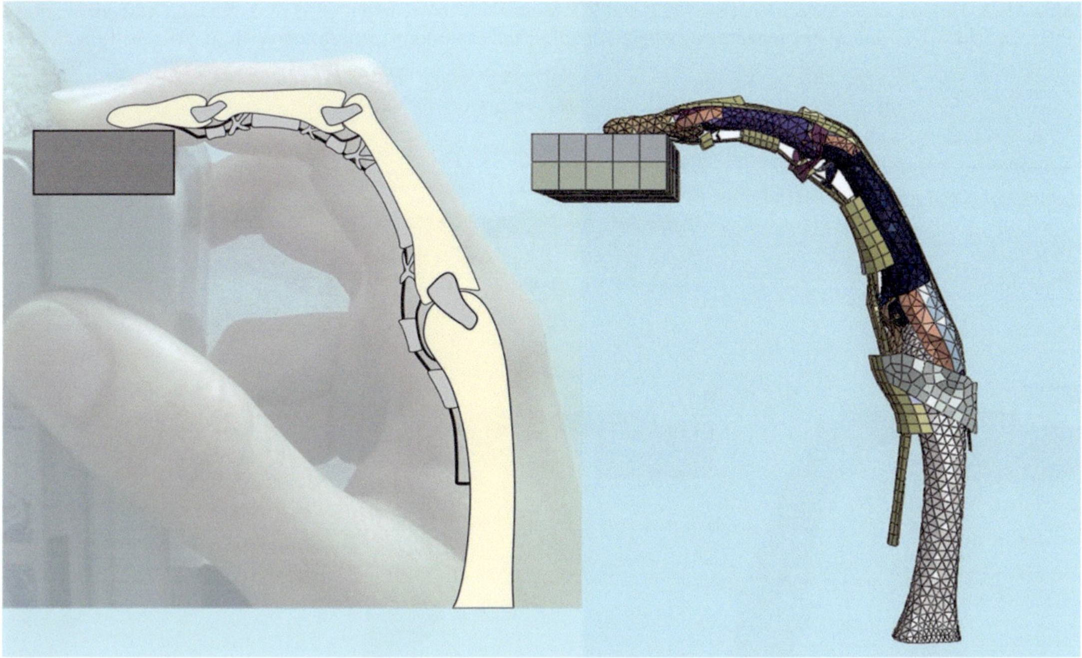

Fig. 15.7 Illustration of the grip posture hypothesized and created using the finite element model CT-based modelling. (From Amirouche et al. [24])

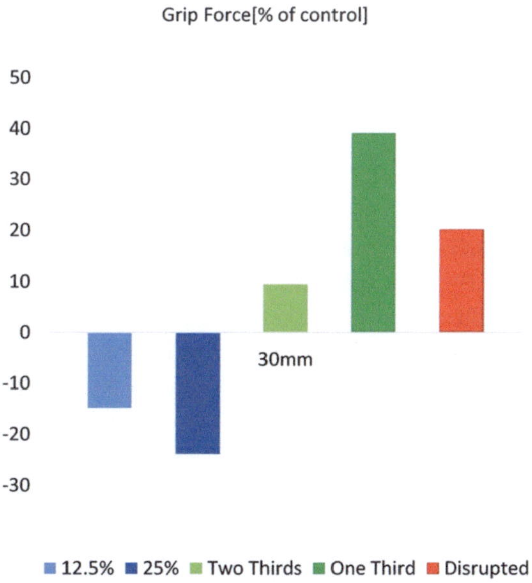

Fig. 15.8 Grip force was achieved with 30 mm of FDS, and the FDP excursion for the four simulated pulley alterations was expressed as a percentage of the value observed for the control configuration. (From Amirouche et al. [24])

stress. Single-loop repairs have been shown to significantly enhance joint flexion, allowing for a more natural range of motion. However, these repairs can place considerable stress on the pulley system, increasing the risk of rerupture. In contrast, double-loop repairs provide a more balanced approach. They effectively manage stress on the pulley while still maintaining optimal functional joint mobility. These findings highlight the critical role of finite element analysis (FEA) as a sophisticated tool in the realm of surgical planning, ultimately aiming to enhance patient outcomes in the delicate process of pulley repair.

Limitations and Considerations

Despite its versatility, FEA has certain limitations that must be considered. The accuracy of simulation results is highly dependent on the quality of input data and the underlying assump-

Table 15.2 Flexion angle measured at the MCP and PIP joints with an excursion of 35 mm of a digit with an intact A2 pulley, excision of A2 pulley, and repair using single-, double-, and triple-loop reconstruction (from Amirouche et al. [24])

	Flexion angle (°)		Percentage change from intact A2 pulley (%)	
	MCP	PIP	MCP	PIP
A2 pulley intact	44.9 ± 11.3	63.8 ± 5.2		
A2 pulley excision	29.4 ± 9.7	44.4 ± 4.8	−34.46 ± 13.75	−30.47 ± 8.60
One loop	36.6 ± 9.7	55.2 ± 4.6	−18.54 ± 14.40	−13.45 ± 11.18
Two loops	39.3 ± 10.3	46.4 ± 4.5	−12.41 ± 9.03	−27.34 ± 14.42
Three loops		43.0 ± 4.7		−32.59 ± 10.53

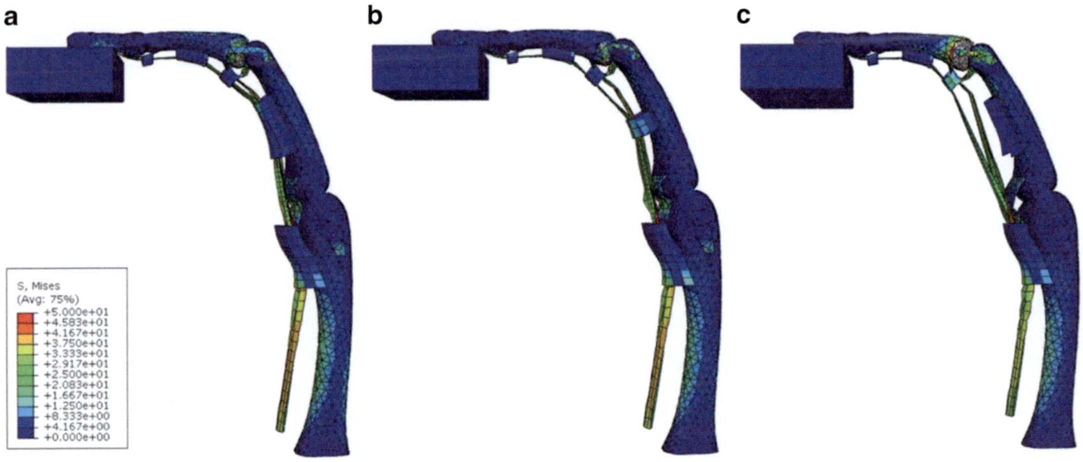

Fig. 15.9 von Mises stress [MPa] distribution in (**a**) intact A2 pulley, (**b**) pulley A2 one-third of the original length, and (**c**) completely excised A2 pulley. (From Amirouche et al. [24])

tions used in model construction. Reliable outcomes in hand biomechanics require validation with experimental data to ensure that the finite element models accurately reflect real-world conditions [4, 24].

Conclusion

Finite element analysis (FEA) has profoundly transformed the investigation of hand pulley biomechanics by establishing a comprehensive and flexible framework for examining the intricate mechanical interactions that occur within the hand. This advanced analytical method facilitates a detailed understanding of how various forces and movements interact in real time, providing critical insights into hand function across a diverse array of activities—from daily tasks to specialized movements in sports or rehabilitation. By identifying stress distribution and deformation patterns, FEA not only enhances our knowledge of normal hand mechanics but also plays a crucial role in refining surgical techniques not addressed in this chapter and developing tailored treatment strategies for hand injuries and disorders. This technique will ultimately contributes to improved patient outcomes and in clinical and therapeutic settings.

References

1. Roberts GL, Pallister I. Finite element analysis in trauma and orthopaedics—an introduction to clinically relevant simulation and its limitations. Orthop

Trauma. 2012;26(6):410–6. https://doi.org/10.1016/j.mporth.2012.10.007.
2. Herrera A, Ibarz E, Cegoñino J, et al. Applications of finite element simulation in orthopedic and trauma surgery. World J Orthop. 2012;3(4):25–41. https://doi.org/10.5312/wjo.v3.i4.25.
3. Scott CEH, Simpson AHRW, Pankaj P. Distinguishing fact from fiction in finite element analysis. Bone Joint J. 2020;102-B(10):1271–3. https://doi.org/10.1302/0301-620X.102B10.BJJ-2020-0827.R1.
4. Huiskes R, Chao EY. A survey of finite element analysis in orthopedic biomechanics: the first decade. J Biomech. 1983;16(6):385–409. https://doi.org/10.1016/0021-9290(83)90072-6.
5. Liu WK, Li S, Park HS. Eighty years of the finite element method: birth, evolution, and future. Arch Computat Methods Eng. 2022;29(6):4431–53. https://doi.org/10.1007/s11831-022-09740-9.
6. Naoum S, Vasiliadis AV, Koutserimpas C, Mylonakis N, Kotsapas M, Katakalos K. Finite element method for the evaluation of the human spine: a literature overview. J Funct Biomater. 2021;12(3):43. https://doi.org/10.3390/jfb12030043.
7. Yoganandan N, Myklebust JB, Ray G, Sances A. Mathematical and finite element analysis of spine injuries. Crit Rev Biomed Eng. 1987;15(1):29–93.
8. Lopez-Espina CG, Amirouche F, Havalad V. Multilevel cervical fusion and its effect on disc degeneration and osteophyte formation. Spine (Phila Pa 1976). 2006;31(9):972–8. https://doi.org/10.1097/01.brs.0000215205.66437.c3.
9. A 3-D dynamic model of human finger for studying free movements—Science Direct. Accessed 12 Feb 2025. https://www.sciencedirect.com/science/article/abs/pii/S0021929001001063
10. Da Cunha FL, et al. Development of anthropomorphic upper limb prostheses with human-like interphalangian and interdigital couplings. Artificial Organs—Wiley Online Library; 2000. Accessed 12 Feb 2025. https://onlinelibrary.wiley.com/doi/abs/10.1046/j.1525-1594.2000.06528.x
11. FADEUP—3D reconstruction of the middle ear for FEM simulation. Accessed 12 Feb 2025. https://sigarra.up.pt/fadeup/en/pub_geral.pub_view?pi_pub_base_id=93358
12. Rocha DN. Development of 3D biomechanical model of the human hand using FEM. Published Online; 2009.
13. The Proven Solution for 3D Design and Product Development | Solidworks. December 7, 2023. Accessed 12 Feb 2025. https://www.solidworks.com/proven-solution-3d-design-and-product-development
14. Netter Atlas of Human Anatomy: Classic Regional—9780323793735. Accessed 12 Feb 2025. https://www.us.elsevierhealth.com/netter-atlas-of-human-anatomy-classic-regional-approach-hardcover-9780323793735.html?srsltid=AfmBOoqsprrtHadeoC0oNQonT3v3yuwidb0mTsQt1GvP21Ukhy8veQoW
15. Wu J, Welcome D, Dong R. Three-dimensional finite element simulations of the mechanical response of the fingertip to static and dynamic compressions. Comput Methods Biomech Biomed Engin. 2006;9:55–63. https://doi.org/10.1080/10255840600603641.
16. Wei Y, Zou Z, Wei G, Ren L, Qian Z. Subject-specific finite element modelling of the human hand complex: muscle-driven simulations and experimental validation. Ann Biomed Eng. 2020;48(4):1181–95. https://doi.org/10.1007/s10439-019-02439-2.
17. Finite element analysis of the effect of meniscal tears and meniscectomies on human knee biomechanics—PubMed. Accessed 12 Feb 2025. https://pubmed.ncbi.nlm.nih.gov/15836937/
18. A numerical-experimental method to characterize the non-linear mechanical behaviour of human skin—Hendriks, 2003. Skin Research and Technology—Wiley Online Library. Accessed 12 Feb 2025. https://onlinelibrary.wiley.com/doi/abs/10.1034/j.1600-0846.2003.00019.x
19. Park WH, Li S. Responses of finger flexor and extensor muscles to transcranial magnetic stimulation during isometric force production tasks. Muscle Nerve. 2013;48(5):739–44. https://doi.org/10.1002/mus.23804.
20. Lawrence JH, De Luca CJ. Myoelectric signal versus force relationship in different human muscles. J Appl Physiol Respir Environ Exerc Physiol. 1983;54(6):1653–9. https://doi.org/10.1152/jappl.1983.54.6.1653.
21. Johansson RS, Birznieks I. First spikes in ensembles of human tactile afferents code complex spatial fingertip events. Nat Neurosci. 2004;7(2):170–7. https://doi.org/10.1038/nn1177.
22. Selective recruitment of single motor units in human flexor digitorum superficialis muscle during flexion of individual fingers—PubMed. Accessed 12 Feb 2025. https://pubmed.ncbi.nlm.nih.gov/15946972/
23. Amirouche F, Gonzalez M, Koldoff J, Tioco J, Ham K. A biomechanical study of the finger pulley system during repair. Technol Health Care. 2002;10(1):23–31. https://doi.org/10.3233/THC-2002-10103.
24. Amirouche F, Solitro GF, Gonzalez MH, Mejia A. Rupture and graft ring repair of finger flexor tendon A2 pulley. Orlando: Presented at: Orthopaedic Research Society Annual Meeting; 2016.

Boutonnière Deformity

16

Seth Roge, Meagan Womack, and Steven Grindel

Introduction

Boutonnière deformity is characterized by flexion of the proximal interphalangeal (PIP) joint and hyperextension of the distal interphalangeal (DIP) joint, as pictured in Fig. 16.1. This deformity may occur acutely after disruption of the extensor mechanism's central slip (CS) via laceration [1], blunt trauma, or burn [2–4]. In the chronic setting, boutonnière deformity commonly occurs due to inflammatory arthropathies like rheumatoid arthritis (RA). RA can cause PIP joint synovitis that degrades the CS over time [5]. The CS may also be attenuated due to congenital abnormality [6].

Pathomechanics of Acute Boutonnière Deformity

The CS is needed for PIP joint extension, and its injury is commonly implicated in boutonnière deformity; however, other extensor structures are involved. The triangular ligament holds the lateral bands in a dorsal position over the PIP joint, and the transverse and oblique fibers of the interosseous hood anchor the lateral bands dorsally to the CS. When these extensor structures are damaged, the intrinsic muscles of the hand contribute to lateral band subluxation, which causes deformity [7–9]. Some clinical and biomechanical papers have examined how progressive damage of these structures creates a boutonnière deformity.

Grundberg et al. found that a surgical tenotomy of the CS to treat mallet finger deformity, known as a Fowler tenotomy, did not result in boutonnière deformity [10]. Mercer et al. used a cadaver model to compare Swanson's method of CS reconstruction to their new technique. Swanson's CS reconstruction technique involved reattaching the CS to its bony insertion on the middle phalanx [11], and Mercer's new technique did not involve CS reattachment to the bone at all. They found that both extensor repair constructs tolerated physiologic loading of 25 N with minimal stiffness. Based on Mercer and her colleagues' experience with this new technique in a cohort of ten patients, they conclude that CS division alone does not necessarily lead to boutonnière deformity [12]. While CS damage alone may not cause boutonnière deformity [13, 14], it can put other extensor structures at risk for degeneration [7].

One study used a cadaveric hand model to determine what combination of structures must be injured to cause a boutonnière deformity. In

S. Roge (✉)
Department of Orthopaedic Surgery, University of Illinois at Chicago, Chicago, IL, USA

M. Womack · S. Grindel
Department of Orthopaedic Surgery, Medical College of Wisconsin, Milwaukee, WI, USA

F. Amirouche et al. (eds.), *Biomechanics of the Hand*,
https://doi.org/10.1007/978-3-032-04688-8_16

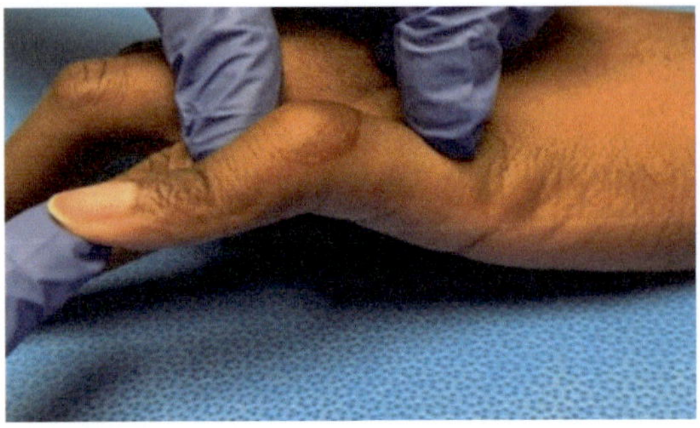

Fig. 16.1 An imaging showing boutonnière deformity

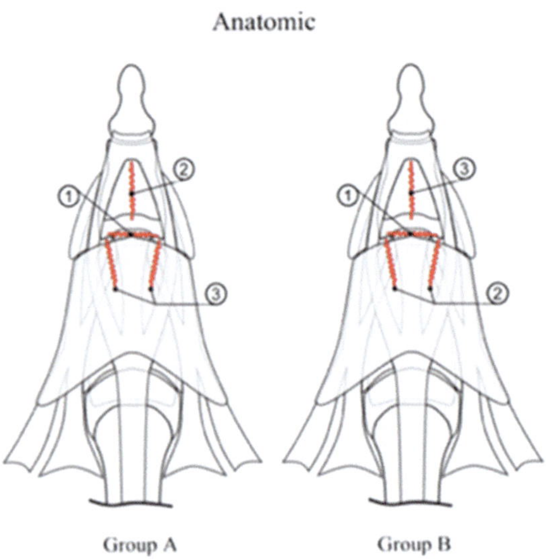

Fig. 16.2 The central slip was divided into digits in groups A and B. The triangular ligament was then divided in group A, followed by the transverse and oblique interosseous hood. In group B, this order was reversed

one aim of a survey by Grau et al. [15], the group dissected ten cadaveric digits and applied constant physiologic loads to the flexor system. Next, the extensor digitorum communis tendon was attached via suture to a computer-controlled servomotor that tensioned the extensor system. At the same time, the PIP joint angle was measured through the digit's range of motion. Figure 16.2 demonstrates how structures were sequentially released to study their contribution to creating a boutonnière deformity. The CS of all ten digits was divided, and their extensor system was tensioned to evaluate for boutonnière deformity. Then, the triangular ligament was divided into

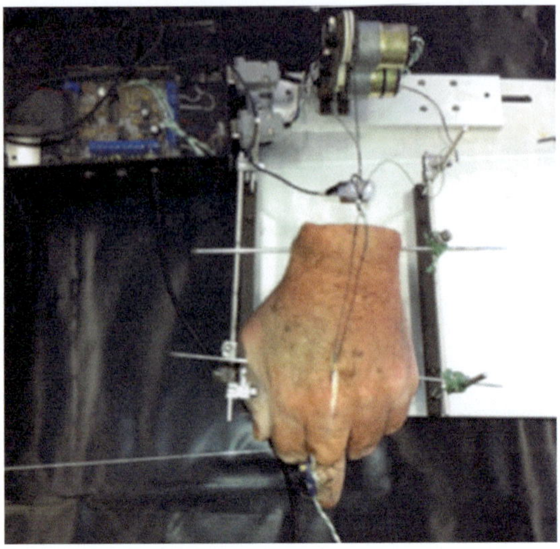

Fig. 16.3 An image of the testing setup used in the first arm of a study by Grau et al. [15]

group A, and the interosseous hood was divided into group B before another round of testing. Finally, the remaining structures were divided between the groups, and the digits were tested again. Figure 16.3 shows the testing apparatus that was used.

The authors' findings are significant in understanding boutonnière deformity. They discovered that the isolated division of the CS, the triangular ligament in group A, or the interosseous hood in group B was not enough to cause a boutonnière deformity. The deformity only occurred when the CS, the triangular ligament, and the interosseous hood were divided into groups A and B. At initial PIP joint flexion angles of 35° or greater, the lateral bands subluxated volar to the PIP joint and acted as paradoxical flexors.

This study had a second aim: eight cadaveric digits were attached to the same testing device. After the sequential division of extensor mechanism structures, the PIP joint angle was measured through the digits' range of motion. After testing the uninjured digit, the CS, the interosseous hood, and the triangular ligament were divided. The extensor mechanism was tested after each division.

Figure 16.4 demonstrates the results of the second experiment. Maximal PIP joint extension was

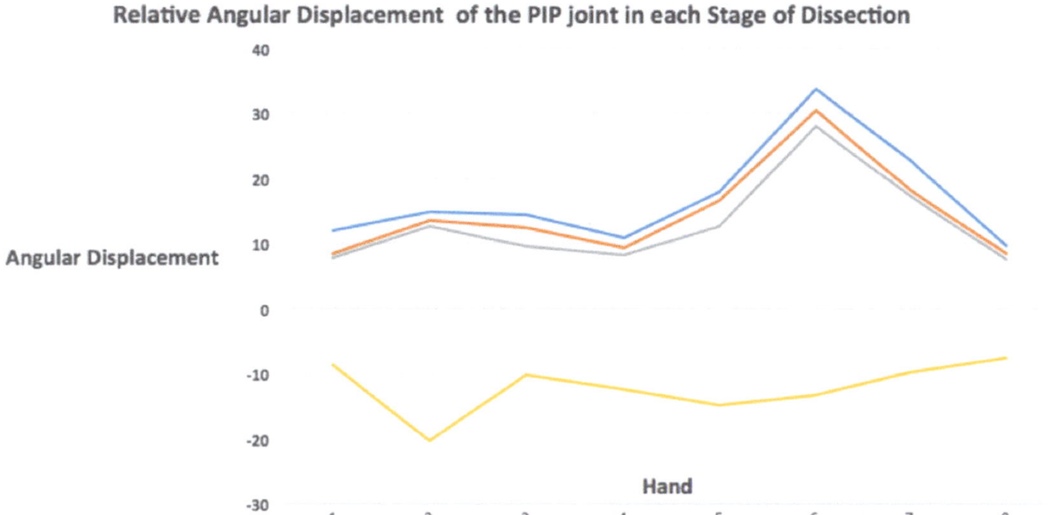

Hand	1	2	3	4	5	6	7	8
Intact	12.1	14.95	14.44	10.88	17.87	33.72	22.86	9.67
Central Slip	8.38	13.45	12.35	9.41	16.66	30.51	18.1	8.49
Central Slip and Interossei Fibers	7.74	12.62	9.54	8.22	12.56	28.02	17.18	7.6
Central Slip and Interossei Fibers and Triangular Ligament	-8.7	-20.37	-10.29	-12.32	-14.83	-13.22	-9.72	-7.62

Fig. 16.4 Results of the second aim of a study by Grau et al. [15]. Positive angles represent PIP joint extension, and negative angles represent PIP joint flexion

observed when all extensor structures were intact. Further division of structures resulted in decreased extension. PIP joint flexion and boutonnière deformity, represented by the yellow line in Fig. 16.4, were observed only after dividing the CS, interosseous hood, and triangular ligament. Through these experiments, the authors conclude that volar displacement of the lateral bands allows the intrinsic and extrinsic digit extensors to act as PIP joint flexors. This kinematic aim further demonstrated that the order of loss of extensor mechanism structures does not affect the development of boutonnière deformity [15].

The results of this study align with other research that shows CS disruption causes mild extensor lag. When the CS is disrupted, the intrinsic muscles of the hand exert excess force on the lateral bands, which attenuates the triangular ligament and other supporting structures. This degeneration of these structures allows volar migration of the lateral bands. This lateral band subluxation subsequently causes the characteristic boutonnière deformity over weeks [16]. Consequently, patients with acute CS injury may not initially present with boutonnière deformity as, initially, extension is maintained through the pull of the lateral bands [3].

Diagnosis of Boutonnière Deformity

These findings have practical implications for diagnosing and treating boutonnière deformity, providing valuable insights for medical professionals in their clinical practice. Boutonnière deformity is essentially a clinical diagnosis, and identifying and treating an acute CS injury can prevent deformity of the PIP joint. Radiographs showing an avulsion of the CS from the base of the middle phalanx may suggest an injury. Additionally, physicians can use two physical exam maneuvers to aid diagnosis.

Elson Test

To perform the Elson test, the patient places their palm flat on a surface with the PIP joint of the

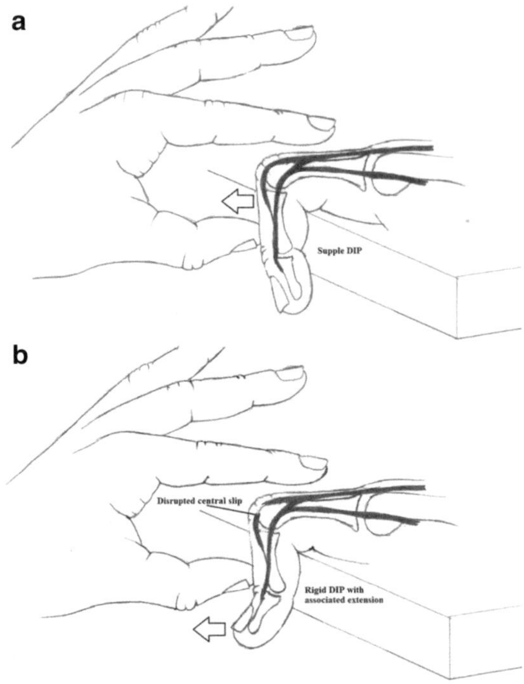

Fig. 16.5 An illustration of the Elson test. (**a**) To perform the Elson test, bend the PIP 90° over the edge of a table and extend middle phalanx against resistance. (**b**) In the presence of central slip injury, there will be weak PIP extension, and the DIP will go rigid. In the absence of a central slip injury, the DIP remains supple because the extension force is now directed entirely on maintaining extension of the PIP joint. The lateral bands are not activated (Adopted from https://www-sciencedirectcom.proxy.cc.uic.edu/science/article/pii/S0278591919300985)

affected digit flexed 90° over the edge. The examiner then has the patient attempt to extend the PIP joint against the examiner's flexion force applied to the middle phalanx. A patient with an intact CS will have supple DIP joint motion during this test, while a patient with a damaged CS will have a rigid DIP joint [17] (Fig. 16.5).

Boyes Test

To perform the Boyes test, the examiner holds the PIP joint of the affected digit in extension. A patient with an intact CS can flex their DIP joint. A patient with a damaged CS may have lateral band subluxation that extends the DIP joint and prevents full DIP joint flexion [18].

Rubin et al. used a cadaver model to evaluate the relative accuracy of these exam maneuvers. The group simulated preboutonnière deformity by cutting the CS. They also simulated passively correctable boutonnière deformity by cutting the interosseous hood's CS, triangular ligament, and oblique fibers. They conducted the above maneuvers in both groups and found that the Elson test could identify both deformities. Still, the Boyes test could not locate either deformity [13]. Therefore, many authors recommend using the Elson test instead of the Boyes test, especially in the acute setting [1].

Houston et al. further questioned using the Elson test in the immediate postinjury setting. Their experiment transected the CS in cadaveric hands to simulate preboutonnière deformity. They then put the digits through a total of 1000 flexion-extension cycles. They simulated an Elson test by measuring the force required to flex the distal phalanx 2.5 and 5 millimeters after 100, 200, 300, and 1000 cycles. The group found that after each set of cycles, the force required to cause the necessary displacement in the distal phalanx increased statistically significantly. However, they state that the change in force required for a positive Elson test between zero and 100 cycles was likely clinically imperceptible, challenging the use of this test immediately after CS injury [19] (Fig. 16.6).

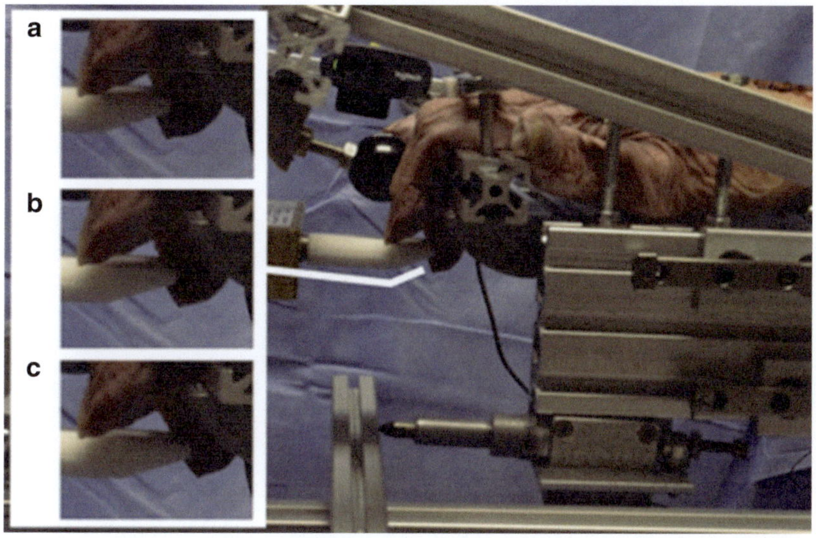

Fig. 16.6 The testing apparatus used to evaluate the reliability of the Elson test in the acute setting. Sequential displacements at the DIP joint to impose flexion during the instrumented Elson test (**a**) at the beginning of the displacement, (**b**) at 2.5 mm, and (**c**) at 5 mm in the experiment performed by Houston et al

Treatment of Boutonnière Deformity

There are various treatment options for boutonnière deformity. PIP joint extension splinting may be used for acute CS injury, while surgical management may be required for more advanced or chronic cases. Furthermore, RA destruction of the PIP joint may require arthrodesis.

Nonoperative Treatment

Traditionally, acute boutonnière deformity or acute CS injuries are treated with splints extending the PIP joint while allowing DIP motion to avoid contracture of the extensor hood. Recent research has focused on using relative motion splints with immediate active motion for CS injuries. One group found excellent clinical results using this technique for acute CS injuries and chronic boutonnière deformities after a period of serial static casting in extension. They argue that splinting the MCP joint of the injured digit in 15–20° of flexion relative to its neighbors relaxes the FDP tendon. This change allows the digit's lumbrical to relax and prevents it from causing volar displacement of the lateral bands [20, 21].

Operative Treatment

There are a handful of published methods to correct boutonnière deformity. However, there is no clear best option due to heterogeneity and limited sample sizes used in clinical reports. Maslow et al. transected the CS in cadaveric hands. They performed a direct CS repair with suture to the bone, a lateral band tenorrhaphy to hold the lateral bands dorsally or a distally based FDS reconstruction. Biomechanical testing of the extensor mechanism showed that all repair constructs had a lower load to failure than an uninjured control but that all constructs tolerated physiologic loading. The FDS slip repair had the highest load to failure at 82 N compared to the direct repair at 28 N or the lateral band tenorrhaphy at 32 N [22].

Models of Boutonnière Deformity

The study of boutonnière deformity has been aided by the cadaveric models mentioned previously. Computer models of human tissues have not been used extensively in boutonnière deformity. Still, general developments in modeling the extensor mechanism may be applied to this field. Past efforts have used finite-element [23,

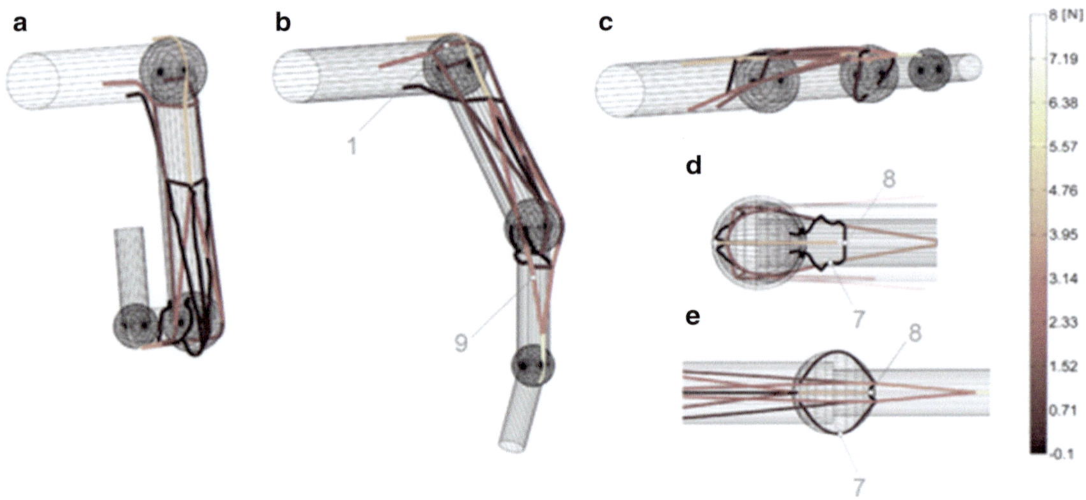

Fig. 16.7 An image of an extensor system model in varying degrees of metacarpophalangeal flexion. The extensor mechanism simulation results in the experiment by Dogadov et al. during three finger postures. Left to right: (**a**) full flexion [DIP = 90°; PIP = 90°; MCP = 90°], (**b**) mid-flexion [DIP = 30°; PIP = 45°; MCP = 45°], (**c**) full extension [DIP = 0°; PIP = 0°; MCP = 0°]; the scaled-up retinacular and triangular ligaments during full flexion (**d**) and full extension (**e**)

24] or rigid-body methods [25, 26]. To model the extensor mechanism, the different mechanical properties of each portion complicate the creation of accurate biomechanical models. Specifically, more proximal areas of the extensor hood tend to be thicker and have greater ultimate load and strain than more distal areas [27]. Furthermore, the behavior of the intrinsic and extrinsic extensors varies with the degree of flexion present [28]. Dogadov et al. created a spring-based model that considered the extensor mechanism's differing mechanical properties along its length. They validated this model for varying degrees of digit flexion [29] (Fig. 16.7).

Future Directions in Boutonnière Deformity Biomechanics

The biomechanical studies presented in this chapter have well established the development of boutonnière deformity due to acute injury or chronic degeneration. Cadaver-based studies are limited in accurately representing live tissue with healing potential. Future studies in this area should focus on the in vivo evaluation of different extensor repair constructs and the application of computer-based models to treat boutonnière deformity.

Acknowledgments The authors of this chapter would like to thank Dr. Farid Amirouche, PhD, for providing edits that were a valuable contribution in the creation of this work.

References

1. McKeon KE, Lee DH. Posttraumatic boutonnière and swan neck deformities. J Am Acad Orthop Surg. 2015;23(10):623–32. https://doi.org/10.5435/JAAOS-D-14-00272.
2. Spinner M, Choi BY. Anterior dislocation of the proximal interphalangeal joint. A cause of rupture of the central slip of the extensor mechanism. Journal of Bone and Joint Surgery. 1970;52(7):1329–36.
3. Strauch R. Extensor tendon injury. In: Wolfe S, Hotchkiss R, Pederson W, Kozin S, Cohen M, editors. Operative hand surgery, vol. 1. 7th ed. Elsevier; 2017. p. 152–82.
4. Sabapathy SR, Bajantri B, Bharathi RR. Management of post-burn hand deformities. Indian J Plast Surg. 2010;43(1 Suppl. 1) https://doi.org/10.4103/0970-0358.70727.
5. Ferlic D. Boutonniere deformities in rheumatoid arthritis. Hand Clin. 1989;5(2):215–22.

6. Carneiro RS. Congenital attenuation of the extensor tendon central slip. J Hand Surg. 1993;18(6):1004–7. https://doi.org/10.1016/0363-5023(93)90391-F.
7. Smith R. Balance and kinetics of the fingers under normal and pathological conditions. Clin Orthop Relat Res. 1974;104:92–111.
8. Williams K, Terrono AL. Treatment of boutonniere finger deformity in rheumatoid arthritis. J Hand Surg. 2011;36(8):1388–93. https://doi.org/10.1016/j.jhsa.2011.05.029.
9. Park J, Lee CH, Choi Y, Joo IH, Lee KH, Kim SJ. Extension mechanism of the proximal interphalangeal joint of the human phalanx: a cadaveric biomechanical study. Biomed Res Int. 2020; https://doi.org/10.1155/2020/7585976.
10. Grundberg AB, Reagan DS. Central slip tenotomy for chronic mallet finger deformity. J Hand Surg. 1987;12(4):545–7. https://doi.org/10.1016/S0363-5023(87)80205-8.
11. Swanson A, de Groot Swanson G. Flexible implant resection arthroplasty of the proximal interphalangeal joint. Hand Clin. 1994;10(2):261–6.
12. Mercer D, FitzPatrick J, Firoozbakhsh K, Carvalho A, Moneim M. Extensor tendon repair with and without central slip reattachment to bone: a biomechanical study. J Hand Surg. 2009;34(1):108–11. https://doi.org/10.1016/j.jhsa.2008.08.021.
13. Rubin J, Bozentka DJ, Bora FW. Diagnosis of closed central slip injuries. A Eadaverie analysis of non-invasive tests. J Hand Surg Br. 1996;21(5):614–6.
14. Harris C, Rutledge G. The functional anatomy of the extensor mechanism of the finger. J Bone Joint Surg. 1972;54(4):713–26.
15. Grau L, Baydoun H, Chen K, Sankary ST, Amirouche F, Gonzalez MH. Biomechanics of the acute boutonniere deformity. J Hand Surg. 2018;43(1):80.e1–6. https://doi.org/10.1016/j.jhsa.2017.07.011.
16. Clavero J, Golanó P, Fariñas O, Alomar X, Monill J, Esplugas M. Extensor mechanism of the fingers: MR imaging–anatomic correlation. Radiographics. 2003;23(3):593–611.
17. Elson R. Rupture of the central slip of the extensor hood of the finger. A test for early diagnosis. J Bone Joint Surg Br Vol. 1986;68(2):229–31.
18. Boyes J. Bunnell's surgery of the hand. 5th ed. Lippincott; 1970.
19. Houston T, Shipley T, Bilderback K, et al. Experimental evaluation of the Elson test efficiency following central slip injury. J Hand Surg Glob Online. 2021;3(6):335–42. https://doi.org/10.1016/j.jhsg.2021.08.004.
20. Merritt WH, Jarrell K. A paradigm shift in managing acute and chronic boutonniere deformity: anatomic rationale and early clinical results for the relative motion concept permitting immediate active motion and hand use. Ann Plast Surg. 2020;84(3):S141–50. https://doi.org/10.1097/SAP.0000000000002307.
21. van Strien G, van Zwieten KJ. An in-depth look at zone III and IV anatomy of the finger extensor mechanism and some clinical implications for using the relative motion flexion orthosis. J Hand Ther. 2023;36(2):280–93. https://doi.org/10.1016/j.jht.2023.01.002.
22. Maslow JI, Posey SL, Habet N, Duemmler M, Odum S, Gaston RG. Central slip reconstruction with a distally based flexor Digitorum Superficialis slip: a biomechanical study. J Hand Surg. 2022;47(2):145–50. https://doi.org/10.1016/j.jhsa.2021.09.010.
23. Jadelis CT, Ellis BJ, Kamper DG, Saul KR. Co-simulation of the index finger extensor apparatus with finite element and musculoskeletal models. J Biomech. 2023:157. https://doi.org/10.1016/j.jbiomech.2023.111725.
24. Elyasi N, Taheri KK, Narooei K, Taheri AK. A study of hyperelastic models for predicting the mechanical behavior of extensor apparatus. Biomech Model Mechanobiol. 2017;16(3):1077–93. https://doi.org/10.1007/s10237-017-0874-x.
25. Vaz A, Singh K, Dauphin-Tanguy G. Bond graph model of the extensor mechanism of the finger based on hook-string mechanism. Mech Mach Theory. 2015;91:187–208. https://doi.org/10.1016/j.mechmachtheory.2015.03.011.
26. Valero-Cuevas FJ. An integrative approach to the biomechanical function and neuromuscular control of the fingers. J Biomech. 2005;38(4):673–84. https://doi.org/10.1016/j.jbiomech.2004.04.006.
27. Qian K, Traylor K, Lee SW, Ellis B, Weiss J, Kamper D. Mechanical properties vary for different regions of the finger extensor apparatus. J Biomech. 2014;47(12):3094–9. https://doi.org/10.1016/j.jbiomech.2014.06.035.
28. Lee SW, Chen H, Towles JD, Kamper DG. Effect of finger posture on the tendon force distribution within the finger extensor mechanism. J Biomech Eng. 2008;130(5) https://doi.org/10.1115/1.2978983.
29. Dogadov A, Alamir M, Serviere C, Quaine F. The biomechanical model of the long finger extensor mechanism and its parametric identification. J Biomech. 2017;58:232–6. https://doi.org/10.1016/j.jbiomech.2017.04.030.

Advancing Understanding of Boutonniere Deformity Through Finite Element Analysis: Insights and Innovations in Hand Biomechanics

17

Farid Amirouche, Ali Rteil, Giovanni Solitro, Diego Barragan, and Mark Gonzalez

Introduction

Boutonniere deformity is a specific type of finger deformity that manifests as hyperextension of the distal interphalangeal (DIP) joint, coupled with an involuntary flexion of the proximal interphalangeal (PIP) joint when the finger is attempted to be straightened. This condition typically results from an injury to the extensor mechanism of the finger, most commonly manifested as a tear of the central slip. This tear causes a volar subluxation of the lateral bands [1], leading to the characteristic joint alignment abnormality.

The disruption of the extensor mechanism can stem from various causes, including acute traumatic injuries to the hand, such as lacerations or fractures that impact the extensor tendons [2, 3]. Additionally, congenital disorders [4] may predispose individuals to develop this deformity from birth, while inflammatory arthropathies, particularly rheumatoid arthritis, can also contribute to its onset. In cases of rheumatoid arthritis [5], the degenerative changes and inflammation can weaken the tendons and joints, increasing the risk for boutonniere deformity. It is crucial to recognize and intervene promptly to restore finger function and prevent permanent disability in individuals affected by this condition.

Multiple studies have well established the development of boutonniere deformity [6, 7]. Still, cadaveric studies are limited by challenges in experimental design or the inability to measure disease processes in an in vivo model. Since cadaveric studies cannot directly quantify the stresses acting upon finger structures during the onset of a boutonniere deformity, finite element analysis (FEA) represents a promising alternative method.

F. Amirouche (✉)
Department of Orthopaedic Surgery, Orthopaedic and Spine Institute, Endeavor HealthSystem (Formerly Northshore University HealthSystem), an Affiliate of the University of Chicago Pritzker School of Medicine, Skokie, IL, USA

Department of Orthopaedic Surgery, University of Illinois at Chicago, Chicago, IL, USA
e-mail: famirouche@northshore.org; amirouch@uic.edu

A. Rteil
Department of Orthopaedic Surgery, Orthopaedic and Spine Institute, Northshore University Health System, An Affiliate of the University of Chicago Pritzker School of Medicine, Skokie, Chicago, IL, USA
e-mail: arteil@northshore.org

G. Solitro
LSU Department of Orthopaedics, Shreveport, LA, USA

D. Barragan · M. Gonzalez
Department of Orthopaedic Surgery, University of Illinois at Chicago, Chicago, IL, USA

F. Amirouche et al. (eds.), *Biomechanics of the Hand*,
https://doi.org/10.1007/978-3-032-04688-8_17

Researchers have increasingly turned to FEA and computational models to better understand and treat complex injuries. These tools enable the simulation of the finger's biomechanics in both healthy and pathological states, which helps to understand the mechanisms of deformity. FEA has become a powerful technique in orthopedic biomechanics, used to reproduce anatomical detail and predict biomechanical changes due to injuries, surgery, or splints [8–10]. Investigators can build anatomically accurate digital finger models and experiment with different injury scenarios and treatments.

This chapter focuses on the intricate world of finite element analysis (FEA) as it applies to boutonniere deformity. It offers a comprehensive examination of recent advancements in the computational modeling of finger biomechanics. It highlights the pivotal role of FEA in elucidating the complex pathogenesis and progression of this specific deformity, offering profound insights into the mechanical factors at play. Additionally, the chapter discusses how FEA optimizes conservative treatment strategies, such as splinting, by simulating various scenarios and outcomes, ultimately enhancing patient care. Furthermore, it sheds light on how these sophisticated models inform meticulous surgical planning for deformity correction, ensuring that interventions are effective and tailored to patients' needs.

Biomechanics of Boutonniere Deformity: Insights from FEA

Boutonniere deformity results from disrupted interactions between the central slip (CS) and lateral bands (LBs) of the extensor digitorum communis (EDC) tendon, along with compromised supporting structures associated with the lateral bands. These structures include the triangular ligament (TL) and the transverse and oblique fibers of the interosseous hood, all of which significantly contribute to the pathogenesis of the deformity [11–13]. It evolves in stages, as defined by Nalebuff and Millender [7], and the deformity becomes visible only in the third stage (Fig. 17.1).

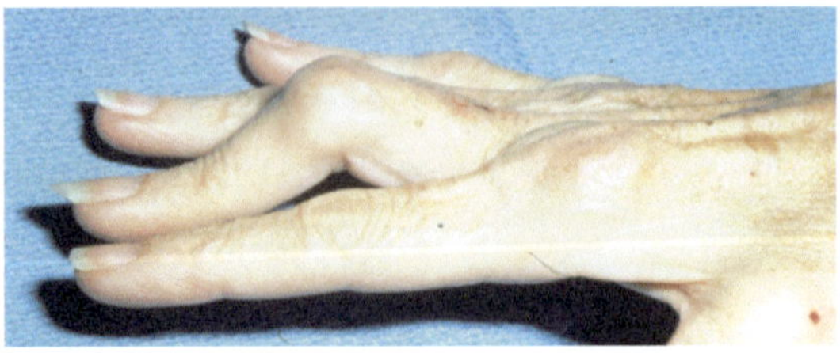

Fig. 17.1 Boutonniere deformity with the characteristic zigzag deformation

A cadaveric biomechanical study by Grau et al. [1] demonstrated that an isolated central slip injury is insufficient to create a complete boutonniere deformity; only a combined disruption of the central slip and the stabilizing transverse, oblique, and triangular ligament fibers produces the classic proximal interphalangeal (PIP) flexion contracture. This finding underscores the importance of modeling the entire extensor apparatus.

Because the finger has a complex network of tendons and ligaments, finite element models (FEA) are especially suited to studying the behavior of its extensor mechanism. FEA can accurately represent structural complexity and heterogeneous material properties, facilitating the simulation of tendon injuries and resultant finger deformities. However, computational modeling of human tissues remains relatively unexplored in the context of boutonniere deformity.

Solitro et al. [14] utilized finite element analysis to examine the biomechanical changes occurring in the progression of boutonniere deformity, focusing on stress redistribution among ligamentous structures. They developed a three-dimensional computational model of the index finger using computed tomography (CT) scans to reconstruct accurate three-dimensional geometry of the bones. In contrast, tendons and ligamentous structures were modeled according to anatomical literature [15–22]. They assigned tissue-specific isotropic elastic properties based on previously reported biomechanical data [23–25] and realistic friction conditions at tendon interfaces. This enables the accurate simulation of finger flexion and extension, as well as pathological conditions associated with boutonniere deformity (Fig. 17.2).

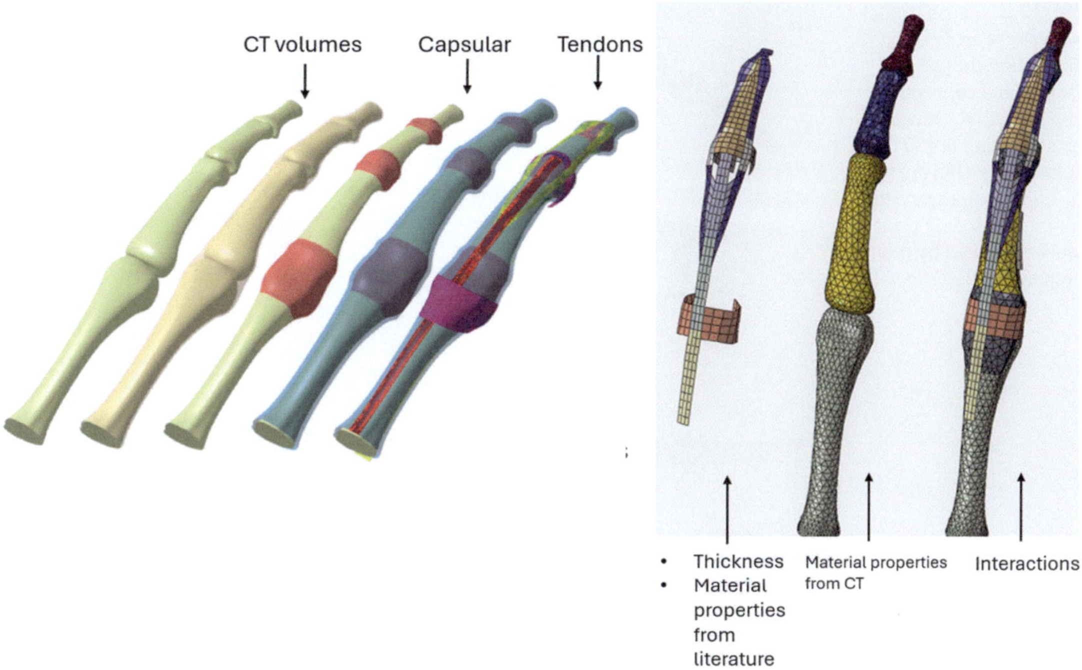

Fig. 17.2 Model of the index finger. (Image created using ANSYS software by University of Illinois Chicago. The figure is based on prior work presented by the authors [14])

The finger model was validated against cadaveric experiments [14] and used to analyze the stress responses of the lateral bands, triangular ligament, and oblique fibers under multiple simulated injury conditions. Their simulations included isolated central slip transection, partial (50%) and complete (100%) attenuation of oblique fibers, and progressive attenuation (25%, 50%, 75%, and 100%) of the triangular ligament. The results showed that the initial central slip injury significantly elevated the stress in the lateral bands, which intensified as the supporting structures, especially the triangular ligament, became progressively attenuated (Fig. 17.3).

This sequence of events led to volar displacement of the lateral bands, which occurred at progressively smaller degrees of PIP joint flexion as the attenuation became more severe. The volar displacement of the LB changes their center of rotation, turning them into paradoxical PIP flexors and DIP extensors when 50% of the triangular ligaments are attenuated, clearly delineating the biomechanical stages involved in the early evolution of the deformity (Fig. 17.4). The study highlights the potential of FEA modeling in providing insights beyond traditional cadaveric studies, underscoring its utility in better understanding the pathomechanics and informing clinical interventions for boutonniere deformity.

Recent work by Jadelis et al. [26] developed an advanced computational work integrating finite element analysis with multibody dynamic musculoskeletal simulations [27] to investigate the biomechanics of the extensor mechanism of the index finger. The developed three-dimensional finite element model adeptly incorporates heterogeneous material properties of the extensor apparatus, demonstrating a robust validation against a comprehensive set of experimental data. This validation highlights the model's impressive capability to closely replicate tissue strain patterns and fingertip endpoint forces across various dynamic finger postures. Notably, the model effectively mirrors experimentally identified strain distributions in critical anatomical structures, such as the central slip and terminal slip, which are essential for the intricate functioning of the finger during various movements.

In addition to strain patterns, the model predicts fingertip force magnitudes and directional

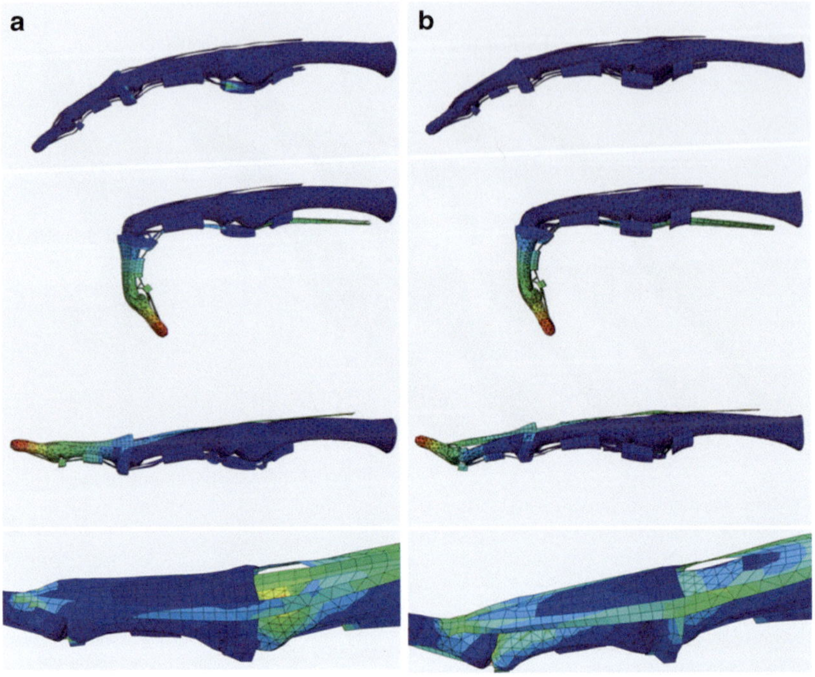

Fig. 17.3 Finger model with stress analysis simulation for (**a**) control and (**b**) central slip alteration. (Image created using ANSYS software by University of Illinois Chicago. The figure is based on prior work presented by the authors [14])

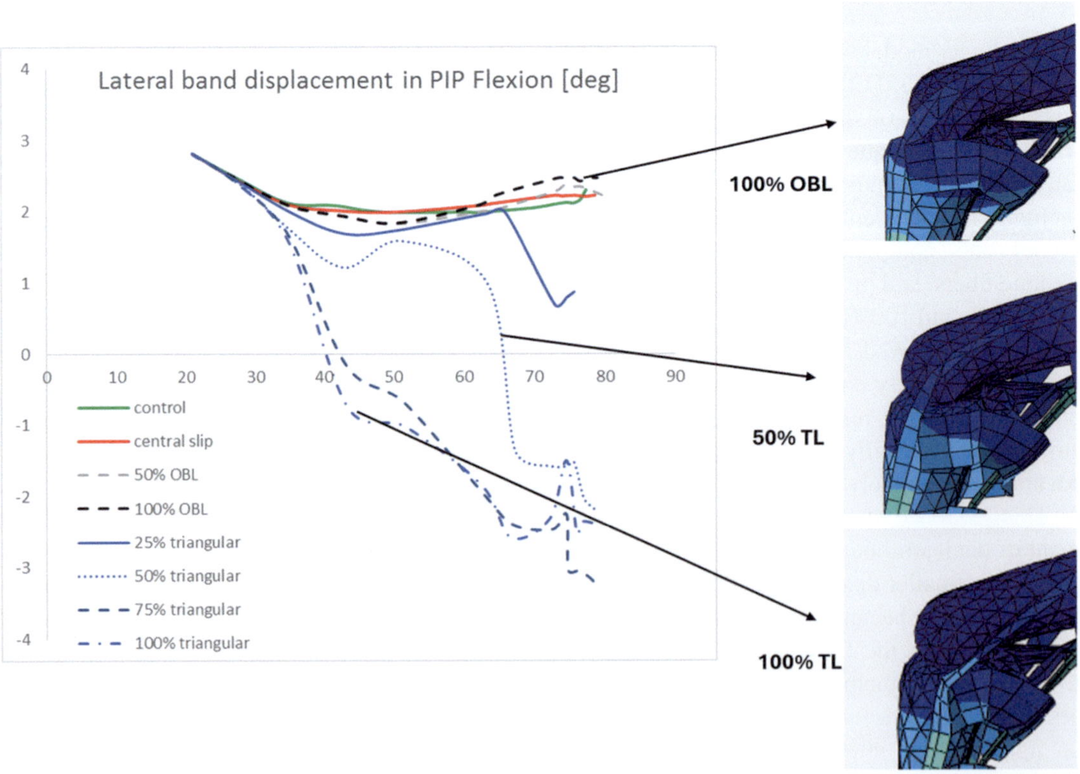

Fig. 17.4 Volar displacement of the lateral band (degrees) with paradoxical PIP flexion when 100% of the oblique ligament (OBL), 50%, and 100% of the triangular ligaments (TL) are attenuated. (Image created using ANSYS software by University of Illinois Chicago. The figure is based on prior work presented by the authors [14])

vectors through a detailed cosimulation process. This nuanced approach enhances the model's utility, allowing for an accurate representation of biomechanical forces at play during functional tasks. Furthermore, rigorous sensitivity analyses demonstrated that variations in tissue modulus and adhesions have a minimal influence on the overall biomechanics of the finger, thereby affirming the model's reliability.

This integrated modeling approach provides invaluable biomechanical insights by capturing the complexities of multiinsertion behavior exhibited by the extensor mechanism. Such insights are pivotal in informing clinical practices, including surgical planning, rehabilitation protocols, and orthotic interventions tailored for various conditions, such as boutonniere deformity. By bridging theoretical modeling with practical application, this research significantly advances the understanding of finger mechanics and the development of targeted treatments.

Faudot et al. [23] also developed a hybrid finite element-musculoskeletal model to estimate in vivo joint contact pressures in the index finger during maximal pinch grip tasks. Their model incorporates realistic anatomical structures and physiological tendon actuation derived from musculoskeletal simulations. It was validated and accurately reproduced experimentally measured fingertip forces and tendon load distributions. They observed significant intersubject variability in joint contact pressures, which was attributed to differences in individual neuromuscular strategies. The model identified peak joint pressures exceeding 30 MPa, potentially surpassing the threshold for cartilage injury, and highlighted the mechanical risk factors associated with pinch grip activities. This approach highlights the promising capabilities of hybrid modeling techniques in guiding precise interventions to mitigate abnormal joint loading. Such loading significantly contributes to the progressive degeneration of cartilage, particularly in the context of deformities like boutonniere. By leveraging these advanced modeling methods, we can gain a deeper understanding of the complex dynamics and implement targeted strategies that address the symptoms and underlying causes of joint abnormalities.

Elyasi et al. [28] developed and validated hyperelastic constitutive models to predict the mechanical behavior of different regions of the finger extensor apparatus, with a primary focus on the central and terminal slips—key structures involved in boutonniere and mallet finger deformities. They compared multiple hyperelastic strain energy density functions. They determined that the invariant-based model provided superior stability and accuracy in various deformation modes, including simple tension, pure shear, and biaxial tension. This study underscores the importance of selecting appropriate constitutive models in FEA to ensure accurate stress and strain predictions during simulations of finger deformities.

Dogadov et al. [29] developed and validated a biomechanical fine element model of the finger extensor mechanism, emphasizing a realistic representation of tendon and ligament interactions. Their model employed a parametric identification approach, enabling the accurate prediction of force distributions within the extensor apparatus during various finger postures. They demonstrated that tendon forces and ligament strains vary significantly with changes in finger posture, highlighting the limitations of traditional models that assume constant biomechanical coefficients.

Wei et al. [30] developed and validated a three-dimensional complex patient-specific human hand for modeling the human hand. This was achieved by combining medical image-based finite element modeling, individualized muscle force, and kinematic measurements. As is the case in FEA first, a subject-specific human hand CT scan is performed from which a finite element (FE) model was developed. The geometries of the phalanges, carpal bones, wrist bones, ligaments, tendons, subcutaneous tissue, and skin were all included except the fluidic part and nerve system. The material properties with some limitations were extracted through empirical data from in vivo and in vitro experiment results available in the literature [31–39]. The boundary and loading conditions were defined based on the kinematic data and muscle forces of a specific subject captured from the in vivo grasping tests making use of optotrack systems and markers as in the case of gait analysis (Fig. 17.5).

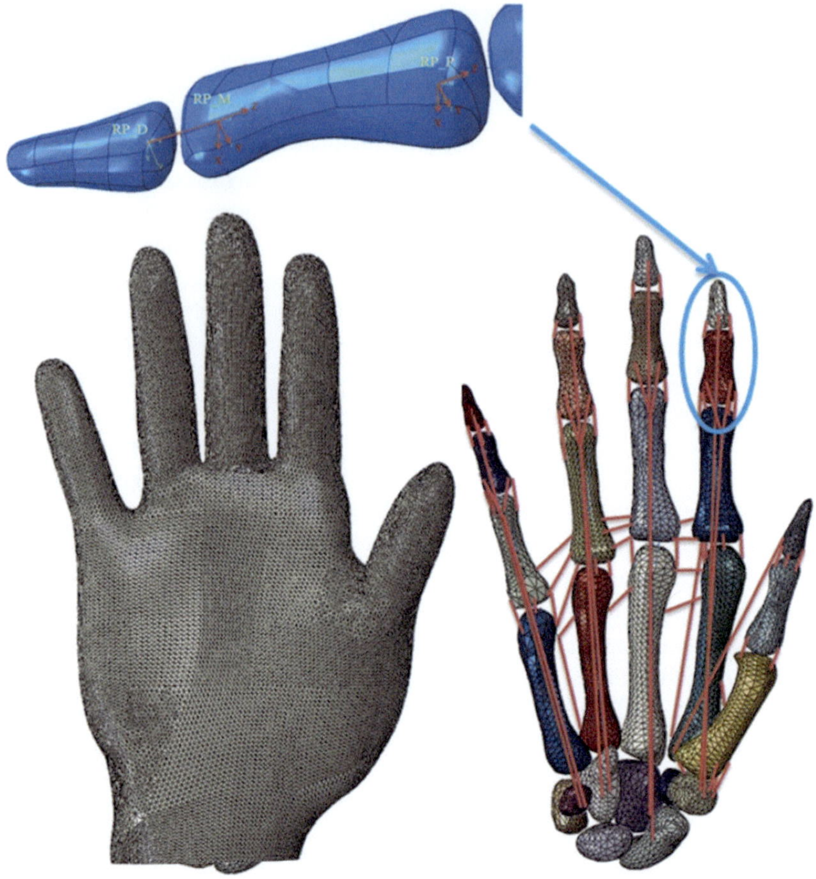

Fig. 17.5 Kinematic motion definition and anatomical view of the FE hand model. The red lines represent ligaments and tendons. The diagram on the top shows the boundary conditions defined at each interphalangeal joint. (Reproduced from Wei et al. [30] under the Creative Commons Attribution 4.0 International License (CC BY 4.0). No changes have been made)

Treatment Simulations and Splint Design Using FEA

Several treatment modalities exist for boutonniere deformity. Acute central slip (CS) injuries can typically be managed effectively through the use of an extension splint for the proximal interphalangeal (PIP) joint. However, surgical intervention might become necessary in advanced or chronic cases. Additionally, severe joint damage caused by rheumatoid arthritis (RA) at the PIP joint may necessitate arthrodesis.

Splints are designed to apply controlled mechanical forces that facilitate the correction of joint stiffness and malalignment, which may arise due to traumatic injuries or pathological conditions, such as arthritis or congenital deformities. Finite element analysis (FEA) and advanced computational modeling techniques are crucial in simulating treatment outcomes and designing orthopedic splints for conditions such as boutonniere deformity.

FEA enables a detailed characterization of the splint's mechanical behavior, allowing researchers and practitioners to generate accurate force displacement and angle-moment curves. These curves are essential for evaluating splint performance across the full range of joint motion, ensuring that the splint is effective in restoring proper function and alignment [8].

Moreover, computational simulations empower clinicians to maintain a consistent corrective force across varying flexion angles, which is crucial for effective deformity correction. This not only enhances the precision of the treatment but also has the potential to significantly improve clinical outcomes, facilitating a more effective recovery process for patients with these complex conditions. By leveraging these cutting-edge technologies, the design and application of orthopedic splints can be optimized to achieve better alignment, increased mobility, and enhanced patient satisfaction.

Chhikara et al. [40] presented a design approach that integrates finite element analysis and computational optimization to create customized three-dimensional printed finger splints for patients with rheumatoid arthritis who have boutonniere or swan-neck deformities. They utilized finite element methods to optimize various splint designs by evaluating their structural rigidity, stress distributions, and deformation under different loading conditions. This approach emphasized precise, patient-specific measurements and customization, utilizing polylactic acid (PLA) as the splint material to enhance comfort, aesthetic acceptance, and patient compliance. The optimized three-dimensional printed splint effectively restored joint alignment and significantly improved finger dexterity, as objectively measured by the Nine-Hole Peg Test, while achieving high patient satisfaction scores as evaluated by the QUEST 2.0 assessment. These findings demonstrate how FEA can aid in treatment simulation by predicting the force-displacement behavior of orthopedic splints for correcting finger deformities.

Surgical Planning and Virtual Repair Evaluation

Finite element analysis has been effectively applied in orthopedic surgery to simulate treatment outcomes and guide surgical planning across various pathologies, including hip prosthesis and lumbar spine interventions [8]. Computational models offer detailed insights into biomechanics, enabling accurate predictions of tissue stresses and implant performance. Similar FEA application and modeling techniques are promising for conditions such as boutonniere deformity, where models could facilitate patient-specific simulations for optimal repair.

The work by Jadelis et al. [26] highlights the potential for finite element analysis as a practical tool in surgical planning for hand surgeries. They showed that their model's fingertip forces remained robust even when varying tissue stiffness or simulating scar adhesions, suggesting that it could be used to predict outcomes of various extensor tendon repair strategies. This capability opens the door to applying FEA-guided surgical planning for hand injuries. By combining high-fidelity finite element models with medical imaging of a patient's hand, surgeons can evaluate different repair techniques to achieve the best functional outcome before making a single incision.

Finite element simulations were also used to compare different PIP joint implant designs for arthritic joints, which is relevant in severe boutonniere deformities caused by rheumatoid arthritis (RA). Kim et al. [41] recently conducted a finite element analysis comparing the biomechanical characteristics of a conventional PIP flexible hinge implant with those of a novel rolling contact joint (RCJ) implant design—their model precisely quantified stress and strain distributions, including moment reactions, during simulated finger motion. The study demonstrated that the RCJ implant exhibited significantly lower stress and strain concentration and substantially reduced moment reactions compared to the traditional flexible hinge implant. The novel RCJ implant showed approximately 6.4-fold lower maximum stress and up to 49.6-fold lower moment reaction within the functional finger range of motion. These findings indicate the potential of advanced FEA methods in designing more durable and functionally practical implants, which could enhance clinical outcomes in patients with severe boutonniere deformities associated with rheumatoid arthritis (Fig. 17.6).

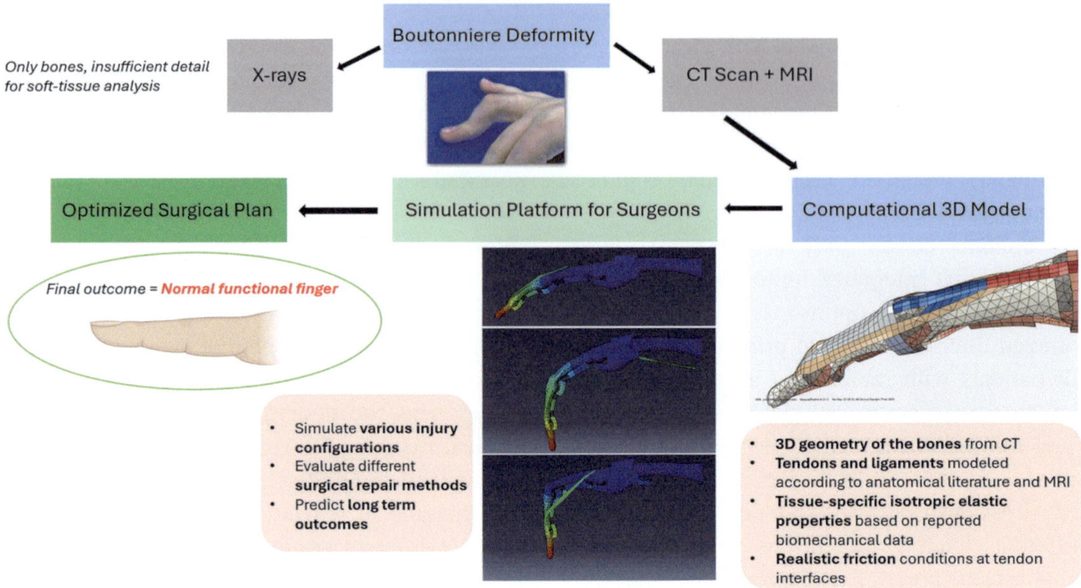

Fig. 17.6 FEA and computational three-dimensional model in the treatment of boutonniere deformity. (Includes embedded images created using ANSYS software by University of Illinois Chicago)

Conclusion

Finite element analysis (FEA) has emerged as a valuable tool for exploring the intricacies of biomechanics associated with boutonniere deformity, effectively addressing many of the limitations inherent in traditional cadaveric and experimental studies. Through sophisticated computational models, researchers can gain profound insights into the intricate stages and mechanical mechanisms that drive the progression of this deformity. These models not only elucidate the underlying pathology but also play a pivotal role in designing customized, patient-specific splints, offering a tailored approach to treatment.

Moreover, FEA facilitates the virtual exploration of various surgical repair strategies, allowing for a thorough comparison of potential outcomes without subjecting patients to immediate surgical interventions. As FEA techniques evolve and advance, we anticipate a shift toward more personalized treatment paradigms that incorporate detailed, patient-specific anatomical data. This approach promises to significantly enhance conservative and surgical outcomes for individuals grappling with boutonniere deformity.

Looking ahead, future research endeavors should concentrate on honing the accuracy of these computational models. There is also a pressing need to expand clinical validation to ensure that FEA can be seamlessly integrated into orthopedic practice, maximizing its practical utility and improving the quality of care for affected patients.

References

1. Grau L, Baydoun H, Chen K, Sankary ST, Amirouche F, Gonzalez MH. Biomechanics of the acute boutonniere deformity. J Hand Surg. 2018;43(1):80.e1–6. https://doi.org/10.1016/j.jhsa.2017.07.011.
2. McKeon KE, Lee DH. Posttraumatic boutonnière and swan neck deformities. J Am Acad Orthop Surg. 2015;23(10):623–32. https://doi.org/10.5435/JAAOS-D-14-00272.
3. Spinner M, Choi BY. Anterior dislocation of the proximal interphalangeal joint. A cause of rupture of the central slip of the extensor mechanism. J Bone Joint Surg Am. 1970;52(7):1329–36.

4. Carneiro RS. Congenital attenuation of the extensor tendon central slip. J Hand Surg. 1993;18(6):1004–7. https://doi.org/10.1016/0363-5023(93)90391-F.
5. Ferlic DC. Boutonniere deformities in rheumatoid arthritis. Hand Clin. 1989;5(2):215–22.
6. Matev I. The boutonniere deformity. Hand. 1969;1(2):90–5. https://doi.org/10.1016/0072-968X(69)90070-9.
7. Nalebuff EA, Millender LH. Surgical treatment of the boutonniere deformity in rheumatoid arthritis. Orthop Clin North Am. 1975;6(3):753–63.
8. Herrera A, Ibarz E, Cegoñino J, et al. Applications of finite element simulation in orthopedic and trauma surgery. World J Orthop. 2012;3(4):25–41. https://doi.org/10.5312/wjo.v3.i4.25.
9. Brekelmans WAM, Poort HW, Slooff TJJH. A new method to analyse the mechanical behaviour of skeletal parts. Acta Orthop Scand. 1972;43(5):301–17. https://doi.org/10.3109/17453677208998949.
10. Prendergast PJ. Finite element models in tissue mechanics and orthopaedic implant design. Clin Biomech Bristol Avon. 1997;12(6):343–66. https://doi.org/10.1016/s0268-0033(97)00018-1.
11. Fox A, Kang N. Reinserting the central slip – a novel method for treating boutonniere deformity in rheumatoid arthritis. J Plast Reconstr Aesthet Surg. 2009;62(5):e91–2. https://doi.org/10.1016/j.bjps.2008.08.055.
12. Littler JW, Eaton RG. Redistribution of forces in the correction of boutonniere deformity. J Bone Joint Surg Am. 1967;49(7):1267–74.
13. Rubin J, Bozentka DJ, Bora FW. Diagnosis of closed central slip injuries: a cadaveric analysis of non-invasive tests. J Hand Surg. 1996;21(5):614–6. https://doi.org/10.1016/S0266-7681(96)80142-2.
14. Solitro GF, Amirouche F, Gonzalez M. Boutonniere deformity mechanics captured and defined using finite element method. Published online 2015. https://doi.org/10.13140/RG.2.1.3151.4724.
15. Sato J, Ishii Y, Noguchi H, Takeda M. Sonographic appearance of the flexor tendon, volar plate, and A1 pulley with respect to the severity of trigger finger. J Hand Surg. 2012;37(10):2012–20. https://doi.org/10.1016/j.jhsa.2012.06.027.
16. Walbeehm ET, McGrouther DA. An anatomical study of the mechanical interactions of flexor Digitorum Superficialis and profundus and the flexor tendon sheath in zone 2. J Hand Surg. 1995;20(3):269–80. https://doi.org/10.1016/S0266-7681(05)80077-4.
17. Ward SR, Loren GJ, Lundberg S, Lieber RL. High stiffness of human digital flexor tendons is suited for precise finger positional control. J Neurophysiol. 2006;96(5):2815–8. https://doi.org/10.1152/jn.00284.2006.
18. Chepla KJ, Goitz RJ, Fowler JR. Anatomy of the flexor digitorum profundus insertion. J Hand Surg. 2015;40(2):240–4. https://doi.org/10.1016/j.jhsa.2014.11.004.
19. Fiorini HJ, Santos JBG, Hirakawa CK, Sato ES, Faloppa F, Albertoni WM. Anatomical study of the A1 pulley: length and location by means of cutaneous landmarks on the palmar surface. J Hand Surg. 2011;36(3):464–8. https://doi.org/10.1016/j.jhsa.2010.11.045.
20. Fok KS, Chou SM. Development of a finger biomechanical model and its considerations. J Biomech. 2010;43(4):701–13. https://doi.org/10.1016/j.jbiomech.2009.10.020.
21. Lee SW, Kamper DG. Modeling of multiarticular muscles: importance of inclusion of tendon–pulley interactions in the finger. IEEE Trans Biomed Eng. 2009;56(9):2253–62. https://doi.org/10.1109/TBME.2009.2019119.
22. Saito S, Suzuki Y. Biomechanics of the volar plate of the proximal interphalangeal joint: a dynamic ultrasonographic study. J Hand Surg. 2011;36(2):265–71. https://doi.org/10.1016/j.jhsa.2010.10.034.
23. Faudot B, Milan JL, Goislard de Monsabert B, Le Corroller T, Vigouroux L. Estimation of joint contact pressure in the index finger using a hybrid finite element musculoskeletal approach. Comput Methods Biomech Biomed Engin. 2020;23(15):1225–35. https://doi.org/10.1080/10255842.2020.1793965.
24. Huang H, Zhang J, Sun K, Zhang X, Tian S. Effects of repetitive multiple freeze–thaw cycles on the biomechanical properties of human flexor digitorum superficialis and flexor pollicis longus tendons. Clin Biomech. 2011;26(4):419–23. https://doi.org/10.1016/j.clinbiomech.2010.12.006.
25. Qian K, Traylor K, Lee SW, Ellis B, Weiss J, Kamper D. Mechanical properties vary for different regions of the finger extensor apparatus. J Biomech. 2014;47(12):3094–9. https://doi.org/10.1016/j.jbiomech.2014.06.035.
26. Jadelis CT, Ellis BJ, Kamper DG, Saul KR. Cosimulation of the index finger extensor apparatus with finite element and musculoskeletal models. J Biomech. 2023;157:111725. https://doi.org/10.1016/j.jbiomech.2023.111725.
27. Amirouche FML. Fundamentals of multibody dynamics. Boston: Birkhäuser; 2006. https://doi.org/10.1007/b137682.
28. Elyasi N, Taheri KK, Narooei K, Taheri AK. A study of hyperelastic models for predicting the mechanical behavior of extensor apparatus. Biomech Model Mechanobiol. 2017;16(3):1077–93. https://doi.org/10.1007/s10237-017-0874-x.
29. Dogadov A, Alamir M, Serviere C, Quaine F. The biomechanical model of the long finger extensor mechanism and its parametric identification. J Biomech. 2017;58:232–6. https://doi.org/10.1016/j.jbiomech.2017.04.030.
30. Wei Y, Zou Z, Wei G, Ren L, Qian Z. Subject-specific finite element modelling of the human hand complex: muscle-driven simulations and experimental validation. Ann Biomed Eng. 2020;48(4):1181–95. https://doi.org/10.1007/s10439-019-02439-2.
31. Adams MJ, Johnson SA, Lefèvre P, et al. Finger pad friction and its role in grip and touch. J R Soc

Interface. 2013;10(80):20120467. https://doi.org/10.1098/rsif.2012.0467.
32. Ali A. Review of constitutive models for rubber-like materials. Am J Eng Appl Sci. 2010;3(1):232–9. https://doi.org/10.3844/ajeassp.2010.232.239.
33. André T, Lévesque V, Hayward V, Lefèvre P, Thonnard JL. Effect of skin hydration on the dynamics of fingertip gripping contact. J R Soc Interface. 2011;8(64):1574–83. https://doi.org/10.1098/rsif.2011.0086.
34. Berger DJ, d'Avella A. Effective force control by muscle synergies. Front Comput Neurosci. 2014;8:46. https://doi.org/10.3389/fncom.2014.00046.
35. Butler TJ, Kilbreath SL, Gorman RB, Gandevia SC. Selective recruitment of single motor units in human flexor digitorum superficialis muscle during flexion of individual fingers. J Physiol. 2005;567(Pt 1):301–9. https://doi.org/10.1113/jphysiol.2005.089201.
36. Carrigan SD, Whiteside RA, Pichora DR, Small CF. Development of a three-dimensional finite element model for carpal load transmission in a static neutral posture. Ann Biomed Eng. 2003;31(6):718–25. https://doi.org/10.1114/1.1574027.
37. Chamoret D, Bodo M, Roth S. A first step in finite-element simulation of a grasping task. Comput Assist Surg. 2016;21(sup1):22–9. https://doi.org/10.1080/24699322.2016.1240294.
38. Chamoret D, Roth S, Feng ZQ, Yan XT, Gomes S, Peyraut F. A novel approach to modelling and simulating the contact behaviour between a human hand model and a deformable object. Comput Methods Biomech Biomed Engin. 2013;16(2):130–40. https://doi.org/10.1080/10255842.2011.608662.
39. Dallard J, Merlhiot X, Duprey S, Wang X, Micaelli A. Fingertip finite element modelling–on choosing the right material property. Comput Methods Biomech Biomed Engin. 2014;17(Suppl 1):30–1. https://doi.org/10.1080/10255842.2014.931084.
40. Chhikara K, Gupta S, Saharawat S, Sarkar S, Chanda A. Design, manufacturing, and trial of a 3D printed customized finger splint for patients with rheumatoid arthritis. Rheumato. 2023;3(1):51–62. https://doi.org/10.3390/rheumato3010004.
41. Kim YJ, Bae HA, Hong SW. Biomechanical comparative finite element analysis between a conventional proximal interphalangeal joint flexible hinge implant and a novel implant design using a rolling contact joint mechanism. J Orthop Surg. 2023;18(1):976. https://doi.org/10.1186/s13018-023-04477-y.

18 Swan Neck Deformity

Seth Roge, Farid Amirouche,
Mohammad Zalzaleh, and Steven Grindel

Introduction

Swan neck deformity (SND) is a condition characterized by hyperextension of the proximal interphalangeal (PIP) joint coupled with flexion of the distal interphalangeal (DIP) joint (Fig. 18.1). The etiology of SND can be categorized into three primary types: extrinsic, intrinsic, and articular. Understanding these types is crucial for diagnosing and treating SND in your practice.

Extrinsic factors contributing to SND include disruptions in the terminal extensor tendon and contractures resulting from trauma to the wrist or metacarpophalangeal (MCP) joint [1]. Intrinsic causes often involve tightness in the intrinsic muscles, volar subluxation of the MCP joint, and tendinous adhesions affecting the intrinsic muscles. Articular-related causes primarily include damage to the volar plate or the flexor digitorum superficialis (FDS) tendon at the PIP joint.

A frequent underlying cause of SND is inflammatory arthropathy, particularly rheumatoid arthritis. In this context, synovitis of the MCP joint can result in volar subluxation, while synovitis of the PIP joint may lead to deterioration of the volar plate or FDS tendon [1]. This degeneration can create an imbalance in the joint, resulting in hyperextension at the PIP joint [2].

It is important to note that this list of causes is not exhaustive; other factors may also lead to SND, including cases stemming from medical procedures such as iatrogenic SND following a trigger finger release [3]. Additionally, conditions that cause generalized joint laxity, such as Ehlers-Danlos syndrome, can contribute to developing SND [4].

Diagnosis of SND typically occurs through clinical evaluation, and the treatment approach is tailored based on the specific etiology and severity of the deformity.

Pathomechanics of Swan Neck Deformity

In our comprehensive exploration of biomechanical factors related to swan neck deformity (SND), we will thoroughly analyze the intricate etiologies contributing to this condition. Our investiga-

S. Roge (✉)
Department of Orthopaedic Surgery, University of Illinois at Chicago, Chicago, IL, USA

F. Amirouche
Department of Orthopaedic Surgery, Orthopaedic and Spine Institute, Endeavor HealthSystem (Formerly Northshore University HealthSystem), an Affiliate of the University of Chicago Pritzker School of Medicine, Skokie, IL, USA

Department of Orthopaedic Surgery, University of Illinois at Chicago, Chicago, IL, USA

M. Zalzaleh · S. Grindel
Medical College of Wisconsin Department of Orthopaedic Surgery, Milwaukee, WI, USA

F. Amirouche et al. (eds.), *Biomechanics of the Hand*,
https://doi.org/10.1007/978-3-032-04688-8_18

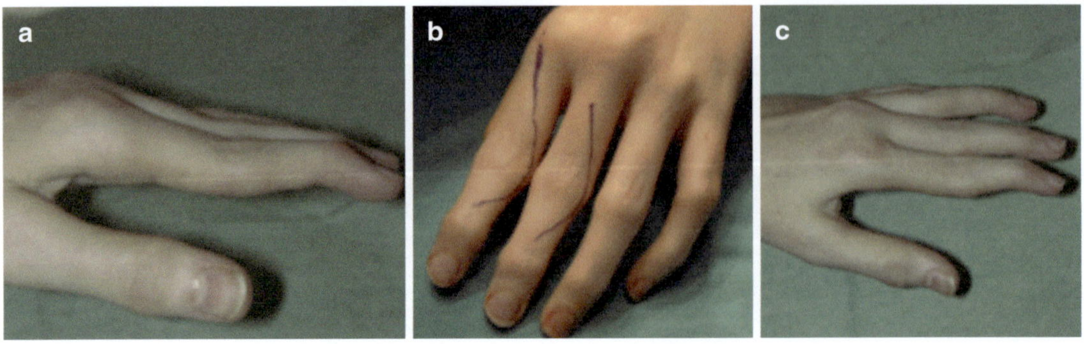

Fig. 18.1 A patient with swan neck deformity, showing the hyperextension of the PIP joint and the flexion of the DIP joint, which are the key features of this condition

tion will begin by closely examining the flexion occurring at the distal interphalangeal (DIP) joint, contrasted with the hyperextension that typically develops at the proximal interphalangeal (PIP) joint. This dual examination is crucial, as it underscores the complexity of the interplay between these joint movements, which is central to understanding the deformity.

Given the myriad of underlying causes that can lead to SND—ranging from rheumatoid arthritis and other inflammatory joint diseases to trauma and congenital factors—it is essential to identify and articulate the overarching biomechanical principles that unify these diverse etiologies. Our analysis will delve into the mechanical aspects of joint function, exploring how ligamentous laxity, tendon dynamics, and muscle imbalances can contribute to the characteristic deviations seen in SND.

Moreover, we will consider the anatomical structures involved, including the roles of the volar plate and the intrinsic and extrinsic musculature of the hand. By synthesizing these insights, we aim to comprehensively understand how these biomechanical principles contribute to the initial onset of SND and influence its progression over time. This understanding can guide your clinical assessments and therapeutic interventions for individuals affected by swan neck deformity.

PIP Joint Hyperextension Etiologies

Causes of SND that begin with PIP joint hyperextension include FDS or volar plate disruption. Since these structures resist hyperextension at the PIP joint, loss of their integrity allows for deformity to occur at this joint. Similarly, volar subluxation of the MCP joint tightens the extensor digitorum communis (EDC) tendon and causes PIP joint hyperextension [1, 2]. Chronic radial collateral ligament (RCL) degeneration in the little digit has been implicated in PIP joint hyperextension [3]. Regardless of the cause, PIP joint hyperextension leads to dorsal subluxation of the lateral bands. Rheumatoid synovitis can exacerbate lateral band subluxation by leading to the degradation of the transverse retinacular ligament (TRL), a crucial structure that secures these bands in their proper position. When the lateral bands become displaced dorsally, they participate in a hyperextension of the proximal interphalangeal (PIP) joint, effectively imparting an extension moment destabilizing this critical joint. In the final stages of developing the syndrome known as SND, this dorsal displacement causes the lateral bands to become slack and lose their effectiveness in facilitating the extension of the distal interphalangeal (DIP) joint, further compromising hand function. FDS flexion predominates, and the deformity is present in the PIP and DIP joints [4].

This PIP joint-first development of SND may be particularly relevant to the hand surgeon performing two-stage flexor tendon repair. Chinchalkar et al. performed biomechanical evaluations of patients undergoing two-stage flexor tendon repair, collecting data preoperatively and then after each stage. They found that FDS resection in stage one can produce PIP joint hyperextension, and subsequent flexor digitorum profundus (FDP) repair in stage two can produce DIP joint flexion [5].

DIP Joint Flexion Etiologies

Causes of SND that begin with DIP joint flexion include mallet finger or DIP joint flexion contractures. In the mallet finger, disruption of the terminal extensor tendon causes proximal migration of the lateral bands and better force distribution on the central slip. Over time, this altered extensor mechanism leads to PIP joint hyperextension [6]. In DIP joint flexion contracture, lateral band extension creates an extension moment at the PIP joint. Again, this extension moment causes the volar plate to undergo creep over time, and PIP joint hyperextension ensues to complete the deformity [7]. A similar pattern of SND development has been noted in interphalangeal (IP) joint flexion contracture of the thumb after flexor pollicis longus (FPL) repair in zone T1 complicated by adhesions. Under these circumstances, IP joint flexion causes increased tension in the extensor pollicis longus (EPL), and MCP joint extension occurs in response [8].

Haddara et al. used a cadaver model to study volar plate strain in a patient progressing from mallet finger to SND. Mallet's finger was simulated by transecting the terminal tendon, and SND progression was simulated by releasing the TRL to allow dorsal subluxation of the lateral bands. A strain gauge was surgically attached in series with the volar plate to measure strain, and PIP joint angle measurements were made using electromagnetic trackers. Twenty-one cadaver models were taken through a full flexion-extension range of motion cycle at baseline, then with a mallet finger, and then with SND conditions using a testing apparatus.

Maximum volar plate strain and PIP joint angle were measured in each condition. The group found nonsignificant increases in PIP joint angle over progression from baseline to mallet finger to SND conditions. Volar plate strain increased significantly between baseline and SND conditions and between mallet finger and SND conditions (Fig. 18.2). Finally, there was a strong positive correlation between the PIP joint angle and volar plate strain over the modeled progression of SND. This was the first study to successfully model a progression of volar plate strain in the development of SND [9].

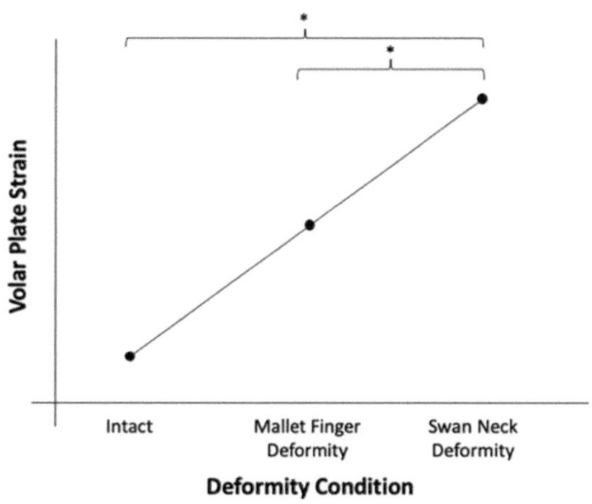

Fig. 18.2 A visualization of the differences in volar plate strain according to the progression of swan neck deformity according to the results published by Haddara et al.

Treatment of Swan Neck Deformity

Depending on the etiology and severity of SND, treatment may involve corrective splinting or surgical interventions, including tenodermodesis, spiral oblique retinacular ligament reconstruction, and arthrodesis [10]. Literature on the biomechanics of different treatment options is relatively limited. Haddara et al. evaluated the biomechanical effects of FDS hemitenodesis with or without DIP joint arthrodesis. In this procedure, one slip of FDS is looped around the A1 or A2 pulley to prevent PIP joint hyperextension (Fig. 18.3) [11].

Using a cadaveric testing apparatus similar to that in their prior study, the group tested maximum PIP joint extension, volar plate strain, and difference in FDS and FDP loading under healthy, SND, FDS hemitenodesis, and added DIP joint fusion conditions. The group found that FDS hemitenodesis significantly decreases maximum PIP joint extension and volar plate strain com-

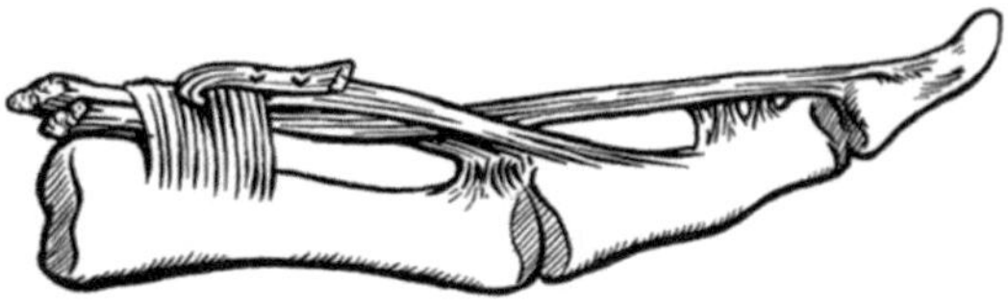

Fig. 18.3 An illustration of FDS hemitenodesis

pared to the SND model. This procedure also induced substantially higher FDP loading than FDS loading. The addition of DIP joint fusion decreased maximum PIP joint extension by an average of 0.5° but did not significantly alter volar plate strain. If DIP joint arthrodesis aims to prevent PIP joint hyperextension, the authors recommend against the added procedure. While these results are interesting, the lack of other biomechanical studies limits comparison to other treatment options [12]. FDS hemitenodesis may have been used in the previously discussed cases of SND caused by RCL degeneration at the little digit [3]. For SND caused by mallet finger deformity, one case study examined the use of terminal extensor tendon reconstruction to correct the deformity, as shown in Fig. 18.4. After this procedure and 4 weeks of splinting in full extension, the patient achieved a full, painless range of motion of the affected digit [13]. Future studies may study the biomechanical properties of such a repair.

Models of Swan Neck Deformity

The cadaveric models [9, 12] and biomechanical vector analyses discussed previously [7, 8, 14] have aided the study of SND. Leijnse et al. conducted an intricate study utilizing motion-tracking cameras to investigate the complicated role of the flexor digitorum profundus (FDP) in coordinating the angles of phalangeal joints during the complex motion of digit flexion closely. Their kinematic analysis, which meticulously examined motion capture data from 68 different digits, uncovered a specific subset of subjects exhibiting locking in the SND condition.

In these individuals, attempts at achieving full-finger flexion led to notable DIP flexion; however, the proximal interphalangeal (PIP) joint stubbornly remained fixed in a position of hyperextension. This state persisted until the surrounding soft tissues accumulated sufficient energy to finally overcome the mechanical resistance of the "locked" PIP joint. This locking phenomenon might be linked to the distinctly felt "snapping" that characterizes SND.

During the "snapping" event, flexion at the PIP joint produces a remarkable snapping sensation. This occurs as the dorsally displaced lateral bands navigate over the PIP joint, shifting in the volar direction. The interplay of these anatomical and biomechanical elements offers valuable insights into the complexities of digit movement and the nuances of joint interactions in the context of SND.

Future Directions in Research

The biomechanical studies outlined in this chapter establish the development of SND, irrespective of its underlying cause. However, cadaver-based studies often fall short of accurately reflecting the dynamics of live tissue with its inherent healing potential. There is a notable need for a more comprehensive kinematic assessment of SND, mainly as it manifests differently depending on its etiology and severity. Ultrasound-based kinematic evaluation of volar plate dynamics has been used to differentiate volar plate injury versus fracture at the PIP joint in vivo, and it is possible that this ultrasound-based method could have clinical applications in the future [15].

Previous research into boutonnière deformity has offered valuable insights through finite-element analysis (FEA) and rigid-body models focused on uninjured digits. These existing models could be adapted to enhance our understanding of SND. Future research should emphasize computational modeling and in vivo evaluations of various surgical interventions to treat SND effectively.

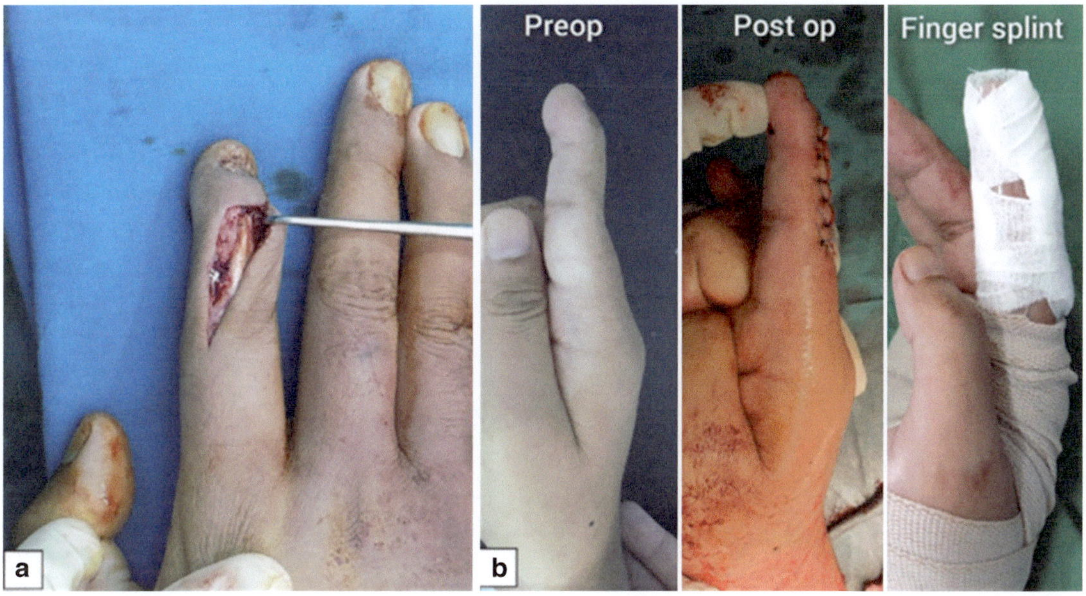

Fig. 18.4 Repair of the terminal extensor tendon to treat swan neck deformity caused by mallet finger. The procedure successfully restored the affected digit's full, active range of motion

References

1. Morrell N, Weiss AP. Biomechanics of rheumatoid finger deformities. In: Clinical management of the rheumatoid hand, wrist, and elbow. Cham: Springer; 2016.
2. Lutter C, Kuerten S, Geppert C, et al. Dynamic study of the finger interphalangeal joint volar plate—motion analysis with magnetic resonance cinematography and histologic comparison. Skeletal Radiol. 2023;52(8):1493–501. https://doi.org/10.1007/s00256-023-04288-6.
3. Hamada Y, Takai H, Satoh R, Hibino N, Ueda Y, Minamikawa Y. Swan neck deformity due to chronic radial collateral ligament injury of the little finger proximal interphalangeal joint. J Hand Surg Eur Vol. 2018;43(5):513–7. https://doi.org/10.1177/1753193417739248.
4. Gonzalez M, Mohan V, Elhassan B, Amirouche F. Biomechanics of the digit. J Hand Surg. 2005;5(1):P48–60.
5. Chinchalkar S, Dunning C, Pipicelli J. Videographic, fluoroscopic, and biomechanical analysis of swan-neck deformity in two-stage flexor tendon graft. J Hand Ther. 2011;24(4):389–90.
6. McKeon KE, Lee DH. Posttraumatic boutonnière and swan neck deformities. J Am Acad Orthop Surg. 2015;23(10):623–32. https://doi.org/10.5435/JAAOS-D-14-00272.
7. Chinchalkar SJ, Lanting BA, Ross D. Swan neck deformity after distal interphalangeal joint flexion contractures: a biomechanical analysis. J Hand Ther. 2010;23(4):420–5. https://doi.org/10.1016/j.jht.2009.11.005.
8. Chinchalkar S, Pereira I, Pipicelli J. Biomechanical analysis of secondary swan neck deformity following IP joint flexion contracture of the thumb. J Hand Ther. 2011;24(4):P389.
9. Haddara MM, Fan S, Matache BA, Chinchalkar SJ, Ferreira LM, Suh N. Development of an in vitro swan neck deformity biomechanical model. Hand. 2022;17(5):833–8. https://doi.org/10.1177/1558944720966736.
10. Elzinga K, Chung K. Managing swan neck and boutonniere deformities. Clin Plast Surg. 2019;46(3):329–37.
11. Wei DH, Terrono AL. Superficialis sling (flexor digitorum superficialis tenodesis) for swan neck reconstruction. J Hand Surg. 2015;40(10):2068–74. https://doi.org/10.1016/j.jhsa.2015.07.018.
12. Haddara MM, Kadar A, Ferreira LM, Suh N. Effect of a Flexor Digitorum Superficialis Hemitenodesis on Reducing Volar Plate Strains for Swan Neck Deformities. Hand. 2023;18(3):421–9. https://doi.org/10.1177/15589447211040877.
13. Sandjaya G, Prabowo I, Patih ID. Terminal extensor tendon reconstruction as a reliable option for chronic mallet finger with swan neck deformity of index finger: a case report. Ann Med Surg. 2022;78:103924. https://doi.org/10.1016/j.amsu.2022.103924.
14. Chinchalkar S, Dunning C, Pipicelli J. Videographic, fluoroscopic, and biomechanical analysis of swan-neck deformity in two stage flexor tendon graft. J Hand Ther. 2011;24(4):P389–90.
15. Saito S, Sawabe K, Suzuki Y, Suzuki S. Ultrasonographic characteristics of volar-lateral ligament constrains after proximal interphalangeal joint injuries. J Plast Surg Hand Surg. 2016;50(4):216–21.

Biomechanic Influences in Choices and Treatment of Pulley Repairs

19

Ali Rteil, Mark Gonzalez, and Farid Amirouche

Introduction

The flexor pulley system is a crucial anatomical structure that plays a vital role in the efficient functioning of finger and thumb movements. This intricate system is designed to prevent tendon bowstringing—an issue where the tendons would spring away from the bones during flexion—and to optimize force transmission necessary for gripping and executing fine motor tasks. The flexor pulley system consists of two main pulleys: annular and cruciate. These pulleys work in concert to stabilize the flexor tendons as they traverse the finger joints, converting the linear excursion of these tendons into the necessary angular motion required for flexion.

A. Rteil
Department of Orthopaedic Surgery, Orthopaedic and Spine Institute, Northshore University Health System, An Affiliate of the University of Chicago Pritzker School of Medicine, Skokie, IL, USA

M. Gonzalez
Department of Orthopaedic Surgery, University of Illinois at Chicago, Chicago, IL, USA

F. Amirouche (✉)
Department of Orthopaedic Surgery, Orthopaedic and Spine Institute, Endeavor HealthSystem (Formerly Northshore University HealthSystem), an Affiliate of the University of Chicago Pritzker School of Medicine, Skokie, IL, USA

Department of Orthopaedic Surgery, University of Illinois at Chicago, Chicago, IL, USA
e-mail: famirouche@northshore.org; amirouch@uic.edu

Among these pulleys, the A2 and A4 pulleys in the fingers are particularly crucial for preserving biomechanical efficiency [1, 2]. Their anatomical positioning and structure effectively enhance the flexor tendons' mechanical advantage, allowing for greater force application without excessive strain. While injuries to the pulley system are relatively uncommon in the general population, they are frequently observed in specific activities, such as rock climbing [3]. Climbers often experience acute injuries to the pulley system due to the high tensile force exerted on the fingers while gripping and pulling on holds.

Damage to the flexor pulleys can significantly disrupt the mechanics of the tendons, resulting in impaired hand function, decreased grip strength, and reduced finger mobility. This disruption can lead to various consequences, from limited finger movement to chronic pain and instability in the hand [4, 5].

This chapter delves into a comprehensive review of the anatomy and biomechanics of the flexor pulley system, emphasizing its indispensable role in maintaining functional hand mechanics. Pathological conditions affecting the pulley system, such as ruptures or tears, are examined in detail, including their classification, the impact on biomechanical function, and the potential complications from such injuries.

In addition, various reconstruction techniques are discussed, highlighting the critical

F. Amirouche et al. (eds.), *Biomechanics of the Hand*,
https://doi.org/10.1007/978-3-032-04688-8_19

need to restore the pulley system's biomechanical function through surgical intervention. This chapter underscores that achieving successful outcomes in hand surgery hinges on a thorough understanding of the pulley system's architecture and its mechanical roles, ensuring patients regain optimal hand function postinjury.

Anatomy and Biomechanics of the Flexor Pulley System

Dynamics of Flexor Tendons

Flexor tendons are crucial in transmitting force from the muscle belly to the fingers, enabling motion. This process relies on tendon excursion, which refers to the sum of the length the muscle can be stretched from its resting length plus the length it can contract from its resting state [6]. The muscle's maximum shortening capacity constrains the extent of this excursion.

Tension within a tendon stays consistent along its length, meaning every part experiences the same force. However, the torque a tendon applies to a joint depends on the vector distance from the line of action of the force to the center of its axis, known as the moment arm (see Fig. 19.1). A larger moment arm increases the torque at the joint. This is achieved by either increasing the moment arm while the force is maintained constant or increasing both force and moment arm. The intact pulleys keep the flexor tendons close to the axis of rotation of the finger joints, thereby maximizing total finger flexion.

This intricate mechanism of pulley-tendon function is critical for understanding motion loss in pathological conditions, such as pulley injuries [7]. The loss of a pulley increases the moment arm (r) by displacing the tendon farther from the joint axis. Since the muscle excursion is physiologically limited, an increase in the moment arm leads to a proportional reduction in joint motion, causing maximum tendon excursion before full finger flexion can be obtained [4].

Components and Biomechanics of the Flexor Pulley System

Pulley Anatomy

The digital flexor pulley system is a key anatomical structure consisting of two main layers: a deep synovial layer and a fibrous pulley layer. Together, these layers form a sheath that stabilizes the tendons and allows them to slide smoothly. The pulleys are dense bands of fibrous tissue that encircle the tendons, creating a tunnel-like structure. This system keeps the tendon close to the bones, which is necessary for transferring force from the muscles into rotational movement at the joints and supporting the efficient function of the flexor tendons.

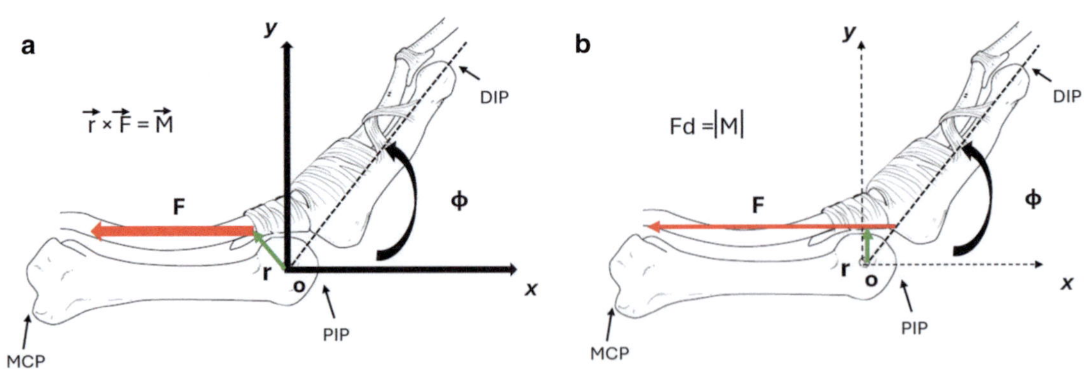

Fig. 19.1 Biomechanical dynamics of the finger pulley system. (**a**) The torque (M) generated at the joint is determined by the cross-product of the force (F) exerted by the tendon and the moment arm (r). (**b**) When the moment arm (r) is perpendicular to the force axis (F), the torque reaches its maximum (|M| = Fd)

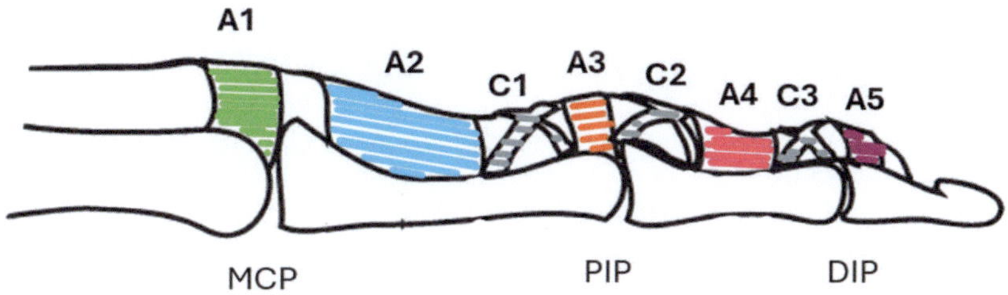

Fig. 19.2 Anatomy of the finger flexor tendon pulley system

The flexor pulley system in the fingers comprises five annular pulleys (A1–A5) and three cruciate pulleys (C1–C3) arranged in order from the base of the finger to the tip. The A1, A3, and A5 pulleys are attached to the volar plates of the metacarpophalangeal (MCP) joint, proximal interphalangeal (PIP), and distal interphalangeal (DIP) joints, respectively. The A2 pulley is attached to the proximal phalanx, while the A4 pulley is attached to the middle phalanx. Three cruciate pulleys are located between the second and fifth pulleys, C1, C2, and C3, located between A2–A3, A3–A4, and A4–A5 pulleys, respectively (Fig. 19.2). The average lengths of the A1, A2, A3, and A4 pulleys are 11 mm, 17 mm, 5 mm, and 8 mm, respectively. Proximal to the A1 pulley is the palmar aponeurotic (PA) pulley, formed by the palmar aponeurosis transverse ligament attached to the connective tissue known as the septa of Legueu and Juvara [8]. Biomechanical studies have shown that removing the PA pulley does not affect tendon function [9].

Biomechanics of the Flexor Pulley System

The efficiency of the flexor system depends on maintaining a short moment arm, which is influenced by several factors: the distance of the pulleys from the joint axis, the perpendicular distance between the pulley edge and the bone's longitudinal axis, and the angle of the joint motion [10]. The closer the pulleys are to the bone, the tighter the radius they form around it, increasing the force acting on the tendon sheath. This tight configuration ensures a short moment arm, allowing the muscle to generate force efficiently while minimizing the required tendon excursion for joint flexion. By sustaining a short moment arm, the pulley system optimizes the uses of tendons' limited excursion and power. This anatomical arrangement enables a 3 cm tendon excursion to produce a 260-degree arc of motion [5].

This positioning prevents the tendons from bowstringing, a condition where the tendons are displaced away from the joint. When bowstringing occurs, the moment arm of the tendon across the joints increases, leading to greater torque at the joint. However, the increased moment arm requires a more extensive tendon excursion to achieve the same degree of flexion. Since the tendon excursion is physiologically limited, damage or rupture to the pulleys results in reduced joint range of motion, stiffness, and contractures [5]. Additionally, the loss of pulley function places the tendon near its maximal excursion during midflexion, which diminishes grip strength, a position where strength is typically optimal [11].

Pulley Injuries

Flexor pulley injuries are most frequently observed in elite rock climbers, but many cases are reported in nonclimbers [3, 12, 13]. Pulley ruptures typically occur due to eccentric loading. At the same time, the finger is in a "crimp" position, characterized by the proximal interphalangeal (PIP) joint being flexed at 90 degrees or more and the distal interphalangeal (DIP) joint hyperextended. Apart from traumatic causes, pulley ruptures have also been identified as complications associated with corticosteroid injections in the treatment of stenosing tenosynovitis [14].

The A2 is typically injured in isolation, but A4 pulley disruption can also occur in isolation or as part of a complex injury [15, 16]. Schöffl et al. introduced a grading system to classify pulley injuries into four grades. Grade I represents pulley strain, grade II indicates complete rupture of

Table 19.1 Pulley injury grading system

Grade	Description	Pulley involved	Treatment
I	Pulley strain	None ruptured	Conservative
II	Complete rupture of A4 or partial rupture of A2	A4 (complete) or A2 (partial)	Conservative
III	Complete rupture of A2 or A3	A2 or A3	Conservative
IV	Multiple ruptures (e.g., A2/A3 or A2/A3/A4) or single rupture (e.g., A2 or A3) with associated trauma	A2, A3, A4, lumbricalis muscle, or collateral ligament	Surgical

A4 or partial rupture of A2, grade III involves complete rupture of A2 or A3, and grade IV includes multiple ruptures (as A2/A3, A2/A3/A4) or single rupture (as A2 or A3) combined with lumbricalis muscle or collateral ligament trauma. This grading system helps determine the appropriate therapeutic approach, with conservative treatment for grades I–III and surgical therapy for grade IV pulley injuries [17] (Table 19.1).

Trigger Finger

Trigger finger or stenosing tenosynovitis is characterized by locking or clicking the affected finger or thumb during movement. This condition results from repetitive friction between the tendon and the enclosing sheath, causing a reactive intratendinous nodule [18]. The obstruction is typically located at the first annular pulley A1, with a mismatch between the volume of the flexor tendon sheath and its content, leading to an impaired excursion of the FPL tendon through the pulley. Surgical release of the A1 pulley is recommended when conservative treatment and corticoid injections fail, with the outcomes being generally favorable [19–23].

Biomechanics of Pulley Reconstruction

Effect of Pulley Excision and Repairs on Hand Biomechanics

The function of the individual pulley has been studied extensively in early research, establishing the critical role of the flexor tendon pulley system in maintaining joint function. As research progressed, the effects of partial pulley excision were explored to determine what components of the pulley system need to be intact for near-normal function to occur. The isolated loss of the A1 or A5 pulley does not significantly impact work efficiency or tendon excursion [24, 25]. In particular, the A2 and A4 pulleys were recognized as essential components for preventing tendon bowstringing and ensuring efficient force transmission during finger flexion [1, 2]. Mitsionis et al. and Tomaino et al. studied the kinematic effects of venting portions of the A2 and A4 pulleys and found that sacrificing 50% or more of either pulley had a detrimental impact on finger motion [26, 27]. They suggested partial A2 and A4 pulley venting may facilitate tendon gliding after surgery, with no clinical implications for finger motion. Areas of controversy exist in the effects of partial proximal versus partial distal pulley excisions. Chow et al. found that the proximal part of both pulleys is more essential to maintain natural finger kinematics, with the PIP joint being the most affected by this alteration, decreasing total ROM during finger flexion [28]. Additionally, Tanaka et al. investigated the effect of venting the proximal portion of the A2 pulley and observed an increase in peak tendon gliding resistance [29]. In contrast, Leeflang et al. found that partial distal venting of the A2 pulley resulted in greater bowstringing than proximal venting [30]. This body of work may support the intentional partial proximal A2 and/or A4 pulley excision as a component for successful treatment.

Pulley reconstruction is advised in case of pulley rupture and flexor tendon bowstringing [31–33]. Studies have shown that reconstructing the distal portion of the A2 pulley significantly improves PIP flexion compared to the reconstruction of the A4 pulley and is more critical in preventing bowstringing [25, 30, 34]. The release of the A4 pulley during flexor tendon repair is often biomechanically tolerable [26, 27, 34, 35]. Conversely, a recent study by Soulii et al. compared the mechanical characteristics of A2 and combined A2–A4 pulley repair. Their findings indicate that a three-ring A2–A4 configuration repair restored the mechanical characteristics of the finger at the PIP joint more than a one-ring or two-ring A2 repair, noting that the difference between the results was statistically significant but not yet clinically relevant [36].

Reconstruction Techniques

Reconstruction surgery of the pulley system helps prevent flexion contractures and decreased range of motion caused by the flexor tendon bowstringing. The principles to restore proper pulley function consist of replicating the length of the native pulley, appropriate tensioning of the construct, and use of a synovial graft [37]. Several techniques for reconstructing the flexor pulley system have been described in the literature. These techniques fall into two broad categories of reconstruction: encircling and no encircling reconstruction (Fig. 19.3).

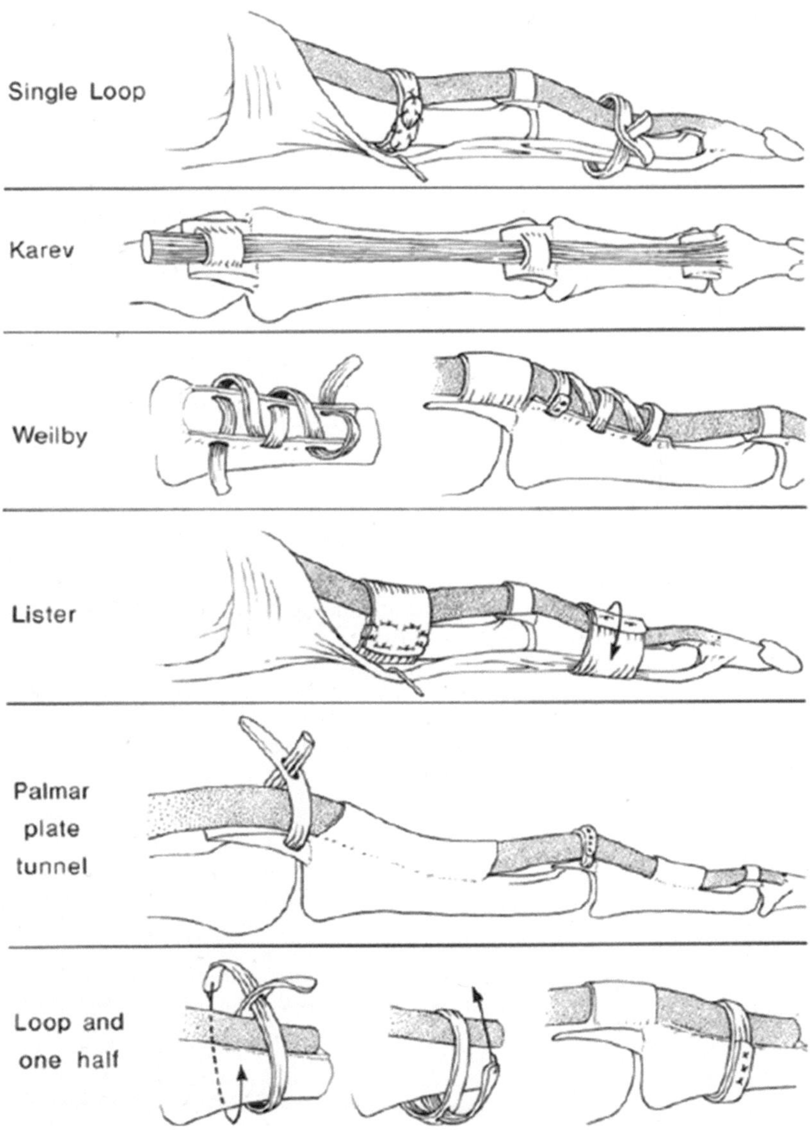

Fig. 19.3 Reconstruction techniques of pulley repair. (From Widstrom et al. [41])

Okutsu's Triple-Loop Technique

This technique is the modification of the original Bunnell technique that uses a single loop of free tendon graft that is overlapped and sutured to itself. Okutsu et al. developed a triple-loop reconstruction using three tendon grafts that are individually passed around the phalanx, tensioned, and sutured to themselves [38]. He suggested that this reconstruction is more expansive and stronger than the Bunnell technique. This technique has been demonstrated to be the most robust among the other reconstruction methods, with a loading capacity comparable to the normal pulley [39]. Additionally, its excursion resistance surpasses the Kleinert/Weilby technique but remains inferior to Lister's and Karev's techniques [40].

Widstrom's Loop-and-a-Half Technique

Widstrom performed this technique by wrapping the tendon graft around the phalanx and passing it through a surgically opened operculum in one graft segment. The two loose ends are then sutured on both sides of the loop and positioned away from the flexor tendon [41]. His biomechanical studies found that this technique has the highest breaking strength and comparable mechanical effectiveness to most other techniques [41, 42].

Kleinert/Weilby Technique

Kleinert and Weilby proposed a reconstruction technique that involves passing the flexor digitorum superficialis (FDS) graft through the fibrous rim of the damaged pulley and suturing it at the rim on each end, resulting in a shoelace-like reconstruction [43]. One notable advantage of this approach is the ability to give the surgeon precise control over tension adjustment during pulley repair. However, this technique's limitation is the lack of strength compared to other techniques [42].

Karev's Belt Loop Technique

This technique requires sliding the flexor digitorum profundus (FDP) tendon through a belt loop made by two transverse parallel incisions in the palmar plates of the joints [44]. Karev's method was the most effective compared to the intact pulley system, requiring less tendon excursion to produce flexion [41]. This finding emphasized placing reconstructed pulleys close to the joint axis to optimize biomechanical efficiency. It is stronger than the Kleinert/Weilby and Lister techniques but weaker than the loop-and-a-half and triple-loop techniques [42].

Lister's Technique

Lister's technique uses an extensor retinaculum graft wrapped around the phalanx and sutured to itself [45]. Nishida et al. found that the retinaculum has a smooth gliding surface, producing the lowest resistance among the other reconstructive techniques [40]. However, this technique has the lowest mechanical efficiency and low strength [41, 42].

Doyle's Palmar Plate Tunnel

Doyle described this technique by creating a transverse tunnel through the MP and PIP volar plates and passing a tendon graft through the tunnel with a constructed loop [41]. Widstrom et al. found this technique to have low mechanical efficiency and strength [41, 42].

Limited research is done on the clinical outcomes of pulley reconstruction techniques. Surgical treatment consistently produced favorable results, with good functional outcomes, significant grip and pinch strength recovery, and increased PIP joint flexion [32, 38, 46, 47]. Additional outcome studies are needed to determine the optimal pulley reconstruction technique and the effect of pulley venting during repair.

Pulley System of the Thumb

Anatomy of the Thumb Pulley System

The thumb pulley system represents a unique anatomical arrangement critical for the precision and power of the thumb movements. This structure stabilizes the flexor pollicis longus (FPL) tendon and facilitates efficient tendon gliding, enabling complex functional activities ranging from pinching to grasping. Unlike the pulley systems of the fingers, the thumb's pulley configuration reflects distinct biomechanical requirements for mobility and strength.

The main pulleys of the thumb include A1, A2, and oblique pulleys, with an additional variable annular (Av) pulley identified in multiple studies [48–51]. A1 pulley is found at the metacarpophalangeal (MCP) joint level, the proximal two-thirds originating from the volar plate. At the same time, the distal one-third attaches to the base of the proximal phalanx. Av pulley is situated between A1 and oblique pulleys, covering most of the proximal half of the proximal phalanx. It exhibits significant variability, appearing as transverse, oblique, or a combination of both orientations, and in some cases, it may be fused with the A1 pulley [49, 51]. The oblique pulley extends obliquely from the ulnar aspect of the proximal phalanx base to the radial side, spanning between the A1 and A2 pulleys. A2 pulley is found at the level of the interphalangeal (IP) joint (Fig. 19.4). Sulivan et al. recently identified a proximal A0 pulley (or palmar aponeurosis), previously described only in fingers, found in 62.5% of thumb specimens (with only 41.6% contained bilaterally). It is a proximal, transverse structure distinct from the A1 pulley [52].

Biomechanics and Surgical Repair of the Thumb's Pulley System

The primary biomechanical role of the thumb pulley system mirrors that of the fingers. It enables the smooth gliding of the flexor pollicis longus (FPL) tendon while maintaining it close to the bone and reducing the risk of bowstringing. However, due to its unique anatomy and functional demands, the thumb pulley system exhibits several distinct biomechanical characteristics.

Zissimos et al. conducted a cadaveric study and found that isolated sectioning of either the A1 or the oblique pulley did not significantly alter thumb biomechanics. Only when both the A1 and the oblique pulleys were disrupted did significant tendon bowstringing occur [53]. Similarly, Espin et al. further highlighted that maintaining the oblique or A1 pulley was sufficient to preserve normal tendon excursion [54]. Conversely, other investigations found that removal of only the oblique pulley caused minimal loss of motion if the A1 and A2 pulleys remained intact and that the oblique pulley alone cannot prevent bowstringing of the FPL tendon. Still, an intact A1 or Av pulley was sufficient to

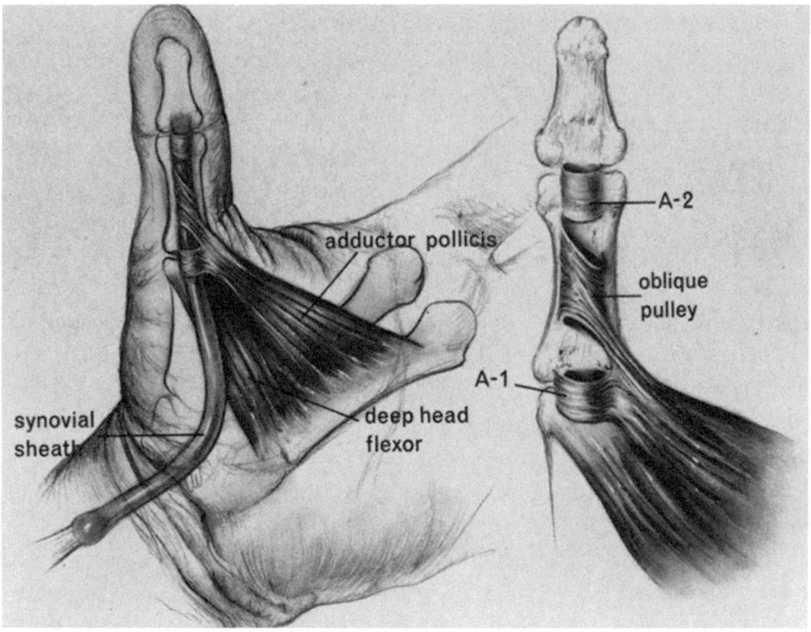

Fig. 19.4 Pulley system of the thumb. (Reproduced from Doyle et al. [48] with permission. No changes have been made)

maintain proper tendon alignment and function [48, 49].

Reconstruction techniques of the thumb's pulley system have been described primarily for the A1 and oblique pulleys. The oblique pulley is typically reconstructed with Okutsu's three-loop technique using palmaris longus tendon [38]. Another method has been described by Slesarenko et al. for the reconstruction of the oblique pulley with distally based extensor pollicis brevis tendon (EPB) [55]. This surgical method involves harvesting the extensor pollicis brevis (EPB) tendon through an incision over the first dorsal wrist compartment. Then a second small incision on the thumb retrieves the distally based EPB tendon. The EPB tendon is passed around the digit volarly to the flexor tendons and extensor pollicis longus tendon, dorsally to the neurovascular bundle, and then sutured to itself. It can be done as a single-, double-, or triple-tendon loop reconstruction. This technique was found to have an ultimate load before failure comparable to an intact oblique pulley and decreased stiffness [55]. Contrary to the pulley venting concept in the fingers, Kutsumi et al. found that the gliding resistance of the flexor pollicis longus tendon is increased after its repair with distal venting of the oblique pulley, suggesting that no partial excision should be made for the oblique pulley to prevent any mechanical and clinical disadvantage [56]. A1 thumb pulley reconstruction includes a free extensor retinaculum graft technique. Guelmi et al. described a modified Lister's method with a fixation procedure similar to Kleinert's technique [57].

Conclusion

The flexor pulley system exhibits a significant biomechanical complexity, underscoring its vital importance in hand function. Injuries to the pulley structures can compromise normal tendon mechanics, leading to joint motion and strength alterations. Surgical reconstruction is often necessary to restore functionality, particularly in instances of high-grade pulley ruptures. This chapter aims to consolidate contemporary knowledge regarding the anatomy, biomechanics, and repair techniques associated with the pulley system. The primary objective is to enhance the understanding of the mechanical constraints inherent to the pulley system, thereby improving surgical interventions and ultimately benefiting patient outcomes.

References

1. Doyle JR. Anatomy of the flexor tendon sheath and pulley system: a current review. J Hand Surg Am. 1989;14(2 Pt 2):349–51. https://doi.org/10.1016/0363-5023(89)90110-x.
2. Peterson WW, Manske PR, Bollinger BA, Lesker PA, McCarthy JA. Effect of pulley excision on flexor tendon biomechanics. J Orthop Res. 1986;4(1):96–101. https://doi.org/10.1002/jor.1100040112.
3. Logan AJ, Makwana N, Mason G, Dias J. Acute hand and wrist injuries in experienced rock climbers. Br J Sports Med. 2004;38(5):545–8. https://doi.org/10.1136/bjsm.2002.003558.
4. Doyle JR. Palmar and digital flexor tendon pulleys. Clin Orthop Relat Res. 2001;383:84–96. https://doi.org/10.1097/00003086-200102000-00011.
5. Goodman HJ, Choueka J. Biomechanics of the flexor tendons. Hand Clin. 2005;21(2):129–49. https://doi.org/10.1016/j.hcl.2004.11.002.
6. Elftman H. Biomechanics of muscle with particular application to studies of gait. J Bone Joint Surg Am. 1966;48(2):363–77.
7. Amirouche F, Gonzalez M, Koldoff J, Tioco J, Ham K. A biomechanical study of the finger pulley system during repair. Technol Health Care. 2002;10(1):23–31.
8. Manske PR, Lesker PA. Palmar aponeurosis pulley. J Hand Surg Am. 1983;8(3):259–63. https://doi.org/10.1016/s0363-5023(83)80154-3.
9. Phillips C, Mass D. Mechanical analysis of the palmar aponeurosis pulley in human cadavers. J Hand Surg Am. 1996;21(2):240–4. https://doi.org/10.1016/S0363-5023(96)80107-9.
10. Idler RS. Anatomy and biomechanics of the digital flexor tendons. Hand Clin. 1985;1(1):3–11.
11. Gonzalez MH, Mohan V, Elhassan B, Amirouche F. Biomechanics of the digit. J Am Soc Surg Hand. 2005;5(1):48–60. https://doi.org/10.1016/j.jassh.2004.11.009.
12. Lourie GM, Hamby Z, Raasch WG, Chandler JB, Porter JL. Annular flexor pulley injuries in professional baseball pitchers: a case series. Am J Sports Med. 2011;39(2):421–4. https://doi.org/10.1177/0363546510387506.
13. Schöffl VR, Jüngert J. Closed flexor pulley injuries in nonclimbing activities. J Hand Surg Am. 2006;31(5):806–10. https://doi.org/10.1016/j.jhsa.2006.02.006.

14. Hamano H, Motomiya M, Iwasaki N. Adverse effect of repeated corticosteroid injections for trigger finger on flexor pulley system. J Hand Surg Eur. 2013;38(3):326–7. https://doi.org/10.1177/1753193412465728.
15. Cockenpot E, Lefebvre G, Demondion X, Chantelot C, Cotten A. Imaging of sports-related hand and wrist injuries: sports imaging series. Radiology. 2016;279(3):674–92. https://doi.org/10.1148/radiol.2016150995.
16. Freilich AM. Evaluation and treatment of jersey finger and pulley injuries in athletes. Clin Sports Med. 2015;34(1):151–66. https://doi.org/10.1016/j.csm.2014.09.001.
17. Schöffl V, Hochholzer T, Winkelmann HP, Strecker W. Pulley injuries in rock climbers. Wilderness Environ Med. 2003;14(2):94–100. https://doi.org/10.1580/1080-6032(2003)014[0094:piirc]2.0.co;2.
18. Hueston J, Wilson W. The aetiology of trigger finger explained on the basis of intratendinous architecture. Hand. 1972;4(3):257–60. https://doi.org/10.1016/S0072-968X(72)80010-X.
19. Ryzewicz M, Wolf JM. Trigger digits: principles, management, and complications. J Hand Surg Am. 2006;31(1):135–46. https://doi.org/10.1016/j.jhsa.2005.10.013.
20. Thorpe AP. Results of surgery for trigger finger. J Hand Surg Br. 1988;13(2):199–201. https://doi.org/10.1016/0266-7681_88_90138-6.
21. Gilberts ECAM, Wereldsma JCJ. Long-term results of percutaneous and open surgery for trigger fingers and thumbs. Int Surg. 2002;87(1):48–52.
22. Lange-Riess D, Schuh R, Hönle W, Schuh A. Long-term results of surgical release of trigger finger and trigger thumb in adults. Arch Orthop Trauma Surg. 2009;129(12):1617–9. https://doi.org/10.1007/s00402-008-0802-8.
23. Saldana MJ. Trigger digits: diagnosis and treatment. J Am Acad Orthop Surg. 2001;9(4):246–52. https://doi.org/10.5435/00124635-200107000-00004.
24. Lin GT, Amadio PC, An KN, Cooney WP. Functional anatomy of the human digital flexor pulley system. J Hand Surg Am. 1989;14(6):949–56. https://doi.org/10.1016/s0363-5023(89)80043-7.
25. Rispler D, Greenwald D, Shumway S, Allan C, Mass D. Efficiency of the flexor tendon pulley system in human cadaver hands. J Hand Surg Am. 1996;21(3):444–50. https://doi.org/10.1016/S0363-5023(96)80361-3.
26. Tomaino M, Mitsionis G, Basitidas J, Grewal R, Pfaeffle J. The effect of partial excision of the A2 and A4 pulleys on the biomechanics of finger flexion. J Hand Surg Br. 1998;23(1):50–2. https://doi.org/10.1016/s0266-7681(98)80218-0.
27. Mitsionis G, Bastidas JA, Grewal R, Pfaeffle HJ, Fischer KJ, Tomaino MM. Feasibility of partial A2 and A4 pulley excision: effect on finger flexor tendon biomechanics. J Hand Surg Am. 1999;24(2):310–4. https://doi.org/10.1053/jhsu.1999.0310.
28. Chow JC, Sensinger J, McNeal D, Chow B, Amirouche F, Gonzalez M. Importance of proximal A2 and A4 pulleys to maintaining kinematics in the hand: a biomechanical study. Hand (NY). 2014;9(1):105–11. https://doi.org/10.1007/s11552-013-9547-0.
29. Tanaka T, Amadio PC, Zhao C, Zobitz ME, An KN. The effect of partial A2 pulley excision on gliding resistance and pulley strength in vitro. J Hand Surg Am. 2004;29(5):877–83. https://doi.org/10.1016/j.jhsa.2004.04.014.
30. Leeflang S, Coert JH. The role of proximal pulleys in preventing tendon bowstringing: pulley rupture and tendon bowstringing. J Plast Reconstr Aesthet Surg. 2014;67(6):822–7. https://doi.org/10.1016/j.bjps.2014.01.041.
31. Hume EL, Hutchinson DT, Jaeger SA, Hunter JM. Biomechanics of pulley reconstruction. J Hand Surg Am. 1991;16(4):722–30. https://doi.org/10.1016/0363-5023(91)90201-l.
32. Moutet F, Forli A, Voulliaume D. Pulley rupture and reconstruction in rock climbers. Tech Hand Up Extrem Surg. 2004;8(3):149–55. https://doi.org/10.1097/01.bth.0000128545.94060.99.
33. Odobescu A, Radu A, Brutus JP, Gilardino MS. Modified flexor digitorum superficialis slip technique for A4 pulley reconstruction. J Hand Surg Eur. 2010;35(6):464–8. https://doi.org/10.1177/1753193410371031.
34. Tang JB. Indications, methods, postoperative motion and outcome evaluation of primary flexor tendon repairs in Zone 2. J Hand Surg Eur. 2007;32(2):118–29. https://doi.org/10.1016/J.JHSB.2006.12.009.
35. Franko OI, Lee NM, Finneran JJ, et al. Quantification of partial or complete A4 pulley release with FDP repair in cadaveric tendons. J Hand Surg Am. 2011;36(3):439–45. https://doi.org/10.1016/j.jhsa.2010.11.041.
36. Soulii L, Amirouche F, Solitro G, et al. Evaluation of A2 and A4 hand pulley repair using tendon graft rings. J Hand Surg Am. 2021;46(7):626.e1–6. https://doi.org/10.1016/j.jhsa.2020.11.025.
37. Dy CJ, Daluiski A. Flexor pulley reconstruction. Hand Clin. 2013;29(2):235–42. https://doi.org/10.1016/j.hcl.2013.02.005.
38. Okutsu I, Ninomiya S, Hiraki S, Inanami H, Kuroshima N. Three-loop technique for A2 pulley reconstruction. J Hand Surg Am. 1987;12(5 Pt 1):790–4. https://doi.org/10.1016/s0363-5023(87)80071-0.
39. Lin GT, Amadio PC, An KN, Cooney WP, Chao EY. Biomechanical analysis of finger flexor pulley reconstruction. J Hand Surg Br. 1989;14(3):278–82. https://doi.org/10.1016/0266-7681_89_90081-8.
40. Nishida J, Amadio PC, Bettinger PC, An KN. Flexor tendon-pulley interaction after pulley reconstruction: a biomechanical study in a human model in vitro. J Hand Surg Am. 1998;23(4):665–72. https://doi.org/10.1016/S0363-5023(98)80053-1.
41. Widstrom CJ, Johnson G, Doyle JR, Manske PR, Inhofe P. A mechanical study of six digital pulley reconstruction techniques: Part I. Mechanical effec-

tiveness. J Hand Surg Am. 1989;14(5):821–5. https://doi.org/10.1016/s0363-5023(89)80082-6.
42. Widstrom CJ, Doyle JR, Johnson G, Manske PR, McGee R. A mechanical study of six digital pulley reconstruction techniques: Part II. Strength of individual reconstructions. J Hand Surg Am. 1989;14(5):826–9. https://doi.org/10.1016/s0363-5023(89)80083-8.
43. Klinert HE, Bennett JB. Digital pulley reconstruction employing the always present rim of the previous pulley. J Hand Surg Am. 1978;3(3):297–8. https://doi.org/10.1016/s0363-5023(78)80098-7.
44. Karev A. The "belt loop" technique for the reconstruction of pulleys in the first stage of flexor tendon grafting. J Hand Surg Am. 1984;9(6):923–4. https://doi.org/10.1016/s0363-5023(84)80082-9.
45. Lister G. Indications and techniques for repair of the flexor tendon sheath. Hand Clin. 1985;1(1):85–95.
46. Arora R, Fritz D, Zimmermann R, et al. Reconstruction of the digital flexor pulley system: a retrospective comparison of two methods of treatment. J Hand Surg Eur. 2007;32(1):60–6. https://doi.org/10.1016/j.jhsb.2006.09.019.
47. Gabl M, Reinhart C, Lutz M, Bodner G, Angermann P, Pechlaner S. The use of a graft from the second extensor compartment to reconstruct the A2 flexor pulley in the long finger. J Hand Surg Br. 2000;25(1):98–101. https://doi.org/10.1054/jhsb.1999.0278.
48. Doyle JR, Blythe WF. Anatomy of the flexor tendon sheath and pulleys of the thumb. J Hand Surg Am. 1977;2(2):149–51. https://doi.org/10.1016/s0363-5023(77)80101-9.
49. Bayat A, Shaaban H, Giakas G, Lees VC. The pulley system of the thumb: anatomic and biomechanical study. J Hand Surg Am. 2002;27(4):628–35. https://doi.org/10.1053/jhsu.2002.34008.
50. Schmidt HM, Fischer G. Pulleys of the tendon sheath of the flexor pollicis longus muscle. Handchir Mikrochir Plast Chir. 1999;31(6):362–6. https://doi.org/10.1055/s-1999-13553.
51. Schubert MF, Shah VS, Craig CL, Zeller JL. Varied anatomy of the thumb pulley system: implications for successful trigger thumb release. J Hand Surg Am. 2012;37(11):2278–85. https://doi.org/10.1016/j.jhsa.2012.08.005.
52. Sullivan JM, Edwards S, Waetjen E, Doğar S, Walker ME. Back to the anatomy lab to redefine the pulley system of the thumb: is there an A0 pulley? Hand (NY). Published online December 11, 2024:15589447241302356. https://doi.org/10.1177/15589447241302356.
53. Zissimos AG, Szabo RM, Yinger KE, Sharkey NA. Biomechanics of the thumb flexor pulley system. J Hand Surg Am. 1994;19(3):475–9. https://doi.org/10.1016/0363-5023(94)90066-3.
54. Esplin VS, Tencer AF, Hanel DP, Cosio MQ. Restoration of function of the thumb flexor apparatus requires repair of the oblique and one adjacent flexor tendon pulley. J Orthop Res. 1996;14(1):152–6. https://doi.org/10.1002/jor.1100140124.
55. Slesarenko YA, Baruch A, Mallo GC, Serra-Hsu F. Thumb flexor pulley reconstruction with distally based extensor pollicis brevis tendon: technique and biomechanical analysis. Tech Hand Up Extrem Surg. 2006;10(2):114–7. https://doi.org/10.1097/00130911-200606000-00011.
56. Kutsumi K, Amadio PC, Zhao C, Zobitz ME, An KN. Gliding resistance of the flexor pollicis longus tendon after repair: does partial excision of the oblique pulley affect gliding resistance? Plast Reconstr Surg. 2006;118(6):1423–8. https://doi.org/10.1097/01.prs.0000239595.97189.54.
57. Guelmi K, Barbato B, Kolb F. Reconstruction of a metacarpophalangeal pulley in the thumb with a free extensor retinaculum graft. J Hand Surg Br. 1997;22(2):274–6. https://doi.org/10.1016/s0266-7681(97)80081-2.

Wrist Joint Biomechanics and Arthroplasty Design

20

Jorge L. Orbay, Peter A. Falgiano, and Ali Rteil

Introduction

The wrist provides a mobile yet stable base for hand function. It is responsible for placing the hand in space to optimize its gripping function and for transmitting to the hand forces generated by the forearm musculature. The wrist joint is a pivot for bending moments generated by forearm muscles whose tendons are located across it. The wrist also transmits to the hand torsional forces (moments) generated by the forearm muscles that drive pronation-supination and are located between the radius and the ulna. Torsion transmission from the forearm to the hand occurs passively, as pronsupinatory muscles do not cross the wrist joint, and torsion transmission occurs regardless of the angle between the hand and the forearm [30].

Anatomically, the wrist comprises the carpus, the distal radius, and the distal ulna. The carpus consists of two transversely aligned rows of bones, the proximal and distal carpal rows, which consist of three and five bones, respectively [5]. These structures form three joints: the radiocarpal, the intercarpal, and the distal radioulnar joints. The latter is mostly involved in forearm rotation and becomes less important in load transmission after wrist arthroplasty. Therefore, we will focus on the first two joints in our analysis. The radiocarpal and intercarpal joints are arranged in tandem. The proximal radiocarpal joint is located between the forearm and the proximal row and the intercarpal joint between the proximal and distal rows. Each joint functions similarly to a hinge joint (ginglymus). They have axes of motion that are orthogonal or essentially perpendicular to each other. This configuration allows the wrist to behave as a universal joint, a mechanical structure that allows motion in two planes and simultaneous transmission of axial rotation (Fig. 20.1).

The eight bones forming the carpus are short, roughly as long as they are wide, primarily cancellous, and covered by large areas of hyaline cartilage. Flat joint surfaces and short intercarpal ligaments allow minimal movement within each carpal row. Wrist motion occurs primarily at the radiocarpal and intercarpal joints, which are controlled by longer extrinsic ligaments [28]. Generally, ligaments prevent abnormal wrist motion and ensure correct joint alignment and direct forces at right angles to neighboring articular surfaces. Hyaline cartilage and subchondral

J. L. Orbay (✉)
Miami Bone and Joint Institute, Miami, FL, USA

P. A. Falgiano
Miami Orthopaedic Research Foundation, Miami, FL, USA

A. Rteil
Department of Orthopaedic Surgery, Orthopaedic and Spine Institute, Northshore University Health System, An Affiliate of the University of Chicago Pritzker School of Medicine, Skokie, IL, USA

F. Amirouche et al. (eds.), *Biomechanics of the Hand*,
https://doi.org/10.1007/978-3-032-04688-8_20

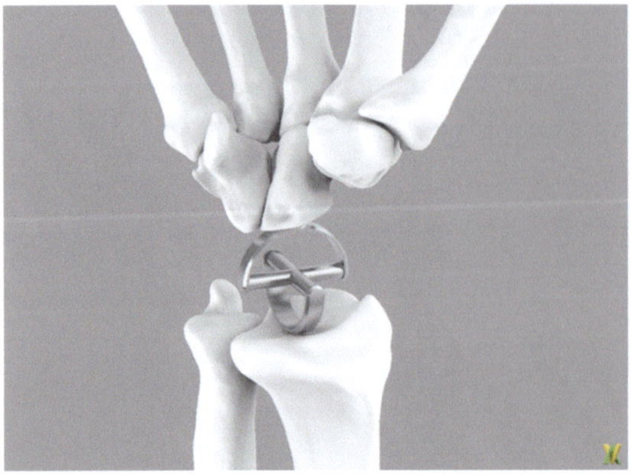

Fig. 20.1 The wrist is a universal joint, having two axes of motion placed in an orthogonal arrangement and permitting flexion-extension, radial-ulnar deviation, and the simultaneous transmission of torsion from the forearm to the hand

bone support these compressive loads while allowing motion. Axial and torsional loads both generate compressive forces on the wrist joint surfaces. Forearm muscles, whose tendons cross the wrist joint, actively apply compressive forces to articular surfaces. Ligaments supporting the wrist joint are arranged in a helical manner. Therefore, active forearm rotation induces torsion across the wrist, producing compressive forces on articular surfaces [24]. Active loads on wrist-crossing tendons also help helical ligaments transmit wrist axial rotation, as they become oriented in a helical arrangement during passive wrist axial rotation [7, 16].

Physiologic wrist motion is traditionally described as flexion-extension and radial-ulnar deviation. Combining these two results in a three-dimensional motion called circumduction. Maximal circumduction describes a cone stretched in the direction of radial extension and ulnar flexion. Simple motion from radial extension to ulnar flexion is known as the "dart thrower's motion." This motion occurs primarily along the axis of the intercarpal joint and is fundamental to many basic activities of the human hand (Fig. 20.2) [4, 28, 29]. Motion along the orthogonally oriented radiocarpal joint axis is of lesser magnitude yet essential for the wrist's function of torque transmission which requires two axis of rotation placed in an orthogonal arrangement [9, 11, 25].

The average range of normal wrist motion is extension 70°, flexion 70°, radial deviation 20°, and ulnar deviation 35°. The minimal motion needed for most activities of daily living is extension 30°, flexion 5°, radial deviation 10°, and ulnar deviation 15° [23]. A sound prosthetic replacement should provide close to 80° of flexion-extension and 40° of radial-ulnar deviation arc. The average passive axial rotation of the wrist is 30°.

Regarding force transmission, the wrist serves a dual purpose, acting as a pivot or fulcrum for active bending and as a universal joint for transmitting torsional forces. This all complicates the design of wrist prostheses. A functional and durable wrist prosthesis should minimize joint surface stress while simultaneously performing both functions. Like hyaline cartilage on the native wrist, the prosthetic joint surfaces are composed of materials that must withstand compressive loading while minimizing wear from movement.

History of Wrist Arthroplasty

Themistocles Gluck, a Moldavian surgeon, designed and implanted the first wrist prosthesis in 1890 in Berlin. It featured a ball-and-socket design with a metacarpal and radial stem. Carved from ivory, it was unsuccessful due to infection. In 1922, he was nominated for a Nobel Prize for this and other surgical innovations.

The introduction of the Swanson® implant (Dow-Corning Corp., Michigan, USA), a stemmed flexible silicone spacer, marked the beginning of the first-generation implant in the 1960s. It was utilized to treat patients with rheumatoid arthritis. However, problems such as implant fractures and silicone synovitis resulted in its retirement [14, 22, 26].

In the 1970s, Meuli and Volz introduced second-generation implants (Fig. 20.3). These implants featured metal-on-polyethylene articulations, with carpal fixation achieved by two long stems cemented into the second and third meta-

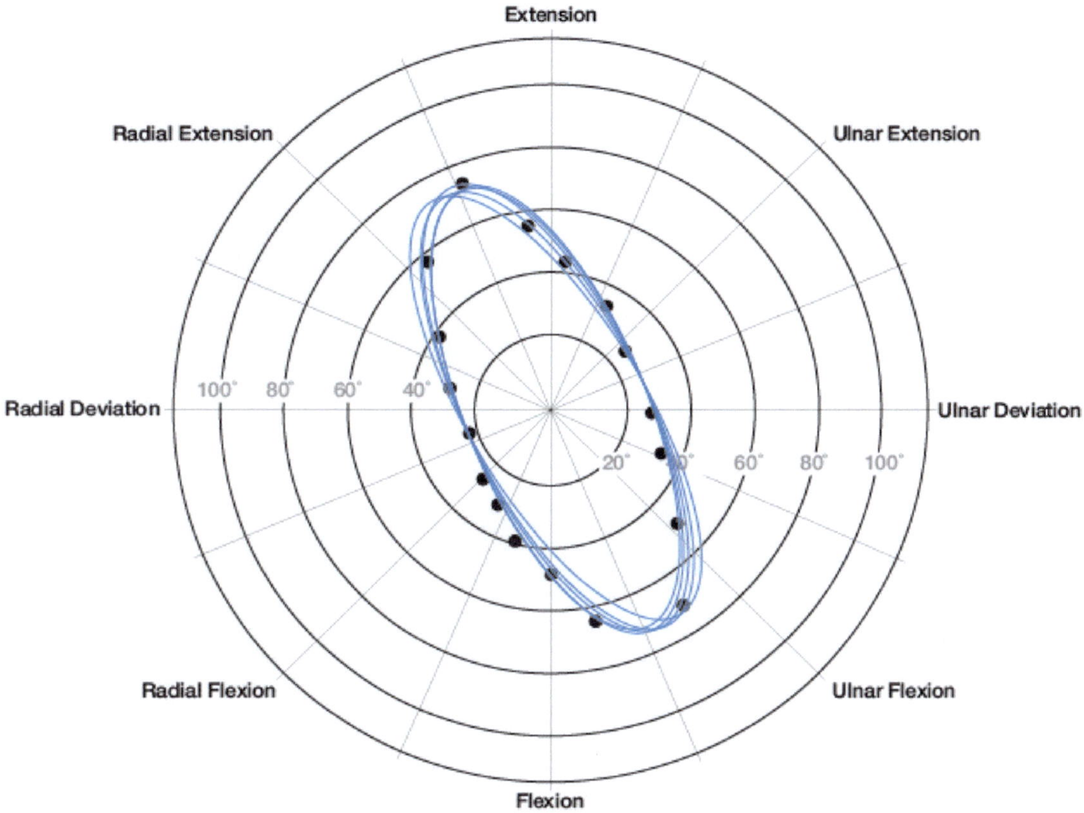

Fig. 20.2 Circumduction is a three-dimensional motion described by the hand as it moves simultaneously along the axis of flexion-extension and radial-ulnar deviation. The extremes of this motion define a cone in space stretched from radial extension to ulnar flexion, a path known as dart throwers

carpals. At the same time, the radial component was secured to the radial shaft.

The Meuli® implant (Sulzer Orthopaedics, Texas, USA) used a polyethylene sphere on the radial component and a metal cup on the carpal side. Failures occurred due to loosening and soft tissue imbalance [15]. In contrast, the Volz® implant (Howmedica, New Jersey, USA) featured a toroidal metal surface on the carpal component and a polyethylene concave surface on the radial component. This toroidal articulation was designed to allow movement in two primary wrist planes: flexion-extension and radial-ulnar deviation. Loosening on the carpal side often resulted in failure [18].

The third-generation TWA implants all relied on cement fixation. The carpal side featured a long third metacarpal stem and a shorter second metacarpal stem to control rotation, and all were provided with a nonspherical polyethylene-on-metal articulation (Fig. 20.4).

The trispherical implant had titanium radial and carpal components, an ultrahigh molecular weight polyethylene (UHMWPE) concave bearing on the carpal component, and a matching head on the radial stem. A nonload-bearing shaft was intended to prevent dislocation. Loosening from both carpal and radial sides was the leading cause of failure [6].

The Biaxial® total wrist implant (DePuy Synthes, New Jersey, USA) was the first to incorporate an ellipsoidal carpal component surface with a matching UHMWPE radial surface. The implant was made of cobalt chrome alloy and had a porous coating. Although it was designed to be cemented, surgeons often used it without cement.

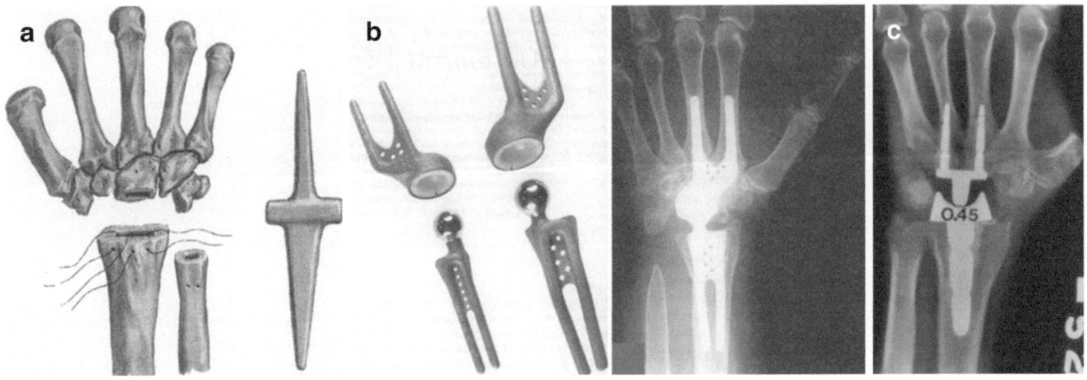

Fig. 20.3 (**a**) Swanson implant, (**b**) Mueli implant, (**c**) Volz implant

Fig. 20.4 (**a**) Trispherical Implant [15]. (**b**) BIAX® Total Wrist System (Depuy) [19]. (**c** and **d**) Universal® Total Wrist Implant (KMI) [20]

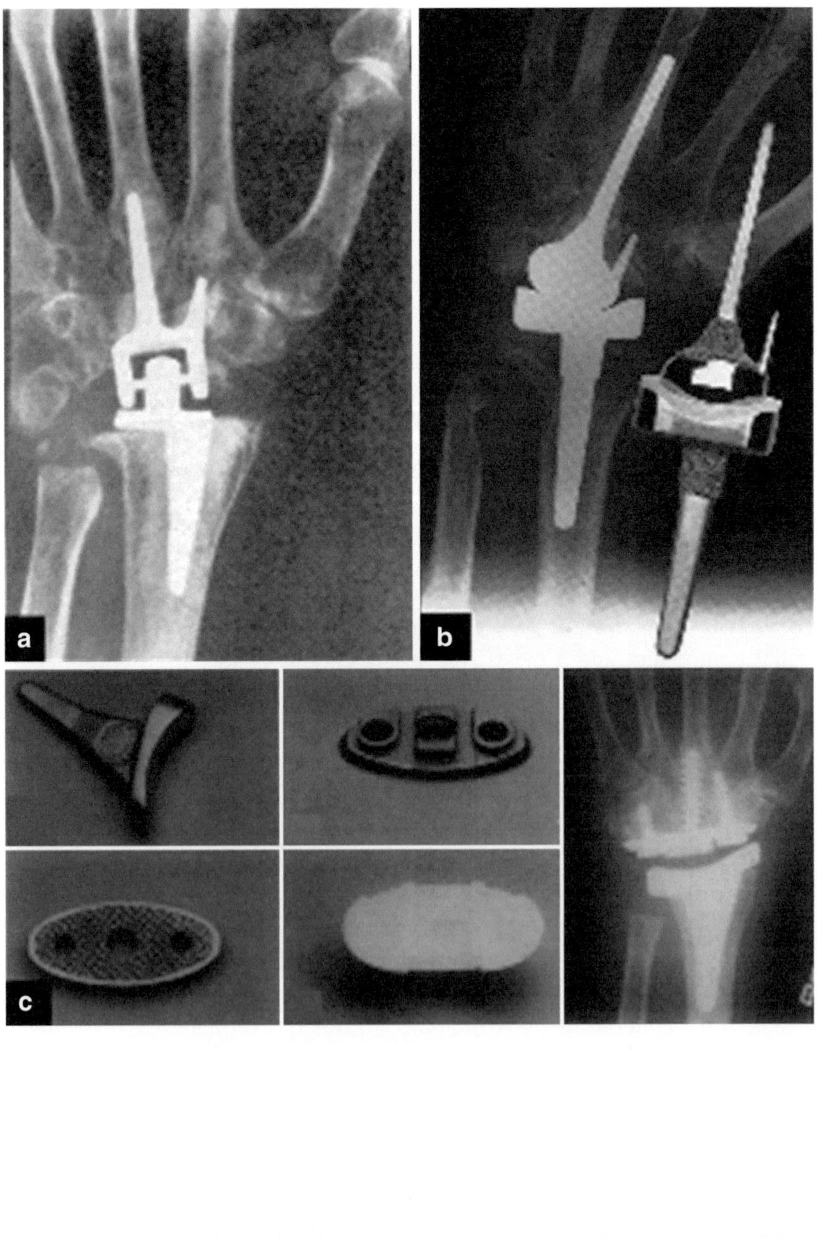

Failures commonly occurred due to loosening of the metacarpal component [3].

The Universal® Total Wrist Implant (Integra Life Sciences, New Jersey, USA) introduced a toroidal joint surface. The carpal component was made of UHMWPE on a titanium base with a porous ingrowth surface, while the radial component was crafted from cobalt chrome. To prevent loosening, the carpal component was designed without a metacarpal stem. The carpus was prepared using a transverse saw cut. Failures resulted from dislocation and loosening on both the carpal and radial sides [3].

Fourth-generation implants show minimal fundamental innovation, as they all utilize the basic design of the Universal wrist, which features a nonspherical joint with carpal rather than metacarpal fixation [19, 21]. Several progressive modifications were introduced to address common complications associated with this basic design, such as loosening and instability [1] (Fig. 20.5).

These implants feature a central stem for fixation into the capitate rather than the metacarpal. Stability issues with the universal wrist were addressed by modifying the articular design from a toroidal-on-toroidal to a toroidal-on-ellipsoidal articulation in the Universal 2® (Integra Life Sciences, New Jersey, USA). The ReMotion® wrist (Small Bone Innovations, Pennsylvania, USA) incorporates 10° of rotation in its carpal polyethylene component, allowing for some axial rotation. These two articular design modifications also reduced stress concentration on the polyethylene components, minimizing wear, debris generation, osteolysis, and loosening. Enhanced wear performance was achieved by utilizing ellipsoidal designs that closely match the greater and lesser diameters, thereby making them more tolerant of axial rotation. This was

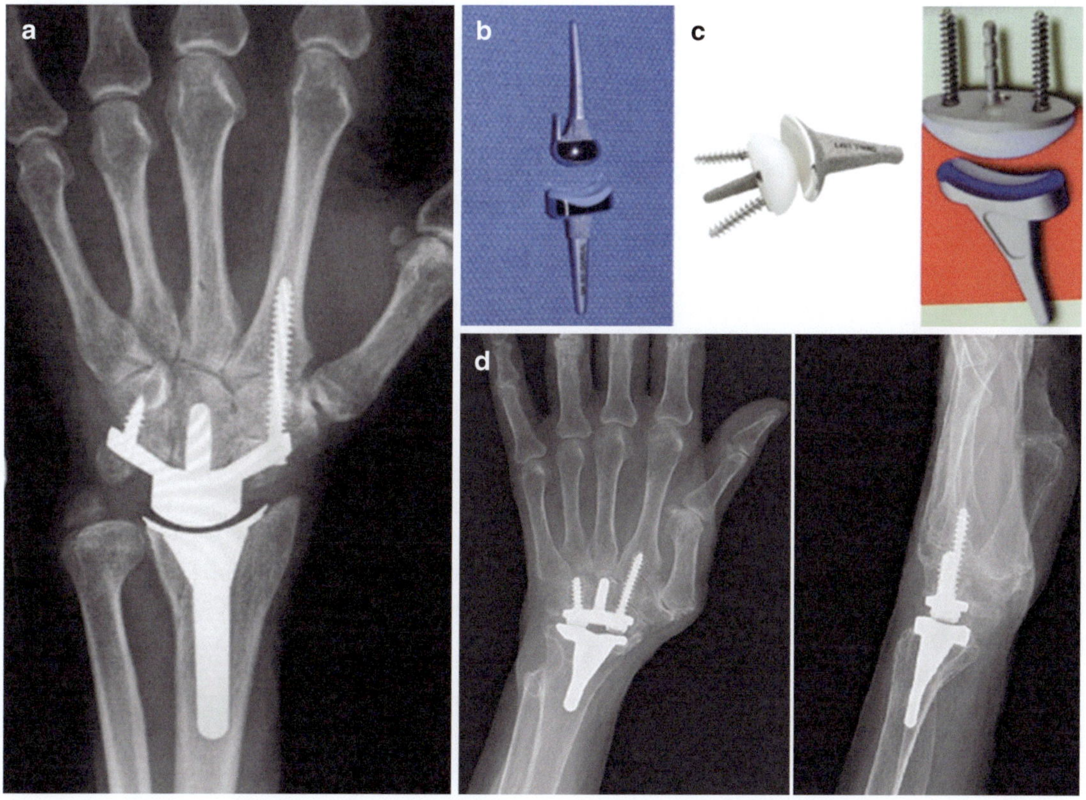

Fig. 20.5 (**a**) Maestro® Total Wrist Implant (Biomet) [31]. (**b**) ReMotion® Total Wrist Implant (Small Bone Innovations). (**c**) Universal 2® Total Wrist Implant (Integra LifeSciences). (**d**) Freedom® Total Wrist Implant (Integra LifeSciences)

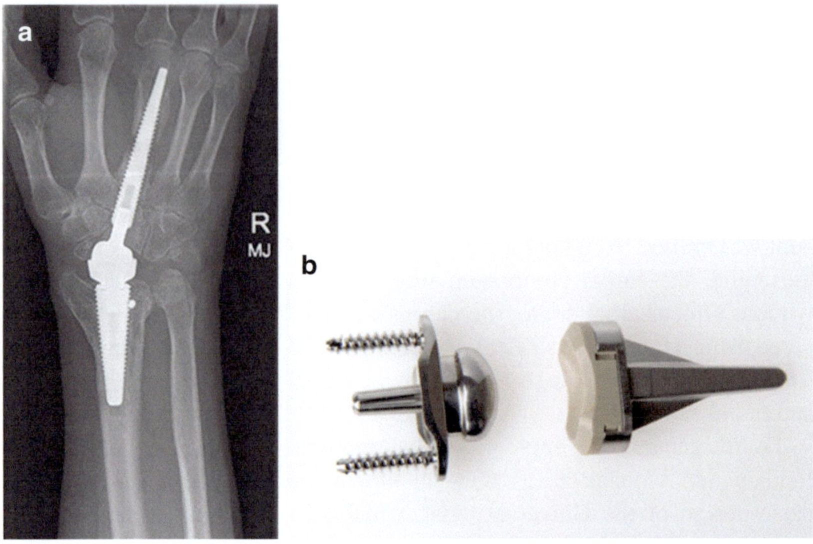

Fig. 20.6 The newer, fourth-generation implants: (**a**) WristMotion®, (**b**) Motec®, (**c**) KinematX®

demonstrated in the Maestro® wrist (Zimmer Biomet, Indiana, USA), which positions the polyethylene component on the concave radial surface. There was a reduced reliance on cement, along with the use of locking screw technology and increased modularity to improve the adjustment of soft tissue tension. New implants were later introduced, some of which are further refinements of the Universal® device, while one returns to the second-generation ball-and-socket concept (Fig. 20.6).

Like the Maestro®, both KinematX® (Extremity Medical, New Jersey, USA) and WristMotion® (Anika Therapeutics, Massachusetts, USA) implants feature an ellipsoidal articular design along with a concave polyethylene component on the radial side. Both are intended to replicate the dart thrower's motion by aligning the long diameter of the ellipsoidal surface with the axis of the intercarpal joint. The Motec® (Swemac Innovation, Linköping, Sweden) wrist implant features a spherical articular design. Initially designed as a metal-on-metal prosthesis, it was modified to include a radial cup with a carbon fiber-reinforced polyether ether ketone (PEEK) surface. Carpal and radial fixation is achieved using hydroxyapatite-coated titanium screws to promote osseointegration. The implant can also be converted to an arthrodesis. The ball joint and the metacarpal screw in the third metacarpal are arranged linearly, creating challenges in aligning the center of rotation with the center of the capitate. Subsequent malalignment may result in soft tissue imbalance. Nevertheless, the implant has the highest reported survival rate of 80% at ten years [2, 10].

Design Considerations for a Wrist Prosthesis

Currently, we cannot wholly replicate the functions of a human wrist using available biomaterials. Nonetheless, with careful engineering, we should aim to provide satisfactory functionality and durability. The key challenge in wrist arthroplasty remains durability, compared to other prosthetic joints, such as the hip, knee, and shoulder, which can perform successfully for 20 years. Currently, only a few wrist implants consistently reach ten years of survivorship [17]. Wrist arthroplasties mostly fail due to loosening and osteolysis [1, 20, 27]. Retrieved failed wrist implants exhibit significant polyethylene wear and shedding of metal debris (Fig. 20.7) [12, 13].

The materials used in these retrieved implants include UHMWPE, cobalt-chrome alloys, and titanium alloys. Interestingly, these have a proven history of success in other prosthetic joints. The

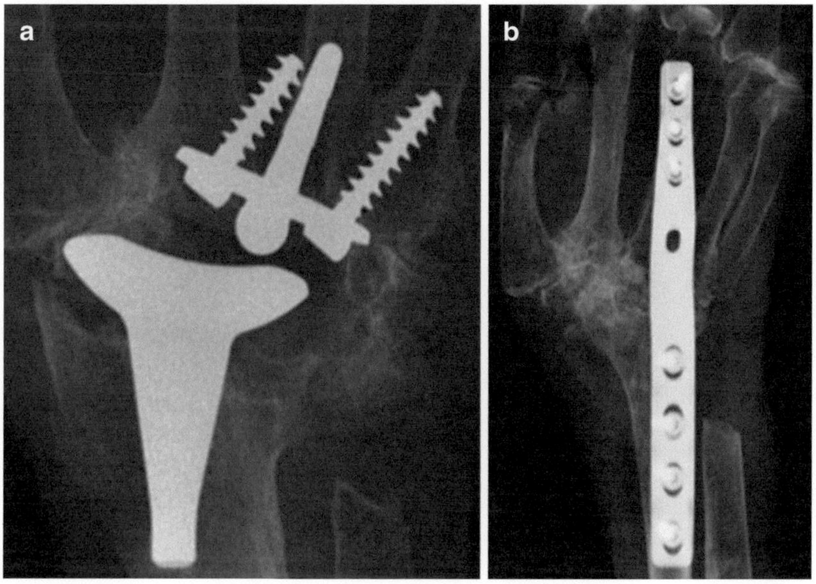

Fig. 20.7 (**a**) Posteroanterior (PA) radiograph of the wrist demonstrating a failed fourth-generation TWA 10 years after implantation on a 57-year-old rheumatoid patient. (**b** and **c**) Retrieved carpal component of this ellipsoidal wrist design. The polyethylene component shows severe wear in the area prone to high stress during torsion transmission (arrows). Wear-through and resulting metal-on-metal contact resulted in severe osteolysis and loosening of both components

issue appears to lie in design inadequacy, as the magnitude of loads experienced by wrist implants is not greater than those faced by other joints successfully treated with arthroplasty. First-generation wrist prostheses were constructed from silicone polymer, a material recognized for having insufficient mechanical properties. Second-generation implants were made from standard materials that had proven effective in other applications and were secured on the carpal side using long stems inserted into the second and third metacarpals. Although it is often said that the second and third carpometacarpal joints are immobile, a brief examination of the hand reveals that at least a few degrees of angulation are possible between these two metacarpals. While this may not have clinical significance for hand function, movement between these metacarpals could negatively affect the fixation of carpal components with stems rigidly attached to both bones. Additionally, a carpal implant with a fixation stem that crosses a CMC joint could experience motion and stress between its metacarpal and carpal fixation points. Unsurprisingly, loosening on the carpal side and stem cutout at the metacarpals have been challenging problems in total wrist arthroplasty [31]. Third-generation implants feature shorter metacarpal stems with a goal of increased fixation on the carpal bones. The carpal components of the Biaxial® and Universal® implants were widened to maximize contact with the carpus, and their articular surfaces were simultaneously expanded, taking on ellipsoidal and toroidal shapes. This design is more visually appealing as it mimics the shape of the native wrist. However, toroidal-shaped wrist joints remain congruent only during radial-ulnar deviation around the central axis of the toroid, which corresponds to the center of curvature of the capitate [8]. They become incongruous with only minimal axial rotation or angular deviation in flexion-extension, leading to high stresses on the polyethylene surface. As a result, this causes wear and failure. In contrast, ellipsoidal articular surfaces, such as those of the biaxial implant, offer better congruity during flexion-extension.

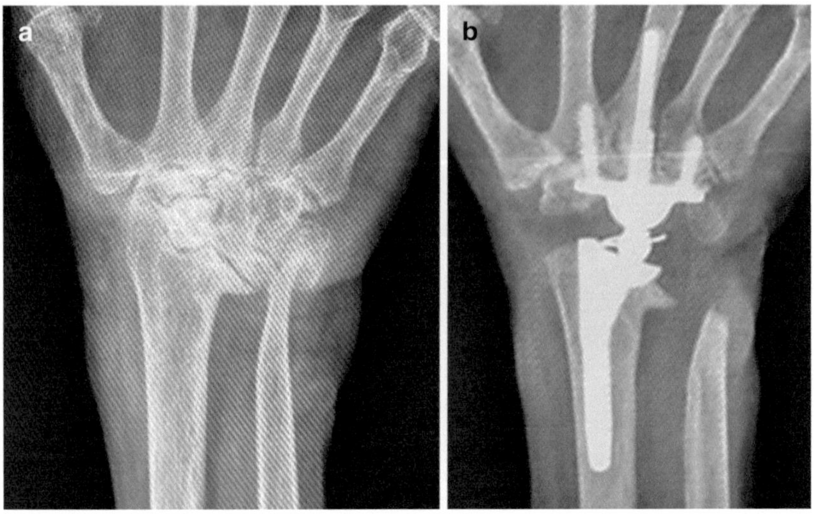

Fig. 20.8 (**a**) Preoperative PA radiograph of the wrist demonstrating rheumatoid arthritis with severe wrist dysfunction. (**b**) Postoperative PA radiograph of the wrist demonstrating a total wrist arthroplasty utilizing a dual-mobility spherical design

However, they exhibit incongruity during slight degrees of axial rotation, resulting in high contact stresses and early loosening.

As previously discussed, a fundamental function of the wrist is to transmit torsion. Torque transmission requires the wrist prosthesis to allow passive rotation along its longitudinal axis. Toroidal and ellipsoidal prosthetic surfaces do not allow rotation without becoming incongruent [8]. Only a wrist prosthesis with a spherical joint surface, such as the Mueli® and Motec®, or one with a secondary axial joint can maintain congruency during axial rotation [10]. Ultimately, minimizing wear on a prosthesis requires achieving a delicate balance. The contact surface must be large enough to reduce contact stress while still being small enough to limit the swept area and decrease abrasive wear. The ideal wrist arthroplasty should feature a spherical articulation to safely transmit torsional loads, be made of well-proven materials, and provide sufficient carpal fixation while minimizing the need for metacarpal fixation (Fig. 20.8).

References

1. Boeckstyns MEH, Herzberg G. Complications after total wrist arthroplasty. J Hand Surg Eur. 2024;49(2):177–87. https://doi.org/10.1177/17531934231203297. PMID: 38315136.
2. Clough OT, Lee G, Hayter E, Hardman J, Anakwe RE. Surgery with the Motec total wrist replacement: learning from earlier designs. J Surg Case Rep. 2021;2021(1):rjaa560. https://doi.org/10.1093/jscr/rjaa560. PMID: 33542810; PMCID: PMC7850115.
3. Cooney W, Manuel J, Froelich J, Rizzo M. Total wrist replacement: a retrospective comparative study. J Wrist Surg. 2012;1(2):165–72. https://doi.org/10.1055/s-0032-1326728. PMID: 24179722; PMCID: PMC3658675.
4. Crisco JJ, Coburn JC, Moore DC, Akelman E, Weiss AC, Wolfe SW. In vivo radiocarpal kinematics and the dart thrower's motion. J Bone Joint Surg Am. 2005;87(12):2729–40.
5. Erwin J, Varacallo MA. Anatomy, shoulder and upper limb, wrist joint. Treasure Island: StatPearls; 2025.
6. Figgie MP, Ranawat CS, Inglis AE, Sobel M, Figgie HE 3rd. Trispherical total wrist arthroplasty in rheumatoid arthritis. J Hand Surg Am. 1990;15(2):217–23. https://doi.org/10.1016/0363-5023(90)90098-c. PMID: 2324447.
7. Genda E, Horii E. Theoretical stress analysis in wrist joint–neutral position and functional position. J Hand Surg Br. 2000;25(3):292–5.
8. Grosland NM, Rogge RD, Adams BD. Influence of articular geometry on prosthetic wrist stability. Clin Orthop Relat Res. 2004;(421):134–42.
9. Hara T, Horii E, An KN, Cooney WP, Linscheid RL, Chao EY. Force distribution across wrist joint: application of pressure-sensitive conductive rubber. J Hand Surg Am. 1992;17(2):339–47.
10. Hooke AW, Pettersson K, Sagerfors M, An KN, Rizzo M. An anatomic and kinematic analysis of a new total wrist arthroplasty design. J Wrist Surg. 2015;4(2):121–7. https://doi.org/10.1055/s-0035-1549288. PMID: 25945297; PMCID: PMC4408131.
11. Horii E, Garcia-Elias M, Bishop AT, Cooney WP, Linscheid RL, Chao EY. Effect on force transmission

across the carpus in procedures used to treat Kienbock's disease. J Hand Surg Am. 1990;15(3):393–400.
12. Joyce TJ, Kandemir G, McGuire D, Solomons M, Brown DJ. Explant analysis of a chromium nitride coated metal-on-metal total wrist replacement: a case study. Proc Inst Mech Eng H. 2024;238(10):985–91.
13. Karjalainen T, Pamilo K, Reito A. Implant failure after Motec wrist joint prosthesis due to failure of ball and socket-type articulation-two patients with adverse reaction to metal debris and polyether ether ketone. J Hand Surg Am. 2018;43(11):1044 e1041–4.
14. Kistler U, Weiss AP, Simmen BR, Herren DB. Long-term results of silicone wrist arthroplasty in patients with rheumatoid arthritis. J Hand Surg Am. 2005;30(6):1282–7.
15. Kretschmer F, Fansa H. BIAX total wrist arthroplasty: management and results after 42 patients. Handchir Mikrochir Plast Chir. 2007;39(4):238–48.
16. Majima M, Horii E, Matsuki H, Hirata H, Genda E. Load transmission through the wrist in the extended position. J Hand Surg Am. 2008;33(2):182–8.
17. Nydick JA, Greenberg SM, Stone JD, Williams B, Polikandriotis JA, Hess AV. Clinical outcomes of total wrist arthroplasty. J Hand Surg Am. 2012;37(8):1580–4. https://doi.org/10.1016/j.jhsa.2012.05.016. Epub 2012 Jul 3. PMID: 22763052.
18. Radmer S, Andresen R, Sparmann M. Total wrist arthroplasty in patients with rheumatoid arthritis. J Hand Surg Am. 2003;28(5):789–94.
19. Reigstad O, Rokkum M. Wrist arthroplasty: where do we stand today? A review of historic and contemporary designs. Hand Surg. 2014;19(2):311–22.
20. Reiser D, Fischer P, Pettersson K, Wretenberg P, Sagerfors M. Total wrist arthroplasty with a new design: 20 cases with 8-year follow-up. J Hand Surg Am. 2025;50(3):377 e371–8.
21. Rossello MI, Zotta I, Rossello C, Formica M, Zoccolan A. Total wrist arthroplasty with Integra freedom® implants: a pilot study with a new evaluation system. Indian J Orthop. 2022;56(6):1040–7. https://doi.org/10.1007/s43465-022-00618-3. PMID: 35669017; PMCID: PMC9123129.
22. Schill S, Thabe H, Mohr W. Long-term outcome of Swanson prosthesis management of the rheumatic wrist joint. Handchir Mikrochir Plast Chir. 2001;33(3):198–206.
23. Schriever T, Olivecrona H, Wilcke M. There is motion between the scaphoid and the lunate during the dart-throwing motion. J Plast Surg Hand Surg. 2021;55(5):294–6.
24. Schuind F, Cooney WP, Linscheid RL, An KN, Chao EY. Force and pressure transmission through the normal wrist. A theoretical two-dimensional study in the posteroanterior plane. J Biomech. 1995;28(5):587–601.
25. Short WH, Werner FW, Fortino MD, Palmer AK. Distribution of pressures and forces on the wrist after simulated intercarpal fusion and Kienbock's disease. J Hand Surg Am. 1992;17(3):443–9.
26. Swanson AB, Swanson G. Flexible implant arthroplasty of the radiocarpal joint. Semin Arthroplast. 1991;2(2):78–84.
27. Townsend CB, Paladino J, Ilyas AM. The evolution & outcomes of total wrist arthroplasty: current concepts. SurgiColl. 2024;2(1)
28. Werner FW, Short WH, Fortino MD, Palmer AK. The relative contribution of selected carpal bones to global wrist motion during simulated planar and out-of-plane wrist motion. J Hand Surg Am. 1997;22(4):708–13.
29. Wolfe SW, Crisco JJ, Orr CM, Marzke MW. The dart-throwing motion of the wrist: is it unique to humans? J Hand Surg Am. 2006;31(9):1429–37.
30. Wolfe SW, Pederson WC, Kozin SH, Cohen MS. Green's operative hand surgery E-book: 2-volume set. Elsevier Health Sciences; 2021.
31. Zijlker HJA, Ritt M, Beumer A. Fourth-generation Total wrist arthroplasty: a systematic review of clinical outcomes. J Wrist Surg. 2022;11(5):456–64.

21 Degenerative Arthritis: Diagnosis and Treatment

Nezar Abunnur, Mark Gonzalez, and Farid Amirouche

Introduction

Definition and Overview of Degenerative Arthritis (Osteoarthritis)

Degenerative arthritis, also known as osteoarthritis (OA), is a joint disorder characterized by the progressive destruction of articular cartilage, subchondral bone remodeling, and the formation of osteophytes [63]. Unlike systemic inflammatory arthropathies such as rheumatoid arthritis, OA follows a localized disease course driven by mechanical stress, metabolic factors, and local inflammatory mediators. In the intricate anatomy of the hand and wrist, specific structures are notably vulnerable to degenerative changes, particularly the carpometacarpal (CMC) joint of the thumb, as well as the proximal interphalangeal (PIP) and distal interphalangeal (DIP) joints. Additionally, the scaphoid bone in the wrist is a notable area of concern due to its significant biomechanical demands. According to Berger [12], these joints are not only critical for movement but also prime examples of regions where osteoarthritis (OA) can have a substantial impact.

The manifestations of OA in the hand and wrist can be particularly distressing, as they fundamentally compromise fine motor control and hinder everyday activities. As noted by Kjeken et al. [48], joint deterioration over time will cause patients to experience pain, stiffness, deformity, and functional impairment. The combination of repetitive loading, the precision necessary for fine motor tasks, and the demands of grip strength all contribute cumulatively to the deterioration of these vital joints over time.

Patients afflicted by this condition typically suffer from a range of debilitating symptoms, including persistent pain, stiffness that restricts movement, visible deformities of the fingers and joints, and significant functional impairment. These challenges profoundly affect their quality of life, as simple actions such as gripping a pen, buttoning a shirt, or even shaking hands become increasingly arduous.

Although often referred to as "wear and tear" arthritis in popular discourse, degenerative arthritis encompasses complex biochemical and bio-

N. Abunnur · M. Gonzalez
Department of Orthopaedic Surgery, University of Illinois at Chicago, Chicago, IL, USA

F. Amirouche (✉)
Department of Orthopaedic Surgery, Orthopaedic and Spine Institute, Endeavor HealthSystem (Formerly Northshore University HealthSystem), an Affiliate of the University of Chicago Pritzker School of Medicine, Skokie, IL, USA

Department of Orthopaedic Surgery, University of Illinois at Chicago, Chicago, IL, USA
e-mail: famirouche@northshore.org; amirouch@uic.edu

F. Amirouche et al. (eds.), *Biomechanics of the Hand*,
https://doi.org/10.1007/978-3-032-04688-8_21

mechanical processes. Articular cartilage essentially comprises a matrix of collagen and proteoglycans that maintain smooth, low-friction joint surfaces. This matrix undergoes enzymatic degradation in OA, leading to fibrillation, cartilage thinning, and loss of protective cushioning [63]. Subchondral bone responds adaptively by remodeling, often resulting in sclerosis and the development of osteophytes, further compromising joint congruency and mobility. The interplay between genetic predisposition, age-related degeneration, and mechanical stress dictates the rate of disease progression, with altered load distribution and microtrauma playing critical roles in cartilage breakdown.

Epidemiology and Global Burden

Osteoarthritis is the most prevalent form of arthritis worldwide, afflicting about 528 million people globally, an increase of 113% since 1990, including over 142.5 million people affected with hand osteoarthritis (HOA) [56]. HOA and wrist osteoarthritis (WOA) are particularly common in the aging population, with an age-standardized prevalence of 44.2% in women and 37.7% in men. While the crude incidence over 9 years is similar (34.6% in women, 33.7% in men), disease progression is common in both groups (96.4% vs. 91.4%), emphasizing its degenerative nature [28].

In the United States, HOA is estimated to be present in one in seven individuals, and WOA has been reported in approximately 13.6% of the population [3]. While less common than HOA, primary WOA develops due to age-related cartilage degeneration and altered load distribution across the radiocarpal and midcarpal joints. Given the wrist's integral role in supporting hand function and weight-bearing activities, WOA contributes to substantial disability, often coexisting with HOA to further impair dexterity and grip function. Given the integral role of the hands and wrists in daily activities, both HOA and WOA impose significant limitations on quality of life. Patients often struggle with tasks that require fine motor control, grip strength, and load-bearing, such as turning doorknobs, fastening buttons, or writing, which can lead to a progressive loss of independence [48].

The impact of hand osteoarthritis (HOA) and wrist osteoarthritis (WOA) extends far beyond mere discomfort, imposing a significant economic burden on individuals and the healthcare system alike. This burden manifests in lost productivity, mounting healthcare costs, and reliance on supportive devices or caregivers to assist in daily activities [38, 53]. Data reveals that patients suffering from HOA face startlingly higher annual medical expenses—averaging $14,521—compared to their counterparts without osteoarthritis, whose costs average only $3629 [96].

Focusing specifically on thumb carpometacarpal (CMC) arthritis, surgical intervention plays a pivotal role in alleviating both pain and functional limitations. Preoperatively, the average costs stand at $20,451, but following surgery, these costs rise to $24,374. Despite this increase, the surgical procedure yields a remarkable enhancement in quality-adjusted life years (QALYs). The incremental cost effectiveness ratio of $25,370 per QALY illustrates the long-term value of this intervention, underscoring its importance for those affected [26].

As global life expectancy continues to rise, the prevalence of osteoarthritis is projected to escalate dramatically, by an estimated 48.2% by the year 2050. This alarming trend highlights the urgent need for improved diagnostic techniques and effective treatment strategies to manage this rapidly growing public health challenge more effectively [86].

Risk Factors (Age, Genetics, Occupation, Trauma, etc.)

Risk factors for OA stem from biological, mechanical, and environmental interactions, with susceptibility shaped by genetics, lifestyle, and medical history. While aging, genetic predisposition, trauma, occupational stress, and hormonal influences are all recognized contributors, OA is not an inevitable consequence of these factors. Instead, they often act synergistically, accelerating disease onset and progression when combined [15, 30].

1. Aging: Increasing age strongly correlates with the onset and severity of OA, reflecting the cumulative burden of mechanical stress and age-related changes in chondrocyte senescence [22].
2. Genetic susceptibility: Certain genetic polymorphisms, particularly those affecting collagen synthesis and the regulation of the cartilage matrix, increase susceptibility to OA [57, 73].
3. Mechanical overload and occupational stress: Manual laborers and individuals in occupations that require repetitive gripping, twisting, or exposure to vibration are at a higher risk of developing hand and wrist osteoarthritis (OA) due to chronic overuse and microtrauma [19, 87].
4. Trauma and joint instability: Injuries, such as distal radius fractures or carpal ligament tears, can destabilize joint structures, cause post-traumatic malalignment, and markedly increase the likelihood of subsequent degenerative changes [5, 64].
5. Gender and hormonal factors: Women are disproportionately affected by hand OA, particularly postmenopausal individuals. This disparity suggests a role for hormonal influences, particularly estrogen deficiency, in disease development [39] (Fig. 21.1).

Importance of Early Diagnosis and Effective Treatment

Timely identification of OA in the hand and wrist is paramount, as early intervention can attenuate disease progression and preserve joint function [32]. Subtle biomechanical alterations, such as increased shear forces or joint malalignment, accelerate cartilage degradation and amplify local inflammation [55]. In its early stages, OA may present with mild pain and transient stiffness, which are often underrecognized or attributed to routine "aches and pains." These early biomechanical derangements escalate without intervention, resulting in persistent pain, pronounced stiffness, and irreversible joint damage.

Advances in biomarker research offer the potential for early identification of individuals at risk of rapid disease progression, allowing for earlier intervention [7]. Management strategies encompass a spectrum of options, from conservative management, such as splinting, targeted exercises, and pharmacological pain relief, to advanced surgical interventions like arthroplasty or arthrodesis [41]. Emerging regenerative therapies, such as mesenchymal stem cell treatments, aim to alter the disease trajectory rather than merely manage symptoms [18]. Across all modalities, treatment success hinges on restoring optimal load distribution and joint mechanics, thereby reinforcing the critical role of biomechanics in disease progression and mitigation.

Relevance to Hand Biomechanics

The intricate structure of the hand and wrist, joints, ligaments, and tendons coordinates precise movements and transmits force [42]. Even minor alterations in carpal alignment or metacarpophalangeal (MCP) joint mechanics can significantly impact force distribution, exacerbating cartilage stress and perpetuating a cycle of degeneration [20].

Pathological conditions such as scapholunate advanced collapse (SLAC), scaphoid nonunion advanced collapse (SNAC), and Kienböck's disease illustrate how disruptions in carpal kinematics promote uneven contact pressures, cartilage breakdown, and osteoarthritic changes [83]. Similarly, repetitive microtrauma from occupational or sporting activities concentrates forces on vulnerable joint regions, accelerating degenerative changes. Understanding these biomechanical underpinnings is crucial for developing targeted interventions, including bracing, rehabilitative exercises, and precisely engineered surgical procedures.

Degenerative arthritis of the hand and wrist is a multifactorial condition in which biomechanical factors play a central role in its pathogenesis and progression. An in-depth understanding of

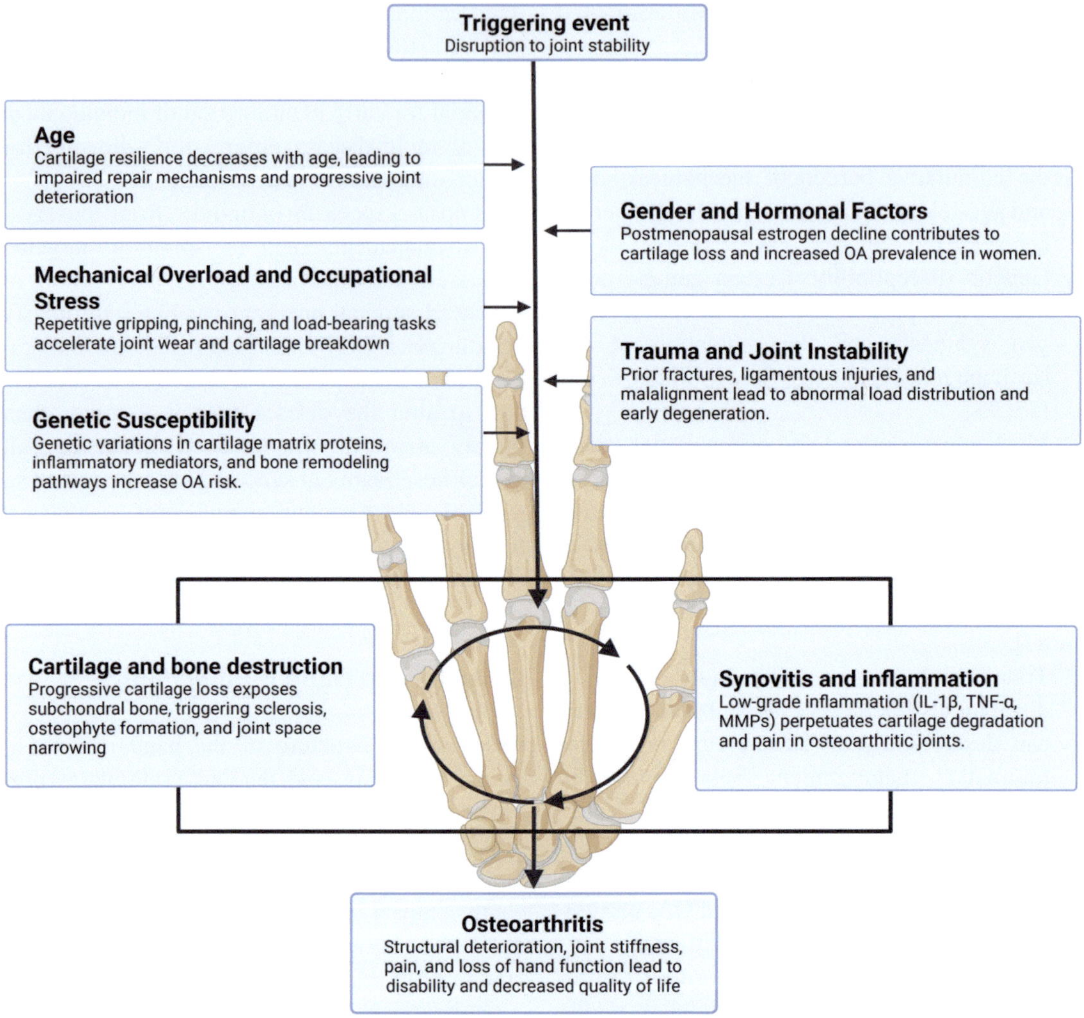

Fig. 21.1 Illustrates the key factors driving hand osteoarthritis, beginning with a triggering event that initiates cartilage and bone destruction, as well as synovitis and inflammation. Risk factors like age, genetic susceptibility, mechanical overload, and hormonal influences contribute to disease progression. The combined effects of biomechanical stress and inflammatory mediators lead to joint deterioration, ultimately resulting in pain, stiffness, and functional impairment characteristic of OA

joint mechanics is essential for refining diagnostic criteria and optimizing treatment strategies to mitigate functional impairment. This chapter examines the pathophysiology, clinical assessment, and management of hand and wrist osteoarthritis, highlighting established and emerging interventions that utilize biomechanical principles to preserve joint function and enhance patient outcomes.

Pathophysiology of Degenerative Arthritis

Cartilage Degeneration and Joint Changes

Articular cartilage is a vital load-bearing and friction-reducing tissue in synovial joints, composed primarily of water, type II collagen, and proteoglycans [20]. In degenerative arthritis, the degradation process initiates with the breakdown of the cartilaginous matrix precipitated by stress, such as excessive mechanical stress or injury. Altered chondrocyte metabolism contributes to an imbalance between anabolic and catabolic activities, favoring matrix metalloproteinases (MMPs) and aggrecanases that degrade collagen and proteoglycans [25, 62]. Cartilage fibrillation occurs as the structural support declines, leading to vertical clefts that extend down to the subchondral bone.

Concomitant changes in the subchondral bone, including sclerosis and reactive bone remodeling, exacerbate cartilage deterioration. Osteophyte formation occurs in periarticular regions, initially as an adaptive response to redistribute mechanical loads, but ultimately results in abnormal joint contours [21]. These changes are especially detrimental in the small joints of the hand, where even minor perturbations in surface congruity can significantly disrupt load transmission, amplify friction, and potentiate further degeneration. High-resolution peripheral quantitative computed tomography studies provide direct evidence of these pathological changes, demonstrating that trabecular bone disorganization, increased bone size, and decreased volumetric bone mineral density are correlated with the severity of osteophyte formation and joint space narrowing. These findings reinforce the role of subchondral bone alterations in disease progression, illustrating that bone remodeling and cartilage loss are closely interconnected. Rather than being confined to cartilage degradation alone, structural deterioration in hand osteoarthritis (OA) extends to bone microarchitecture, further disrupting joint mechanics and perpetuating disease progression [58] (Fig. 21.2).

Inflammatory and Mechanical Factors

Although often regarded as a "wear and tear" disease, osteoarthritis (OA) is increasingly recognized as a condition in which inflammatory

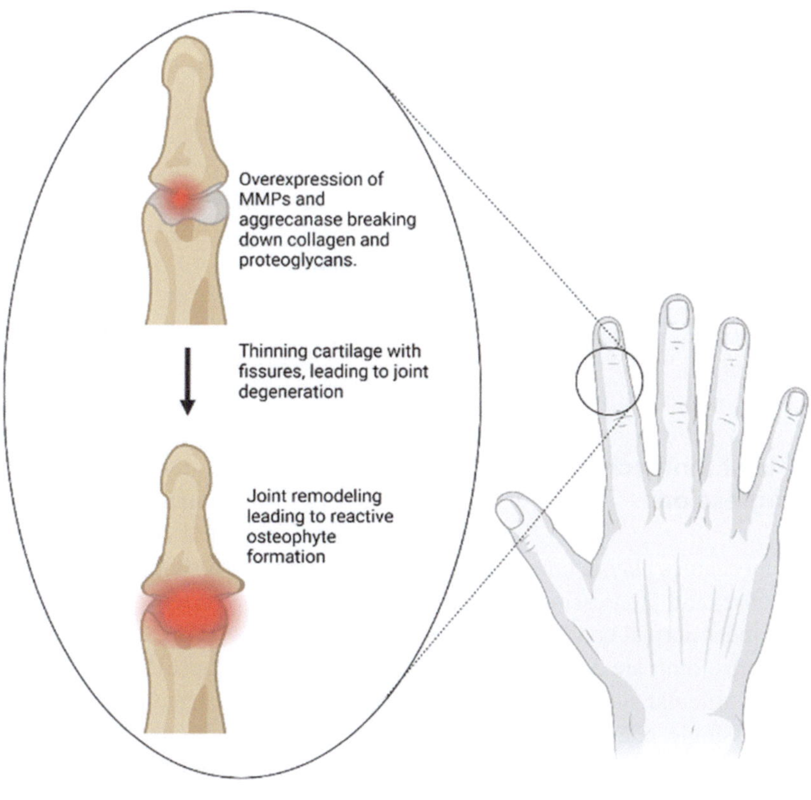

Fig. 21.2 Progressive cartilage degeneration in hand osteoarthritis. Overexpression of matrix metalloproteinases (MMPs) and aggrecanases leads to the breakdown of collagen and proteoglycans, resulting in cartilage thinning and fissuring. As degeneration progresses, joint remodeling triggers the formation of reactive osteophytes, contributing to pain and functional impairment

cascades play a crucial role in disease progression [68]. The OA joint environment becomes enriched in proinflammatory mediators, including interleukin-1β (IL-1β), tumor necrosis factor-α (TNF-α), and prostaglandin E2 (PGE2), all of which enhance proteolytic enzyme activity and contribute to cartilage matrix degradation. MMPs and a disintegrin-like metalloproteinase with thrombospondin motif (ADAMTS) play a central role in extracellular matrix breakdown, disrupting the structural integrity of cartilage and periarticular tissues [68]. Chronic, low-grade inflammation perpetuates a cycle of tissue destruction and suboptimal healing, driving cartilage loss, synovial hyperplasia, subchondral bone sclerosis, and osteophyte formation.

While inflammation is a key driver, mechanical overload or malalignment often serves as the primary trigger for this inflammatory response. Repetitive shear or compressive forces, such as those experienced during sustained gripping, lifting, or weight-bearing activities, induce microfractures in subchondral bone, prompting the release of cytokines and growth factors that fuel further degeneration [29]. This synergy between mechanical stress and local inflammation accelerates the degenerative trajectory, where structural deterioration fuels further inflammation, ultimately culminating in clinically apparent osteoarthritis (OA). Given the interplay between mechanical and inflammatory mechanisms, targeted interventions aimed at reducing joint stress, modulating inflammatory pathways, and inhibiting MMP activity may hold promise in mitigating disease progression and preserving joint function.

Role of Biomechanics in Disease Progression

Biomechanics plays a crucial role in determining the onset and progression of degenerative arthritis. Under normal conditions, forces within hand and wrist joints are distributed evenly across congruent articular surfaces, minimizing peak stress at any single point. Degenerative changes concentrate forces on these load-bearing pathways, concentrating forces on smaller, more vulnerable regions of the cartilage [20]. On the other hand, certain joints are naturally subjected to high stress due to frequent usage and intricate movement patterns. For instance, the thumb CMC joint endures substantial axial loads during pinching and grasping. Minor changes in CMC alignment or stability can thus precipitate substantial biomechanical deviations, magnifying local forces and prompting cartilage damage [20]. Once cartilage deterioration initiates, the subsequent maladaptive remodeling of osteophyte formation and subchondral sclerosis further skews joint forces, fostering a vicious cycle of degeneration. Assessment of 36 trapezial specimens revealed three distinct wear patterns on the metacarpal articular surface: saddle (47% of cases), preserving overall anatomy with cartilage loss; dish (33%), exhibiting concavity and loss of the usual standard saddle shape; and cirque (19%), showing preferential volar concavity that disrupts the typical volar-dorsal arc. Notably, the Eaton stage, sex, handedness, and age differed among these morphologies, with advanced (IV) disease predominantly associated with the dish and cirque configurations [90]. These findings highlight the characteristic topographical changes in advanced thumb carpometacarpal arthritis, providing insight into progression and potential surgical considerations.

Several pathological conditions exemplify how biomechanical disruptions contribute to progressive OA in the hand and wrist:

1. SLAC (scapholunate advanced collapse) [83]:

   SLAC occurs when the scapholunate ligament is disrupted or attenuated, leading to abnormal motion between the scaphoid and lunate bones in the wrist. The ensuing instability concentrates loading on the capitulohumeral joints, provoking chondral wear and subchondral collapse. Osteophyte formation and sclerosis follow, often progressing to a predictable pattern of degenerative changes in the wrist.
2. SNAC (scaphoid nonunion advanced collapse) [83]:

   Similar to SLAC, SNAC arises from a chronic nonunion of a scaphoid fracture, typically in the waist region of the scaphoid. Persistent instability leads to articular surface incongruity, which promotes abnormal motion and increased contact pressures. The resultant malalignment accelerates cartilage erosion and advances osteoarthritis in a pattern closely paralleling SLAC.

3. Kienböck's disease [14, 59]:

   Kienböck's disease is a form of avascular necrosis affecting the lunate. The etiology is multifactorial, and factors such as lunate anatomy, variations in blood supply, and radiocarpal alignment have been identified as predisposing factors to this condition. As the lunate collapses, carpal biomechanics are substantially disrupted, altering load transmission across the radiocarpal joint. Inevitably, these mechanical imbalances invite degenerative changes and osteoarthritic sequelae, often requiring surgical salvage procedures if untreated.

4. Posttraumatic OA following distal radius fractures [5, 64]:

   Distal radius fractures are a common cause of posttraumatic wrist osteoarthritis (PTOA), often resulting from malalignment, articular stepoffs, and altered load transmission across the radiocarpal joint. Minor incongruities can increase contact pressures, accelerating cartilage wear and OA development. Beyond mechanical disruption, joint trauma induces oxidative stress, leading to chondrocyte senescence, DNA damage, and impaired cartilage repair, which may explain why PTOA can develop even in younger individuals. Intraarticular fractures and residual dorsal tilt deformities further increase PTOA risk, particularly in untreated or improperly reduced fractures.

Clinical Presentation and Diagnosis

Symptoms and Signs

Pain is the defining symptom of hand osteoarthritis (HOA) and wrist osteoarthritis (WOA), typically manifesting as a dull, persistent ache associated with physical activity. This discomfort intensifies during gripping, pinching, or weight-bearing movements, making everyday tasks increasingly challenging. Research by Kjeken et al. [48] highlights that while pain during joint use shows excellent reliability in assessments, its ability to detect osteoarthritis (OA) is somewhat limited. This is particularly evident as some individuals with radiographic evidence of OA may not report significant pain.

Throughout the day, pain intensity tends to fluctuate, often peaking in the evening as cumulative joint stress builds up or following extended use of the affected limb. Patients may also endure unpredictable pain flares, where discomfort suddenly escalates, leading to further restrictions in joint mobility and affecting their ability to engage in daily activities. In contrast to inflammatory arthropathies, which often present with persistent morning stiffness, individuals with OA experience only brief bouts that typically improve with movement. This characteristic of OA highlights the importance of activity in managing symptoms, as noted by Huskisson et al. [35].

Physical Examination

The hand and wrist are highly intricate structures, comprising 27 bones, including 8 carpal bones, 5 metacarpals, and 14 phalanges for each finger. Each finger contains three phalanges, except for the thumb, which has only two. A thorough physical examination assesses joint mobility, structural deformities, and pain, guiding diagnosis and treatment.

1. Joint Deformities and Swelling
2. Hand and wrist OA primarily affects the DIP, PIP, thumb CMC, and radiocarpal joints. DIP involvement is the most prevalent, followed by the thumb base, PIP, and MCP joints [17, 27, 71, 75]. Hand osteoarthritis (OA) often affects both hands symmetrically, particularly in radiographic joint space narrowing. It also tends to cluster within the same row or ray, making it more likely for multiple joints in the same finger or neighboring fingers to develop OA [71, 76]. Osteophyte formation and capsular thickening at the PIP level develop Bouchard's nodes, and formation at the DIP level develops Heberden's nodes. These bony

enlargements can impair grip strength, dexterity, and precision pinch, contributing to significant functional limitations [9]. The thumb CMC joint, frequently affected in OA, develops adduction deformity and subluxation, resulting in a characteristic squared appearance [10].

3. Range of Motion and Joint Stability
   (a) In the early stages of OA, range of motion (ROM) deficits in the hand and wrist may be subtle or absent. However, as the disease progresses, flexion, extension, and rotational movements become increasingly restricted ([CSL STYLE ERROR: reference with no printed form]). The hand and wrist should be assessed through active and passive range of motion (ROM) testing, with careful observation for tenderness, stiffness, and limitations in movement, particularly at the end range of motion. Joint instability or crepitus may be palpable during motion, indicating cartilage degradation and surface irregularities.
4. Specific Tests for Hand Involvement
5. A series of provocative maneuvers can isolate pathology within specific joints and help differentiate OA from other conditions ([CSL STYLE ERROR: reference with no printed form]):
   (a) CMC grind and shear tests: These tests are used for thumb CMC OA. The grind test applies axial compression and circumduction to reproduce pain and crepitus. In contrast, the shear test involves circumduction and translation of the CMC joint to assess joint instability. Tenderness over the CMC joint of the thumb further supports an OA diagnosis.
   (b) Watson test: In wrist OA with scapholunate involvement, dorsal wrist pain is elicited by applying pressure to the scaphoid. Moving the wrist from ulnar to radial deviation suggests ligamentous instability or early degenerative changes.
   (c) Scissor technique for MCP joints: Palpate each MCP joint using a scissor grip, flexing the joints to 90°, and applying thumb pressure to the joint line to assess for effusion, swelling, or tenderness.
   (d) Four-finger technique for PIP/DIP joints: Stabilize the PIP or DIP joint from the sides using the thumb and index finger of one hand while applying anteroposterior pressure with the other hand to detect effusion, osteophytes, or joint line tenderness.
   (e) Capitate pressure test: This test assesses wrist joint effusion by applying firm dorsal pressure at the capitate level while palpating the radiocarpal joint line for fluid displacement or tenderness.

Imaging Techniques

1. Radiography

   Plain radiographs remain the primary imaging modality for diagnosing hand and wrist osteoarthritis (OA) due to their widespread availability, cost effectiveness, and standardized interpretation [60]. Standard posteroanterior (PA), lateral, and oblique views enable assessment of joint space narrowing, osteophyte formation, subchondral sclerosis, and subchondral cysts. Weight-bearing or functional views can cause subtle carpal malalignment, which may not be apparent in static images [40]. Several radiographic grading systems exist, with the Kellgren-Lawrence scale being the most commonly used to quantify disease severity. This system categorizes OA based on the presence and progression of osteophytes, joint space narrowing, and subchondral changes. However, other classification methods, such as the Verbruggen-Veys scoring system for erosive hand OA, provide additional insight into disease progression and joint remodeling patterns [92]. Despite its advantages, radiography has limitations in assessing early-stage OA, as it primarily detects structural changes rather than soft tissue or inflammatory involvement.

2. MRI and CT Scans

   Magnetic resonance imaging (MRI) provides superior visualization of soft tissues and can detect early osteoarthritic changes that may precede radiographic abnormalities [8, 50]. Contrast-enhanced MRI can further delineate synovial inflammation and vascular supply, aiding in the early detection of conditions such as Kienböck's disease. Although MRI has demonstrated high sensitivity in detecting OA features, its clinical value is limited due to additional costs and adverse contraindications for patients, such as the use of contrast [80]. Computed tomography (CT) provides a detailed visualization of bony structures, making it valuable for assessing complex carpal alignments, subtle fractures, and advanced osteoarthritic remodeling. High-resolution peripheral quantitative computed tomography (HR-pQCT) offers high sensitivity for detecting erosions, osteophytes, and periarticular bone changes, surpassing plain radiography in resolution [69]. Although a CT scan can be performed more quickly, it involves exposure to ionizing radiation [13]. Emerging evidence suggests that combining HR-pQCT and MRI may enhance the evaluation of joint inflammation and bone damage, offering a more comprehensive assessment of joint pathology [52].
3. Ultrasound

   Although ultrasound has not traditionally been a primary tool in diagnosing osteoarthritis (OA), advancements in high-resolution transducers have enabled the identification of synovitis, osteophytes, and subtle cartilage irregularities in superficial joints [44, 45]. Ultrasound also facilitates dynamic assessment, permitting real-time observation of tendon gliding and joint subluxation during functional maneuvers. However, operator dependency represents one of the most significant limitations of ultrasound assessment [36]. While research has expanded into its use in hand OA, ultrasound is more commonly employed in inflammatory osteoarthritis, where it is valuable and beneficial for detecting gray-scale synovitis and assessing inflammatory activity in the finger joints. Studies have demonstrated a high prevalence of synovitis on ultrasound in inflammatory osteoarthritis (OA), reinforcing its value in distinguishing between the structural and inflammatory components of the disease [94].

Laboratory Tests

Laboratory evaluation is generally used to exclude systemic or inflammatory arthritis rather than to confirm a diagnosis of OA, which remains predominantly clinical and mainly imaging-based [3, 27]. Rheumatoid factor (RF), anticyclic citrullinated peptide (anti-CCP) antibodies, erythrocyte sedimentation rate (ESR), and C-reactive protein (CRP) are typically within normal ranges in OA. However, they may be elevated in inflammatory subtypes of OA [79, 91]. These markers can help differentiate OA from rheumatoid arthritis, psoriatic arthritis, or crystal arthropathies.

Ultimately, a comprehensive diagnostic approach to hand and wrist OA integrates clinical history, physical examination, and imaging findings, with selective laboratory studies reserved for cases where differential diagnosis remains unclear. Early and accurate diagnosis allows for tailored treatment strategies to prevent functional decline and irreversible joint damage.

Treatment Strategies

Nonpharmacological Management

1. Lifestyle Modifications

   While hand OA does not correlate as strongly with body mass index (BMI) as knee OA does, optimizing weight remains beneficial for overall joint health [55]. Reducing excess weight can mitigate systemic inflammatory mediators and indirectly lessen the load on upper extremities during weight-bearing tasks, such as using a cane or walker. Activity adjustments aim to reduce repetitive, high-impact maneuvers. Patients with frequent pinching, twisting, or gripping motions may benefit from scheduled rest breaks and modified work routines [87]. Engaging in low-impact exercises, such as swimming or light resistance training, helps maintain musculoskeletal fitness without exacerbating joint stress.

2. Physical and Occupational Therapy

   Hand therapists implement range-of-motion exercises to maintain flexibility and prevent contractures alongside isometric and isotonic strengthening exercises that target intrinsic and extrinsic hand muscles [46]. Neuromuscular reeducation and fine motor training help improve coordination and reduce joint overload in daily tasks. Orthotics, joint protection programs, and exercise regimens are tailored based on biomechanical principles, even in the absence of clear evidence-based guidelines. Expanding the effectiveness of hand therapy requires early intervention, improved access, multidisciplinary collaboration, and strategies to enhance adherence to therapy programs.

3. Assistive Devices and Orthotics

   Splints or braces provide external stabilization, particularly in joints subject to substantial mechanical stress. In the thumb, a rigid or semirigid thumb spica splint, such as the CMC OA, can reduce pain by limiting excessive motion and offloading compressive forces at the trapeziometacarpal joint. Studies show that splinting is well tolerated and effective, with 76% of early-stage patients reporting symptom improvement and 54% of advanced-stage patients reporting. While it does not fully eliminate symptoms, splinting offers a viable conservative treatment option to reduce inflammation and enhance function over time ([CSL STYLE ERROR: reference with no printed form]). Ring splints may correct hyperextension deformities in the PIP joints, while custom orthotics provide carpal support and stress relief at the wrist [77]. Custom splints for distal interphalangeal (DIP) osteoarthritis (OA) have demonstrated a 66% reduction in pain over 6 months, although functional improvements remain limited [37]. Proper fit and wear schedules are essential for striking a balance between joint protection and maintaining mobility for optimal rehabilitation.

Pharmacological Treatment

1. Analgesics and NSAIDs

   Given that joint pain or discomfort upon activity is reported by patients as the primary symptom of HOA and WOA, first-line pharmacological management typically involves pain control through the use of acetaminophen or nonsteroidal antiinflammatory drugs (NSAIDs) [51, 85]. However, chronic use of these medications warrants caution due to gastrointestinal, renal, and cardiovascular risks [81]. Topical NSAIDs, such as diclofenac gel and topical capsaicin, may offer localized symptom relief with fewer systemic side effects. For severe or refractory pain, opioids have historically been used in OA of large and small joints [34]. Recent guidelines have stated that opioids such as tramadol are only conditionally recommended due to modest benefits [51]. Opioids carry significant risks, including constipation, nausea, dizziness, somnolence, and addiction [23].

2. Corticosteroid Injections

   Intraarticular corticosteroid injections can relieve and suppress localized inflammation in osteoarthritis (OA) flares. However, recent evidence shows intraarticular injections are generally not recommended for thumb base arthritis [33, 66]. On the other hand, a study has demonstrated the efficacy of interphalangeal therapy. The small size and anatomical complexity of hand joints make accurate needle placement challenging, and the procedure is conditionally recommended over other forms of intraarticular injection. Repeated injections should be approached judiciously, as excessive corticosteroid exposure may weaken cartilage and other periarticular tissues [65]

3. Disease-Modifying Agents

Although DMARDs (disease-modifying antirheumatic drugs) exist for rheumatoid arthritis, definitive OA-specific DMARDs remain under investigation. Early trials on strontium ranelate, diacerein, and WNT inhibitors show modest potential in slowing progression but lack large-scale validation. Several existing systemic agents, including hormones (PTH, calcitonin), bisphosphonates, monoclonal antibodies, statins, supplements (glucosamine, chondroitin, hyaluronic acid), antibiotics, and nonsteroidal antiinflammatory drugs (NSAIDs), have been investigated for their cartilage-protective effects, with mixed clinical outcomes [6]. PTH analogs may enhance chondrocyte activity, while bisphosphonates may reduce subchondral bone turnover. Additionally, statins, monoclonal antibodies, and doxycycline show potential antiinflammatory or cartilage-preserving effects. However, none of these are standard OA therapies, and further trials are needed to confirm their efficacy and safety.

Surgical Interventions

Surgical treatment of hand and wrist OA is tailored to the affected joints, severity of degeneration, and patient-specific functional demands. Options range from arthroscopic debridement in the early stages of disease to arthrodesis and arthroplasty in advanced cases, with various reconstructive or salvage procedures available depending on the degree of joint involvement [2]:

1. Arthroscopy

   Wrist arthroscopy serves both diagnostic and therapeutic roles, allowing for direct visualization of cartilage surfaces, assessment of ligament integrity, and debridement of loose bodies or osteophytes. It is most effective in early to moderate OA, providing symptomatic relief by removing degenerative tissue and smoothing articular surfaces. However, its efficacy in advanced OA remains controversial, as structural joint instability and widespread cartilage loss often necessitate more definitive surgical options [1].

2. Joint Fusion (Arthrodesis)

   Arthrodesis remains the gold standard for pain relief and stability in end-stage OA, particularly in the DIP joints, thumb MCP joints, and selected wrist fusions. By eliminating joint movement, arthrodesis provides durable pain relief, though at the cost of mobility loss. The surgical technique involves the removal of articular cartilage, osteophytes, and sclerotic bone, with fixation via K-wires, tension-band constructs, compression screws, or plates. A cup-and-cone preparation is often preferred to enhance fusion stability and bone contact. In the wrist, partial or total arthrodesis is considered in cases where motion preservation is less critical than pain relief and stability. A complete wrist fusion is generally reserved for patients with severe pain from carpal arthritis, often yielding high satisfaction rates, though functional tradeoffs must be discussed.

3. Joint Replacement (Arthroplasty)

   For patients seeking to preserve joint mobility, arthroplasty is an alternative to fusion in select joints, particularly metacarpophalangeal (MCP) and thumb carpometacarpal (CMC) joints. Implant options include silicone, pyrocarbon, or metal-plastic joint replacements, each with varying durability and complication profiles [93]. While advances in implant materials and design have improved functional outcomes, concerns remain regarding implant loosening, wear, and long-term revision rates.

   Thumb CMC arthroplasty: Trapeziectomy with ligament reconstruction and tendon interposition (LRTI) remains widely used for thumb CMC osteoarthritis, providing pain relief and joint stability [99]. A refined trapeziectomy with suture suspension arthroplasty has shown improved grip strength, key pinch, and pain reduction, with over 50% of patients reporting no postoperative pain [74]. This technique preserves ligamentous support, minimizing proximal migration and enabling an early return to motion. Surgical modifications further help reduce complications and enhance outcomes, particularly in advanced stage III or IV of OA.

PIP and MCP arthroplasty: Although silicone implants are commonly used, they exhibit high rates of fracture over time, often resulting in asymptomatic conditions. Pyrocarbon and metal-plastic implants offer theoretical durability advantages; however, studies show comparable revision rates [67, 89].

4. Salvage and Motion-Preserving Procedures

   Several motion-preserving procedures are available for select wrist OA cases, including the following:

   Proximal row carpectomy (PRC): PRC involves the removal of the scaphoid, lunate, and triquetrum, allowing the capitate to articulate with the lunate facet of the radius [95]. This procedure can relieve pain while preserving partial wrist motion, making it suitable for patients with mild to moderate midcarpal arthritis. PRC is generally favored in patients over 35 years old or those in whom limited wrist fusion (LWF) is not viable due to poor bone healing potential. While postoperative radiographic changes are common by 10 years, many remain asymptomatic, and failure rates tend to plateau after 3 years. Surgical modifications, such as capitate reshaping or interpositional capsular grafting, have been proposed to improve outcomes in patients with mild cartilage changes on the capitate or lunate facet.

   Limited wrist fusion (LWF) involves removing the scaphoid and fusing selected carpal bones, with common constructs including four-corner and two-corner fusions [97]. The four-corner fusion involves the arthrodesis of the capitate, lunate, hamate, and triquetrum, thereby eliminating pain associated with scapholunate instability while retaining some wrist motion. The two-column fusion fuses the capitate to the lunate and the triquetrum to the hamate, providing stability while optimizing motion.

   LWF is generally favored in younger, high-demand patients or active individuals aged 35–55 who require greater wrist durability and stability [88]. However, it carries higher costs (up to 425% more than PRC), increased complication rates, and longer recovery times [43]. Potential complications include nonunion, dorsal impingement, hardware issues, and adjacent joint arthritis, though the latter is often asymptomatic ([CSL STYLE ERROR: reference with no printed form]).

   Denervation procedures: Selective denervation of the posterior interosseous nerve (PIN) and anterior interosseous nerve (AIN) can provide pain relief without altering wrist mechanics, making it an option for patients who are not surgical candidates for more invasive procedures [98] (Fig. 21.3).

Emerging Therapies and Research

1. Stem Cell Therapy

   Mesenchymal stem cells (MSCs) have garnered considerable interest in regenerative medicine, mainly due to their promising applications in chondral repair and their capacity to modulate inflammatory processes. Emerging preliminary studies indicate that intraarticular injections of either autologous or allogeneic MSCs could potentially provide significant pain relief and encourage modest cartilage regeneration in patients suffering from osteoarthritis (OA). Research conducted by Mahla [61], Geng et al. [24], and Kim et al. [47] highlights these encouraging findings. However, the quest for consistent and reproducible outcomes remains challenging, underscoring the urgent need for more comprehensive clinical trials to thoroughly investigate the efficacy and safety of these innovative treatments.

2. Biologics and Regenerative Medicine

   Platelet-rich plasma (PRP), along with various growth factors and other biological agents, shows significant potential in

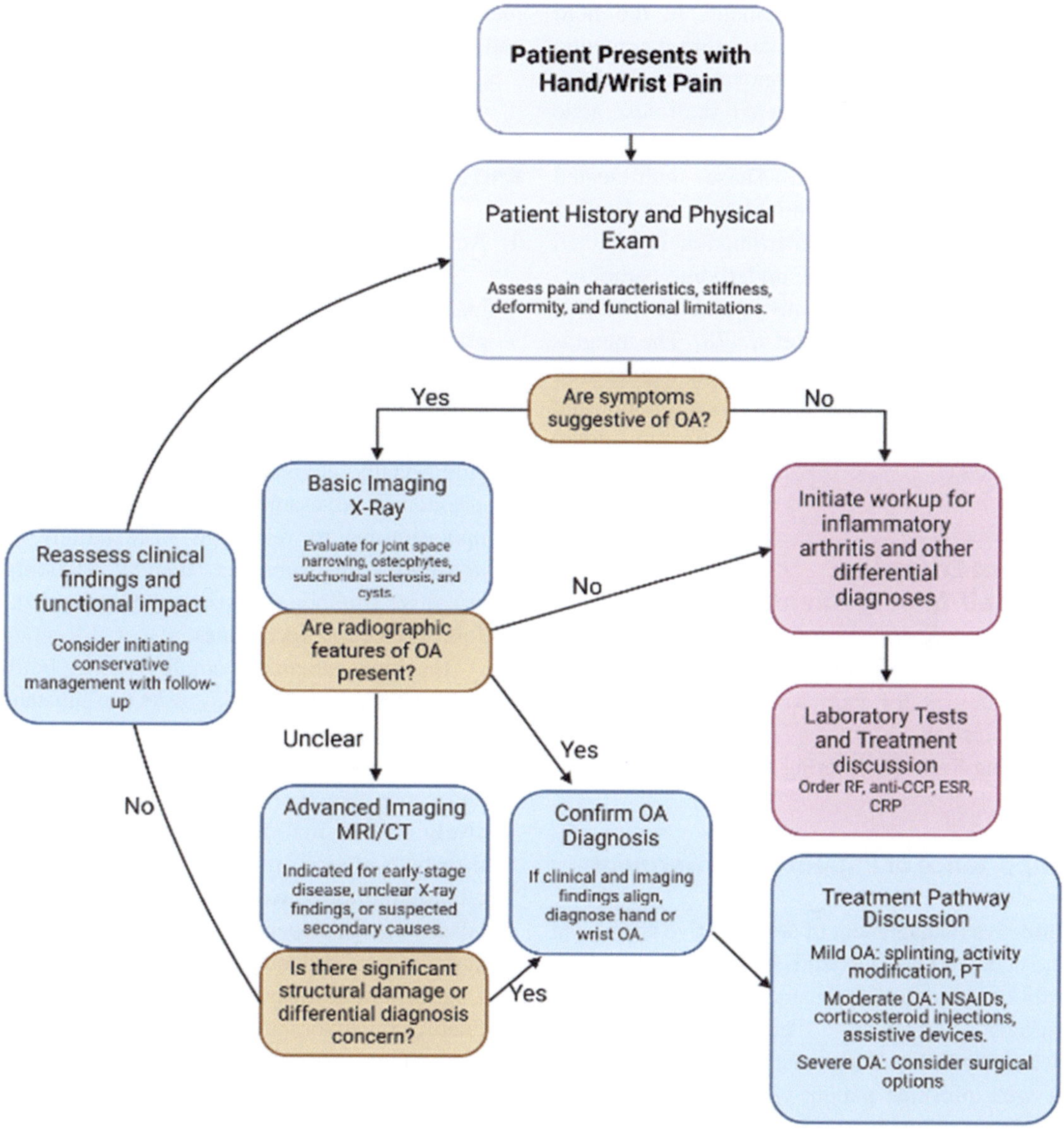

Fig. 21.3 This flowchart outlines the stepwise approach to diagnosing hand and wrist OA. The process begins with patient history and physical examination, assessing pain characteristics, stiffness, and functional limitations. If OA is suspected, essential imaging is performed to evaluate for hallmark radiographic features, including joint space narrowing and osteophytes. If the findings are unclear, advanced imaging is considered. Laboratory tests are reserved for cases where inflammatory arthritis or other differential diagnoses are suspected. Once OA is confirmed, treatment discussions focus on symptom severity, guiding the selection of conservative or surgical management options. The algorithm ensures a structured and evidence-based approach to diagnosis and clinical decision-making

modulating the inflammatory environment present in arthritic joints, which can lead to improved clinical outcomes. Recent studies have suggested that these interventions may offer transient symptomatic relief for patients with knee osteoarthritis (OA), prompting further investigation into their efficacy for hand and wrist OA as well. Notably, research by Bennell et al. [11] underscores the regenerative properties inherent in these treatments.

Furthermore, as advancements in the field continue, it is anticipated that future therapeutic strategies will incorporate innovative methodologies such as tissue-engineered scaffolds, gene therapy, and advanced biomaterials to enhance targeted cartilage repair. These multifaceted approaches have the potential to enhance healing processes and restore joint function in affected areas by addressing the underlying pathology associated with degenerative joint diseases, as highlighted by Raeissadat et al. [78]. The integration of regenerative medicine and engineering holds promise for delivering more durable and practical solutions for osteoarthritis and related conditions.

Patient Education and Self-Management

A. Importance of Patient Empowerment
B. Strategies for Pain Management and Joint Protection
C. Long-Term Monitoring and Follow-Up

Importance of Patient Empowerment

Effective management of degenerative arthritis in the hand and wrist extends beyond clinical interventions. It requires a nuanced approach encompassing patient education, shared decision-making, and proactive self-management. A well-informed patient who understands their condition and its biomechanical underpinnings is likely to adhere to treatment regimens, implement lifestyle modifications, and engage in long-term joint protection strategies [72]. The NICE guidelines emphasize the importance of patient participation in their care, collaborative decision-making, and strategies that account for multiple morbidities [70]. Integrating these principles into clinical practice strengthens patient engagement, ensuring that treatment aligns with individual needs and functional goals. Furthermore, patients should be directed to reliable sources of information on principal management strategies that include flare management, exercise modalities, and general education about OA [72].

Strategies for Pain Management and Joint Protection

1. Activity Pacing and Rest-Activity Cycles

   Patients should be encouraged to alternate high-demand tasks, like prolonged typing or gripping, with rest periods that minimize cumulative stress on arthritic joints. Strategic scheduling should help control inflammation and reduce mechanical overload [87]. Additionally, advising the patient to perform ergonomic assessments can help identify opportunities to rearrange workstations or household environments, thereby reducing repetitive strain on the wrist and fingers [19].
2. Joint Protection Techniques

   Teaching patients to employ proper body mechanics during everyday tasks can substantially lessen peak stress. For instance, using the larger joints of the upper extremity to lift and carry objects, rather than relying exclusively on the small joints of the hand, helps distribute loads more evenly. Likewise, employing tools with enlarged handles or adaptive grips disperses force across a broader surface area, reducing localized pressure on fragile joint surfaces. These include built-up handles on silverware, rocker knives, zip and button adaptors, and adaptive jar openers. Studies have shown that these assistive devices have improved occupational performance and are tolerated by OA patients [4, 49].
3. Heat, Cold, and Other Modalities

   Applying local heat through a warm compress can alleviate muscle tension and improve blood flow around arthritic joints, thereby enhancing flexibility before therapeutic exercise. Conversely, cold therapy, commonly through cold packs, helps decrease acute inflammation and pain flare-ups. Paraffin wax baths for the hands are another commonly used modality that combines warmth with gentle stretching, facilitating joint range of motion [31, 82].

4. Self-Monitoring in Osteoarthritis Management

   Self-monitoring has shown promise in inflammatory arthritis conditions, such as rheumatoid arthritis, where tracking symptoms and activity levels can improve disease management and treatment adherence [54, 84]. Regular self-assessment of pain levels, joint stiffness, and functional capacity allows patients to recognize patterns in symptom fluctuations and adjust their daily activities accordingly. Keeping a pain and activity diary helps facilitate data-driven discussions during clinical follow-ups, enabling clinicians to tailor interventions based on individualized disease progression and functional limitations. By actively engaging in self-monitoring, patients can develop greater awareness of modifiable triggers and make informed decisions that support joint protection and symptom control.

Long-Term Monitoring and Follow-Up

Periodic consultations enable clinicians to monitor radiographic changes, functional capacity, and the progression of pain patterns [63]. These visits provide an opportunity to adjust treatment plans, addressing new patient concerns while minimizing disease progression and associated discomfort. The patient learning process is inherently iterative, highlighting the crucial need for continuous reinforcement of biomechanical principles.

This entails a thorough evaluation of splint use and a diligent focus on ensuring proper exercise techniques that align with individual needs. Additionally, addressing ergonomic adjustments plays a pivotal role in optimizing daily activities. A comprehensive assessment of orthotic fitment is essential, as making the necessary modifications to adapt to joint alignment changes or disease progression can significantly enhance comfort and functionality.

Moreover, persistent pain and functional decline can lead to mood disturbances, social withdrawal, and a notable decrease in overall quality of life [16]. To counter these challenges, integrating mental health support—whether through personalized counseling, engaging support groups, or structured self-management programs—can empower patients to cope more effectively with their circumstances.

Furthermore, educating family members and caregivers is paramount to creating a nurturing home environment. This education reinforces joint-protective strategies and adaptive techniques that can lead to meaningful improvements in daily function, ultimately fostering a holistic approach to patient care and well-being.

Conclusion

Hand and wrist OA is a complex, multifactorial disorder driven by biomechanical stress, inflammatory processes, and progressive structural deterioration. OA imposes significant functional and economic burdens with increasing global prevalence, particularly among aging populations. Early diagnosis and targeted interventions are crucial in mitigating pain, preserving mobility, and delaying disease progression. A comprehensive approach to managing osteoarthritis (OA) encompasses conservative therapies, pharmacological interventions, and surgical options, each tailored to the severity of degeneration and the specific functional needs of the patient. Nonsurgical strategies, including splinting, activity modification, and structured rehabilitation programs, are crucial for joint protection and symptom alleviation. Pharmacological treatments, such as nonsteroidal antiinflammatory drugs (NSAIDs) and intraarticular injections, provide additional symptomatic relief while emerging biological and regenerative therapies offer potential disease-modifying opportunities. For advanced cases, surgical interventions such as arthrodesis, arthroplasty, and motion-preserving procedures remain viable surgical options.

As healthcare professionals, we must stress the importance of early detection in our approach to osteoarthritis management. Patient education and self-management remain cornerstones of long-term OA care, emphasizing activity pacing, ergonomic modifications, and joint protection techniques to optimize function and quality of life. As biomechanical advancements and regenerative medicine continue to evolve, the future of OA treatment is shifting toward more personalized, multimodal interventions to restore joint integrity and enhance patient outcomes. By integrating early detection, individualized treatment plans, and emerging innovations, we can more effectively address the challenges of hand and wrist osteoarthritis (OA), ultimately improving functional independence and patient well-being. Continued advancements in regenerative medicine and personalized treatment strategies will shape the next era of OA management.

References

1. Adams JE. Does arthroscopic Débridement with or without interposition material address carpometacarpal arthritis? Clin Orthop Relat Res. 2014;472:1166. https://doi.org/10.1007/s11999-013-2905-y.
2. Adams JE. Surgical management of osteoarthritis of the hand and wrist. J Hand Ther. 2022;35:418–27. https://doi.org/10.1016/j.jht.2022.01.001.
3. Akhondi H, Panginikkod S. Wrist arthritis. In: StatPearls. Treasure Island (FL): StatPearls Publishing; 2025.
4. Amaral DS, Duarte ALBP, Barros SS, et al. Assistive devices: an effective strategy in non-pharmacological treatment for hand osteoarthritis-randomized clinical trial. Rheumatol Int. 2018;38:343–51. https://doi.org/10.1007/s00296-017-3892-1.
5. Anderson DD, Chubinskaya S, Guilak F, et al. Post-traumatic osteoarthritis: improved understanding and opportunities for early intervention. J Orthop Res Off Publ Orthop Res Soc. 2011;29:802–9. https://doi.org/10.1002/jor.21359.
6. Apostu D, Lucaciu O, Mester A, et al. Systemic drugs with impact on osteoarthritis. Drug Metab Rev. 2019;51(4):498–523.
7. Attur M, Krasnokutsky-Samuels S, Samuels J, Abramson SB. Prognostic biomarkers in osteoarthritis. Curr Opin Rheumatol. 2013;25:136–44. https://doi.org/10.1097/BOR.0b013e32835a9381.
8. Backhaus M, Kamradt T, Sandrock D, et al. Arthritis of the finger joints: a comprehensive approach comparing conventional radiography, scintigraphy, ultrasound, and contrast-enhanced magnetic resonance imaging. Arthritis Rheum. 1999;42:1232–45. https://doi.org/10.1002/1529-0131(199906)42:6<1232::AID-ANR21>3.0.CO;2-3.
9. Bagis S, Sahin G, Yapici Y, et al. The effect of hand osteoarthritis on grip, pinch strength, and hand function in postmenopausal women. Clin Rheumatol. 2003;22:420–4. https://doi.org/10.1007/s10067-003-0792-4.
10. Bakri K, Moran SL. Thumb carpometacarpal arthritis. Plast Reconstr Surg. 2015;135:508. https://doi.org/10.1097/PRS.0000000000000916.
11. Bennell KL, Paterson KL, Metcalf BR, et al. Effect of intra-articular platelet-rich plasma vs placebo injection on pain and medial tibial cartilage volume in patients with knee osteoarthritis: the RESTORE randomized clinical trial. JAMA. 2021;326:2021–30. https://doi.org/10.1001/jama.2021.19415.
12. Berger RA. The anatomy and basic biomechanics of the wrist joint. J Hand Ther Off J Am Soc Hand Ther. 1996;9:84–93. https://doi.org/10.1016/s0894-1130(96)80066-4.
13. Bousson V, Lowitz T, Laouisset L, et al. CT imaging for investigating subchondral bone in knee osteoarthritis. Osteoporos Int. 2012;23:S861–5. https://doi.org/10.1007/s00198-012-2169-5.
14. Camus EJ, Van Overstraeten L. Kienböck's disease in 2021. Orthop Traumatol Surg Res OTSR. 2022;108:103161. https://doi.org/10.1016/j.otsr.2021.103161.
15. Chaganti RK, Lane NE. Risk factors for incident hip and knee osteoarthritis. Curr Rev Musculoskelet Med. 2011;4:99–104. https://doi.org/10.1007/s12178-011-9088-5.
16. da Silva MD, Guginski G, Sato KL, et al. Persistent pain induces mood problems and memory loss by involving cytokines, growth factors, and supraspinal glial cells. Brain Behav Immun Health. 2020;7:100118. https://doi.org/10.1016/j.bbih.2020.100118.
17. Dahaghin S, Bierma-Zeinstra SMA, Ginai AZ, et al. Prevalence and pattern of radiographic hand osteoarthritis and association with pain and disability (the Rotterdam study). Ann Rheum Dis. 2005;64:682–7. https://doi.org/10.1136/ard.2004.023564.
18. Diekman BO, Guilak F. Stem cell-based therapies for osteoarthritis: challenges and opportunities. Curr Opin Rheumatol. 2013;25:119–26. https://doi.org/10.1097/BOR.0b013e32835aa28d.
19. Dillon C, Petersen M, Tanaka S. Self-reported hand and wrist arthritis and occupation: data from the U.S. National Health Interview Survey-Occupational Health Supplement. Am J Ind Med. 2002;42:318–27. https://doi.org/10.1002/ajim.10117.
20. Eschweiler J, Li J, Quack V, et al. Anatomy, biomechanics, and loads of the wrist joint. Life (Basel). 2022;12:188. https://doi.org/10.3390/life12020188.

21. Felson DT, Neogi T. Osteoarthritis: is it a cartilage disease or bone disease? Arthritis Rheum. 2004;50:341–4. https://doi.org/10.1002/art.20051.
22. Felson DT, Zhang Y, Hannan MT, et al. Risk factors for incident radiographic knee osteoarthritis in the elderly: the Framingham Study. Arthritis Rheum. 1997;40:728–33. https://doi.org/10.1002/art.1780400420.
23. Fuggle N, Curtis E, Shaw S, et al. Safety of opioids in osteoarthritis: outcomes of a systematic review and meta-analysis. Drugs Aging. 2019;36:129–43. https://doi.org/10.1007/s40266-019-00666-9.
24. Geng Y, Chen J, Alahdal M, et al. Intra-articular injection of hUC-MSCs expressing miR-140-5p induces cartilage self-repairing in the rat osteoarthritis. J Bone Miner Metab. 2020;38:277–88. https://doi.org/10.1007/s00774-019-01055-3.
25. Goodman C, Fuller K. Pathology: implications for the physical therapist. Third ed. St. Louis: Elsevier; 2009.
26. Grobet C, Audigé L, Eichler K, et al. Cost-utility analysis of thumb carpometacarpal resection arthroplasty: a health economic study using real-world data. J Hand Surg. 2022;47:445–53. https://doi.org/10.1016/j.jhsa.2022.01.013.
27. Hart DJ, Spector TD, Brown P, et al. Clinical signs of early osteoarthritis: reproducibility and relation to x-ray changes in 541 women in the general population. Ann Rheum Dis. 1991;50:467–70. https://doi.org/10.1136/ard.50.7.467.
28. Haugen IK, Englund M, Aliabadi P, et al. Prevalence, incidence and progression of hand osteoarthritis in the general population: the Framingham Osteoarthritis Study. Ann Rheum Dis. 2011;70:1581–6. https://doi.org/10.1136/ard.2011.150078.
29. He Y, Li Z, Alexander PG, et al. Pathogenesis of osteoarthritis: risk factors, regulatory pathways in chondrocytes, and experimental models. Biology. 2020;9:194. https://doi.org/10.3390/biology9080194.
30. Heidari B. Knee osteoarthritis prevalence, risk factors, pathogenesis, and features: part I. Caspian J Intern Med. 2011;2:205–12.
31. Helfand AE, Bruno J. Therapeutic modalities and procedures. Part I: cold and heat. Clin Podiatry. 1984;1:301–13.
32. Hermann W, Lambova S, Muller-Ladner U. Current treatment options for osteoarthritis. Curr Rheumatol Rev. 2018;14:108–16. https://doi.org/10.2174/1573397113666170829155149.
33. Heyworth BE, Lee JH, Kim PD, et al. Hylan versus corticosteroid versus placebo for treatment of basal joint arthritis: a prospective, randomized, double-blinded clinical trial. J Hand Surg. 2008;33:40–8. https://doi.org/10.1016/j.jhsa.2007.10.009.
34. Hochberg MC, Altman RD, April KT, et al. American College of Rheumatology 2012 recommendations for the use of nonpharmacologic and pharmacologic therapies in osteoarthritis of the hand, hip, and knee. Arthritis Care Res. 2012;64:465–74. https://doi.org/10.1002/acr.21596.
35. Huskisson EC, Dieppe PA, Tucker AK, Cannell LB. Another look at osteoarthritis. Ann Rheum Dis. 1979;38:423–8. https://doi.org/10.1136/ard.38.5.423.
36. Iagnocco A, Conaghan PG, Aegerter P, et al. The reliability of musculoskeletal ultrasound in the detection of cartilage abnormalities at the metacarpophalangeal joints. Osteoarthr Cartil. 2012;20:1142–6. https://doi.org/10.1016/j.joca.2012.07.003.
37. Ikeda M, Ishii T, Kobayashi Y, et al. Custom-made splint treatment for osteoarthritis of the distal interphalangeal joints. J Hand Surg. 2010;35:589–93. https://doi.org/10.1016/j.jhsa.2010.01.012.
38. Jin X, Liang W, Zhang L, et al. Economic and humanistic burden of osteoarthritis: an updated systematic review of large sample studies. PharmacoEconomics. 2023;41:1453–67. https://doi.org/10.1007/s40273-023-01296-1.
39. Kalichman L, Hernández-Molina G. Hand osteoarthritis: an epidemiological perspective. Semin Arthritis Rheum. 2010;39:465–76. https://doi.org/10.1016/j.semarthrit.2009.03.001.
40. Kani KK, Mulcahy H, Chew FS. Understanding carpal instability: a radiographic perspective. Skeletal Radiol. 2016;45:1031–43. https://doi.org/10.1007/s00256-016-2390-y.
41. Karl Spies C, Langer M, Hahn P, et al. The treatment of primary arthritis of the finger and thumb joint. Dtsch Ärztebl Int. 2018;115:269–75. https://doi.org/10.3238/arztebl.2018.0269.
42. Kaufmann RA, Pfaeffle HJ, Blankenhorn BD, et al. Kinematics of the midcarpal and radiocarpal joint in flexion and extension: an in vitro study. J Hand Surg. 2006;31:1142–8. https://doi.org/10.1016/j.jhsa.2006.05.002.
43. Kazmers NH, Stephens AR, Presson AP, et al. Comparison of direct surgical costs for proximal row corpectomy and four-corner arthrodesis. J Wrist Surg. 2019;8:66–71.
44. Keen HI, Wakefield RJ, Grainger AJ, et al. An ultrasonographic study of osteoarthritis of the hand: synovitis and its relationship to structural pathology and symptoms. Arthritis Rheum. 2008;59:1756–63. https://doi.org/10.1002/art.24312.
45. Keen HI, Wakefield RJ, Hensor EMA, et al. Response of symptoms and synovitis to intra-muscular methylprednisolone in osteoarthritis of the hand: an ultrasonographic study. Rheumatology. 2010;49:1093–100. https://doi.org/10.1093/rheumatology/keq010.
46. Kerkhof F, Kenney D, Ogle M, et al. The biomechanics of osteoarthritis in hand: implications and prospects for hand therapy. J Hand Ther. 2022;35:367–76. https://doi.org/10.1016/j.jht.2022.11.007.
47. Kim YG, Choi J, Kim K. Mesenchymal stem cell-derived exosomes for effective cartilage tissue repair and treatment of osteoarthritis. Biotechnol J. 2020;15:e2000082. https://doi.org/10.1002/biot.202000082.

48. Kjeken I, Dagfinrud H, Slatkowsky-Christensen B, et al. Activity limitations and participation restrictions in women with hand osteoarthritis: patients' descriptions and associations between dimensions of functioning. Ann Rheum Dis. 2005;64:1633–8. https://doi.org/10.1136/ard.2004.034900.
49. Kjeken I, Darre S, Smedslund G, et al. Effect of assistive technology in hand osteoarthritis: a randomised controlled trial. Ann Rheum Dis. 2011;70:1447–52. https://doi.org/10.1136/ard.2010.148668.
50. Klarlund M, Ostergaard M, Jensen KE, et al. Magnetic resonance imaging, radiography, and scintigraphy of the finger joints: one year follow up of patients with early arthritis. The TIRA Group. Ann Rheum Dis. 2000;59:521–8. https://doi.org/10.1136/ard.59.7.521.
51. Kolasinski SL, Neogi T, Hochberg MC, et al. 2019 American College of Rheumatology/Arthritis Foundation guideline for the management of osteoarthritis of the hand, hip, and knee. Arthritis Care Res. 2020;72:149–62. https://doi.org/10.1002/acr.24131.
52. Lee CH, Srikhum W, Burghardt AJ, et al. Correlation of structural abnormalities of the wrist and metacarpophalangeal joints evaluated by high-resolution peripheral quantitative computed tomography, 3 Tesla magnetic resonance imaging and conventional radiographs in rheumatoid arthritis. Int J Rheum Dis. 2015;18:628–39. https://doi.org/10.1111/1756-185X.12495.
53. Leifer VP, Katz JN, Losina E. The burden of OA-health services and economics. Osteoarthr Cartil. 2022;30:10–6. https://doi.org/10.1016/j.joca.2021.05.007.
54. Li LC, Xie H, Feehan LM, et al. Effect of digital monitoring and counselling on self-management ability in patients with rheumatoid arthritis: a randomised controlled trial. Rheumatology (Oxford). 2025;64:310–20. https://doi.org/10.1093/rheumatology/kead709.
55. Loeser RF. Age-related changes in the musculoskeletal system and the development of osteoarthritis. Clin Geriatr Med. 2010;26:371–86. https://doi.org/10.1016/j.cger.2010.03.002.
56. Long H, Liu Q, Yin H, et al. Prevalence trends of site-specific osteoarthritis from 1990 to 2019: findings from the global burden of disease study 2019. Arthritis Rheumatol. 2022;74:1172–83. https://doi.org/10.1002/art.42089.
57. Loughlin J. Genetic epidemiology of primary osteoarthritis. Curr Opin Rheumatol. 2001;13:111–6. https://doi.org/10.1097/00002281-200103000-00004.
58. Ma C, Aitken D, Wu F, et al. Association between radiographic hand osteoarthritis and bone microarchitecture in a population-based sample. Arthritis Res Ther. 2022;24:223. https://doi.org/10.1186/s13075-022-02907-6.
59. MacLean SBM, Bain GI. Lunate fracture morphology in Kienbock's Disease—A computed tomography study abstract. J Wrist Surg. 2023;12(02):128–34. https://doi.org/10.1055/s-0042-1756497.
60. Maheu E, Altman RD, Bloch DA, et al. Design and conduct of clinical trials in patients with osteoarthritis of the hand: recommendations from a task force of the Osteoarthritis Research Society International. Osteoarthr Cartil. 2006;14:303–22. https://doi.org/10.1016/j.joca.2006.02.010.
61. Mahla RS. Stem cells applications in regenerative medicine and disease therapeutics. Int J Cell Biol. 2016;2016:6940283. https://doi.org/10.1155/2016/6940283.
62. Mansell JP, Tarlton JF, Bailey AJ. Biochemical evidence for altered subchondral bone collagen metabolism in osteoarthritis of the hip. Rheumatology. 1997;36:16–9. https://doi.org/10.1093/rheumatology/36.1.16.
63. Martel-Pelletier J, Barr AJ, Cicuttini FM, et al. Osteoarthritis. Nat Rev Dis Primers. 2016;2:1–18. https://doi.org/10.1038/nrdp.2016.72.
64. Martin JA, Brown T, Heiner A, Buckwalter JA. Post-traumatic osteoarthritis: the role of accelerated chondrocyte senescence. Biorheology. 2004;41:479–91.
65. McAlindon TE, LaValley MP, Harvey WF, et al. Effect of intra-articular triamcinolone vs. saline on knee cartilage volume and pain in patients with knee osteoarthritis: a randomized clinical trial. JAMA. 2017;317:1967–75. https://doi.org/10.1001/jama.2017.5283.
66. Meenagh GK, Patton J, Kynes C, Wright GD. A randomized controlled trial of intra-articular corticosteroid injection of the carpometacarpal joint of the thumb in osteoarthritis. Ann Rheum Dis. 2004;63:1260–3. https://doi.org/10.1136/ard.2003.015438.
67. Mora AN, Earp BE, Blazar PE. Midterm clinical and radiographic follow-up of pyrolytic carbon PIP arthroplasty. J Hand Surg. 2020;45:253.e1–6. https://doi.org/10.1016/j.jhsa.2019.06.008.
68. Mukherjee A, Das B. The role of inflammatory mediators and matrix metalloproteinases (MMPs) in the progression of osteoarthritis. Biomater Biosyst. 2024;13:100090. https://doi.org/10.1016/j.bbiosy.2024.100090.
69. Nagaraj S, Finzel S, Stok KS, Barnabe C. High-resolution peripheral quantitative computed tomography imaging in the assessment of periarticular bone of metacarpophalangeal and wrist joints. J Rheumatol. 2016;43:1921–34. https://doi.org/10.3899/jrheum.160647.
70. National Institute for Health and Care Excellence: Guidelines. National Institute for Health and Care Excellence (NICE); 2003.
71. Niu J, Zhang Y, LaValley M, et al. Symmetry and clustering of symptomatic hand osteoarthritis in elderly men and women: the Framingham Study. Rheumatology (Oxford). 2003;42:343–8. https://doi.org/10.1093/rheumatology/keg110.
72. Osteoarthritis in over 16s: diagnosis and management. London: National Institute for Health and Care Excellence (NICE); 2022.
73. Palotie A, Väisänen P, Ott J, et al. Predisposition to familial osteoarthrosis linked to type II collagen

gene. Lancet. 1989;1:924–7. https://doi.org/10.1016/s0140-6736(89)92507-5.
74. Pistorio AL, Moore JB. Lessons learned: trapeziectomy and suture suspension arthroplasty for thumb carpometacarpal osteoarthritis. J Hand Microsurg. 2020;14:233–9. https://doi.org/10.1055/s-0040-1716607.
75. Plato CC, Norris AH. Osteoarthritis of the hand: age-specific joint-digit prevalence rates. Am J Epidemiol. 1979;109:169–80. https://doi.org/10.1093/oxfordjournals.aje.a112672.
76. Poole J, Sayer AA, Hardy R, et al. Patterns of interphalangeal hand joint involvement of osteoarthritis among men and women: a British cohort study. Arthritis Rheum. 2003;48:3371–6. https://doi.org/10.1002/art.11339.
77. Rabinovich RV, Rahman OF, Nasra MH, et al. Midcarpal instability. J Am Acad Orthop Surg. 2023;31:834–44. https://doi.org/10.5435/JAAOS-D-22-00777.
78. Raeissadat SA, Ghazi Hosseini P, Bahrami MH, et al. The comparison effects of intra-articular injection of Platelet Rich Plasma (PRP), Plasma Rich in Growth Factor (PRGF), Hyaluronic Acid (HA), and ozone in knee osteoarthritis; a one year randomized clinical trial. BMC Musculoskelet Disord. 2021;22:134. https://doi.org/10.1186/s12891-021-04017-x.
79. Raza K, Breese M, Nightingale P, et al. Predictive value of antibodies to the cyclic citrullinated peptide in patients with very early inflammatory arthritis. J Rheumatol. 2005;32:231–8.
80. Rundback JH, Nahl D, Yoo V. Contrast-induced nephropathy. J Vasc Surg. 2011;54:575–9. https://doi.org/10.1016/j.jvs.2011.04.047.
81. Sanders D, Krause K, O'Muircheartaigh J, et al. Pharmacologic modulation of hand pain in osteoarthritis: a double-blind placebo-controlled functional magnetic resonance imaging study using naproxen. Arthritis Rheumatol. 2015;67:741–51. https://doi.org/10.1002/art.38987.
82. Schmidt KL, Ott VR, Röcher G, Schaller H. Heat, cold and inflammation. Z Rheumatol. 1979;38:391–404.
83. Shah CM, Stern PJ. Scapholunate advanced collapse (SLAC) and scaphoid nonunion advanced collapse (SNAC) wrist arthritis. Curr Rev Musculoskelet Med. 2013;6:9–17. https://doi.org/10.1007/s12178-012-9149-4.
84. Shao J-H, Yu K-H, Chen S-H. Feasibility and acceptability of a self-management program for patients with rheumatoid arthritis. Orthop Nurs. 2020;39:238–45. https://doi.org/10.1097/NOR.0000000000000676.
85. Sharma CV, Mehta V. Paracetamol: mechanisms and updates. Contin Educ Anaesth Crit Care Pain. 2014;14:153–8. https://doi.org/10.1093/bjaceaccp/mkt049.
86. Steinmetz JD, Culbreth GT, Haile LM, et al. Global, regional, and national burden of osteoarthritis, 1990–2020 and projections to 2050: a systematic analysis for the Global Burden of Disease Study 2021. Lancet Rheumatol. 2023;5:e508–22. https://doi.org/10.1016/S2665-9913(23)00163-7.
87. Sun HB. Mechanical loading, cartilage degradation, and arthritis. Ann N Y Acad Sci. 2010;1211:37–50. https://doi.org/10.1111/j.1749-6632.2010.05808.x.
88. Tang P, Wei DH, Ueba H, et al. Scaphoid excision and 4-bone arthrodesis versus proximal row carpectomy: a comparison of contact biomechanics. J Hand Surg. 2012;37:1861–7. https://doi.org/10.1016/j.jhsa.2012.05.040.
89. Trail IA, Martin JA, Nuttall D, Stanley JK. Seventeen-year survivorship analysis of silastic metacarpophalangeal joint replacement. J Bone Joint Surg Br. 2004;86:1002–6. https://doi.org/10.1302/0301-620X.86B7.15061.
90. Van Nortwick S, Berger A, Cheng R, et al. Trapezial topography in thumb carpometacarpal arthritis. J Wrist Surg. 2013;2:263–70. https://doi.org/10.1055/s-0033-1350088.
91. van Venrooij WJ, van Beers JJBC, Pruijn GJM. Anti-CCP antibodies: the past, the present, and the future. Nat Rev Rheumatol. 2011;7:391–8. https://doi.org/10.1038/nrrheum.2011.76.
92. Verbruggen G, Goemaere S, Veys EM. Systems to assess the progression of finger joint osteoarthritis and the effects of disease-modifying osteoarthritis drugs. Clin Rheumatol. 2002;21:231–43. https://doi.org/10.1007/s10067-002-8290-7.
93. Vitale MA, Fruth KM, Rizzo M, et al. Prosthetic arthroplasty versus arthrodesis for osteoarthritis and posttraumatic arthritis of the index finger proximal interphalangeal joint. J Hand Surg. 2015;40:1937–48. https://doi.org/10.1016/j.jhsa.2015.05.021.
94. Vlychou M, Koutroumpas A, Alexiou I, et al. High-resolution ultrasonography and 3.0 T magnetic resonance imaging in erosive and nodal hand osteoarthritis: high frequency of erosions in nodal osteoarthritis. Clin Rheumatol. 2013;32:755–62. https://doi.org/10.1007/s10067-013-2166-x.
95. Wall LB, Stern PJ. Proximal row carpectomy. Hand Clin. 2013;29:69–78. https://doi.org/10.1016/j.hcl.2012.08.022.
96. Wang SX, Ganguli AX, Bodhani A, et al. Healthcare resource utilization and costs by age and joint location among osteoarthritis patients in a privately insured population. J Med Econ. 2017;20:1299–306. https://doi.org/10.1080/13696998.2017.1377717.
97. Watson HK, Weinzweig J, Guidera PM, et al. One thousand intercarpal arthrodeses. J Hand Surg Br. 1999;24:307–15. https://doi.org/10.1054/jhsb.1999.0066.
98. Wu CH, Strauch RJ. Wrist denervation: techniques and outcomes. Orthop Clin North Am. 2019;50:345–56. https://doi.org/10.1016/j.ocl.2019.03.002.
99. Yuan F, Aliu O, Chung KC, Mahmoudi E. Evidence-based practice in the surgical treatment of thumb carpometacarpal joint arthritis. J Hand Surg. 2017;42:104–112.e1. https://doi.org/10.1016/j.jhsa.2016.11.029.

Three-Dimensional Printing and Scaffolding in Hand Mechanics

22

Angela Hussain and Farid Amirouche

Fundamentals of Three-Dimensional Printing and Scaffolding

Three-dimensional printing has revolutionized the field of biomechanics, offering unprecedented customization and precision. When combined with developments in scaffolding techniques, it opens new frontiers in tissue engineering and prosthetic development.

Principles of Three-Dimensional Printing

Additive manufacturing, the foundation of three-dimensional printing, involves constructing objects layer by layer using computer-aided design (CAD) models. Key techniques include the following:

- Fused deposition modeling (FDM): FDM is an additive manufacturing technology within the material extrusion-based three-dimensional printing class. It constructs parts by depositing a range of materials in layers of thermoplastic material, and it is commonly used to create low-cost prototypes and preliminary models (Kim, Acierno). For example, in the clinical setting, FDM has created lightweight, anatomically tailored braces for patients recovering from surgery (Van Leishout et al.).
- Stereolithography (SLA): This process employs a laser to cure resin, producing high-resolution parts ideal for medical models. SLA is often used to create precise bone models for presurgical planning in patients with complex fractures [5].
- Selective laser sintering (SLS): Uses a laser to fuse powdered materials, creating durable and complex structures. SLS has been used to produce custom-fit finger splints that optimize joint stability while maintaining the range of motion in patients with rheumatoid arthritis [8].

These techniques rely on various materials, such as biocompatible polymers, metals, and ceramics.

In the case of fused deposition modeling three-dimensional printing, a wide range of mate-

A. Hussain
Department of Orthopaedic Surgery, University of Illinois College of Medicine at Chicago, Chicago, IL, USA

F. Amirouche (✉)
Department of Orthopaedic Surgery, Orthopaedic and Spine Institute, Endeavor HealthSystem (Formerly Northshore University HealthSystem), an Affiliate of the University of Chicago Pritzker School of Medicine, Skokie, IL, USA

Department of Orthopaedic Surgery, University of Illinois at Chicago, Chicago, IL, USA
e-mail: famirouche@northshore.org; amirouch@uic.edu

F. Amirouche et al. (eds.), *Biomechanics of the Hand*,
https://doi.org/10.1007/978-3-032-04688-8_22

rials are available. Still, the most commonly utilized filament materials are polylactic acid (PLA), acrylonitrile butadiene styrene (ABS), polyethylene terephthalate (PETG), and thermoplastic polyurethane (TPU). The choice of material used for three-dimensional printing depends on obtaining the right properties that allow for the functional use they are intended for. Other than simply considering the cost and speed of production of the materials available for use, other key criteria need to be considered: heat resistance, layer adhesion, impact resistance, max stress, visual quality, and ease of printing (Kantarose et al.; [23]).

PLA is among the easiest polymers to print. It offers good visual quality and is strong; however, it is also a very brittle material with little impact resistance. When considering the application of PLA in hand prototypes to study biomechanics, it is essential to assess the limitations the brittle quality of the material poses. ABS is usually favored when higher temperature resistance is desired. It provides more impact resistance than PLA and offers good abrasion resistance. However, ABS is much more difficult to print and potentially has high fume emissions. PET is considered a softer polymer than both PLA and ABS. It offers high resistance to humidity and chemicals and is safe to be in contact with foods. Its noteworthy limitation to its use is that it is significantly heavier than PLA and ABS. TPU is chosen to be used when the application requires flexibility. It offers strong support and impact resistance but lacks printing and postprinting processing ease.

When using SLA printing, we can produce high-resolution products that are ideal for medical models. We can create precise, accurate models made from resins with biocompatibility certifications. The liquid resins used in SLA three-dimensional printing come in various blends that offer different properties, each chosen based on the intended application of a given part or model. SLA resins can create intricate details while providing lots of rigidity and stability. SLA-produced items used for medical purposes are made with class I biocompatible resins. They often are utilized in the construction of custom medical equipment. Parts made using class I biocompatible resins can withstand sterilization from an autoclave, which allows them to be used in sterile environments such as medical operating rooms.

SLS technology allows for low costs per part and high-productivity manufacturing. The most commonly used material for SLS is nylon. It offers profound environmental stability and is ideal for creating durable and complex assemblies. The nylon material is impact-resistant and biocompatible; thus, it is ready for use in multiple applications. SLS three-dimensional printers can also create TPU parts, allowing for greater tear strength and versatility. TPU is preferred for producing flexible, skin-safe prototypes that withstand everyday wear and tear.

Scaffolding in Biomechanics

Bioscaffolds are critical in tissue engineering, acting as temporary frameworks to support cell attachment, proliferation, and differentiation. The effectiveness of a scaffold depends on several key factors:

- Material properties: Scaffolds must be biocompatible and have mechanical properties that mimic native tissues. For instance, polylactic acid (PLA) has been used to create scaffolds for tendon repair in flexor tendon injuries, ensuring adequate tensile strength during the healing process [7, 13].
- Porosity and architecture: Scaffolds' porous structure is essential for nutrient diffusion and vascularization. In trials, collagen-based scaffolds with interconnected pores have shown promise in regenerating tendon tissue [14]. A combination of synthetic and natural hydrogels is typically used to achieve optimal clinical outcomes.
- Degradation and integration: Scaffolds must degrade at a rate that matches tissue regeneration, allowing native tissues to replace the scaffold seamlessly (Chen et al. 2023). For example, hydrogel-based scaffolds with controlled degradation rates have been developed to support ligament repair in hand injuries,

enabling gradual load transfer to the healing tissue.

Advanced scaffolding technologies now incorporate bioactive molecules to enhance cell behavior and employ multimaterial printing for heterogeneous tissue structures [22]. The development of novel scaffolds that respond to environmental stimuli, such as load or pH changes, is being explored for their potential to accelerate bone healing in complex hand fractures. Bone tissue engineering most commonly incorporates porous three-dimensional scaffolds with bioactive growth factors and cells, providing structural support for cells to migrate and differentiate in forming new tissue. Multipotent stromal cells (MSCs) can differentiate into osteoblasts, chondrocytes, and adipocytes and thus are favored for their osteogenic potential in bone formation [17]. For instance, bioinks have been employed to create three-dimensional printed scaffolds seeded with chondrocytes for cartilage regeneration in patients with degenerative joint diseases in the hand [11].

Applications in Hand Mechanics

Integrating three-dimensional printing and scaffolding has transformative implications for hand mechanics, particularly in prosthetics, tissue repair, and surgical planning.

Three-Dimensional Printed Prosthetics

Custom prosthetics tailored using three-dimensional printing offer enhanced comfort and functionality. Case studies highlight the success of patient-specific designs, which outperform traditional prosthetics in adaptability and aesthetics. For example, the SoftHand Pro, a robotic hand prosthesis, has demonstrated remarkable functionality in tasks requiring fine motor control, such as buttoning a shirt, typing, or holding a coffee cup. Patients fitted with this device during competition events reported improved usability due to its adaptive grip technology and natural movement compared to standard prosthetic models [3, 26].

Another significant application includes myoelectric prosthetics, which use sensors to detect muscle signals, enabling precise control. Three-dimensional printed myoelectric hands have been personalized for individuals with partial hand amputations, allowing them to regain dexterity for cooking and writing [10]. This approach reduces the time and cost of creating these devices compared to traditional manufacturing methods.

Three-dimensional printed prosthetics also enable affordable solutions in low-resource settings. For instance, a low-cost, three-dimensional printed hand prosthetic provided to a young patient in a rural community restored basic grasping and fine motor skills, improving their ability to attend school and participate in daily activities.

Tissue Engineering for Hand Repair

Bioprinting enables the creation of cartilage, tendons, and ligaments, which are crucial for hand repair. For instance, bioprinted cartilage scaffolds seeded with stem cells have been successfully used to reconstruct damaged metacarpophalangeal joints in patients with osteoarthritis [19]. These scaffolds help restore joint functionality and decrease pain with movement, significantly improving quality of life [27] (Fig. 22.1).

Tendon regeneration has seen notable advancements with new tendon tissue engineering techniques. Research conducted by Dechovi et al. looked at strategies for enhancing tendon repair by applying growth factors, natural biomaterials, and stem cells. In a series of studies, it was demonstrated that MSC-seeded collagen sponges, compared to native tendon tissue, could improve mechanical characteristics of the headed tendon in vivo and that the MSC-seeded collagen constructs used produced better results than tendon allografts [1, 9, 15, 25] (Fig. 22.2).

Additionally, dual-material scaffolds that combine soft hydrogel layers with more substantial synthetic materials have been employed in

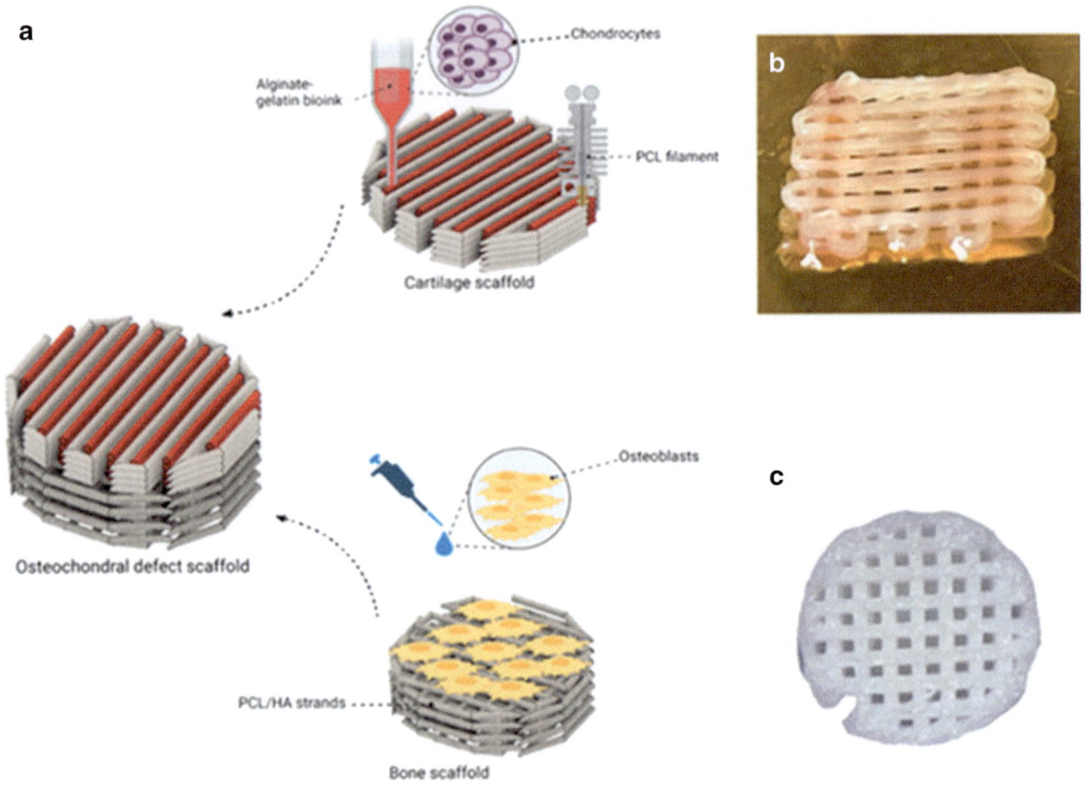

Fig. 22.1 (**a**) Schematics demonstrating the incorporation of the co-printed alginate–gelatin hydrogel and PCL scaffold and DIW-printed PCL/HA composite scaffold into a hybrid osteochondral scaffold for bone-cartilage interface regeneration. (**b**) Coprinted alginate/gelatin hydrogel and PCL scaffold for cartilage regeneration. (**c**) Three-dimensional (3D) printed PCL/30HA composite scaffold with the DIW technique for subchondral bone regeneration [4]

reconstructive surgeries for patients with complex injuries involving both tendon and ligament damage. For instance, scaffold implants designed to stabilize the scapholunate ligament have shown improved outcomes in cases of wrist instability.

Surgical Planning and Simulation

Three-dimensional printed models replicate patient-specific anatomy, aiding presurgical planning and improving accuracy. In one case, surgeons treating a comminuted distal radius fracture used a three-dimensional printed model to visualize the fracture pattern and practice the surgical approach [18]. This reduced operation time and improved fracture alignment, minimizing postoperative complications.

In surgical training, simulation models created via three-dimensional printing have been invaluable. For instance, tendon transfer procedures, such as using the palmaris longus to restore thumb movement, have been practiced on realistic, patient-specific models. These simulations enable surgeons to refine their techniques without risking patient safety.

Beyond orthopedic applications, these models have been used in microsurgical training for nerve repair, where precise suturing of tiny nerve fibers is required. This has significantly improved surgical outcomes and has helped decrease costs associated with surgical training models [20].

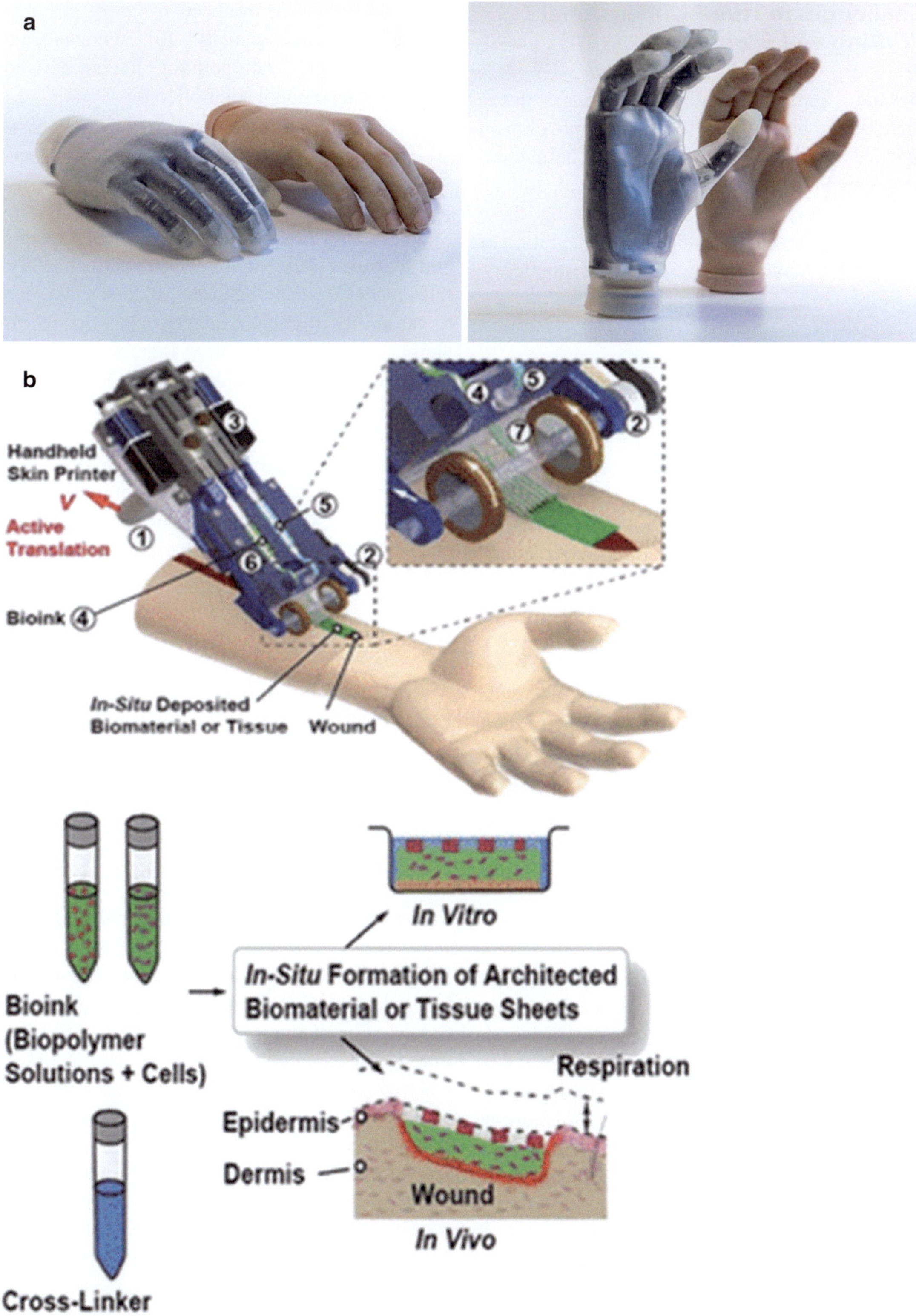

Fig. 22.2 Three-dimensional bioprinting of skin tissue: (b) employing a handheld bioprinter to generate skin cell-laden sheets with controllable thickness, width, and composition via incorporating dermal and epidermal cells into various cross-linkable hydrogels [2]

Challenges in Three-Dimensional Printing and Scaffolding

Despite its potential, integrating three-dimensional printing and scaffolding into clinical practice faces significant hurdles. These challenges include technological, biological, ethical, and regulatory limitations. Overcoming these challenges is key to the widespread integration of three-dimensional printing in hand prosthetics.

Technical Limitations

Current printers struggle with resolution and scalability. For example, creating scaffolds with micron-level precision required for regenerating intricate structures like the flexor tendons in the hand remains challenging. Multimaterial printing, critical for replicating composite structures such as the tendon-bone interface, often requires costly and sophisticated equipment that is not widely accessible.

Material constraints also hinder progress. For example, while silicone-based materials provide excellent flexibility for prosthetic fingers, their durability under constant mechanical stress remains limited. Developing hybrid materials that combine flexibility with long-term wear resistance is an ongoing area of research.

Biological Challenges

Vascularization remains a significant obstacle for three-dimensional printed scaffolds. In reconstructive surgeries following severe burns or trauma to the hand, insufficient blood vessel formation within scaffolds can lead to tissue necrosis [12, 28]. Efforts are underway to bioprint vascularized scaffolds using endothelial cell-laden bioinks, which could revolutionize tissue regeneration.

Another challenge is immune compatibility. A recent study noted that synthetic scaffolds used for tendon repair elicited mild inflammatory responses in patients with autoimmune conditions, underscoring the need for materials that minimize immunogenicity [6]. Personalized scaffold designs incorporating patient-derived cells offer a potential solution to this issue.

Ethical and Regulatory Issues

Standardizing three-dimensional printed implants and scaffolds is essential to ensure patient safety. Differences in manufacturing processes can lead to variability in implant quality, raising concerns about long-term outcomes. Regulatory bodies must establish clear guidelines to address these issues while ensuring accessibility, particularly in underserved populations.

Beyond regulation, there are ethical considerations related to equity. For example, while advanced prosthetics and regenerative therapies offer life-changing benefits, their high costs limit availability for many patients. Addressing this disparity requires collaborative efforts among researchers, clinicians, and policymakers.

Future Directions and Innovations

Emerging technologies promise to address current limitations, paving the way for advanced applications in hand biomechanics.

Emerging Materials and Techniques

Innovative materials with adaptive properties are being developed to mimic the behavior of natural tissues. For example, shape-memory polymers that adjust to temperature changes are tested in adaptive finger splints, enabling dynamic support for conditions such as Dupuytren's contracture.

Self-healing materials are another exciting innovation. These materials can repair minor tears or damage autonomously, extending the lifespan of prosthetics or scaffold implants. For example, a three-dimensional printed wrist brace using self-healing elastomers has been tested for long-term use in repetitive strain injuries [16].

Advances in Bioprinting

Incorporating stem cells into bioinks enables the creation of living tissues. Recent advances have led to bioprinted finger cartilage for patients with traumatic injuries, showing promising integration and functional restoration in early trials [28]. Furthermore, fully printed composite tissues—combining the bone, cartilage, and ligament—have been explored to reconstruct complex injuries like those seen in severe crush injuries to the hand.

Real-time monitoring during printing ensures precision in constructing complex structures, such as the interphalangeal joints. Advanced bioreactors are being integrated with these technologies to simulate in vivo conditions, promoting better maturation of bioprinted tissues.

Artificial intelligence optimizes three-dimensional-printed designs for specific biomechanical needs. For example, AI algorithms have been used to analyze patient-specific movement patterns and customize prosthetics to enhance grip strength or reduce fatigue [26]. One notable success involved using AI to design a robotic hand with improved finger articulation for musicians who suffered partial hand amputations, allowing them to play instruments again [21].

Robotics integrated with three-dimensional printed components have enabled more intuitive control systems, improving rehabilitation outcomes for patients with partial hand amputations. A recent example includes a bionic hand with tactile feedback sensors, allowing patients to feel pressure or texture, significantly enhancing their environmental interaction [24].

Conclusion

Three-dimensional printing and scaffolding represent a paradigm shift in hand biomechanics, offering innovative prosthetics, tissue repair, and surgical planning solutions. Addressing current challenges requires interdisciplinary collaboration, continuous research, and technological advancement. The future of hand mechanics involves integrating transformative technologies into clinical practice.

References

1. AlAli AB, Griffin MF, Butler PE. Three-dimensional printing surgical applications. Eplasty. 2015;15:e37. PMID: 26301002; PMCID: PMC4539849.
2. Askari M, Afzali Naniz M, Kouhi M, Saberi A, Zolfagharian A, Bodaghi M. Recent progress in extrusion 3D bioprinting of hydrogel biomaterials for tissue regeneration: A comprehensive review focusing on advanced fabrication techniques. Biomater Sci. 2020; https://doi.org/10.1039/D0BM00973C.
3. Capsi-Morales P, Piazza C, Grioli G, Bicchi A, Catalano MG. The SoftHand Pro platform: a flexible prosthesis with a user-centered approach. J Neuroeng Rehabil. 2023;20(1):20. Published 2023 Feb 8. https://doi.org/10.1186/s12984-023-01130-x.
4. Chen H, Gonnella G, Huang J, Di-Silvio L. Fabrication of 3D bioprinted bi-phasic scaffold for bone–cartilage interface regeneration. Biomimetics. 2023;8(1):87. https://doi.org/10.3390/biomimetics8010.
5. Cornejo J, Cornejo-Aguilar JA, Vargas M, et al. Anatomical engineering and 3D printing for surgery and medical devices: international review and future exponential innovations. Biomed Res Int. 2022;2022:6797745. Published 2022 Mar 24. https://doi.org/10.1155/2022/6797745.
6. Eugenis I, Wu D, Rando TA. Cells, scaffolds, and bioactive factors: engineering strategies for improving regeneration following volumetric muscle loss. Biomaterials. 2021;278:121173. https://doi.org/10.1016/j.biomaterials.2021.121173.
7. Govindaraju DT, Chen CH, Shalumon KT, Kao HH, Chen JP. Bioactive nanostructured scaffold-based approach for tendon and ligament tissue engineering. Nanomaterials (Basel). 2023;13(12):1847. Published 2023 Jun 12. https://doi.org/10.3390/nano13121847.
8. Guven E, Karpat Y, Cakmakci M. Improving the dimensional accuracy of micro parts 3D printed with projection-based continuous vat photopolymerization using a model-based grayscale optimization method, 102954., ISSN 2214-8604. Additive Manuf. 2022;57 https://doi.org/10.1016/j.addma.2022.102954.
9. Kinneberg KR, Galloway MT, Butler DL, Shearn JT. Effect of implanting a soft tissue autograft in a central-third patellar tendon defect: biomechanical and histological comparisons. J Biomech Eng. 2011;133(9):091002. https://doi.org/10.1115/1.4004948.
10. Lee J, Kim HC, Choi JW, et al. A review on 3D printed smart devices for 4D printing. Int J Precis Eng Manuf

Green Tech. 2017;4:373–83. https://doi.org/10.1007/s40684-017-0042-.

11. Liang Q, Ma Y, Yao X, Wei W. Advanced 3D-printing bioinks for articular cartilage repair. Int J Bioprint. 2022;8(3):511. Published 2022 Apr 22. https://doi.org/10.18063/ijb.v8i3.511.
12. Meyer-Szary J, Luis MS, Mikulski S, et al. The role of 3D printing in planning complex medical procedures and training of medical professionals-cross-sectional multispecialty review. Int J Environ Res Public Health. 2022;19(6):3331. Published 2022 Mar 11. https://doi.org/10.3390/ijerph19063331.
13. Narayanan G, Vernekar VN, Kuyinu EL, Laurencin CT. Poly (lactic acid)-based biomaterials for orthopaedic regenerative engineering. Adv Drug Deliv Rev. 2016;107:247–76. https://doi.org/10.1016/j.addr.2016.04.015.
14. Ning F, Cong W, Qiu J, Wei J, Wang S. Additive manufacturing of carbon fiber reinforced thermoplastic composites using fused deposition modeling. Compos Part B Eng. 2015;80:369–78. ISSN 1359-8368, https://doi.org/10.1016/j.compositesb.2015.06.013. (https://www.sciencedirect.com/science/article/pii/S1359836815003777).
15. Nirmalanandhan VS, Juncosa-Melvin N, Shearn JT, et al. Combined effects of scaffold stiffening and mechanical preconditioning cycles on construct biomechanics, gene expression, and tendon repair biomechanics. Tissue Eng Part A. 2009;15(8):2103–11. https://doi.org/10.1089/ten.tea.2008.0335.
16. Oud TAM, Lazzari E, Gijsbers HJH, Gobbo M, Nollet F, Brehm MA. Effectiveness of 3D-printed orthoses for traumatic and chronic hand conditions: a scoping review. PLoS One. 2021;16(11):e0260271. Published 2021 Nov 18. https://doi.org/10.1371/journal.pone.0260271.
17. Percival KM, Paul V, Husseini GA. Recent advancements in bone tissue engineering: integrating smart scaffold technologies and bio-responsive systems for enhanced regeneration. Int J Mol Sci. 2024;25(11):6012. Published 2024 May 30. https://doi.org/10.3390/ijms25116012.
18. Roelofs LJM, Assink N, Kraeima J, et al. Clinical application of 3D-assisted surgery techniques in treatment of intra-articular distal radius fractures: a systematic review in 718 patients. J Clin Med. 2024;13(23):7296. Published 2024 Nov 30. https://doi.org/10.3390/jcm13237296.
19. Roseti L, Cavallo C, Desando G, et al. Three-dimensional bioprinting of cartilage by the use of stem cells: a strategy to improve regeneration. Materials (Basel). 2018;11(9):1749. Published 2018 Sep 17. https://doi.org/10.3390/ma11091749.
20. Serrano C, Fontenay S, van den Brink H, Pineau J, Prognon P, Martelli N. Evaluation of 3D printing costs in surgery: a systematic review. Int J Technol Assess Health Care. Published online June 3, 2020. https://doi.org/10.1017/S0266462320000331.
21. Shafti A, Haar S, Mio R, Guilleminot P, Faisal AA. Playing the piano with a robotic third thumb: assessing constraints of human augmentation. Sci Rep. 2021;11(1):21375. Published 2021 Nov 1. https://doi.org/10.1038/s41598-021-00376-6.
22. Sindhi R. In vitro C5a generation by synthetic vascular prostheses: implications for graft incorporation in vivo. Biomater Med Devices Artif Organs. 1986;14(3–4):249–55. https://doi.org/10.3109/10731198609117546.
23. Sun Y, Lee D, Wang Y, et al. Thermal decomposition behavior of 3D printing filaments made of wood-filled polylactic acid/starch blend. J Appl Polym Sci. 2021;138:e49944. https://doi.org/10.1002/app.49944.
24. Varaganti P, Seo S. Recent advances in biomimetics for the development of bio-inspired prosthetic limbs. Biomimetics. 2024;9(5):273. https://doi.org/10.3390/biomimetics9050273.
25. Yang G, Rothrauff BB, Tuan RS. Tendon and ligament regeneration and repair: clinical relevance and developmental paradigm. Birth Defects Res C Embryo Today. 2013;99(3):203–22. https://doi.org/10.1002/bdrc.21041.
26. Zahid MJ, Mavani P, Awuah WA, et al. Sculpting the future: a narrative review of 3D printing in plastic surgery and prosthetic devices. Health Sci Rep. 2024;7(6):e2205. Published 2024 Jun 23. https://doi.org/10.1002/hsr2.2205.
27. Zahid MJ, Mavani P, Awuah WA, Alabdulrahman M, Punukollu R, Kundu A, Mago A, Maher K, Adebusoye FT, Khan TN. Sculpting the future: a narrative review of 3D printing in plastic surgery and prosthetic devices. Health Sci Rep. 2024;7(6):e2205. https://doi.org/10.1002/hsr2.2205. PMID: 38915353; PMCID: PMC11194296.
28. Zhang J, Wehrle E, Rubert M, Müller R. 3D bioprinting of human tissues: biofabrication, bioinks, and bioreactors. Int J Mol Sci. 2021;22(8):3971. Published 2021 Apr 12. https://doi.org/10.3390/ijms22083971.

Fractures

23

Barbara Mera, Majd Mzeihem,
and Farid Amirouche

Introduction

The study of hand and wrist fracture biomechanics is a challenging yet critical discipline that delves into the intricate and vital forces and mechanical principles responsible for fractures in these regions. The hand and wrist comprise a sophisticated network of bones, joints, ligaments, and tendons, forming a complex system providing remarkable range of motion and functional versatility.

Fractures in the hand and wrist arise from diverse mechanisms, including high-impact trauma such as falls, vehicle collisions, or sports-related injuries, as well as repetitive stress or pathological conditions like osteoporosis [1, 2]. By delving into the biomechanics underlying these injuries, several key aspects of fracture causation, diagnosis, and management can be better understood:

Various types of fractures, including Colles' (distal radius), Smith's, and scaphoid fractures, exhibit distinct mechanical behaviors [1]. These behaviors are governed by factors such as the magnitude, direction, and point of application of forces [2]. Understanding these differences provides valuable insights for identifying the nature of the injury and tailoring appropriate treatment strategies.

How forces are transmitted through the hand and wrist during an injury is pivotal to fracture outcomes [2]. Parameters such as the impact's angle, velocity, and energy determine the fracture's extent and pattern [2]. By analyzing these force dynamics, we can guide the development of preventive measures and refine intervention techniques to reduce fracture severity.

The structural integrity and alignment of bones and joints play a central role in the occurrence and healing of fractures [3]. Misalignment during or after injury can lead to compromised functionality and long-term complications. Therefore, achieving and maintaining proper anatomical alignment is essential for optimal recovery and minimizing adverse outcomes.

The ligaments, tendons, and other soft tissues surrounding the bones significantly influence fracture risk and healing [4]. These structures contribute to shock absorption during impact and are often injured concurrently [4]. Addressing the condition of these tissues is critical for ensuring

B. Mera (✉) · M. Mzeihem
Department of Orthopaedic Surgery, University of Illinois at Chicago, Chicago, IL, USA
e-mail: bmera2@uic.edu

F. Amirouche
Department of Orthopaedic Surgery, Orthopaedic and Spine Institute, Endeavor HealthSystem (Formerly Northshore University HealthSystem), an Affiliate of the University of Chicago Pritzker School of Medicine, Skokie, IL, USA

Department of Orthopaedic Surgery, University of Illinois at Chicago, Chicago, IL, USA

F. Amirouche et al. (eds.), *Biomechanics of the Hand*,
https://doi.org/10.1007/978-3-032-04688-8_23

comprehensive management and rehabilitation of fractures.

The mechanisms and patterns of fractures vary significantly with age and overall health. For instance, younger individuals are more likely to sustain fractures from high-energy trauma. In contrast, older adults are often vulnerable to fractures from low-energy impacts due to conditions like decreased bone density and fragility. These differences highlight the need for age-specific approaches to prevention and treatment, underscoring the importance of tailored care in fracture management.

By thoroughly analyzing the biomechanics of hand and wrist fractures, we can advance the development of more effective treatment strategies, rehabilitation protocols, and preventive measures. This knowledge facilitates restoring functionality and reduces the risk of recurrent injuries, improving the overall quality of life for affected individuals. Contributions to this field can significantly impact clinical outcomes and patient care.

Biomechanical Principles in Fracture Causation

The biomechanics of the hand and forearm can be categorized into rotational mechanisms and stabilizing systems [5]. The three radioulnar joints (proximal, middle, and distal) collaborate to enable forearm rotation, while the interosseous membrane provides stability and force transfer. This setup allows forearm rotation along a single axis passing through the radial head and ulnar styloid [5, 6]. Stabilization systems include longitudinal stability, which prevents proximal radius migration, and transverse stability, which maintains proper bone spacing and relationships [5]. Understanding these biomechanical principles is crucial in comprehending the forces and mechanisms involved in hand and wrist injuries.

Fractures of the hand and forearm can arise from various applied forces, including axial loading, shear, and torsion. Axial loading refers to forces directed along the bone axis, such as those resulting from a direct impact or a fall on an outstretched hand. This can lead to fractures when the bone's compressive strength is exceeded [8, 9]. Distal radius fractures are commonly caused by falls, whereas direct impacts, such as those in boxing, often result in metacarpal fractures [10–12] (Fig. 23.1).

Shear forces, in contrast, act parallel to the bone surface, causing it to slide over itself. These forces are particularly detrimental to healing, as they can misalign bone fragments [13]. Shear-related injuries may occur during manual labor or industrial accidents involving heavy machinery

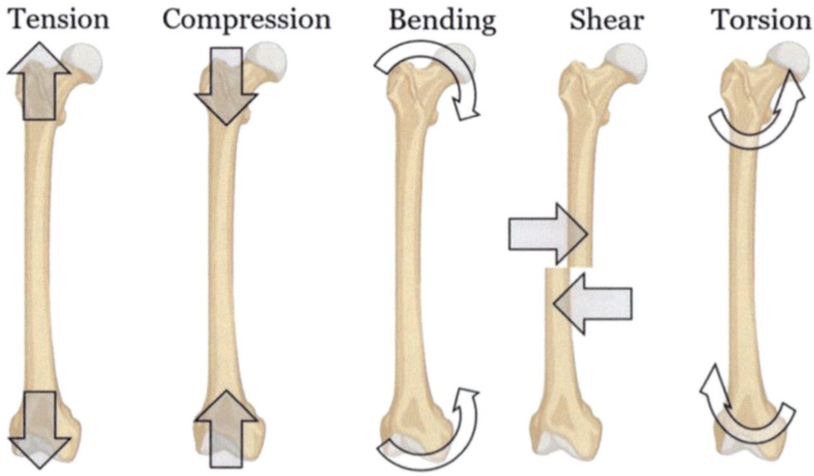

Fig. 23.1 *Biomechanical forces*—five types of mechanical forces acting on bones: *tension*, which pulls the bone apart; *compression*, which pushes it together; *bending*, which applies force along a curve; *shear*, which causes opposing forces along different planes; and *torsion*, which twists the bone. These forces influence fracture patterns and bone strength

[13] and in sports like basketball or football, where lateral impacts to the hand are frequent [14, 15].

Torsion forces create rotational stress on the bone, resulting in spiral fractures characterized by a helical fracture line. These injuries can occur during sports such as snowboarding or skiing, where the hand may become twisted, or during vehicular accidents involving rotational forces [16, 17].

Distal Radius Fractures

Distal radius fractures represent some of the most prevalent injuries affecting the upper extremities, particularly among individuals engaged in high-impact activities or falls, especially in older adults. Fracture typically occurs when a person falls onto an outstretched hand, resulting in a distal fragment of the radius. This mechanism of injury often involves a combination of force and wrist extension, which leads to the wrist's characteristic "dinner fork" appearance when viewed laterally. In contrast, other fractures occur due to a fall on a flexed wrist or a direct blow to the wrist—this type of fracture results in the distal radius fragment.

Understanding the differences between these fracture types is essential, as they dictate specific treatment approaches and rehabilitation strategies. Accurate assessment of the wrist position at injury and the mechanisms involved is crucial for appropriate management and recovery (Fig. 23.2).

Colles' Fracture

A Colles' fracture is a distal radius fracture characterized by dorsal angulation of the distal fragment. It typically occurs from a fall on an outstretched hand, leading to a hyperextension injury, and is most prevalent among older adults, particularly those with osteoporosis. The biomechanical mechanisms underlying this fracture include dorsal angulation, axial load, and wrist hyperextension.

When a person falls with the wrist in a hyperextended position, the axial load transmitted through the carpal bones to the distal radius can exceed the bone's structural integrity, resulting in a fracture. The dorsal displacement of the distal fragment leads to the characteristic dorsal angulation [18]. Research indicates that off-axis loads—those deviating from a purely axial direction—contribute significantly to the failure of the distal radius, as they induce bending, shear, and torsional stresses. These stresses are more detrimental to the bone's structural integrity than purely axial loads, causing fractures at lower forces [8, 19]. Additionally, hyperextension shifts the contact points of the radiocarpal joint dorsally, concentrating stress on the dorsal aspect of the radius. This increased dorsal load creates a "bending moment" that predisposes the radius to fracture in this specific pattern [20, 21].

Smith's Fracture

A Smith's fracture, in contrast, features volar angulation of the distal radius fragment and occurs due to a fall on a flexed wrist, resulting in a hyperflexion injury. While the biomechanical forces—axial loading, shear forces, and torsional stresses—are similar to those seen in Colles' fractures, their application differs due to the wrist's flexed position during the injury.

In a Smith's fracture, axial loading creates compressive forces on the volar side of the distal radius. Shear forces, influenced by the impact angle, contribute to a sliding motion that displaces the radius volarly [22]. Although torsional forces may also be present, they play a less significant role in this injury mechanism. Unlike Colles' fractures, Smith's fractures are less dependent on age and bone density and can occur across all age groups.

Clinical Comparisons of Colles' and Smith's Fractures

Colles' and Smith's fractures differ in their mechanisms, angulation patterns, and clinical

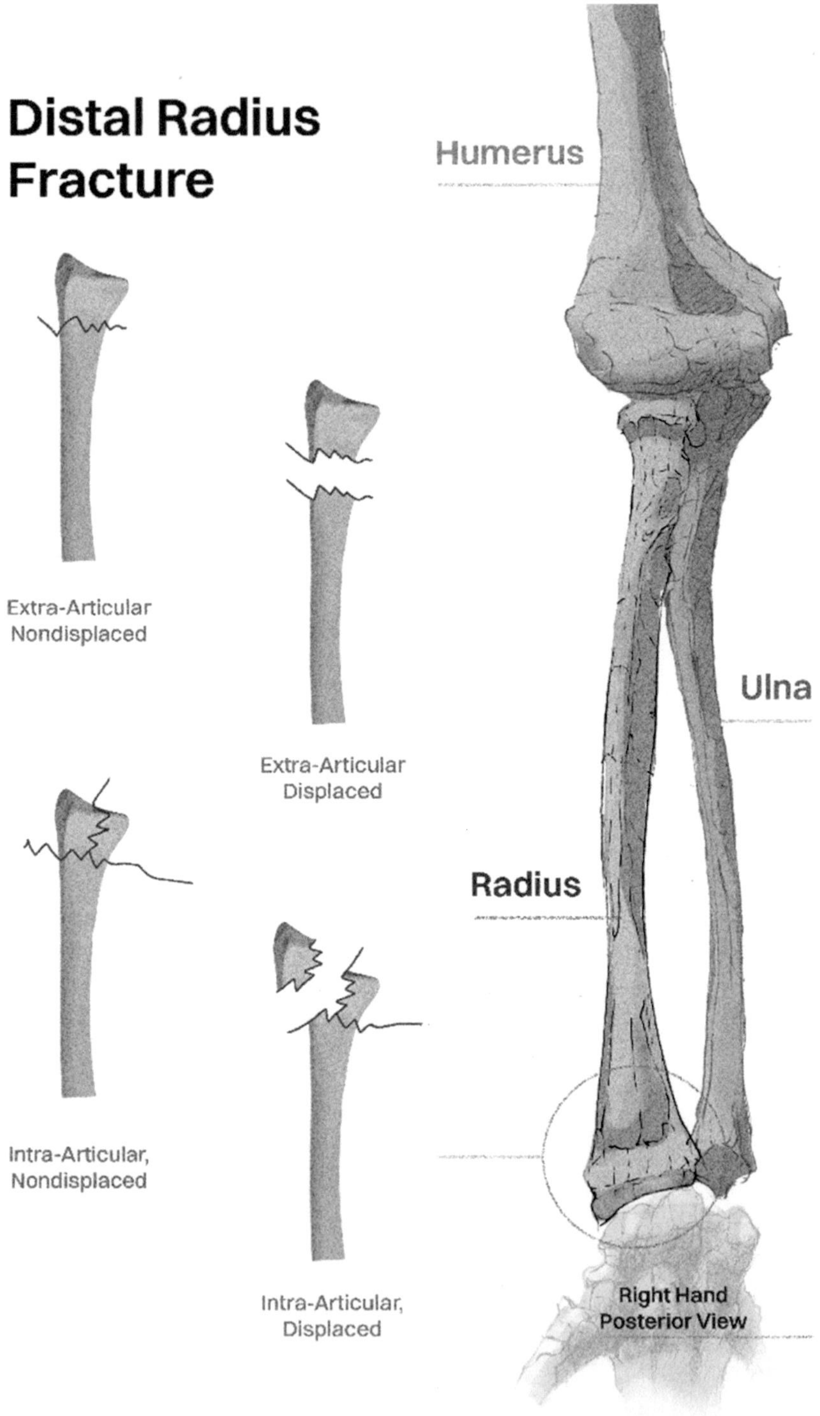

Fig. 23.2 Extraarticular fracture is a break that occurs outside of the joint capsule. Intraarticular fracture is a break in the bone that extends into the joint and can damage the cartilage, making it difficult for the joint to move smoothly. (Reproduced with permission from Cast21, Inc., "Distal Radius Fracture: What Is It?", https://www.cast21.com/blog/distal-radius-fracture-explained. No changes were made)

implications. Colles' fractures are commonly associated with complications such as decreased wrist flexion, diminished radial inclination, and altered dorsal tilt [23, 24]. In contrast, Smith's fractures may lead to reduced grip strength and impaired forearm rotation due to volar angulation, which can also affect distal radioulnar joint kinematics [23, 24].

Understanding these fractures' biomechanical forces and clinical consequences is essential for accurate diagnosis and effective management, ensuring optimal patient functional outcomes (Figs. 23.3, 23.4, and 23.5).

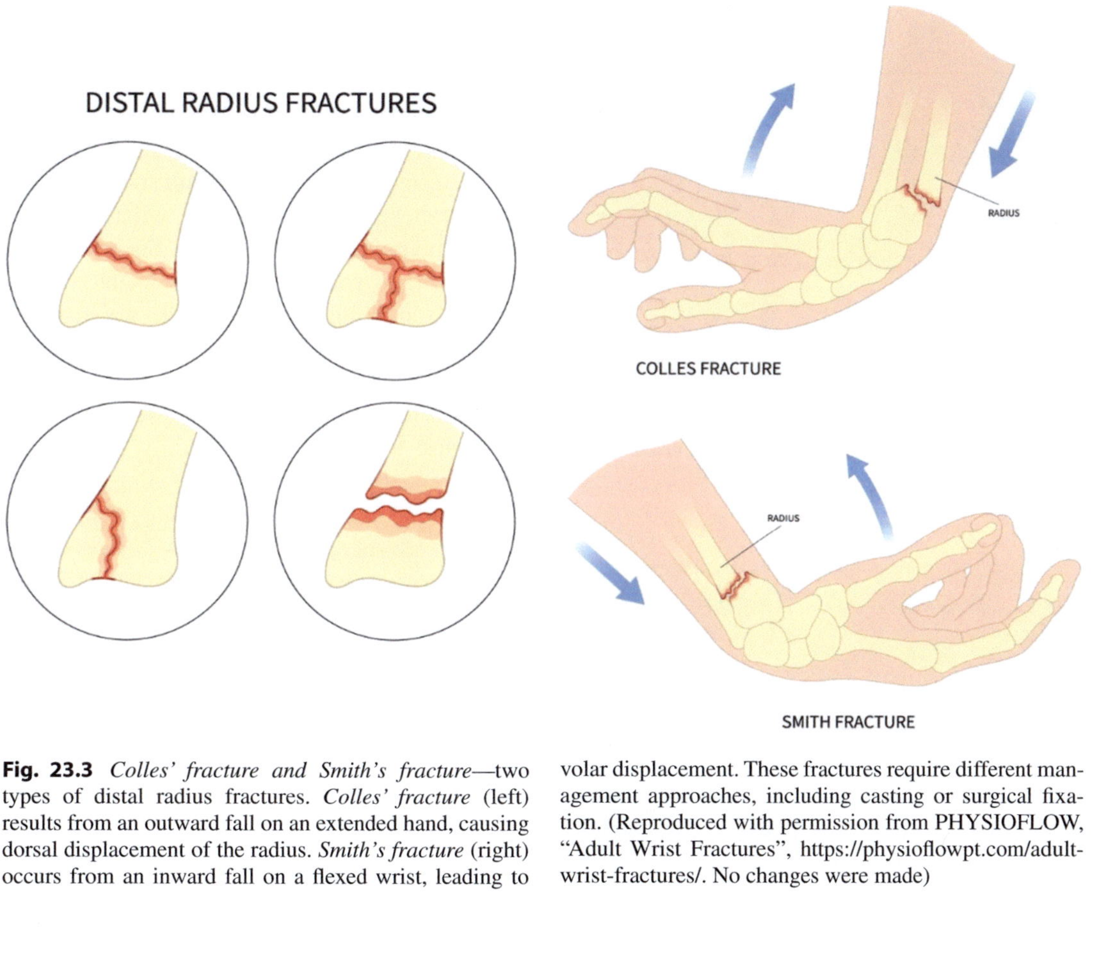

Fig. 23.3 *Colles' fracture and Smith's fracture*—two types of distal radius fractures. *Colles' fracture* (left) results from an outward fall on an extended hand, causing dorsal displacement of the radius. *Smith's fracture* (right) occurs from an inward fall on a flexed wrist, leading to volar displacement. These fractures require different management approaches, including casting or surgical fixation. (Reproduced with permission from PHYSIOFLOW, "Adult Wrist Fractures", https://physioflowpt.com/adult-wrist-fractures/. No changes were made)

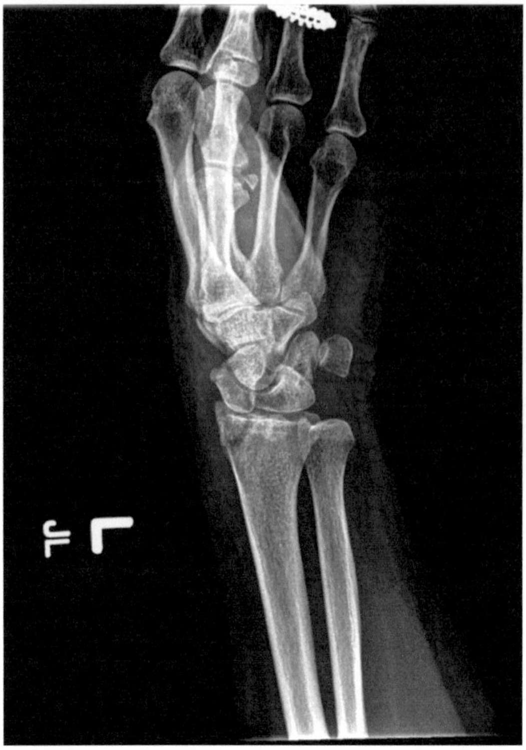

Fig. 23.4 (left): Lateral X-ray view of distal radius fracture

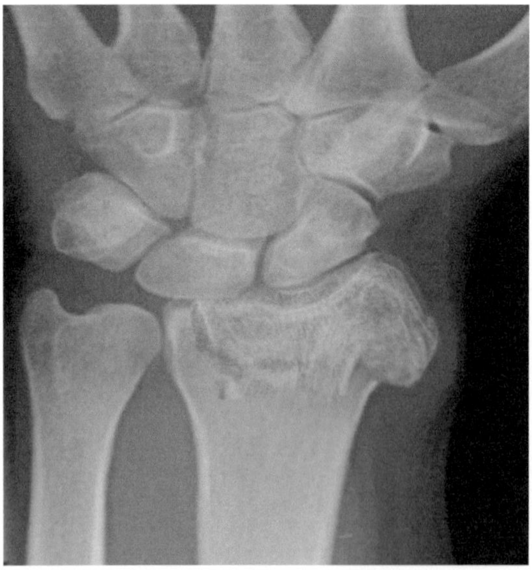

Fig. 23.5 (right): Anteroposterior X-ray of a wrist showing a distal radius fracture

Carpal Bone Fractures

Scaphoid Fracture

Scaphoid fractures are among the most common carpal bone injuries, posing challenges due to the scaphoid's unique anatomy and vascular supply. Spanning the proximal and distal carpal rows, the scaphoid plays a critical role in wrist biomechanics [25, 26]. Its nearly complete articular cartilage coverage limits bony remodeling and healing capacity [25]. Blood supply primarily originates from radial artery branches that enter retrogradely, making the proximal pole especially prone to avascular necrosis during injury [27].

These fractures typically result from a fall on an outstretched hand, with the wrist hyperextended and radially deviated [28]. Such positioning produces compressive and shear forces, frequently causing fractures at the waist of the bone [28]. Limited vascularity and difficulties achieving stable fixation often lead to delayed healing or nonunion [29, 30]. Persistent nonunion may cause a "humpback deformity," characterized by scaphoid collapse, shortening, and disrupted carpal alignment, progressing to scaphoid nonunion advanced collapse (SNAC), a condition marked by degenerative changes and arthritis [31].

Early diagnosis is critical. Nondisplaced fractures generally respond well to immobilization [28, 29, 32], while displaced fractures often require surgical intervention, such as open reduction with internal fixation [31] (Fig. 23.6).

Triquetral Fracture

Triquetral fractures, the second most common carpal fracture, account for 15–18% of all carpal injuries [33]. These fractures typically result from a fall on an outstretched hand with the wrist extended and ulnar deviated [33]. The three main types are dorsal cortical fractures (93–95%, most common), body fractures, and volar cortical fractures (least common) [33].

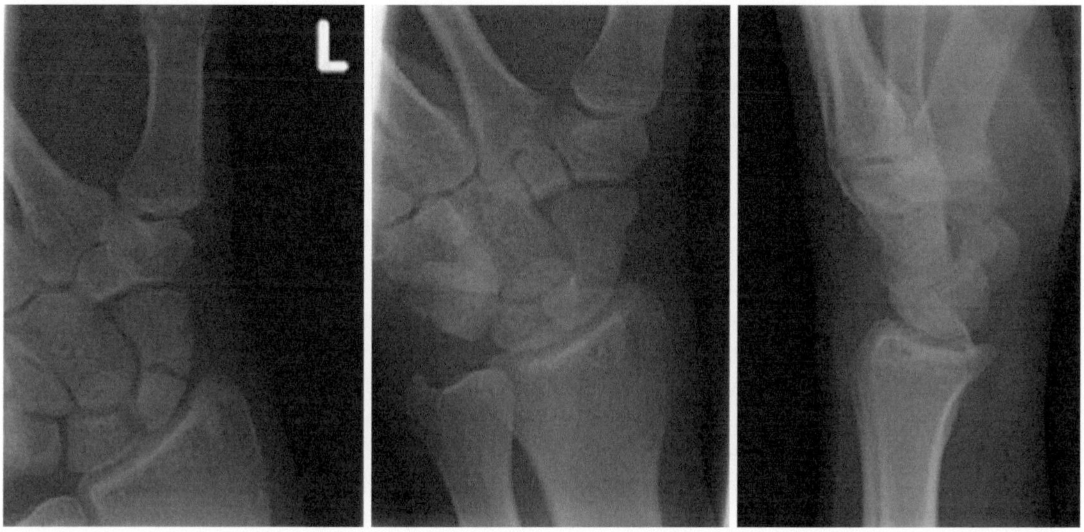

Fig. 23.6 Scaphoid fracture—X-ray series of the left wrist, demonstrating a scaphoid fracture. The images include multiple views to improve visualization of the fracture, which commonly occurs due to a fall on an outstretched hand. Scaphoid fractures are clinically significant due to the risk of avascular necrosis, given the bone's unique blood supply. Early detection through radiographic imaging is crucial for appropriate management, which may include casting or surgical fixation, depending on fracture displacement and location

Dorsal avulsion fractures arise from hyperextension with ulnar deviation, creating tensile forces on the dorsal triquetrum and often involving avulsion of dorsal carpal ligaments, such as the dorsal radiocarpal and lunotriquetral ligaments [34]. These fractures involve a small bone fragment pulled off by attached ligaments and are frequently associated with ligamentous injuries, complicating healing [34].

Body fractures occur due to direct trauma or compressive forces during wrist hyperextension with ulnar deviation [34, 35]. These fractures involve a significant portion of the triquetrum and can disrupt carpal alignment and stability [34, 35].

Biomechanically, wrist hyperextension with ulnar deviation places significant stress on the triquetrum, causing avulsion fractures under ligament tension or body fractures under compressive forces [34–36]. Both types disrupt wrist kinematics, resulting in pain, reduced motion, and complications such as carpal instability or arthritis [34–36]. Dorsal avulsion fractures are associated with chronic pain, instability, delayed healing, nonunion, and reduced grip strength [37, 38].

Lunate Fracture

Lunate fractures are concerning due to the risk of avascular necrosis, also known as Kienböck's disease [39]. These fractures result from axial compression forces during a fall on an outstretched hand, with the wrist hyperextended and ulnarly deviated [39].

Anatomical factors, such as negative ulnar variance and a high lunate uncovering index, exacerbate fracture progression by increasing internal stresses on the lunate under heavy loading conditions [39]. These stresses can cause trabecular microfractures and bone collapse [39, 40]. The lunate's retrograde blood supply from radial artery branches makes it susceptible to avascular necrosis when intraosseous pressure disrupts blood flow [41–43].

Complications include avascular necrosis, carpal instability, posttraumatic arthritis, and median nerve dysfunction [39, 43, 44, 45]. Functional outcomes often involve a reduced range of motion, chronic pain, diminished grip strength, and impaired hand function [46–48].

Hamate Fracture

Hamate fractures can involve either the body or the hook, with differing mechanisms of injury. Body fractures typically result from high-energy trauma, such as a direct blow to the hand or wrist during a fall or punch, where axial compression and shear forces are transmitted through the fourth and fifth metacarpals to the hamate [49]. In contrast, hook fractures are more common in athletes engaged in repetitive gripping or swinging activities, such as golf, tennis, or baseball. These fractures result from direct trauma, such as impact from sports equipment handles or repetitive stress applied by the adjacent deep flexor tendons [51–53]. Biomechanically, body fractures occur from significant impact forces exceeding the bone's structural integrity, whereas hook fractures arise from fatigue or vascular compromise induced by repetitive tendon loads [49–53]. Both fractures can lead to functional impairments, including decreased grip strength and altered tendon mechanics [54, 55].

Hook of Hamate Fracture

Hook of hamate fractures are frequent in athletes performing repetitive gripping and swinging motions, such as golfers, baseball players, and tennis players. These fractures can result from direct trauma (e.g., impact from equipment handles) or repetitive stress from gripping and tendon loading [51–53]. Mechanically, the hook serves as a pulley for the flexor tendons, and its fracture disrupts tendon mechanics, leading to decreased grip strength and altered kinematics. Stress fractures may also arise from vascular compromise due to sustained tendon pressure, which reduces blood flow and weakens the bone [51–53].

Pisiform Fracture

Pisiform fractures, often seen in athletes involved in sports requiring significant wrist loading, result from direct trauma to the hypothenar eminence or repetitive stress [56–58]. The pisiform stabilizes the wrist's ulnar column and is a fulcrum for the flexor carpi ulnaris tendon. Its fracture can lead to chronic pain, instability, and reduced wrist function [56, 57]. Pisiform excision has increased forces required for wrist motion, altering tendon mechanics [59, 60].

Capitate Fracture

Capitate fractures are rare but crucial to wrist stability. These injuries typically result from high-energy axial loadings, such as a fall on an outstretched hand or a direct blow to the wrist [49, 61, 62]. The capitate's central location in the wrist predisposes it to complications like avascular necrosis and nonunion due to limited blood supply [2, 61, 62].

High axial loads transmitted through the carpus generate significant compressive forces, often leading to fractures, particularly when the wrist is extended [7, 11, 63]. Ligament disruptions can further destabilize the capitate and alter wrist kinematics, increasing fracture risk [64]. Rotational dynamics of the proximal carpal row during axial loading also contribute to these injuries [65].

Trapezium Fracture

The trapezium at the thumb base is integral to trapeziometacarpal (TMC) joint stability, facilitating thumb opposition and grip. Its fractures compromise TMC stability and thumb function, causing pain and reduced grip strength [66, 67]. Ligament disruptions further destabilize the joint, increasing the risk of subluxation or dislocation [68].

Trapezium fractures often result from high-energy trauma, such as falls onto an outstretched hand or axial loading through the thumb, causing longitudinal shear fractures or sagittal splits [70, 71]. Inadequate treatment may result in chronic pain, weakness, and impaired thumb function. Early management is essential to prevent complications such as nonunion or osteoarthritis [72].

Thumb and Finger Fractures

A Bennett fracture is an intraarticular fracture at the base of the first metacarpal bone that extends into the thumb's carpometacarpal (CMC) joint. It typically results from axial loading of a flexed thumb, such as during a fall or a direct blow [69, 73]. This fracture is characterized by a large fragment that remains attached to the trapezium via the volar oblique ligament. At the same time, the rest of the metacarpal is displaced dorsally and radially by the pull of the abductor pollicis longus muscle, creating a V-shaped fracture pattern.

The anterior oblique ligament (AOL) and the ulnar collateral ligament (UCL) play key roles in maintaining the stability of the CMC joint. The more robust UCL is primarily attached to the fracture fragment, while the thinner AOL contributes less to the fracture morphology [74]. Displacement forces result from the abductor pollicis longus muscle exerting a dorsomedial pull on the metacarpal shaft, contributing to the fracture's instability [69, 74].

Following a Bennett fracture, the contact area within the trapeziometacarpal joint shifts. The palmar region's contact area decreases, leading to altered joint mechanics and load redistribution [75].

A Rolando fracture is a comminuted intraarticular fracture of the first metacarpal bone that disrupts the biomechanics of the thumb, particularly affecting grip and mobility by compromising the integrity of the CMC joint. This disruption results in joint instability and altered load distribution [73, 76]. The ability to perform opposition and pinching—critical for practical grip function—is notably impaired [73, 76].

The CMC joint supports flexion, extension, abduction, adduction, and opposition, and fractures involving this joint often lead to stiffness, reduced range of motion (ROM), and a higher risk of posttraumatic arthritis [76, 77]. Surgical interventions such as open reduction and internal fixation (ORIF) or external fixation are often required to restore anatomical alignment and achieve stable fixation. External fixation may be preferable in cases of significant comminution or associated soft tissue injury [73, 77].

The fracture typically results from axial loading of the thumb, producing a Y- or T-shaped pattern that disrupts the articular surface of the CMC joint, leading to instability [69, 73]. Multiple fragments complicate joint restoration and stability, significantly impacting grip strength by compromising the mechanical efficiency of the thumb muscles [69, 73] (Fig. 23.7).

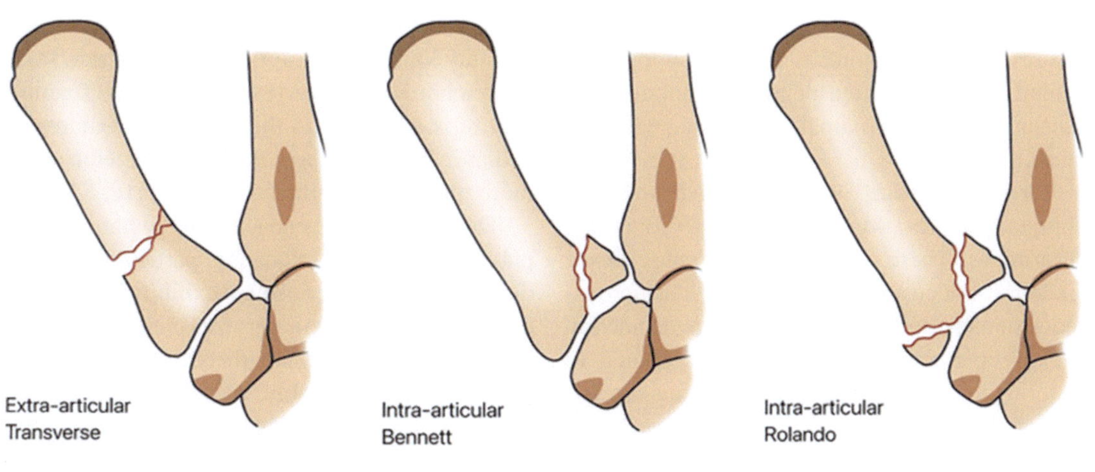

Fig. 23.7 *Comparison of Bennett and Rolando fractures*—three types of first metacarpal base fractures: a transverse extraarticular fracture, an intraarticular Bennett fracture with a single displaced fragment, and a comminuted intraarticular Rolando fracture. Treatment varies based on complexity, often requiring immobilization or surgery

Metacarpal fractures commonly result from punching, axial loading, or direct trauma. Punching injuries frequently cause neck fractures of the little finger metacarpal and shaft fractures of the ring finger metacarpal [78]. Axial loading can produce fractures at different points along the metacarpal, depending on the force and angle of impact. Direct trauma often results in comminuted or transverse fracture patterns [75].

Rotational deformities in metacarpal fractures are critical to assess, as even minimal misalignment can impair hand function, particularly grip strength and dexterity. Clinical evaluation involves checking rotational alignment by having the patient flex their fingers into a fist, noting any overlap or scissoring of the digits. Misalignment affects the customary excursion of flexor and extensor tendons, increasing the workload during finger movements [79, 80]. Surgical intervention is often necessary to correct rotational deformities, ensuring proper alignment and preventing long-term deficits [79, 80]. Studies indicate significant declines in the mechanical efficiency of the flexor system when angulation exceeds 30° [79, 80].

Phalangeal fractures often result from crushing or torsional injuries. Crushing injuries typically cause comminuted fractures with extensive soft tissue damage, leading to tendon adhesions and impaired gliding [81, 82]. This disruption can result in malalignment and shortening, particularly affecting extensor tendon function and causing lag at the proximal interphalangeal (PIP) joint [81, 82].

Torsional injuries usually produce spiral or oblique fractures, which may lead to rotational deformities. These deformities can cause flexor tendons to rotate abnormally within their fibroosseous tunnels, impairing alignment and function [81]. Proper skeletal realignment is crucial to restoring tendon function and preventing complications such as adhesions and deformities [81].

Proximal phalangeal fractures, in particular, can disrupt extensor tendon function, often leading to extensor lag due to adhesions or malunion with angulation or shortening [81]. These disruptions may require compensatory actions by the extrinsic extensors or intrinsic muscles, highlighting the importance of precise realignment during treatment [81].

Pediatric Fractures

Greenstick fractures occur due to the distinctive properties of children's bones, which exhibit greater elasticity and plasticity than adult bones [81]. This elasticity arises from a higher ratio of immature to mature collagen crosslinks in the bone matrix, enabling more significant deformation before fracture [81]. When a bending force is applied, the bone bends rather than breaking completely, disrupting the cortical layer on one side while the other remains intact, resulting in a greenstick fracture [82].

Pediatric bones have a lower modulus of elasticity and reduced bending strength than adult bones, allowing them to absorb more energy and undergo more significant deflection before fracturing [82]. Finite element analysis reveals that the failure moment of pediatric bones under bending and torsional loads increases with age and body mass. Consequently, younger bones, with their lower strength and higher deformability, are more prone to greenstick fractures [83].

Torus, or buckle fractures of the distal radius, are caused by axial compression forces. They are characterized by localized bulging of the cortex without complete disruption, making these fractures inherently stable [84]. This unique fracture pattern is attributed to the distinctive features of pediatric bone, such as a thicker periosteum and a more pliable cortex than adult bone [10, 84].

The remodeling capacity of pediatric bone plays a critical role in healing, particularly in the distal radius. The periosteum exhibits high osteogenic potential, and the growth plate contributes significantly to bone lengthening and structural adaptation [84]. Notably, even substantial angulations in distal radius fractures can remodel over time, with a reported remodeling rate of approximately 2° per month in the coronal plane [84].

The distal radius is especially prone to torus fractures due to the characteristics of the metaphyseal cortex, which is relatively thin and less

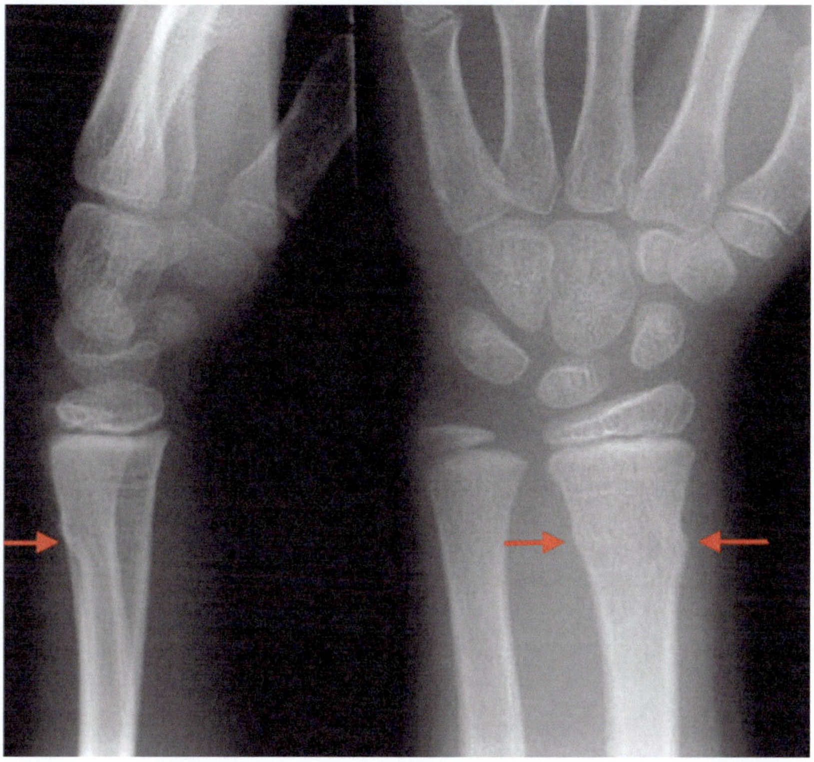

Fig. 23.8 *Comparison of Buckle (torus) fracture and Greenstick fractures*—X-ray image of the distal radius and ulna, illustrating a *torus (buckle) fracture*. The red arrows indicate areas of cortical bulging, a characteristic feature of this type of fracture, which is common in pediatric patients due to the elasticity of their bones. Torus fractures typically occur from axial loading forces, such as a fall on an outstretched hand, and demonstrate subtle, nondisplaced cortical irregularities without complete bone breakage

dense during growth [85]. This susceptibility is further compounded by mechanical challenges, such as increasing body weight and forearm length, which outpace the growth in bone strength during childhood [85]. The metaphyseal cortex undergoes dynamic remodeling during this period, characterized by high rates of periosteal resorption and endocortical apposition, enabling the bone to adapt to mechanical stresses [85].

Healing in torus fractures is primarily driven by the remodeling capacity of pediatric bone. The periosteum's osteogenic potential and the growth at the distal radius physis allow for substantial correction of angulations and deformities over time, with studies confirming remodeling rates of about 2° per month in the coronal plane [84] (Fig. 23.8).

Complex Wrist Injuries

Perilunate dislocation fractures result from high-energy trauma, such as falls from significant heights or motor vehicle accidents [86]. These injuries involve extensive disruption of the lunate and other carpal bones, leading to carpal instability [86]. The mechanism typically starts with a fall on an outstretched hand, disrupting the scapholunate interosseous ligament (SLIL). This initial injury progresses to involve other critical ligaments, such as the dorsal intercarpal (DIC) and radiolunate (RL) ligaments [64, 87, 88]. The resulting ligamentous disruptions alter the kinematics of the carpal bones—particularly the scaphoid, lunate, and capitate—culminating in significant carpal instability [64, 88].

The lunate remains normal in these fractures, while the other carpal bones dislocate dorsally. This dislocation is often accompanied by fractures of the scaphoid, capitate, or other carpal bones [86, 89].

The biomechanics of a Galeazzi fracture involve both load transmission and rotational forces. This injury is characterized by a radial shaft fracture and a dislocation of the distal radio-ulnar joint (DRUJ) [90]. The injury mechanism involves the application of an axial load to the forearm, coupled with a rotational force.

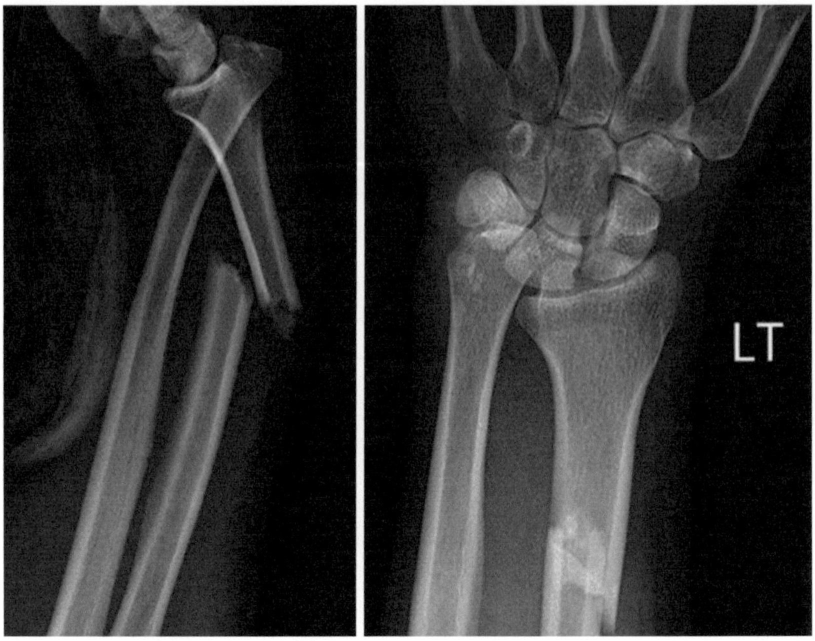

Fig. 23.9 *Galeazzi fracture*—radiographic images of a Galeazzi fracture. The left-side image illustrates a midshaft radial fracture with an intact ulna, while the right-side image highlights the associated DRUJ disruption

Together, these forces result in a radial shaft fracture and disruption of the DRUJ [90]. The axial load is transmitted through the radius, while the rotational force causes DRUJ dislocation by disrupting stabilizing ligamentous structures, particularly the triangular fibrocartilage complex (TFCC) [61, 90]. The TFCC, which includes the central disk and distal radioulnar ligaments and other associated structures, is vital in maintaining DRUJ stability by providing a cushioning effect and preventing excessive movement [61, 91] (Fig. 23.9).

Clinical and Surgical Considerations

Effective management of hand fractures relies on accurate diagnostics, tailored treatment approaches, and a focus on rehabilitation and recovery to ensure optimal functional outcomes. Imaging modalities such as x-rays are the cornerstone for initial fracture assessment, providing essential details on fracture type, alignment, and displacement [61, 90]. They are usually sufficient for initial assessment and treatment planning in many cases.

Advanced imaging, including computed tomography (CT) and magnetic resonance imaging (MRI), is particularly valuable in evaluating complex fractures, intraarticular involvement, or associated soft tissue injuries [91]. CT imaging is recommended when initial radiographs are normal. Still there is a high clinical suspicion of a fracture or when more detailed imaging is needed to assess complex fractures intraarticular involvement or to aid in preoperative planning. The American College of Radiology (ACR) suggests CT scans for better visualization of fracture displacement, depression, and comminution, especially in intraarticular fracture [92].

The severity and nature of the fracture dictate treatment options. Nonoperative strategies, such as casting and splinting, are appropriate for stable fractures and emphasize early initiation of rehabilitation to prevent stiffness and maintain range of motion [61, 93]. Operative interventions, including internal fixation, external fixation, and bone grafting, are reserved for unstable or comminuted fractures, ensuring anatomical alignment and stability while facilitating early mobilization [61, 93]. Surgical options include percutaneous Kirschner wires (K-wires), screws and plates for internal fixation, external fixation for complex or comminuted fractures, and bone grafting in cases of nonunion or significant bone loss [61, 94].

Posttreatment rehabilitation is critical, leveraging biomechanical principles to guide healing and restore function. Combined with structured therapy, early controlled mobilization minimizes complications such as joint stiffness and tendon adhesions, promoting a return to preinjury activities [1, 94]. Professional guidelines from the American Academy of Orthopaedic Surgeons and the American Society for Surgery of the Hand emphasize the importance of individualized treatment plans, not just as a recommendation but as a key strategy and early mobilization to optimize outcomes in hand fracture management [17].

Types of Orthopedic Plates

The type of plates used for surgical correction of these fractures is tailored depending on their desired function: protection (neutralization), compression, bridging, and buttress (antiglide) [95, 96, 111]. The plates are designed for their anatomic location, as they must fit the shape of the bone [95, 96, 111]. They are precontoured to fit the region where they will be placed [95, 96, 111].

Dynamic Compression Plates (DCP)

Dynamic compression plates (DCPs) are designed to compress fracture sites, ensuring absolute stability, a principle in fracture management that aims to eliminate interfragmentary motion and promote primary bone healing [95, 96, 111]. They are effective for two-part fractures, achieving compression via eccentrically positioned screws and a prebend applied to the plate [95, 96, 111]. Research shows that a 1.5 mm prebend offers uniform pressure distribution, while more significant prebends reduce compressive force [97]. External tensioning devices further enhance compression [98].

DCPs are ideal for simple transverse and short oblique fractures requiring absolute stability and primary bone healing. They are instrumental in osteoporotic bone, where controlling interfragmentary motion is crucial [99]. Limitations include reduced compressive force with larger prebends, reliance on precise screw placement, and diminished effectiveness with initial fracture gaps [97]. Excessive rigidity from multiple screw-locking elements may also impair callus formation in osteoporotic bone [99].

Locking Compression Plates (LCP)

Locking plates do not require complete bone contact because their screws lock into the plate, pulling the bone toward it [95, 96, 111]. This fixed-angle construct uses locking screws that securely thread into the plate, creating angular stability and reducing the risk of screw loosening, especially in osteoporotic bone or complex fractures [100].

Compared to traditional plates, locking plates offer enhanced stability by resisting torsional and bending forces, minimizing implant failure risk [100]. Their design preserves the periosteal blood supply by reducing bone contact and promoting better biological healing [95, 96, 111]. Locking plates are versatile, supporting various configurations and decreasing the likelihood of screw loosening, thus lowering the risk of inflammatory complications [95, 96, 111].

Locking compression plates are ideal for osteoporotic fractures due to their superior fixation in poor-quality bone [100], comminuted fractures where bridging enables indirect reduction and callus formation [101], periprosthetic fractures requiring stabilization around implants [100], and metaphyseal or diaphyseal fractures, where stable fixation minimizes soft tissue disruption [102].

Bridge Plates

Bridge plates are typically used in cases where intramedullary nailing or conventional plate fixation is unsuitable, such as in multifragmentary long bone fractures [95, 96, 111]. These plates provide fixation of the two main fragments, ensuring proper length, rotational alignment, and transverse alignment [95, 96, 111]. By leaving the fracture site undisturbed, this technique allows the bone to heal through callus formation [95, 96, 111].

The biomechanics of bridge plating involve the use of long plates that span the fracture zone, with screws placed in the proximal and distal fragments while avoiding the comminuted zone [103]. This approach permits controlled micromotion at the fracture site, a critical factor for callus formation and subsequent fracture healing [103]. Bridge plates are particularly effective in managing highly comminuted fractures where direct anatomical reduction is not feasible. They are commonly employed in treating comminuted fractures of the humerus, femur, and forearm [104].

Specialty Plates

Specialty plates, mainly anatomical plates, have transformed fracture management in regions like the distal radius. These plates are precontoured to match the bone's anatomy, providing a better fit and improved stability while minimizing soft tissue irritation. The Synthes VA-LCP distal two-column-radius plate has shown the best anatomical fit among tested options [105].

The development of three-dimensional contoured plates, customized to a patient's anatomy using advanced imaging and printing, offers further advancements. These plates improve fit, reduce surgical time, and enhance outcomes. Studies on three-dimensional printed plates for distal radius and forearm osteotomies report significant improvements in grip strength, range of motion, and patient-reported outcomes one year postoperatively [105].

Anatomical and three-dimensional contoured plates are particularly effective for complex fractures and malunions, providing stable fixation while preserving fracture site biology. They are commonly used in areas like the distal radius, humerus, and femur, where precise anatomical fit is critical [106]. These innovations have significantly advanced orthopedic surgery, offering tailored solutions for complex fracture patterns and improving patient outcomes.

Types of Orthopedic Screws

Cortical Screws

Cortical screws are primarily designed for fixation in dense cortical bone, characterized by a smaller thread pitch and higher thread density for improved purchase [107]. Their smaller core diameter and higher thread count enhance pullout strength and resistance to torsional forces [107]. Stainless steel cortical screws demonstrate superior mechanical properties, including the highest mean yield point and maximal load at failure in three-point bending tests, outperforming titanium and bioabsorbable screws [107].

These screws are commonly used in orthopedic surgeries for fracture fixation in long bones such as the femur, tibia, and humerus [108]. In spinal surgeries, cortical bone trajectory (CBT) screws are employed to achieve better fixation in the vertebrae's dense cortical bone [108].

Cancellous Screws

Cancellous screws are commonly used in the metaphyseal regions of long bones (e.g., distal femur, proximal tibia, distal radius) and the epiphyseal areas, especially in pediatric orthopedics for procedures like epiphysiodesis to address leg-length discrepancies and angular deformities [109]. Their larger outer diameter, deeper threads, and broader thread pitch maximize engagement with the trabecular structure of softer cancellous bone, providing superior pullout strength and resistance to displacement in these regions [110].

While cortical screws are ideal for denser diaphyseal bone, cancellous screws are better suited for metaphyseal and epiphyseal regions, particularly in osteoporotic bone, where their design ensures more stable fixation [110].

Locking Screws

Locking screws are designed to integrate with locking plates, creating fixed-angle constructs by threading the screw heads into corresponding threaded holes in the plate. This mechanism ensures the screws lock into the plate, providing stability without relying on friction between the plate and bone.

Locking screws offer biomechanical advantages, including fixed-angle stability, which is particularly beneficial in the osteoporotic bone where traditional screw purchase may be inadequate [95, 96, 103, 111]. By locking into the plate, the screws prevent toggling and maintain alignment. Additionally, they distribute the load, reduce the risk of periprosthetic fractures, and allow some angulation during insertion while preserving locking strength. This flexibility is advantageous in complex fracture patterns where optimal screw placement is challenging [112].

Lag Screws

Lag screws are used to achieve interfragmentary compression, providing stable and rigid fixation by compressing fracture fragments together. They are indicated for intraarticular fractures, such as supracondylar distal femoral fractures, to stabilize the intraarticular component and ensure proper joint surface alignment [113]. Oblique and spiral fractures benefit from the compressive forces generated by lag screws, which help maintain reduction and stability [114]. In metaphyseal fractures, lag screws are often used alongside plates to enhance fixation stability in the metaphyseal regions [114].

The lag screw technique involves drilling through both fracture fragments, overdrilling the near cortex to create a gliding hole, and then inserting the screw to compress the fragments. This method significantly increases interfragmentary compression compared to positional screws [115]. The optimal compressive force is achieved when the screw is applied at an angle of 60–90° relative to the fracture plane [115]. Lag screws are often combined with locking plates to minimize shear movements and provide more predictable fracture gap motion, improving overall fixation stability [114]. Reduction clamps are commonly used initially to align and compress the fracture fragments before definitive lag screw fixation [113].

Cannulated Screws

Cannulated screws provide significant advantages due to their minimally invasive approach and precision. Their percutaneous insertion minimizes soft tissue dissection, reducing the risks of infection and wound complications [116]. This technique is associated with shorter operative times and less intraoperative blood loss than traditional open methods ([116, 117]. The cannulated design allows for precise guide wire placement, enhancing the accuracy of screw insertion and reducing the risk of malposition [118]. Moreover, the stability offered by cannulated screws facilitates early mobilization and functional recovery, improving patient outcomes [118].

Cannulated screws have diverse applications in orthopedic surgery. For calcaneal fractures, particularly Sanders II and III types, percutaneous cannulated screws achieve clinical outcomes comparable to traditional plating while reducing complications [116, 118]. They are also effective in the closed reduction and percutaneous fixation of tibial pilon fractures, avoiding extensive soft tissue dissection [117]. In distal radius fractures, these screws allow for percutaneous fixation, enabling an immediate range of motion and rapid functional recovery [119]. Similarly, for proximal humeral fractures, cannulated screws provide stable fixation with minimal soft tissue disruption [120].

These advantages and varied applications highlight the value of cannulated screws in orthopedic procedures, offering reduced morbidity and improved patient outcomes.

Bone Healing and Fracture Fixation

Primary Bone Healing

Primary bone healing occurs without forming a callus, directly bridging the fracture site with bone tissue [121]. It bypasses the cartilaginous phase in secondary healing, regenerating lamellar bone and Haversian systems without remodeling [121]. This process relies on close apposition of bone fragments and rigid fixation, which minimizes interfragmentary strain and enables osteons to bridge the fracture [121]. Coordinated osteoclast and osteoblast activity facilitates this: osteoclasts create cutting cones, and osteoblasts deposit new bone matrix, restoring original bone structure and mechanical properties [121].

Rigid fixation provides absolute stability, essential for primary bone healing without callus formation [106]. This stability minimizes strain, allowing osteoclasts and osteoblasts to directly form new bone tissue and restore mechanical properties [106].

For example, in a distal radius fracture, a volar locking plate can achieve compression and direct healing [122]. Contoured to fit the volar aspect of the distal radius, the plate is fixed with screws, including eccentrically placed ones, to generate interfragmentary compression critical for healing [122].

Secondary Bone Healing

Secondary bone healing, the more common form of bone repair, involves callus formation [121]. It occurs under relative stability at the fracture site, allowing interfragmentary motion [121]. The process includes hematoma formation, inflammation, soft callus formation, complex callus formation, and remodeling [121]. Hematoma triggers an inflammatory response that recruits mesenchymal stem cells, which differentiate into chondrocytes and osteoblasts to form a soft callus of cartilage and fibrous tissue [121]. This soft callus is replaced by a hard callus of woven bone, which is eventually remodeled into lamellar bone [121].

A controlled interfragmentary motion under relative stability stimulates the biological processes for callus formation and remodeling [123]. Bridge plating, used for comminuted fractures, provides relative stability by spanning the fracture with screws in the proximal and distal fragments, preserving blood supply and enabling micromotion for callus formation [123].

Far cortical locking constructs in bridge plating offer flexible fixation and near-parallel interfragmentary motion, enhancing callus formation and healing strength compared to standard locked plating [123]. Dynamic plating techniques, such as near cortex over drilling, can increase axial motion, promote robust callus formation, and reduce nonunion rates [123].

Fixation Techniques in Orthopaedics

Plating Techniques

Neutralization and compression plating are key techniques in fracture fixation within orthopedic surgery. Neutralization plating serves to protect the primary fixation, such as lag screws, from forces like shear, bending, and torsion. While this technique does not apply compression across the fracture site, it effectively "neutralizes" the forces acting on the fracture, allowing the primary fixation to maintain stability. This method is beneficial in comminuted fractures where direct compression is impossible [124]. Neutralization plating is commonly employed to fix comminuted fractures of the metacarpals and phalanges. For instance, in the treatment of metacarpal fractures, a neutralization plate can be used to safeguard lag screws that provide interfragmentary compression.

In contrast, compression plating aims to achieve absolute stability by applying compression across the fracture site, facilitating primary bone healing through direct bone contact [125]. This technique is especially suited for simple transverse or short oblique fractures. The dynamic compression plate (DCP) is a widely used device in this method, enabling controlled compression through eccentric screw placement. Additionally, external devices such as Verbrugge clamps or articulated tensioning devices can augment plate compression, further enhancing the compressive forces applied [98, 124].

Screw Fixation Techniques

The principles of lag screw application in orthopedic hand surgery focus on achieving interfragmentary compression to promote primary bone healing. Key principles include overdrilling the near cortex, achieving compression, optimal screw angle, avoiding overtightening, and ensuring biomechanical stability. Overdrilling the near cortex to match the screw's outer thread diameter allows the screw to glide through this cortex and engage only the far cortex, pulling bone fragments together as the screw is tightened [126]. This compression facilitates primary bone healing through direct contact [115].

The optimal screw angle maximizes compressive force, with lag screws being most effective when inserted perpendicular to the fracture plane, typically at 60°–90° [115]. Overtightening must be avoided to prevent thread stripping and compromising fixation [127]. Maintaining a balance between interfragmentary compressive force (IFCF) and pullout strength (POS) is critical for stable fixation [127]. Lag screws provide superior biomechanical stability, especially at angles of 60° and 90°, aiding fracture reduction and enabling early mobilization [115].

An example of this technique is its use in fixing long oblique fractures of the proximal phalanx, where interfragmentary compression is achieved using lag screws.

Special Technique

Innovations in orthopedic fixation techniques include using bone grafts to enhance stability and promote healing, particularly in bone defects or poor-quality cases. In hand surgery, bone grafts may be autologous (harvested from the patient), allogenic (from a donor), or synthetic. Autologous grafts, commonly taken from the distal radius or iliac crest, provide osteoconductive, osteoinductive, and osteogenic properties critical for bone healing. For instance, autologous bone grafts are frequently used to fill defects and support lag screw fixation in scaphoid nonunions [128].

Synthetic substitutes, such as calcium phosphate cement or hydroxyapatite, improve mechanical stability by enhancing the bone-implant interface [129]. Calcium sulfate/hydroxyapatite (CaS/HA) biomaterials have been shown to significantly increase the peak extraction force of lag screws, making them practical for fragile bone [129]. Cement augmentation with materials like polymethylmethacrylate (PMMA) enhances lag screw anchorage in osteoporotic bone, improving interfragmentary compression and maintaining fixation stability by reducing relaxation effects [130]. These techniques are beneficial for complex fractures, nonunions, and osteoporotic bone.

Conclusion

The future of hand fracture treatment, particularly in bone grafting and fixation, is poised for significant advancements driven by innovations in materials and techniques and a deeper understanding of bone biology. Biomaterial and synthetic graft research is increasingly focused on developing synthetic grafts that mimic natural bone's mechanical and biological properties. These materials aim to enhance osteoconductivity, promote better integration with host bone, and reduce the risks of complications associated with autografts, such as donor site morbidity.

Using stem cells in conjunction with bone grafts holds promise for enhancing healing in complex fractures. Mesenchymal stem cells can differentiate into osteoblasts, facilitating bone regeneration and repair. Combining stem cells with scaffolding materials may improve the effectiveness of grafts and reduce healing time.

Emerging technologies like three-dimensional bioprinting allow customized grafts tailored to patients' anatomical and defect characteristics. This technology could improve the compatibility and functionality of grafts in repairing bone defects.

Advances in surgical instrumentation and techniques have led to the development of minimally invasive approaches. These techniques reduce soft tissue trauma and can enhance recovery times, decrease complications, and improve overall outcomes for patients with hand fractures.

New constructs in internal fixation, such as bioresorbable plates and screws, are gaining traction. These devices gradually dissolve in the body, eliminating the need for removal surgery and promoting natural healing processes without the burden of permanent hardware.

External fixation methods are also evolving. Improved designs for external fixation devices allow for better patient stability and mobility, facilitating early rehabilitation while maintaining fixation. Adjustable and modular external fixators are being developed to adapt to each patient's needs.

Integrating robotic technology in orthopedic surgery is expected to enhance precision in the placement of fixation devices. Robotics can aid in achieving optimal alignment and stabilize complex fractures more reliably than traditional techniques.

The future of hand fracture treatment will heavily emphasize personalized approaches. Advances in imaging and biomarker identification may allow for tailored treatment plans considering individual patient factors such as age, activity level, and specific fracture characteristics.

With improved fixation and grafting strategies, rehabilitation protocols are refined to focus on early mobilization while ensuring stability. Innovations in physical therapy, including virtual reality and biofeedback, may enhance recovery by making rehabilitation more engaging and effective.

In conclusion, the future of hand fracture treatment will benefit from advancements in material science, surgical techniques, and rehabilitation strategies, paving the way for improved patient outcomes. These will enhance the healing process and aim to restore function more efficiently and effectively.

References

1. Jones NF, Jupiter JB, Lalonde DH. Common fractures and dislocations of the hand. Plast Reconstr Surg. 2012;130(5):722e–36e. https://doi.org/10.1097/PRS.0b013e318267d67a. PMID: 23096627
2. Forman J, Perry B, Alai A, Freilich A, Salzar R, Walilko T. Injury tolerance of the wrist and distal forearm to impact loading onto outstretched hands. J Trauma Acute Care Surg. 2014;77(3 Suppl 2):S176–83. https://doi.org/10.1097/TA.0000000000000329. PMID: 25159352
3. Sosa EM, Moure MM. Simulation of low-energy impacts on the human hand for prediction of peak reaction forces and bone fracture. J Biomech. 2023;160:111813. https://doi.org/10.1016/j.jbiomech.2023.111813. Epub 2023 Sep 28. PMID: 37778278.
4. Rahmati M, Haffner M, Lee MA, Leach JK, Saiz AM. The critical impact of traumatic muscle loss on fracture healing: basic science and clinical aspects. J Orthop Res. 2024;42(2):249–58. https://doi.org/10.1002/jor.25746. Epub 2023 Nov 29. PMID: 37990953.
5. Soubeyrand M, Assabah B, Bégin M, Laemmel E, Dos Santos A, Crézé M. Pronation and supination of the hand: anatomy and biomechanics. Hand Surg Rehabil. 2017;36(1):2–11. https://doi.org/10.1016/j.hansur.2016.09.012. Epub 2016 Oct 27. PMID: 28137437.
6. Padmore CE, Chan AHW, Langohr GDG, Johnson JA, Suh N. The effect of forearm position on wrist joint biomechanics. J Hand Surg Am. 2021;46(5):425.e1–425.e10. https://doi.org/10.1016/j.jhsa.2020.10.017. Epub 2020 Dec 7. PMID: 33303325.
7. Badida R, Akhbari B, Vutescu E, Moore DC, Wolfe SW, Crisco JJ. The role of scapholunate interosseous, dorsal intercarpal, and radiolunate ligaments in wrist biomechanics. J Biomech. 2021;125:110567. https://doi.org/10.1016/j.jbiomech.2021.110567. Epub 2021 Jun 12. PMID: 34246909; PMCID: PMC8355209
8. Johnston JD, McDonald MP, Kontulainen SA. Off-axis loads cause distal radius failure at lower magnitudes than axial loads: a side-to-side experimental study. J Orthop Res. 2020;38(8):1688–92. https://doi.org/10.1002/jor.24601. Epub 2020 Jan 31. PMID: 31989687.
9. Buchanan D, Ural A. Finite element modeling of the influence of hand position and bone proper-

ties on the Colles' fracture load during a fall. J Biomech Eng. 2010;132(8):081007. https://doi.org/10.1115/1.4001681. PMID: 20670056

10. Patel DS, Statuta SM, Ahmed N. Common fractures of the radius and ulna. Am Fam Physician. 2021;103(6):345–54. PMID: 33719378
11. Chiu J, Robinovitch SN. Prediction of upper extremity impact forces during falls on the outstretched hand. J Biomech. 1998;31(12):1169–76. https://doi.org/10.1016/s0021-9290(98)00137-7. PMID: 9882050
12. McNemar TB, Howell JW, Chang E. Management of metacarpal fractures. J Hand Ther. 2003;16(2):143–51. https://doi.org/10.1016/s0894-1130(03)80009-1. PMID: 12755165
13. Steiner M, Claes L, Ignatius A, Simon U, Wehner T. Disadvantages of interfragmentary shear on fracture healing--mechanical insights through numerical simulation. J Orthop Res. 2014;32(7):865–72. https://doi.org/10.1002/jor.22617. Epub 2014 Mar 20. PMID: 24648331.
14. Aitken S, Court-Brown CM. The epidemiology of sports-related fractures of the hand. Injury. 2008;39(12):1377–83. https://doi.org/10.1016/j.injury.2008.04.012. Epub 2008 Jul 25. PMID: 18656191.
15. Wang H, Liu H, Wu J, Li C, Zhou Y, Liu J, Ou L, Xiang L. Age, gender, and etiology differences of sports-related fractures in children and adolescents: a retrospective observational study. Medicine (Baltimore). 2019;98(4):e13961. https://doi.org/10.1097/MD.0000000000013961. PMID: 30681556; PMCID: PMC6358360.
16. Sun X, Wu W, Zhang R, Qu H, Wang J, Xu K, Fang L, Xu L, Jiang R. Mechanical response and in-situ deformation mechanism of cortical bone materials under combined compression and torsion loads. PLoS One. 2022;17(7):e0271301. https://doi.org/10.1371/journal.pone.0271301. PMID: 35895673; PMCID: PMC9328520.
17. Kamal RN, Shapiro LM. American Academy of Orthopaedic Surgeons/American Society for Surgery of the hand clinical practice guideline summary management of distal radius fractures. J Am Acad Orthop Surg. 2022;30(4):e480–6. https://doi.org/10.5435/JAAOS-D-21-00719. PMID: 35143462; PMCID: PMC9196973.
18. Wagner DW, Lindsey DP, Beaupre GS. Replicating a Colles fracture in an excised radius: revisiting testing protocols. J Biomech. 2012;45(6):997–1002. https://doi.org/10.1016/j.jbiomech.2012.01.014. Epub 2012 Jan 26. PMID: 22281409.
19. Troy KL, Grabiner MD. Off-axis loads cause distal radius failure at lower magnitudes than axial loads: a finite element analysis. J Biomech. 2007;40(8):1670–5. https://doi.org/10.1016/j.jbiomech.2007.01.018. Epub 2007 Mar 26. PMID: 17368466; PMCID: PMC4254763
20. Tang JB, Chen YR. In vivo changes in contact regions of the radiocarpal joint during wrist hyperextension. J Hand Surg Am. 2012;37(11):2257–62. https://doi.org/10.1016/j.jhsa.2012.08.030. PMID: 23101521
21. Augat P, Iida H, Jiang Y, Diao E, Genant HK. Distal radius fractures: mechanisms of injury and strength prediction by bone mineral assessment. J Orthop Res. 1998;16(5):629–35. https://doi.org/10.1002/jor.1100160517. PMID: 9820289
22. Matsuura Y, Rokkaku T, Kuniyoshi K, Takahashi K, Suzuki T, Kanazuka A, Akasaka T, Hirosawa N, Iwase M, Yamazaki A, Orita S, Ohtori S. Smith's fracture generally occurs after falling on the palm of the hand. J Orthop Res. 2017;35(11):2435–41. https://doi.org/10.1002/jor.23556. Epub 2017 Apr 7. PMID: 28262985.
23. Caldwell RA, Shorten PL, Morrell NT. Common upper extremity fracture eponyms: a look into what they mean. J Hand Surg Am. 2019;44(4):331–4. https://doi.org/10.1016/j.jhsa.2018.07.012. Epub 2018 Sep 18. PMID: 30241976.
24. Erhart S, Toth S, Kaiser P, Kastenberger T, Deml C, Arora R. Comparison of volarly and dorsally displaced distal radius fracture treated by volar locking plate fixation. Arch Orthop Trauma Surg. 2018;138(6):879–85. https://doi.org/10.1007/s00402-018-2925-x. Epub 2018 Apr 25. PMID: 29696363.
25. Sendher R, Ladd AL. The scaphoid. Orthop Clin North Am. 2013;44(1):107–20. https://doi.org/10.1016/j.ocl.2012.09.003. PMID: 23174330
26. Kobayashi M, Garcia-Elias M, Nagy L, Ritt MJ, An KN, Cooney WP, Linscheid RL. Axial loading induces rotation of the proximal carpal row bones around unique screw-displacement axes. J Biomech. 1997;30(11–12):1165–7. https://doi.org/10.1016/s0021-9290(97)00080-8. PMID: 9456385
27. Morsy M, Sabbagh MD, van Alphen NA, Laungani AT, Kadar A, Moran SL. The vascular anatomy of the scaphoid: new discoveries using micro-computed tomography imaging. J Hand Surg Am. 2019;44(11):928–38. https://doi.org/10.1016/j.jhsa.2019.08.001. Epub 2019 Sep 19. PMID: 31543293.
28. Ko JH, Pet MA, Khouri JS, Hammert WC. Management of scaphoid fractures. Plast Reconstr Surg. 2017;140(2):333e–46e. https://doi.org/10.1097/PRS.0000000000003558. PMID: 28746289
29. Kawamura K, Chung KC. Treatment of scaphoid fractures and nonunions. J Hand Surg Am. 2008;33(6):988–97. https://doi.org/10.1016/j.jhsa.2008.04.026. PMID: 18656779; PMCID: PMC4405116
30. Schmidle G, Ebner HL, Klima G, Pfaller K, Fritz J, Hoermann R, Gabl M. Time-dependent changes in the bone healing capacity of scaphoid fractures and non-unions. J Anat. 2018;232(6):908–18. https://doi.org/10.1111/joa.12795. Epub 2018 Feb 27. PMID: 29488208; PMCID: PMC5979627

31. Gray RRL, Halpern AL, King SR, Anderson JE. Scaphoid fracture and nonunion: new directions. J Hand Surg Eur Vol. 2023;48(2_suppl):4S–10S. https://doi.org/10.1177/17531934231165419. PMID: 37704024.
32. Li NY, Dennison DG, Shin AY, Pulos NA. Update to management of acute scaphoid fractures. J Am Acad Orthop Surg. 2023;31(15):e550–60. https://doi.org/10.5435/JAAOS-D-22-01210. Epub 2023 Jun 16. PMID: 37332224.
33. Guo RC, Cardenas JM, Wu CH. Triquetral fractures overview. Curr Rev Musculoskelet Med. 2021;14(2):101–6. https://doi.org/10.1007/s12178-021-09692-w. Epub 2021 Jan 23. PMID: 33483875; PMCID: PMC7991068.
34. Becce F, Theumann N, Bollmann C, Omoumi P, Richarme D, Guerini H, Campagna R, Meuli R, Drapé JL. Dorsal fractures of the triquetrum: MRI findings with an emphasis on dorsal carpal ligament injuries. AJR Am J Roentgenol. 2013;200(3):608–17. https://doi.org/10.2214/AJR.12.8736. Erratum in: AJR Am J Roentgenol. 2013;201(3):698. PMID: 23436851.
35. Weintraub MD, Hansford BG, Stilwill SE, Allen H, Leake RL, Hanrahan CJ, Chan BY, Soltanolkotabi M, Kobes P, Mills MK. Avulsion injuries of the hand and wrist. Radiographics. 2020;40(1):163–80. https://doi.org/10.1148/rg.2020190085. PMID: 31917655
36. Kamal RN, Rainbow MJ, Akelman E, Crisco JJ. In vivo triquetrum-hamate kinematics through a simulated hammering task wrist motion. J Bone Joint Surg Am. 2012;94(12):e85. https://doi.org/10.2106/JBJS.J.01644. PMID: 22717837; PMCID: PMC3368495.
37. Garçon C, Degeorge B, Coulet B, Lazerges C, Chammas M. Perilunate dislocation and fracture dislocation of the wrist: outcomes and long-term prognostic factors. Orthop Traumatol Surg Res. 2022;108(5):103332. https://doi.org/10.1016/j.otsr.2022.103332. Epub 2022 May 21. PMID: 35609818.
38. Spiry C, Bacle G, Marteau E, Charruau B, Laulan J. Radiocarpal dislocations and fracture-dislocations: injury types and long-term outcomes. Orthop Traumatol Surg Res. 2018;104(2):261–6. https://doi.org/10.1016/j.otsr.2017.12.016. Epub 2018 Feb 8. PMID: 29428553.
39. Camus EJ, Van Overstraeten L. Kienböck's disease in 2021. Orthop Traumatol Surg Res. 2022;108(1S):103161. https://doi.org/10.1016/j.otsr.2021.103161. Epub 2021 Nov 30. PMID: 34861414.
40. Owers KL, Scougall P, Dabirrahmani D, Wernecke G, Jhamb A, Walsh WR. Lunate trabecular structure: a radiographic cadaver study of risk factors for Kienböck's disease [corrected]. J Hand Surg Eur Vol. 2010;35(2):120–4. https://doi.org/10.1177/1753193409103732. Epub 2009 Jul 20. Erratum in: J Hand Surg Eur Vol. 2010;35(2):163. PMID: 19620185.
41. Wilke B, Kakar S. Delayed avascular necrosis and fragmentation of the lunate following Perilunate dislocation. Orthopedics. 2015;38(6):e539–42. https://doi.org/10.3928/01477447-20150603-92. PMID: 26091230
42. van Alphen NA, Morsy M, Laungani AT, Kadar A, Vercnocke AJ, Lachman N, Ritman EL, Moran SL. A three-dimensional micro-computed tomographic study of the intraosseous lunate vasculature: implications for surgical intervention and the development of avascular necrosis. Plast Reconstr Surg. 2016;138(5):869e–78e. https://doi.org/10.1097/PRS.0000000000002696. PMID: 27782999
43. Ledoux P, Lamblin D, Wuilbaut A, Schuind F. A finite-element analysis of Kienbock's disease. J Hand Surg Eur Vol. 2008;33(3):286–91. https://doi.org/10.1177/1753193408090757. PMID: 18562358
44. DeGeorge BR Jr, Chawla SS, Lewallen L, Kakar S. Functional and radiographic disease progression in nonoperatively managed Kienböck disease. Plast Reconstr Surg. 2021;147(5):1117–23. https://doi.org/10.1097/PRS.0000000000007838. PMID: 33890893
45. Stahl S, Hentschel PJ, Held M, Manoli T, Meisner C, Schaller HE, Santos SA. Characteristic features and natural evolution of Kienböck's disease: five years' results of a prospective case series and retrospective case series of 106 patients. J Plast Reconstr Aesthet Surg. 2014;67(10):1415–26. https://doi.org/10.1016/j.bjps.2014.05.037. Epub 2014 Jun 2. PMID: 24947083.
46. Shin YH, Kim JK, Han M, Lee TK, Yoon JO. Comparison of long-term outcomes of radial osteotomy and nonoperative treatment for Kienböck disease: a systematic review. J Bone Joint Surg Am. 2018;100(14):1231–40. https://doi.org/10.2106/JBJS.17.00764. PMID: 30020130
47. Allan CH, Joshi A, Lichtman DM. Kienbock's disease: diagnosis and treatment. J Am Acad Orthop Surg. 2001;9(2):128–36. https://doi.org/10.5435/00124635-200103000-00006. PMID: 11281636
48. Goliver JA, Adamow JS, Goliver J. Hamate body and capitate fracture in punch injury. Am J Emerg Med. 2014;32(10):1303.e1–2. https://doi.org/10.1016/j.ajem.2014.03.050. Epub 2014 Apr 3. PMID: 24792935.
49. Yalcinkaya M, Azar N, Dogan A. A rare wrist injury: simultaneous fractures of the hamate body and scaphoid waist. Orthopedics. 2009;32(8) https://doi.org/10.3928/01477447-20090624-23. PMID: 19708624.
50. Campbell FC, Jones SW, Campbell DA. The etiology of fracture and nonunion in the hook of the hamate. J Hand Surg Eur Vol. 2024;49(9):1172–8. https://doi.org/10.1177/17531934241235803. Epub 2024 Mar 15. PMID: 38488623.
51. Guha AR, Marynissen H. Stress fracture of the hook of the hamate. Br J Sports Med. 2002;36(3):224–5.

https://doi.org/10.1136/bjsm.36.3.224. PMID: 12055122; PMCID: PMC1724513.
52. Stark HH, Jobe FW, Boyes JH, Ashworth CR. Fracture of the hook of the hamate in athletes. J Bone Joint Surg Am. 1977;59(5):575–82. PMID: 873952
53. Demirkan F, Calandruccio JH, Diangelo D. Biomechanical evaluation of flexor tendon function after hamate hook excision. J Hand Surg Am. 2003;28(1):138–43. https://doi.org/10.1053/jhsu.2003.50005. PMID: 12563651
54. Kuptniratsaikul V, Luangjarmekorn P, Kerr S, Vinitpairot C, Kitidumrongsook P. Grip strength after hamate hook excision and reconstruction surgery: a biomechanical cadaveric study. J Biomech. 2022;141:111221. https://doi.org/10.1016/j.jbiomech.2022.111221. Epub 2022 Jul 12. PMID: 35908472.
55. Geissler WB. Carpal fractures in athletes. Clin Sports Med. 2001;20(1):167–88. https://doi.org/10.1016/s0278-5919(05)70254-4. PMID: 11227704
56. Henderson CJ, Kobayashi KM. Ulnar-sided wrist pain in the athlete. Orthop Clin North Am. 2016;47(4):789–98. https://doi.org/10.1016/j.ocl.2016.05.017. PMID: 27637665
57. Wallace DR, Floyd WE 3rd. Posttraumatic arthrosis and triquetral nonunion associated with pisotriquetral subluxation in adolescent female softball players. J Hand Surg Am. 2022;47(10):1021.e1–4. https://doi.org/10.1016/j.jhsa.2021.07.032. Epub 2021 Sep 17. PMID: 34538669.
58. Beckers A, Koebke J. Mechanical strain at the pisotriquetral joint. Clin Anat. 1998;11(5):320–6. https://doi.org/10.1002/(SICI)1098-2353(1998)11:5<320::AID-CA5>3.0.CO;2-T. PMID: 9725576
59. O'Keefe KD, Werner FW, Boyette M, Palmer AK, Garcia-Elias M, Harley BJ. Effect of pisiform excision or pisotriquetral arthrodesis as a treatment for pisotriquetral arthritis: a biomechanical study. J Hand Surg Am. 2013;38(10):1913–8. https://doi.org/10.1016/j.jhsa.2013.07.021. Epub 2013 Sep 8. PMID: 24021737.
60. Kadar A, Morsy M, Sur YJ, Akdag O, Moran SL. Capitate fractures: a review of 53 patients. J Hand Surg Am. 2016;41(10):e359–66. https://doi.org/10.1016/j.jhsa.2016.07.099. Epub 2016 Aug 11. PMID: 27524693.
61. Catalano LW 3rd, Minhas SV, Kirby DJ. Evaluation and management of carpal fractures other than the scaphoid. J Am Acad Orthop Surg. 2020;28(15):e651–61. https://doi.org/10.5435/JAAOS-D-20-00062. PMID: 32732656
62. Freeland AE, Pesut TA. Oblique capitate fracture of the wrist. Orthopedics. 2004;27(3):287–90. https://doi.org/10.3928/0147-7447-20040301-13. PMID: 15058451
63. Fan YL, Xu HY, Xia MY, Zhang W, Wen HL, Gao LB, Pei YH. Biomechanical evaluation of axial-loading simulated experiment in wrist fractures: a finite element analysis. J Int Med Res. 2020;48(10):300060520966884. https://doi.org/10.1177/0300060520966884. Erratum in: J Int Med Res. 2022;50(11):3000605221133587. https://doi.org/10.1177/03000605221133587. PMID: 33135534; PMCID: PMC7780565.
64. Majima M, Horii E, Matsuki H, Hirata H, Genda E. Load transmission through the wrist in the extended position. J Hand Surg Am. 2008;33(2):182–8. https://doi.org/10.1016/j.jhsa.2007.10.018. PMID: 18294538
65. Pellegrini VD Jr. The ABJS 2005 Nicolas Andry Award: osteoarthritis and injury at the base of the human thumb: survival of the fittest? Clin Orthop Relat Res. 2005;438:266–76. https://doi.org/10.1097/01.blo.0000176968.28247.5c. PMID: 16131901
66. Bouredoucen H, Abs B, Ferreira Branco D, Buzzi M, Poletti PA, Piguet E, Beaulieu JY, Boudabbous S. Trapeziometacarpal joint imaging: normal high-resolution MRI, US and CT compared with cadaveric specimens and pathological imaging findings. Eur J Radiol. 2024;177:111561. https://doi.org/10.1016/j.ejrad.2024.111561. Epub 2024 Jun 12. PMID: 38897054.
67. Norose M, Nimura A, Tsutsumi M, Fujita K, Okawa A, Akita K. Anatomical study for elucidating the stabilization mechanism in the trapeziometacarpal joint. Sci Rep. 2022;12(1):20790. https://doi.org/10.1038/s41598-022-25355-3. PMID: 36456627; PMCID: PMC9715720.
68. Liverneaux PA, Ichihara S, Hendriks S, Facca S, Bodin F. Fractures and dislocation of the base of the thumb metacarpal. J Hand Surg Eur Vol. 2015;40(1):42–50. https://doi.org/10.1177/1753193414554357. Epub 2014 Oct 13. PMID: 25311936.
69. Wiesler ER, Chloros GD, Kuzma GR. Arthroscopy is used in the treatment of fractures in the trapezium. Arthroscopy. 2007;23(11):1248.e1–4. https://doi.org/10.1016/j.arthro.2006.07.051. Epub 2007 Jan 5. PMID: 17986420.
70. Garneti N, Tuson CE. Sagittally split fracture of trapezium associated with the subluxated carpometacarpal joint of the thumb. Injury. 2004;35(11):1172–5. https://doi.org/10.1016/j.injury.2003.09.004. PMID: 15488511
71. Botte MJ, von Schroeder HP, Gellman H, Cohen MS. Fracture of the trapezial ridge. Clin Orthop Relat Res. 1992;276:202–5. PMID: 1537153
72. Soyer AD. Fractures of the base of the first metacarpal: current treatment options. J Am Acad Orthop Surg. 1999;7(6):403–12. https://doi.org/10.5435/00124635-199911000-00006. PMID: 11505928.
73. Kang JR, Behn AW, Messana J, Ladd AL. Bennett fractures: a biomechanical model and relevant ligamentous anatomy. J Hand Surg Am. 2019;44(2):154.e1–5. https://doi.org/10.1016/j.jhsa.2018.04.024. Epub 2018 Jun 8. PMID: 29891266.

74. Cullen JP, Parentis MA, Chinchilli VM, Pellegrini VD Jr. Simulated Bennett fracture treated with closed reduction and percutaneous pinning. A biomechanical analysis of residual incongruity of the joint. J Bone Joint Surg Am. 1997;79(3):413–20. https://doi.org/10.2106/00004623-199703000-00015. PMID: 9070532
75. Lambi AG, Rowland RJ, Brady NW, Rodriguez DE, Mercer DM. Metacarpal fractures. J Hand Surg Eur Vol. 2023;48(2_suppl):42S–50S. https://doi.org/10.1177/17531934231184119. PMID: 37704027
76. Kontakis GM, Katonis PG, Steriopoulos KA. Rolando's fracture was treated by closed reduction and external fixation. Arch Orthop Trauma Surg. 1998;117(1-2):84–5. https://doi.org/10.1007/BF00703448. PMID: 9457345
77. Soong M, Got C, Katarincic J. Ring and little finger metacarpal fractures: mechanisms, locations, and radiographic parameters. J Hand Surg Am. 2010;35(8):1256–9. https://doi.org/10.1016/j.jhsa.2010.05.013. PMID: 20684925
78. Wahl EP, Richard MJ. Management of metacarpal and phalangeal fractures in the athlete. Clin Sports Med. 2020;39(2):401–22. https://doi.org/10.1016/j.csm.2019.12.002. PMID: 32115091
79. Birndorf MS, Daley R, Greenwald DP. Metacarpal fracture angulation decreases flexor mechanical efficiency in human hands. Plast Reconstr Surg. 1997;99(4):1079–83; discussion 1084–5. https://doi.org/10.1097/00006534-199704000-00024. PMID: 9091906.
80. Berteau JP, Gineyts E, Pithioux M, Baron C, Boivin G, Lasaygues P, Chabrand P, Follet H. Ratio between mature and immature enzymatic cross-links correlates with post-yield cortical bone behavior: An insight into greenstick fractures of the child fibula. Bone. 2015;79:190–5. https://doi.org/10.1016/j.bone.2015.05.045. Epub 2015 Jun 14. PMID: 26079997.
81. Currey JD, Butler G. The mechanical properties of bone tissue in children. J Bone Joint Surg Am. 1975;57(6):810–4. PMID: 1158919
82. Altai Z, Viceconti M, Offiah AC, Li X. Investigating the mechanical response of pediatric bone under bending and torsion using finite element analysis. Biomech Model Mechanobiol. 2018;17(4):1001–9. https://doi.org/10.1007/s10237-018-1008-9. Epub 2018 Mar 10. PMID: 29525976.
83. Lynch KA, Wesolowski M, Cappello T. Coronal remodeling potential of pediatric distal radius fractures. J Pediatr Orthop. 2020;40(10):556–61. https://doi.org/10.1097/BPO.0000000000001580. PMID: 32427798
84. Rauch F, Neu C, Manz F, Schoenau E. The development of metaphyseal cortex--implications for distal radius fractures during growth. J Bone Miner Res. 2001;16(8):1547–55. https://doi.org/10.1359/jbmr.2001.16.8.1547. PMID: 11499878
85. Stanbury SJ, Elfar JC. Perilunate dislocation and perilunate fracture-dislocation. J Am Acad Orthop Surg. 2011;19(9):554–62. https://doi.org/10.5435/00124635-201109000-00006. PMID: 21885701
86. Murray PM, Palmer CG, Shin AY. The mechanism of ulnar-sided perilunate instability of the wrist: a cadaveric study and 6 clinical cases. J Hand Surg Am. 2012;37(4):721–8. https://doi.org/10.1016/j.jhsa.2012.01.015. Epub 2012 Mar 6. PMID: 22397843.
87. Zhang X, Tham S, Ek ET, McCombe D, Ackland DC. Scaphoid, lunate and capitate kinematics in the normal and ligament deficient wrist: a bi-plane X-ray fluoroscopy study. J Biomech. 2023;158:111685. https://doi.org/10.1016/j.jbiomech.2023.111685. Epub 2023 Jun 19. PMID: 37573806.
88. Scalcione LR, Gimber LH, Ho AM, Johnston SS, Sheppard JE, Taljanovic MS. Spectrum of carpal dislocations and fracture-dislocations: imaging and management. AJR Am J Roentgenol. 2014;203(3):541–50. https://doi.org/10.2214/AJR.13.11680. PMID: 25148156
89. Cheah AE, Yao J. Hand fractures: indications, the tried and true and new innovations. J Hand Surg Am. 2016;41(6):712–22. https://doi.org/10.1016/j.jhsa.2016.03.007. Epub 2016 Apr 22. PMID: 27113910.
90. Ganesh Kumar N, Chung KC. An evidence-based guide for managing phalangeal fractures. Plast Reconstr Surg. 2021;147(5):846e–61e. https://doi.org/10.1097/PRS.0000000000007840. PMID: 33890906
91. Bloom JMP, Hammert WC. Evidence-based medicine: metacarpal fractures. Plast Reconstr Surg. 2014;133(5):1252–60. https://doi.org/10.1097/PRS.0000000000000095. PMID: 24776556
92. Expert Panel on Musculoskeletal Imaging, Torabi M, Lenchik L, Beaman FD, Wessell DE, Bussell JK, Cassidy RC, Czuczman GJ, Demertzis JL, Khurana B, Klitzke A, Motamedi K, Pierce JL, Sharma A, Walker EA, Kransdorf MJ. ACR appropriateness criteria® for acute hand and wrist trauma. J Am Coll Radiol. 2019;16(5S):S7–S17. https://doi.org/10.1016/j.jacr.2019.02.029. PMID: 31054760
93. Meals C, Meals R. Hand fractures: a review of current treatment strategies. J Hand Surg Am. 2013;38(5):1021–31; quiz 1031. https://doi.org/10.1016/j.jhsa.2013.02.017. PMID: 23618458.
94. Ring D, Allende C, Jafarnia K, Allende BT, Jupiter JB. Ununited diaphyseal forearm fractures with segmental defects: plate fixation and autogenous cancellous bone-grafting. J Bone Joint Surg Am. 2004;86(11):2440–5. PMID: 15523016
95. AO Foundation. Basic principles of plating. AO Surgery Reference. Accessed 20 Jan 2025. https://surgeryreference.aofoundation.org/orthopedic-trauma/adult-trauma/basic-technique/basic-principles-of-plating.

96. AO Foundation. ORIF - compression plating: principles. AO Surgery Reference. Accessed 20 Jan 2025. https://surgeryreference.aofoundation.org/orthopedic-trauma/adult-trauma/humeral-shaft/simple-fracture-oblique-30/orif-compression-plating#principles.
97. Wee H, Staub J, Koroneos Z, Kunselman A, Reid JS, Lewis GS. Mechanics of dynamic compression plate application in fracture fixation. Clin Biomech (Bristol). 2024;113:106209. https://doi.org/10.1016/j.clinbiomech.2024.106209.
98. Lucas JF, Lee MA, Eastman JG. Optimizing compression: Comparing eccentric plate holes and external tensioning devices. Injury. 2016;47(7):1461–5. https://doi.org/10.1016/j.injury.2016.04.020. Epub 2016 May 3. PMID: 27206844.
99. Cuadrado A, Yánez A, Carta JA, Garcés G. Suitability of DCPs with screw locking elements to allow sufficient interfragmentary motion to promote secondary bone healing of osteoporotic fractures. Med Eng Phys. 2013;35(6):852–9. https://doi.org/10.1016/j.medengphy.2012.08.018.
100. Cronier P, Pietu G, Dujardin C, Bigorre N, Ducellier F, Gerard R. The concept of locking plates. Orthop Traumatol Surg Res. Published online May 4, 2010. https://doi.org/10.1016/j.otsr.2010.03.008
101. Wagner M. General principles for the clinical use of the LCP. Injury. 2003;34 Suppl 2:B31–42. https://doi.org/10.1016/j.injury.2003.09.023.
102. Tan SL, Balogh ZJ. Indications and limitations of locked plating. Injury. 2009;40(7):683–91. https://doi.org/10.1016/j.injury.2009.01.003.
103. Egol KA, Kubiak EN, Fulkerson E, Kummer FJ, Koval KJ. Biomechanics of locked plates and screws. J Orthop Trauma. 2004;18(8):488–93. https://doi.org/10.1097/00005131-200409000-00003.
104. Oppermann J, Wacker M, Stein G, et al. The anatomical fit of seven different palmar distal radius plates. Arch Orthop Trauma Surg. 2014;134(10):1483–9. https://doi.org/10.1007/s00402-014-2072-y.
105. Schindele S, Oyewale M, Marks M, Brodbeck M, Herren DB. Three-dimensionally planned and printed patient-tailored plates for corrective osteotomies of the distal radius and forearm. J Hand Surg Am. 2024;49(3):277.e1–8. https://doi.org/10.1016/j.jhsa.2022.06.021.
106. Augat P, von Rüden C. Evolution of fracture treatment with bone plates. Injury. 2018;49 Suppl 1:S2–7. https://doi.org/10.1016/S0020-1383(18)30294-8.
107. Collinge CA, Stern S, Cordes S, Lautenschlager EP. Mechanical properties of small fragment screws. Clin Orthop Relat Res. 2000;(373):277–84. https://doi.org/10.1097/00003086-200004000-00034.
108. Ueno M, Sakai R, Tanaka K, et al. Should we use cortical bone screws for cortical bone trajectory? J Neurosurg Spine. 2015;22(4):416–21. https://doi.org/10.3171/2014.9.SPINE1484.
109. Normandin BM, Tennent DJ, Baldini TH, Blanchard AM, Rhodes JT. Mechanical testing of epiphysiodesis screws. Orthopedics. 2018;41(2):e240–4. https://doi.org/10.3928/01477447-20180123-01.
110. Wang T, Boone C, Behn AW, Ledesma JB, Bishop JA. Cancellous screws are biomechanically superior to cortical screws in metaphyseal bone. Orthopedics. 2016;39(5):e828–32. https://doi.org/10.3928/01477447-20160509-01.
111. AO Foundation. Insertion of locking head screw and variable angle screw. AO Surgery Reference. https://surgeryreference.aofoundation.org/orthopedic-trauma/adult-trauma/basic-technique/insertion-of-locking-head-screw-and-variable-angle-screw#locking-head-screws. Accessed 21 Jan 2025.
112. Glowacki J, Bartkowiak T, Gapinski B, et al. Biomechanical evaluation of variable angle locking systems. A micro-CT analysis. J Orthop Trauma. Published online October 16, 2023. https://doi.org/10.1097/BOT.0000000000002712
113. Au B, Groundland J, Stoops TK, Santoni BG, Sagi HC. Comparison of 3 methods for maintaining interfragmentary compression after fracture reduction and fixation. J Orthop Trauma. 2017;31(4):210–3. https://doi.org/10.1097/BOT.0000000000000769.
114. Märdian S, Schmölz W, Schaser KD, Duda GN, Heyland M. Locking plate constructs benefit from interfragmentary lag screw fixation with decreased shear movements and more predictable fracture gap motion in simple fracture patterns. Clin Biomech (Bristol). 2019;70:89–96. https://doi.org/10.1016/j.clinbiomech.2019.08.008.
115. Kuzma AL, Luo TD, De Gregorio M, et al. Biomechanical evaluation of interfragmentary compression of lag screw versus positional screw at different angles of fixation. J Orthop Trauma. 2019;33(5):e183–9. https://doi.org/10.1097/BOT.0000000000001429.
116. Kong D, Fan X, Song C, et al. A comparative analysis of between percutaneous cannulated screw fixation and traditional plate internal fixation in treatment of sanders II and III calcaneal fractures. J Foot Ankle Surg. 2024;63(3):327–32. https://doi.org/10.1053/j.jfas.2023.12.004.
117. Syed MA, Panchbhavi VK. Fixation of tibial pilon fractures with percutaneous cannulated screws. Injury. 2004;35(3):284–9. https://doi.org/10.1016/s0020-1383(03)00173-6.
118. El-Azab H, Ahmed KFE, Khalefa AH, Marzouk AR. A prospective comparative study between percutaneous cannulated screws and Kirschner wires in the treatment of displaced intra-articular calcaneal fractures. Int Orthop. 2022;46(11):2667–83. https://doi.org/10.1007/s00264-022-05521-y.
119. Nalbantoglu U, Gereli A, Kocaoglu B, Turkmen M. Percutaneous cannulated screw fixation in the treatment of distal radius fractures. Arch Orthop Trauma Surg. 2012;132(9):1335–41. https://doi.org/10.1007/s00402-012-1541-4.
120. Watford KE, Jazrawi LM, Eglseder WA Jr. Percutaneous fixation of unstable proximal

humeral fractures with cannulated screws. Orthopedics. 2009;32(3):166. https://doi.org/10.3928/01477447-20090301-29.

121. Marsell R, Einhorn TA. The biolDoi of fracture healing. Injury. 2011;42(6):551–5. https://doi.org/10.1016/j.injury.2011.03.031.
122. Willis AA, Kutsumi K, Zobitz ME, Cooney WP 3rd. Internal fixation of dorsally displaced fractures of the distal part of the radius. A biomechanical analysis of volar plate fracture stability. J Bone Joint Surg Am. 2006;88(11):2411–7. https://doi.org/10.2106/JBJS.E.00946.
123. Bottlang M, Lesser M, Koerber J, et al. Far cortical locking can improve healing of fractures stabilized with locking plates. J Bone Joint Surg Am. 2010;92(7):1652–60. https://doi.org/10.2106/JBJS.I.01111.
124. Hernigou P, Pariat J. History of internal fixation with plates (part 2): new developments after World War II; compressing plates and locked plates. Int Orthop. 2017;41(7):1489–500. https://doi.org/10.1007/s00264-016-3379-9. Epub 2016 Dec 30. PMID: 28035429.
125. Ellis E III, Schubert W. Locking plate principles. AO Surgery Reference. https://surgeryreference.aofoundation.org/cmf/basic-technique/locking-plate-principles. Accessed 20 Jan 2025.
126. Nicklin S, Ingram S, Gianoutsos MP, Walsh WR. In vitro comparison of lagged and nonlagged screw fixation of metacarpal fractures in cadavers. J Hand Surg Am. 2008;33(10):1732–6. https://doi.org/10.1016/j.jhsa.2008.07.010. PMID: 19084171
127. Sun P, Xu D, Zhao W, Jiao P, Li Z, Liu C, Ouyang J. Interfragmentary compression and pull-out properties of 6.5-mm AO cancellous lag screws in a uniform synthetic material during tightening procedures. Med Eng Phys. 2017;42:116–20. https://doi.org/10.1016/j.medengphy.2017.01.011. Epub 2017 Jan 31. PMID: 28159451
128. Ayalon O, Rettig SA, Tedesco LJ. Bone graft and fixation options in the surgical management of scaphoid nonunion. J Am Acad Orthop Surg. 2024; https://doi.org/10.5435/JAAOS-D-24-00510. Epub ahead of print. PMID: 39531592
129. Raina DB, Markevičiūtė V, Stravinskas M, Kok J, Jacobson I, Liu Y, Sezgin EA, Isaksson H, Zwingenberger S, Tägil M, Tarasevičius Š, Lidgren L. A new augmentation method for improved screw fixation in fragile bone. Front Bioeng Biotechnol. 2022;10:816250. https://doi.org/10.3389/fbioe.2022.816250. PMID: 35309986; PMCID: PMC8926351.
130. Wähnert D, Hofmann-Fliri L, Schwieger K, Brianza S, Raschke MJ, Windolf M. Cement augmentation of lag screws: an investigation on biomechanical advantages. Arch Orthop Trauma Surg. 2013;133(3):373–9. https://doi.org/10.1007/s00402-012-1676-3. Epub 2012 Dec 21. PMID: 23263012.

24 Advances in Implants for the Distal Radioulnar Joint

Luis R. Scheker

Introduction

The importance of the distal radioulnar joint has not been known for an extended period. It is well known that Galen [1] was the reference for everything in medicine, including anatomy, for 13 centuries until Andreas Vesalius [2], 1514–1564, wrote De Humani Corporis Fabrica Libri Septem, which is considered the reason for the transformation of human anatomy knowledge.

We must remember that in middle age, extremity function was described as pulling on muscles. Still, there were no means to verify that a particular muscle did not work in concert with other muscles or that the muscle reacted to mechano-receptors and not being a voluntary muscle.

For clarification, let us take the lumbrical muscles, for example. The literature says these muscles flex the metacarpal-phalangeal joints and extend the interphalangeal joints. Still, the tendons of these muscles are inserted on the radial side of the distal phalanges to modulate the flexor digitorum profundus function and allow for grip span.

The other example is the brachioradialis, which responds to mechano-receptors; it is supposed to flex the elbow. As you read this chapter, flex your elbow to 90 degrees and put the contralateral hand on the triceps. Now, try to contract the brachioradialis without contracting the triceps. It is impossible because the brachioradialis is not a voluntary muscle; it responds to mechanoreceptors. It is not an elbow flexor but a modulator of the elbow extension when handling weight. It strengthens the biceps and the brachialis to control the elbow extension.

To understand the upper extremity function, specifically when lifting weights until the elbow is flexed to 90 degrees in a neutral position, the brachialis anterior, which inserts distal to the coronoid process of the ulna, is responsible. After full supination, the biceps brachii works in concert with the brachialis anterior. This is why we supinate both forearms when lifting heavy objects.

Now, it starts to be clear that the head of the ulna is an essential factor in lifting weights, no matter how little. The radius is an appendix of the ulna, attached at both ends in a bicondylar joint. In the past, the distal radioulnar joint was irrelevant. The first description of the excision of the head of the ulna appeared in the book *Illustrated Manual of Operative Surgery and Surgical Anatomy* by Claude Bernard and Charles Huette [3].

Many others described similar procedures. The best known is the work of William Darrach [4] in 1913. Other modifications followed,

L. R. Scheker (✉)
University of Louisville, Louisville, KY, USA

Christine M Kleinert Institute for Hand and Microsurgery, Louisville, KY, USA

F. Amirouche et al. (eds.), *Biomechanics of the Hand*,
https://doi.org/10.1007/978-3-032-04688-8_24

including Kapandji [5], Bowers [6], Watson [7], and others.

In the 1970s, the treatment of rheumatoid patients was mainly based on steroids and antiinflammatory medications. The DMARS [8] was not available, and surgeons were the last resort for treatment; it included splinting and surgery with help for the metacarpal-phalangeal joints with replacement with Swanson [9] silicone implant and fusion of the interphalangeal joints to give stability to the fingers. But they had problems with the wrist and the ulnar head (see Fig. 24.1).

Rana and Taylor [10] in 1973 and Jackson et al. [11] in 1974 described the excision of the ulnar head in patients with severe rheumatoid arthritis to relieve pain at rest. These unfortunate people had all their joints affected by the disease, and they had minimal function left. They were unable to lift even a cup of tea. Excision of the ulnar head gave them some relief.

I did my last excision of the ulnar in 1983; at that point, the pendulum was moving to perform excision of the ulnar head in cases of posttraumatic and degenerative arthritides. In our case of an osteoarthritic patient, the result was not optimal, and the patient let me know that the pain was more proximal than before surgery and was more severe. After 12 months of waiting for the result to improve, I tried to show the patient that the two bones were separated, showing her an AP X-ray of the forearm (see Fig. 24.2). The patient asked me to put my hand on her flexed forearm, and she rotated the forearm; the feeling was sickening; she had to grind between the radius and ulna, as was demonstrated with an X-ray taken with her placing the forearm in a neutral position and the X-ray beam horizontally (see Fig. 24.3).

Excision of the ulnar head leads to contact between the radius and the stump of the ulna, with what has been described as ulnar impingement by King et al. [12] and dynamic impingement by Lees and Scheker [13]. But there is a mistake. It is not the ulna that impinges on the radius; it is the radius that falls on the stump of the ulna.

To help with this problem in rheumatoid patients, Swanson [14] created the ulnar silicone

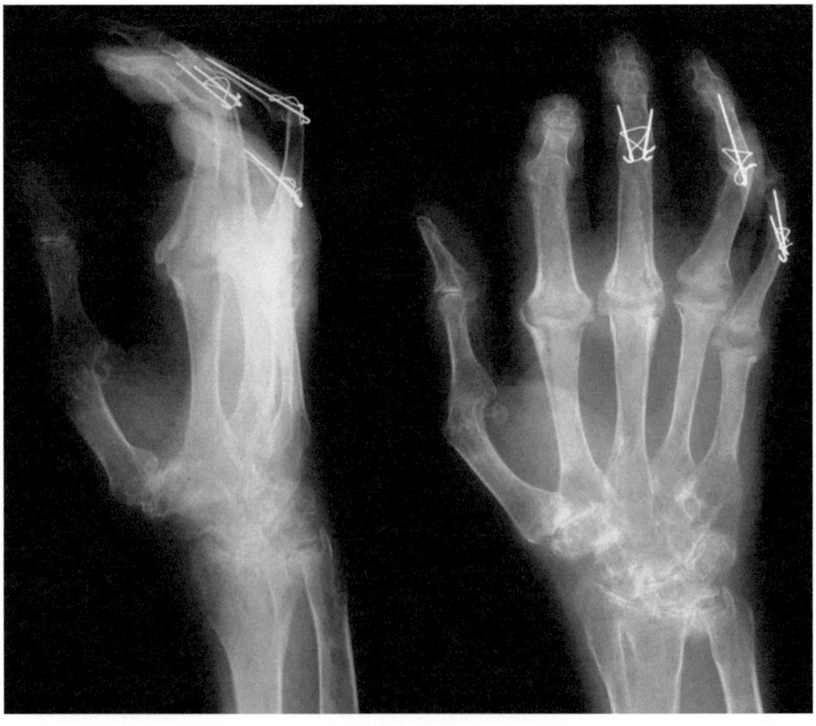

Fig. 24.1 The rheumatoid patients had their metacarpal phalangeal joints replaced with Swanson silicone implants, the extensor tendons centralized, and the proximal interphalangeal joints fused, giving them a functional hand, but the pain of the cartilage depleted sigmoid notch with the deformed ulnar head gave the patients pain even at rest

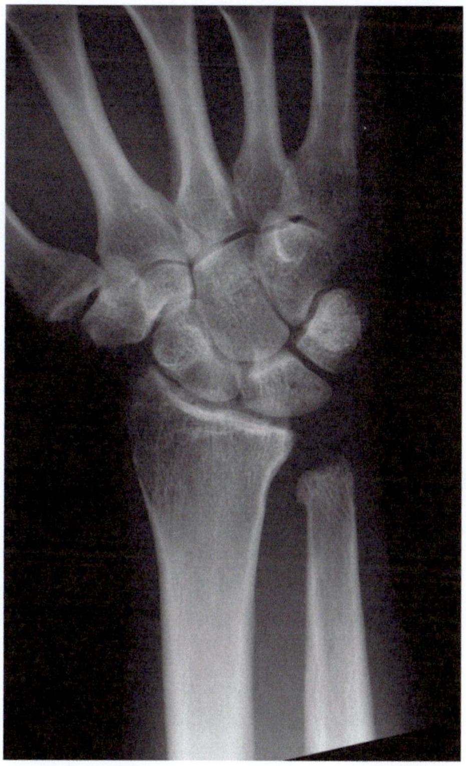

Fig. 24.2 This image was taken with the patient resting the forearm in the X-ray table; the gravity was not approximating the bones

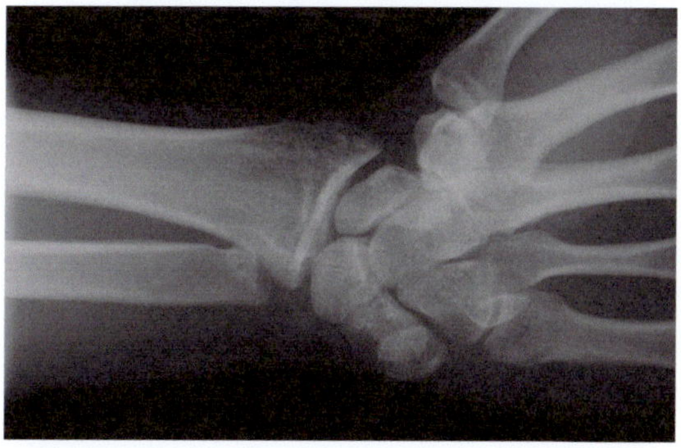

Fig. 24.3 The forearm was held in the air with the elbow flexed at 90 degrees in neutral rotation, with the X-ray beam directed horizontally. Here the brachialis anterior is lifting the ulna while the radius is falling on top of the ulna stump as there is no support for the distal radius

cap. The cap was used to get subluxated because the pressure exerted by the radius falling on the end of the ulna caused it to become subluxated.

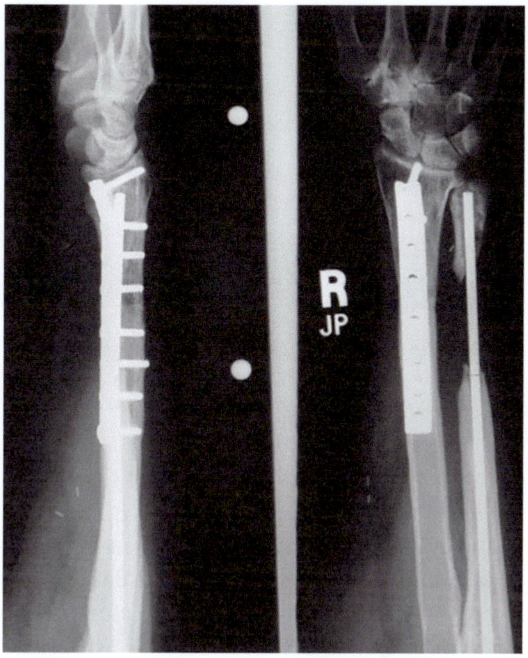

Fig. 24.4 Shortening osteotomy to both segments of the radius allowed us to stabilize this bone, but the ulna defect could not be grafted as the possibility of failure of the replated hand as well as the development of an infection made us delay the possible bone graft

Hand surgeons wanted to create a sturdier implant that could withstand the forces between the radius and ulna at the distal radioulnar joint. They developed metal, ceramic, and carbon fiber unipolar implants. Some stayed in place, but many subluxated because the triangular fibrocartilage could not heal to nonviable materials.

In 1987, while on call, I received a coal mine worker with an avulsion amputation of the right forearm. After thorough debridement, we shortened the radius, but the ulna was missing too much bone. Not knowing if the replant would survive or if there would be no infection, we inserted a Steinmann pin to bridge the gap between the proximal and distal ulna segments (see Fig. 24.4).

We explained to the patient that we would do a bone graft later if he regained protective sensation and some grip. After 2 years of follow-up, he had a protective sensation and some gripping, as well as pronation and supination. He refused our

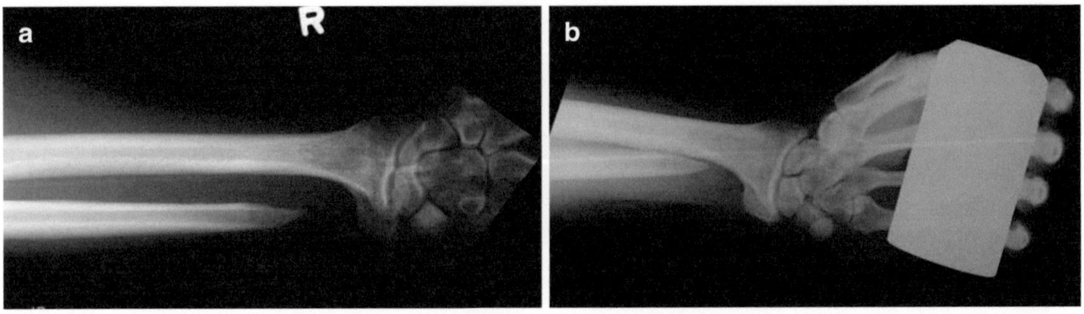

Fig. 24.5 (**a**, **b**) The patient had a Sauve-Kapandji procedure performed previously making the ulna look like a pencil tip, shaven by the radius as it rotated around the ulna. (**a**) was taken on the X-ray table. (**b**) was taken with the elbow in 90 degrees and the forearm in neutral rotation

recommendation and did not want the bone graft, saying he was happy with what he had.

In 1992, a piano player came to see us because of an inability to make ends meet; he complained of pain when rotating the forearm. A stress X-ray showed the two bones touching (see Fig. 24.5). We told him about our patient with the replant and how the Steinmann pin helped him. He was advised to do no lifting but only play the piano. He promised that he was going to follow all the recommendations (see Fig. 24.6).

A year later, he came back complaining that he felt something moving distally; we X-rayed and found that the cavity in the ulnar head was more significant than initially. His wife was very helpful in telling us that he was playing the piano and lifting it with his friends. We measured the cavity defect in an AP view; the radius blocked the lateral; we decided that a metal ball with an Allen screw fixed at the end of a Steinmann pin would help us to solve the problem.

While introducing the ball into the socket, the wall cracked; after the initial shock, we used a six holes 3.5 mm plate bent at the hole before last and secured it to the ulnar side of the radius with three cortical screws. It gave us breathing, and he was happy (see Fig. 24.7).

Three years later, we saw a patient with a hand grenade injury to his right hand; he was a veteran of the Korean and Vietnam wars. He lost his first carpometacarpal joint and damaged the radiocarpal and radioulnar joints. He had an arthroplasty of the base of the thumb with good results and multiple procedures to fuse the radiocarpal joint,

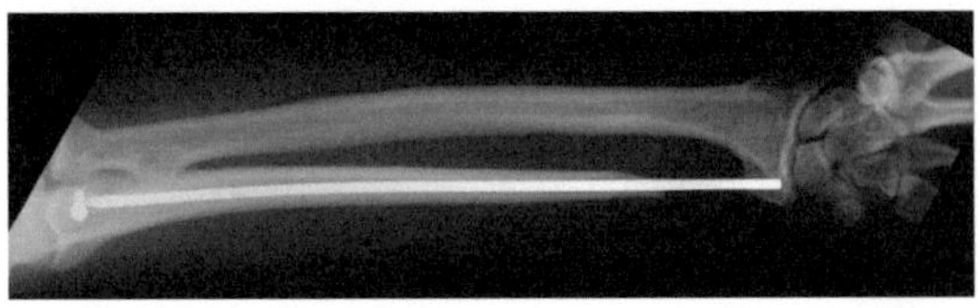

Fig. 24.6 A Steinmann pin was utilized to support the distal radius, by placing the distal pin in the head of the ulna. To prevent pin proximal migration, a screw as introduced at the proximal end of the pin

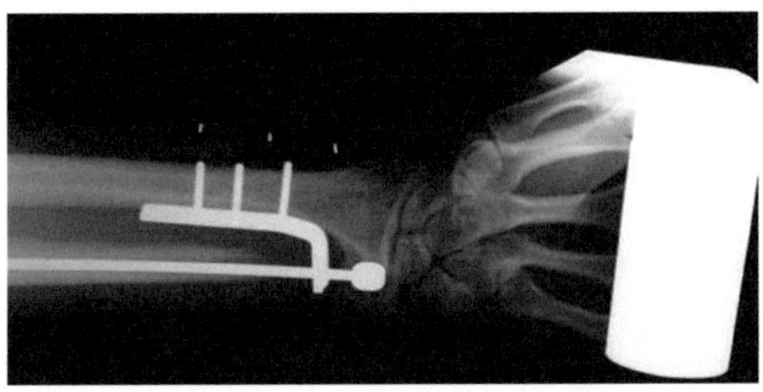

Fig. 24.7 Due to the cavity of the ulnar head not being circular, but ovoid, the wall of the cavity broke, and a bent six-hole-plate helped us to provide support for the distal radius by securing it to the interosseous side of the radius with three screws

which finally was fused when the new radio carpal fusion plate was available, but an attempt to stabilize the radius to the ulna after a Darrach procedure did not work.

His complaint was about the lack of motion and the pain he had as the radioulnar fusion had not consolidated (see Fig. 24.8).

Having had two previous patients' experiences, we told the patient we could create an implant to

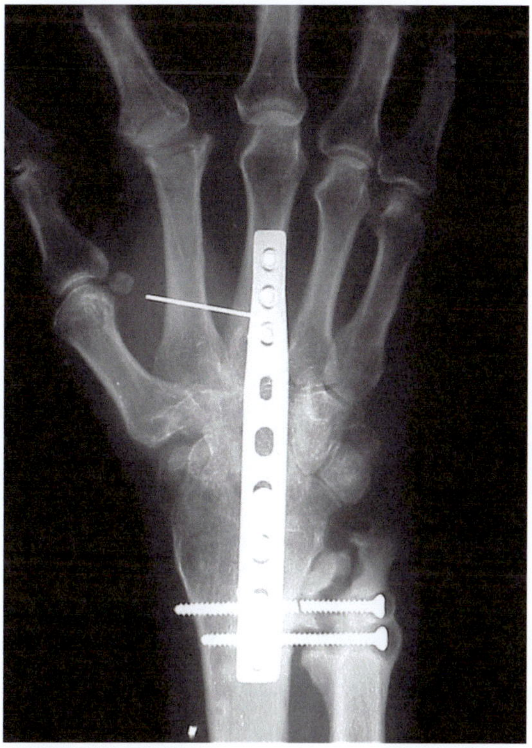

Fig. 24.8 The reconstruction of his grenade blown hand was extensive; he had arthroplasty of the basal joint of the thumb and silicone implant to the index finger metacarpal phalangeal joint. Many attempts to fuse the radio carpal joint until the fusion plate was developed. His main problem was lack of pronation and supination and pain while using his right hand due to nonunion of the radio ulnar distally

help him. He was delighted. We embarked on creating an implant that would mimic the function of the radioulnar joint, allowing full pronation and supination, radial migration, and changing the rotation angle as there would be no radial translation on the seat of the ulna.

The implant design was brought to the IRB of our institution in 1995. Most of the members of the board were people who knew me, and I expected a positive response. Well, not so fast; the kindest comment was that I probably have not read that constraint implants do not work. It did not matter that I explained that this device would have more motion than the original joint and that the forces would be compressing rather than displacing the components.

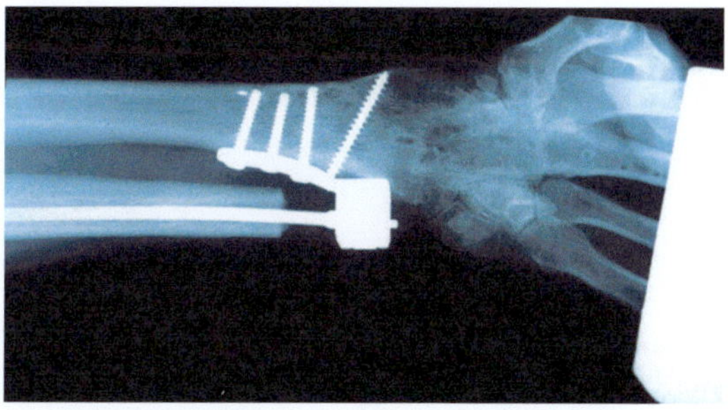

Fig. 24.9 The original implant consisted in a box attached to the radius by four screws, one cancellous and three cortical. Inside of the box, there was an ultrahigh molecular weight polyethylene ball with a tunnel in the center allowing the radius to migrate; the ball permitted the angles of rotation to change as there was no radial translation possible

The idea of making a semiconstrained implant was because the work of DeSmet and Fabry [15] demonstrated that the sigmoid notch has three different angulations, and the work of Tollat et al. [16] showed that there are four shapes of the sigmoid notch. By making it a bipolar implant, we obviate the 12 possibilities needed to accommodate the shape of the patient's sigmoid notch, as it does not require this structure, nor does it need the ligaments. The implant creates a stable joint capable of all the motion of the original distal radio ulnar joint.

Two years later, the board members were replaced, and I got the go-ahead from the president with whom I had worked closely. The implant was made and implanted in the early summer of 1997 (see Fig. 24.9).

The patient came back to follow up with pictures taken by his wife. He was pleased, and I was very worried (see Fig. 24.10). I told him that he exceeded the weight we thought the implant made of stainless steel could tolerate. With a smile, he said, "You told me that it is a modular implant, so I know that if any part breaks, you can fix it. Anyway, lifting is not important to me; now I can shave myself, drink from a glass, and open a doorknob."

That comment sounded to me like an alarm clock; I had many patients with missing distal radio ulnar joints and complaining of impingement (see Fig. 24.11).

Fig. 24.10 (**a–c**) The patient recovered pronation and supination; the picture demonstrates the patient in the first week after surgery starting to rotate the forearm. Four weeks after surgery, he brought pictures taken by his wife while he was working in the yard cutting and lifting a large piece of wood

With the success of this implant, we requested permission from the IRB to perform 30 implants. It was granted, and the implant started to change people's lives. The implant also changed as we moved forward. Eventually, it became made of cobalt chrome, ten times stronger than stainless steel (Figs. 24.12 and 24.13).

The design changed to have the implant cavity covering less than 100% of the ball, as it was determined that it was not needed. We also created a radial peg to facilitate the insertion of the plate. Initially, the stem was a 3 mm three-point fixation, which was changed to a press fit. An extension of the stem allowed us to seal the ulna

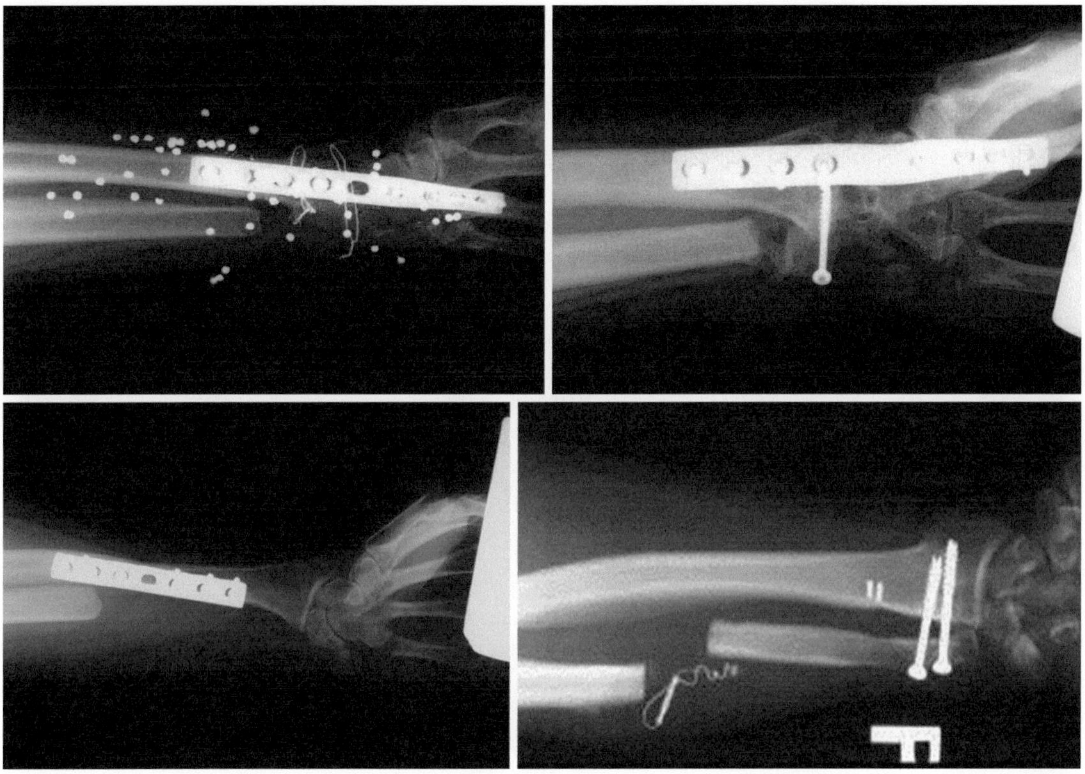

Fig. 24.11 Many of our patients were missing the distal radio ulnar joint due to trauma or surgery that were candidates for this type of self-stabilizing implant

cavity and prevent bone marrow scaping and ectopic bone formation.

The play of the ball inside the cavity and the tunnel on the ball have such a slight difference that no ball had to be changed because of wear in the 20 years that the implant has been in the market.

Because patients' bones vary in size, and some have had excision of the ulna of various lengths, we have three sizes of plates, four diameter stems, and four lengths of the collar to adapt to the missing bones.

Acknowledgments I want to express my sincere gratitude to Dr. Farid Amirouche for his invaluable assistance with this chapter. His insights have greatly enriched the content, and his support throughout the editing process has been instrumental in shaping this work.

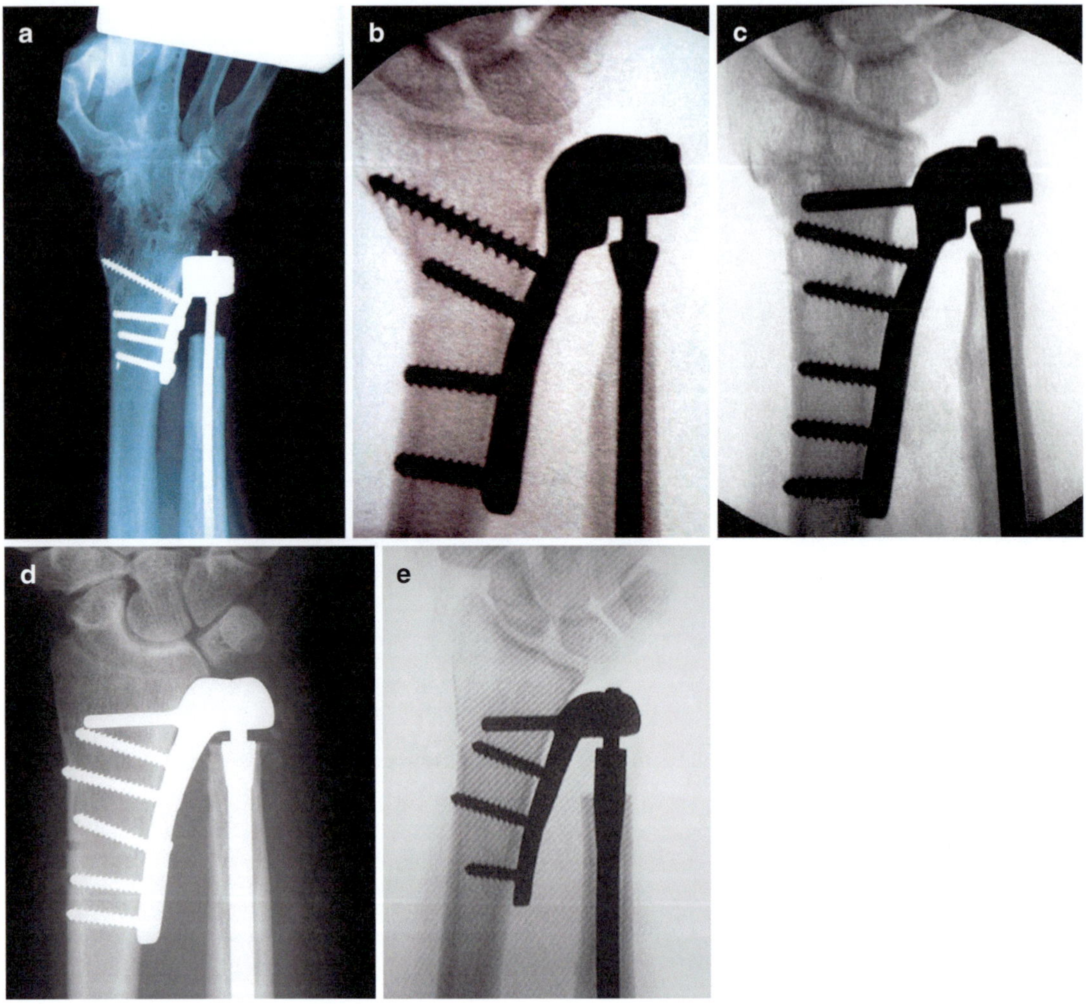

Fig. 24.12 From the beginning, the prosthesis has been modified to improve the implantation and reduced the early complication that included ectopic bone formation and extensor carpi ulnaris tendinitis especially in patients with small radius. From a box, distally the implant has evolved into a smooth head that prevents tendon irritation; also by having an extension at the collar of the stem, the ulna cavity is sealed, and no ectopic bone is now seen. The present implant is the sixth generation

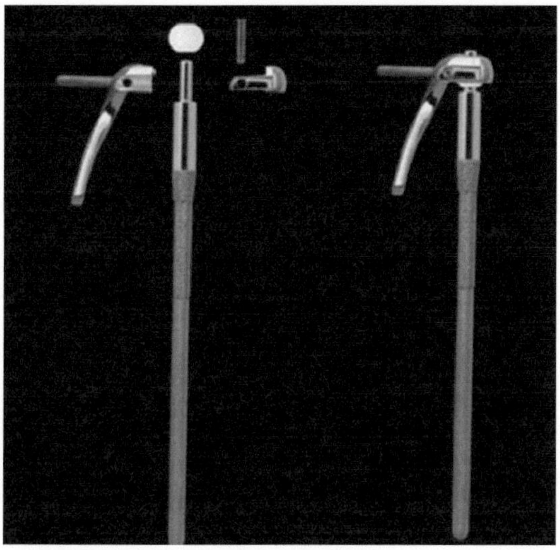

Fig. 24.13 Cobalt chrome implant

References

1. Dunn PM. Galen (AD129-200) of Pergaminum: anatomist and experimental physiologist. Arch Dis Child Fetal Neonatal Ed. 2003;88:441–3.
2. Vesalius A (1514–1564). De Humani Corporis Fabrica Libri Septem. Basel: Johannes Oporinus; 1543.
3. Bernard C, Huette C. In: Norton AT, editor. A textbook of operative surgery and surgical anatomy; tr. from the French. London: Bailliere, Tindall, and Cox; 1878.
4. Darrach W. Partial excision of lower shaft of the ulna for deformity following Colles' fracture. Ann Surg. 1913;57:764–5.
5. Sauvé L, Kapandji M. Nouvelle technique de traitement chirurgical des luxations récidivantes isolées de l'extrémité inférieure du cubitus. J Chir. 1936;7:589.
6. Bowers WH. Distal radioulnar joint arthroplasty: the hemiresection interposition technique e. J Hand Surg. 1985;10A:169–78.
7. Watson HK, Ryu JY, Burgess RC. Matched distal ulnar resection. J Hand Surg. 1986;11A:812–7.
8. Hunneyball IM. Recent developments in disease-modifying antirheumatic drugs. Prog Drug Res. 1980;24:101–216.
9. Swanson AB, de Groot Swanson G. Flexible implant arthroplasty in the rheumatoid metacarpophalangeal joint. Clin Rheum Dis. 1984;10(3):609–29.
10. Rana NA, Taylor AR. Excision of the lower end of the ulna in rheumatoid arthritis. Ann Rheum Dis. 1973;32(3):277.
11. Jackson IT, Milward TM, Lee P, Webb J. Ulnar head resection in rheumatoid arthritis. Hand. 1974;6(2):172–80.
12. Bell MJ, Hill RJ, McMurtry RY. Ulnar impingement syndrome. J Bone Joint Surg. 1985;67B:126–9.
13. Lees VC, Scheker LR. The radiological demonstration of dynamic ulnar impingement. J Hand Surg. 1997;22B:448–50.
14. Swanson AB. Implant arthroplasty for disabilities of the distal radioulnar joint. A silicone rubber capping implant is used following resection of the ulnar head. Orthop Clin North Am. 1973;4(2):373–8211.
15. De Smet L, Fabry G. Orientation of the sigmoid notch of the distal radius: determination of different types of the distal radioulnar joint. Acta Orthop Belg. 1993;59(3):269–72.
16. Tolat AR, Stanley JK, Trail IA. A cadaveric study of the anatomy and stability of the distal radioulnar joint in the coronal and transverse planes. J Hand Surg Br. 1996;21(5):587–94.
17. Scheker LR. Implant arthroplasty for the distal radioulnar joint. J Hand Surg Am. 2008;33(9):1639–44.

Sutures in Hand Surgery

25

Yvon Bogdonoff, Farid Amirouche, and Alfonso Mejia

Introduction

Sutures are an indispensable tool in hand surgery, enabling surgeons to achieve precise alignment of tissues and optimize healing outcomes. The intricate anatomy of the hand, with its dense network of tendons, nerves, and ligaments, demands surgical techniques that are both meticulous and biomechanically sound. Sutures provide mechanical support during the healing process and ensure the restoration of function in this highly specialized and critical body region. A vital review article by Lekic et al. discusses the advantages of different materials, needles, and methods in other contexts [10]. The significance of sutures extends beyond their immediate mechanical role; they reduce infection risks, minimize scarring, and promote optimal recovery, giving patients hope for a better postsurgery life (Table 25.1).

Y. Bogdonoff · A. Mejia
Department of Orthopaedic Surgery, University of Illinois at Chicago, Chicago, IL, USA

F. Amirouche (✉)
Department of Orthopaedic Surgery, Orthopaedic and Spine Institute, Endeavor HealthSystem (Formerly Northshore University HealthSystem), an Affiliate of the University of Chicago Pritzker School of Medicine, Skokie, IL, USA

Department of Orthopaedic Surgery, University of Illinois at Chicago, Chicago, IL, USA
e-mail: famirouche@northshore.org

Historically, the evolution of sutures has paralleled advances in surgical science. Early sutures made from natural materials such as silk and catgut offered rudimentary solutions but often fell short in strength and predictability. However, the development of synthetic and bioengineered materials has revolutionized this field, offering surgeons diverse options tailored to specific needs. Innovations such as absorbable sutures and drug-eluting variants reflect the ever-increasing sophistication of surgical techniques. Looking to the future, potential advancements in suture materials and methods include the development of smart sutures that can monitor the healing process and provide real-time feedback. These smart sutures incorporate a small electronic sensor that can monitor wound integrity, gastric leakage, and tissue micromotions while providing healing outcomes equivalent to medical-grade sutures [7]. Biodegradable materials can be dissolved within the healing process as an alternative to Vycril, which takes 42 days by hydrolysis. These potential advancements underscore the essential role of sutures in modern medicine, particularly in the intricate field of hand surgery, and stimulate critical thinking and innovation among hand surgeons and surgical trainees [24].

Biomechanically, sutures act as temporary scaffolds that hold tissues together while natural

F. Amirouche et al. (eds.), *Biomechanics of the Hand*,
https://doi.org/10.1007/978-3-032-04688-8_25

Table 25.1 Advantages and disadvantages of common suture materials

Suture material	Advantages	Disadvantages
Vicryl (polyglactin 910)	Absorbable, degrades in 2–3 weeks, good for subcutaneous closures	A short degradation time may not provide long-term support
PDS (polydioxanone)	Absorbable, lasts 4–6 months, suitable for long-term healing (e.g., TFCC repairs)	Long absorption time, which can irritate sensitive areas
Nylon	Nonabsorbable, high tensile strength, commonly used in tendon repairs	Requires removal, may cause irritation if left in permanently
Prolene (polypropylene)	Nonabsorbable, minimal tissue reactivity, good for cardiovascular and skin closure	Less flexible, can be difficult to handle for deep suturing
Silk	Good handling, secure knots, but causes higher inflammatory response	Nonabsorbable, induces inflammatory reaction, less commonly used today

healing occurs. They must withstand the forces generated during movement and resist degradation long enough to allow the tissues to regain sufficient strength. This delicate balance between supporting and minimizing tissue stress underscores the importance of selecting appropriate suture materials and techniques. An excellent review of the interplay of different suturing techniques with the biomechanical properties of flexor tendon repair is elucidated by an article by Muslu et al. Comparative analysis of modified Kessler, Bunnel, and Krackow techniques using Monosorb Ti-Cron and V-loc suture demonstrated that the ultimate failure load and 2-mm gap load in the setting of flexor tendon repair varied significantly [13]. The Krackow technique provides the highest failure load due to its multilooped locking structure, making it the strongest. In contrast, the Modified Kessler technique has the lowest strength, offering limited suture-tendon interaction. Bulk at the repair site also varies by technique. Krackow creates more bulk, which can limit tendon gliding, while Kessler minimizes bulk but sacrifices strength. Bunnell provides a balance between the two.

Failure modes differ by technique. The Modified Kessler technique often fails at the knot, where slippage or loosening reduces repair integrity. The Krackow and Bunnell techniques lock more securely into the tendon and are more likely to fail due to suture breakage under excessive tension. Additionally, nonlocking techniques, such as the Modified Kessler, are more prone to suture cutting through the tendon, especially if the suture does not grip the tissue securely. These findings highlight the need to balance strength, bulk, and failure risks when selecting a suture technique (Fig. 25.1).

These findings underscore the critical importance of clinicians' understanding of unique suturing materials and techniques. A detailed knowledge of the biomechanical properties of sutures is, therefore, essential for optimizing outcomes in hand surgery.

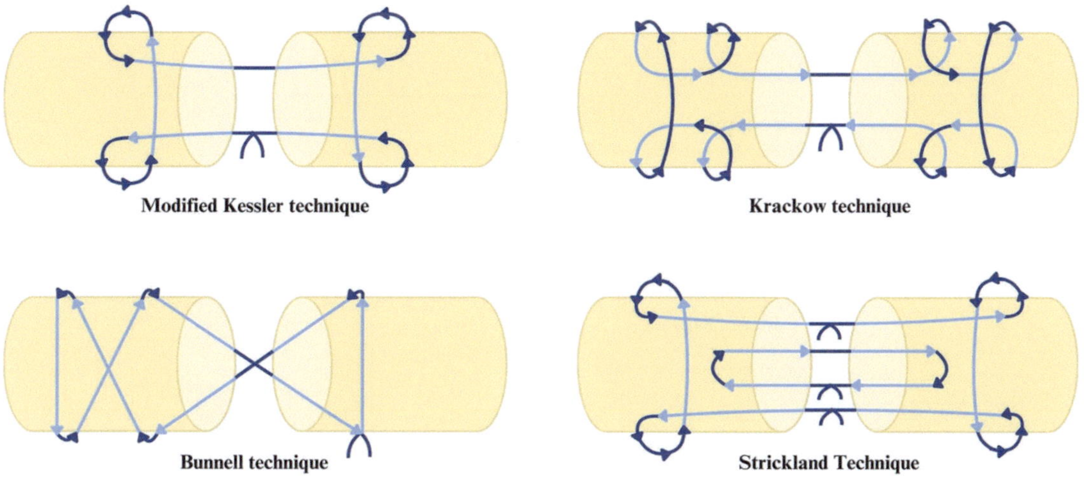

Fig. 25.1 Comparison of the Modified Kessler-Bunnell and Strickland suture techniques in tendon repair. The Modified Kessler-Bunnell method incorporates locking loops and cross-linked suture patterns, optimizing tensile strength and resistance to gapping. The Strickland technique emphasizes core suture integrity with peripheral reinforcement to enhance smooth gliding and reduce adhesions. Both methods are evaluated based on tensile strength, gap resistance, and biomechanical efficiency in tendon healing

Biomechanical Properties of Sutures

The biomechanical properties of sutures play a pivotal role in determining their effectiveness during surgical procedures, particularly in hand surgery. These properties encompass a variety of factors, including the strength and stability of the knot formed with the suture, the specific type of material used for the suture itself, and the selected technique for the repair. Surgeons aim to create a repair that is robust enough to facilitate early movement of the affected area while simultaneously safeguarding the surrounding tissues from injury (Fig. 25.2, Table 25.2).

Among the various characteristics of sutures, the material composition stands out as one of the most significant factors. This composition ultimately dictates whether a suture is classified as absorbable or nonabsorbable and whether synthetic or derived from natural sources. Understanding these distinctions is essential for selecting the appropriate suture for a given surgical procedure, ensuring optimal healing and functional recovery for the patient. Absorbable sutures, like polyglycolic acid and polydioxanone, are designed to degrade naturally within the body over time, eliminating the need for removal. Absorbable sutures degrade at different rates, influencing their use in various surgical applications. Vicryl (polyglactin 910) degrades over 2–3 weeks, making it suitable for subcutaneous closures where short-term support is needed. PDS (polydioxanone) degrades over 4–6 months, making it ideal for longer-healing structures such as TFCC repairs, where extended support is required without the irritation of a permanent suture.

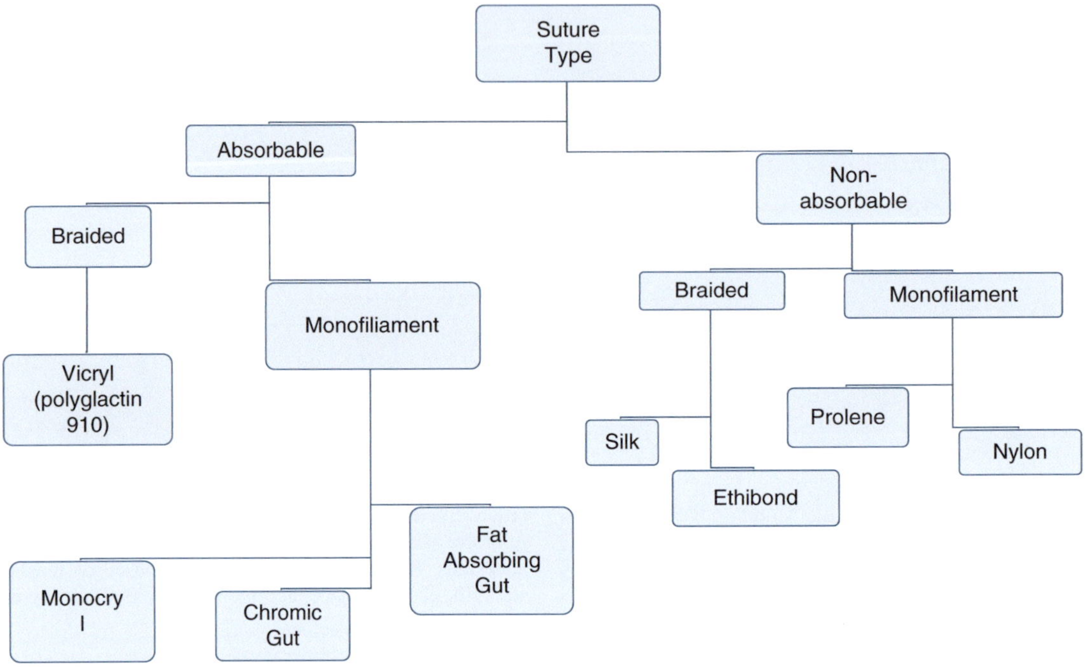

Fig. 25.2 Flowchart comparing absorbable and nonabsorbable sutures. The chart categorizes sutures based on their biodegradability, with absorbable sutures (Vicryl, Monocryl, Chromic Gut, and Fast Absorbing Gut) breaking down naturally in the body over time, reducing the need for removal, and typically used for internal tissues. Nonabsorbable sutures (Silk, Ethibond, Prolene, and Nylon) provide long-term tensile strength, making them suitable for external or high-stress applications where durability is critical

Table 25.2 Comparative biomechanical properties of various 3-0 suture materials commonly used in hand surgery

Suture material	Strength	Durability	Inflammatory reaction	Deformability
Nylon (3-0)	High	Very high	Very low	High
Polypropylene (3-0)	Medium-high	High	Very low	Medium
Silk (3-0)	Medium	Low	High	Low
Polyglactin (Vicryl 3-0)	Medium-high	Medium	Medium	Very high

Nonabsorbable sutures like nylon are often used for structures requiring prolonged tensile strength, particularly in tendon repairs, where healing can take months. PDS provides a middle ground, offering long-term strength without the irritation of a permanent suture, which is particularly useful in areas with minimal overlying soft tissue where a subcutaneous foreign body could cause discomfort. This balance between degradation time and mechanical support is crucial in selecting the optimal suture material for different surgical procedures.

These materials can be particularly beneficial in selective hand surgery for wound closure [14]. Additionally, they are advantageous for internal repairs where the long-term presence of sutures could lead to complications such as the potential of excessive scarring, infection, or a foreign body reaction where the body is not able to break down the suture material. In contrast, nonabsorbable sutures, such as nylon and proline, provide permanent support and are typically used in situations that require prolonged tensile strength.

Tensile strength is another key property, as it determines the suture's ability to hold tissues together under stress. High tensile strength is crucial in hand surgery, where forces generated by even slight movements can strain the repair site.

Table 25.3 Handling properties and surgical use of various suture materials

Suture material	Handling properties	Influence on surgical use
Nylon	High memory, stiff, can cause tangling	Requires stretching to reduce memory; can be harder to handle for precise suturing
Prolene (polypropylene)	Slippery, low friction, high memory	Tends to resist knot tying but glides smoothly through tissue, making it useful for vascular and skin closures
Vicryl (polyglactin 910)	Soft, pliable, minimal memory	Easy to handle and ties securely, making it ideal for subcutaneous closures
PDS (polydioxanone)	Moderate memory, slightly stiff but flexible	Holds knots well but can be more challenging to manage than Vicryl; used when long-term absorbable support is needed
Silk	Very pliable, excellent knot security	Easy to handle but causes more tissue reaction, limiting its use in modern surgeries
Coated sutures (e.g., coated vicryl, coated PDS)	Reduced friction, smoother passage through tissue	Enhances handling, reduces tissue trauma, and improves knot security

Testing methods, such as uniaxial tensile tests, provide insights into how sutures perform under controlled conditions. Research has demonstrated that synthetic sutures generally offer superior tensile strength with better knot security and predictability than natural materials, making them the preferred choice for many applications. However, they have a high risk of infection because they provide better physical conditions for colonization by microorganisms [2].

Elasticity and stress distribution are equally important in minimizing wound tension, especially in delicate hand tissues. Sutures must be flexible enough to accommodate slight movements without causing additional stress to the wound edges. Elasticity also influences how sutures distribute mechanical forces across the repair site, reducing the risk of tissue damage or suture pull-through. Biomechanical studies have shown that synthetic materials with controlled elasticity can significantly improve outcomes in hand surgery by reducing the likelihood of wound dehiscence [3].

Knot security is another critical factor influencing future performance. A secure knot ensures the integrity of the repair and minimizes the risk of slippage. Factors such as the material's friction coefficient and knotting technique play a significant role in achieving optimal knot security. Failure mechanisms, including knot untying and breakage under load, have been extensively studied to develop strategies that maximize reliability. The tensile strength of the suture material heavily influences the number of knots needed for a secure suture. Unraveling is a problem with some sutures, so the number of "throws" and length of the suture tail cut after completing the knot need is essential in suture security. For example, fiber tape used for 360-degree SL ligament reconstruction requires five "throws" to minimize the risk of unraveling. A biomechanical comparative analysis compared a novel higher tensile Dynatape to a more conventional, lower tensile strength, SutureTape. The study demonstrated that the higher tensile strength material only required four knots in wet conditions compared to six for the lower tensile strength suture tape [15].

Finally, handling properties, such as pliability and memory, influence a surgeon's ability to place sutures precisely. Materials with poor handling characteristics can complicate procedures, leading to increased operative times and potential complications (Table 25.3).

These advancements allow for greater surgical precision and reduced tissue trauma, demonstrating the role of technology in enhancing the practice of hand surgery. For instance, suture coatings reduce friction, making it easier to tie knots. Recently developed flexible thermoplastic polyurethane sutures are feasible and safe for midline laparotomy wound closure with promising properties to prevent postoperative complications such as abdominal dissection after abdominal surgery. The elastic joints showed significant material strength; the average elastic elongation was 56 × 10 cycles [1]. At the same time, material engineering enhances the flexibility and memory of sutures, allowing them to conform to the tissue's shape and maintain their position during the healing process.

Suture Techniques in Hand Surgery

Hand surgery is a complex field that requires various sophisticated suture techniques tailored to the hand's unique anatomical and functional requirements. The most commonly employed methods are simple interrupted and running sutures, each offering distinct advantages and optimal applications in various surgical scenarios. While these foundational techniques lay the groundwork for effective closures, hand surgeons utilize specialized approaches for tendon repairs, capsular closures, and reticular closures.

Simple interrupted sutures are characterized by individual stitches that create strong fixation points. This makes them particularly effective for securing the edges of a wound under moderate tension. This technique allows for precise control over the tissue alignment, ensuring that each wound segment is well supported.

In contrast, running sutures are favored for their efficiency in quickly closing linear wounds. This method involves a continuous stitching pattern that accelerates the suturing process, making it a valuable choice in situations where speed is of the essence. However, the success of running sutures hinges on meticulous placement, as uniform tension distribution along the wound is crucial to prevent complications and ensure optimal healing outcomes. Each technique, thus, plays a vital role in achieving the best possible surgical results.

Vertical and horizontal mattress sutures are frequently utilized in surgical procedures when extra support is required to alleviate wound tension. These suturing techniques effectively distribute forces evenly across the edges of the wound, helping to minimize the risk of ischemia—reduced blood flow that can lead to tissue damage—and dehiscence, the separation of a surgical incision. Furthermore, horizontal mattress sutures eliminate cross-hatching on the skin that can lead to a cosmetically inferior scar. Vertical mattress sutures are helpful to move tension away from the skin edge in the skin with decreased vascularity.

Retention sutures leverage the skin's viscoelastic properties, similar to how a "Sure-Close" device functions. Typically, a thicker suture stretches the skin and prevents it from cutting out before the final sutures are placed.

Due to their biomechanical properties, these sutures are particularly advantageous for closing thick or high-tension tissues, such as those in the hand, for very thick skin, such as the palm. Locked horizontal mattress sutures dock tissue edges together. A recent study on porcine models provided insightful findings, revealing that vertical mattress suture configurations outperformed horizontal and cross-stitch configurations. Notably, the vertical mattress stitches demonstrated superior tensile strength, meaning they can withstand greater pulling forces without failing. They were also associated with the least occurrence of suture pullout, making them a reliable choice for surgeons [17].

Advanced techniques, such as figure-of-eight and pulley sutures, are commonly used in tendon repair. The figure-of-eight configuration provides robust fixation by capturing multiple tissue points, while pulley sutures enhance force distribution across the repair site. Comparative studies have demonstrated the superior biomechanical performance of these techniques in withstanding the high tensile forces generated during hand movements.

Specialized applications, such as nerve, tendon, and ligament repairs, require particular suture configurations. One such example was recently highlighted by Mckernan et al. In a recent article, Mckernan demonstrated that SutureTape may provide an alternative technique for arthrodesis of the MCP and PIP joints compared to surgical steel wire [11]. Such advancements in materials demonstrate that sutures may replace more rigid, metal-based anchoring materials in the future. In nerve repairs, the primary goal is to achieve precise alignment without excessive tension, which can impede axonal regeneration. Tendon repairs demand sutures with high tensile strength and knot security to withstand the dynamic forces of movement. Ligament repairs often involve combining suture techniques and augmentation with grafts to restore stability and function.

Innovations in Suture Technologies

The suturing field has witnessed remarkable technological advancements to improve healing outcomes and reduce complications. Bioactive sutures, for instance, incorporate coatings that promote tissue healing and minimize the risk of infection. Antibacterial coatings like triclosan have effectively reduced postoperative infections, particularly in high-risk surgeries [1].

Drug-eluting sutures represent another significant innovation. These sutures release therapeutic agents, such as antiinflammatory or antimicrobial drugs, directly at the repair site [5]. This localized delivery system enhances healing while minimizing systemic side effects. Antimicrobial sutures have the potential to decrease postoperative infection rates significantly. Developing protocols for antimicrobial sutures is as crucial as the sutures themselves. A randomized controlled trial recently outlined the indications and limitations of antimicrobial sutures for hand surgery [23]. Furthermore, emerging research suggests that drug-eluting sutures may be pivotal in reducing adhesion formation and scarring in hand surgery. A recent meta-analysis by Suleiman et al. demonstrated the efficacy of antibiotic sutures. The secondary meta-analysis showed a significant risk reduction in surgical site infections when antibiotic-eluting sutures were employed [19].

Smart sutures, equipped with embedded sensors, are at the forefront of surgical innovation. These sutures can monitor wound tension and temperature parameters, providing real-time data to guide postoperative management [7]. Future trends in suture technologies include biodegradable sensors and sutures that respond dynamically to changes in the wound environment, further enhancing their therapeutic potential.

Challenges and Limitations

Despite their numerous advantages, sutures are not without challenges and limitations. One of the most common issues is the risk of slippage or breakage, which can compromise the repair. High-tension zones, such as flexor tendons, are prone to suture failure due to the significant forces generated during movement. Tissue pull-through, where sutures cut through the tissue instead of holding it securely, is another frequent complication in delicate hand structures.

Moreover, certain suture types have been shown to predispose patients to a hypersensitivity reaction more than others. An exemplary study by Koppa et al. demonstrated that in a sample of 184 patients, patients who had chronic catgut used in their surgeries were 2.82 times more likely to develop a hypersensitivity reaction than those who did not [8]. Understanding the patient's history of atopy, sensitivity to different materials, and prior surgical history are central considerations in clinical decision-making and suture material selection.

Mitigating these risks requires a combination of careful material selection and refined surgical techniques. Strategies such as using reinforced suture configurations and selecting materials with optimal elasticity and tensile strength can significantly reduce failure rates. However, achieving consistent success in high-tension repairs remains challenging, underscoring the need for ongoing research and innovation.

Evidence-based guidelines are essential to maximize the effectiveness of sutures in hand surgery. Surgeons should consider anatomical site, tissue type, and repair complexity when selecting suture materials and techniques. For example, absorbable sutures are ideal for internal repairs where a long-term presence may irritate, while nonabsorbable sutures provide lasting support in high-tension areas. A randomized controlled trial by Theopold et al. demonstrated that absorbable Vicryl Rapid sutures were superior to nonabsorbable Novafil in the context of comfort, scarring, and pain. No significant difference was identified in the infection rate between the two types [20].

Tendon Repair and Pulley Repairs

Tailoring suture techniques to the specific requirements of each repair is equally important. Advanced configurations should be employed, such as pulley sutures for tendons or tension-relieving mattress sutures for thick tissues. Adhering to these recommendations enhances surgical outcomes and minimizes complications, ensuring optimal recovery for hand surgery patients.

These properties play a decisive role in tendon and pulley repairs. Tendon repairs typically demand sutures with high tensile strength to counteract the significant forces generated during finger flexion and extension. Pulley repairs, by contrast, require sutures that provide precise tension and minimize bulk, as excessive material can interfere with tendon gliding and increase the risk of adhesions (Fig. 25.3).

Tendon repair requires sutures that hold the tendon ends securely and distribute stress evenly across the repair site to prevent gapping. Sutures with superior tensile strength and controlled elasticity, such as braided polyester or high-tensile synthetic monofilaments, are preferred. A biomechanical analysis of four-strand flexor tendon

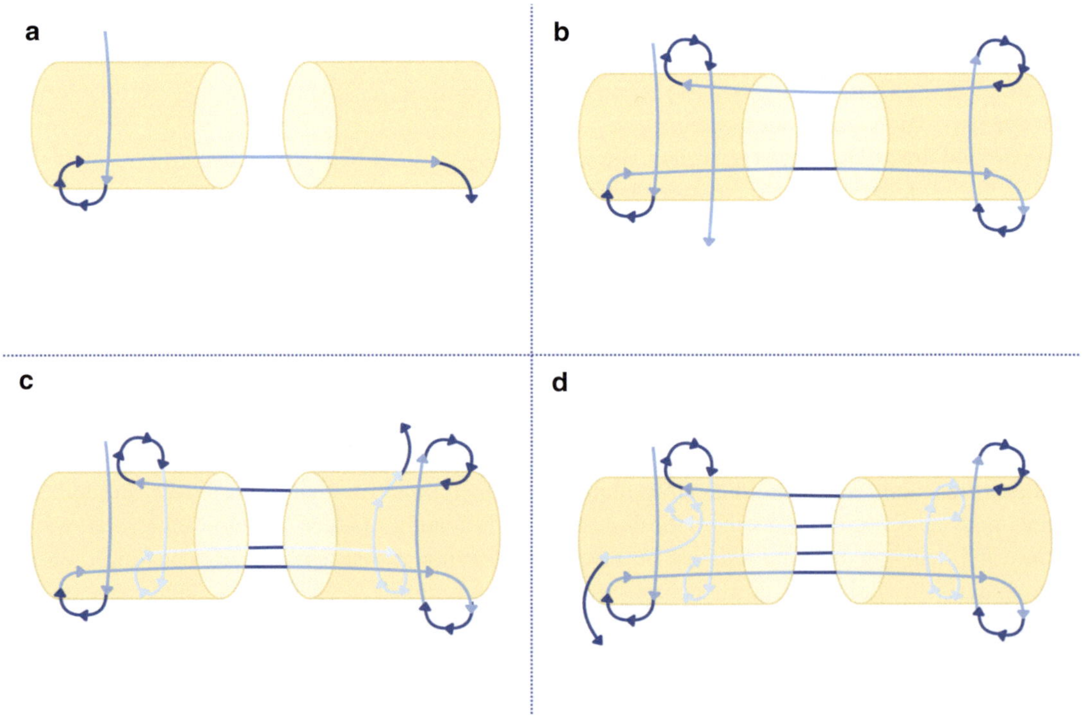

Fig. 25.3 Successive knotless flexor tendon repair steps using a four-strand modified Kessler suture technique (**a**–**d**). This technique demonstrates precise sequential steps to achieve a secure and efficient repair. Barbs are strategically positioned contrary to the direction of the tension force to enhance grip and stability. These barbs are carefully buried within the tendon structure to minimize surface irregularities, promote smooth gliding during tendon movement, and reduce the risk of adhesions. The method ensures optimal alignment and tensile strength, making it highly effective for restoring tendon functionality while minimizing bulk at the repair site [6]

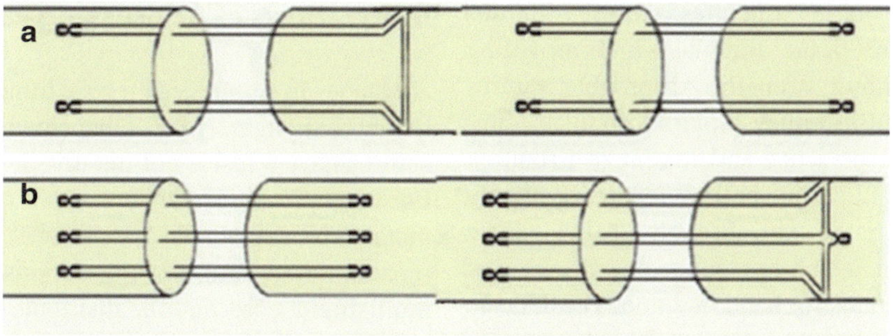

Fig. 25.4 Line diagram comparison of (**a**) conventional two-strand suture and (**b**) conventional four-strand suture techniques in tendon repair. The two-strand suture demonstrates a simple configuration with limited tensile strength and gap resistance, which is suitable for low-load scenarios. In contrast, the four-strand suture provides enhanced tensile strength and superior gap resistance due to the increased number of core strands, offering more excellent stability under higher load conditions

repair demonstrated that nylon suture repairs consistently performed inferiorly to stiffer firewire and proline-based sutures [9]. Several core suture strands, suture caliber, and suture configuration were also critical determinants in the strength of the repair. Intuitively, the number of core suture strands used demonstrated improved yield force and stiffness [21] (Fig. 25.4).

However, failure can occur due to improper technique, excessive tension on the repair, or insufficient knot security. Tendon suture failures often manifest as gapping or rupture, especially in high-tension repairs. These failures are exacerbated in cases where early mobilization protocols are initiated to improve functional outcomes. One such study demonstrating the difference in repair protocols underscored the importance of appropriate protocol selection. The study comparing grasping type, nonlocking repair (Strickland repair) to locked flexor tendon repair (MGH repair) demonstrated that Strickland failed 74% of the time from suture pullout compared to a 99% failure rate of MGH via suture breakage [12]. Different repair techniques' advantages, failure modes, and limitations are needed to guide clinical decision-making.

Pulley repairs, on the other hand, present unique challenges. The annular pulleys in the hand are thin and delicate structures critical for maintaining tendon alignment during motion. Pulley reconstruction with sutures has been proven in some instances to improve force production in flexor tendons postoperatively. A study demonstrated a new technique of pulley reconstruction involving a tendon graft pulled through a tunnel of the proximal phalanx [16]. The repair utilized Ethicon sutures for their absorbable properties and tensile strength. Failure of flexor tendon surgery can subsequently lead to further pulley incompetence, requiring additional surgeries (Fig. 25.5).

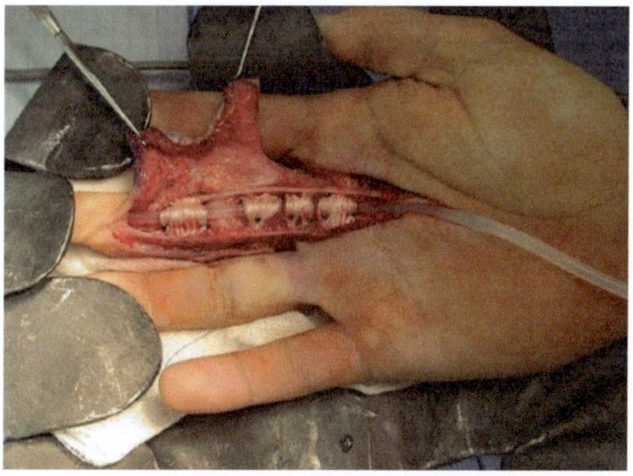

Fig. 25.5 Reconstruction of multiple pulleys over a rod using material harvested from discarded flexor tendon. This technique demonstrates the application of surplus flexor tendon tissue to reconstruct pulley systems to restore tendon alignment and functional biomechanics during flexor tendon repair

Repairing these structures requires sutures that minimize tissue trauma while providing adequate holding strength. Absorbable sutures are often used in pulley repairs to reduce long-term material presence and minimize irritation. However, failures in pulley repairs typically result from tissue pull-through or inadequate suture tension, which can lead to bowstringing of the tendon and loss of hand function. The delicate balance between achieving sufficient strength and preserving smooth tendon gliding is a key consideration in pulley repair techniques.

Suture Failures and Contributing Factors

Several factors contribute to the failure of sutures in both tendon and pulley repairs. One common issue is tissue pull-through, where the suture cuts through the repaired tissue rather than holding it securely. This failure is often observed when using inadequate elastic sutures or repairing friable tissues under high tension. They can add bulk in tight areas such as the flexor sheath, leading to friction and less ultimate motion. The choice of suture material and configuration can mitigate this risk; for instance, broader or reinforced suture patterns distribute forces more evenly and reduce the likelihood of pull-through.

Knot security is another critical aspect. Inadequately secured knots can unravel under load, compromising the repair. Research has shown that certain materials, such as monofilament nylon, may require additional throws to ensure knot stability compared to braided alternatives. However, excessive knot bulk can interfere with tissue healing and function, particularly in confined spaces like the digital pulleys.

Suture degradation is another potential source of failure, particularly with absorbable materials. While these sutures are designed to degrade predictably, factors such as infection, inflammation, or poor vascularity at the repair site can accelerate degradation, leading to premature loss of tensile strength.

Innovations in Addressing Failures

Advances in suture design have aimed to address these challenges. Like other medical devices, sutures have a history of iterative improvements. For example, suture anchors have evolved from metal to biodegradable materials [22]. For tendon repairs, newer suture configurations, such as the multistrand core suture, distribute stress more evenly across the repair site, reducing the likelihood of gapping. Similarly, coated sutures with enhanced knot-holding properties have been developed to improve security without increasing bulk. New devices to support surgery in tendon repair also prove to be promising. A novel stainless steel and ultrahigh tenacity polyethylene implant named CoNextions TR Tendon Repair System provides a stronger, faster, and thinner alternative to traditional suture repair techniques and can be used for repairs of flexor and extensor tendons in the hand, wrist, and forearm [4]. Continued advancements in material science and biotechnology hold promise for cheaper, more reliable devices in the future.

The development of thinner, high-strength materials has allowed for effective pulley repairs that minimize interference with tendon gliding. Additionally, biologic scaffolds or adjuncts, such as dermal grafts, have been explored to reinforce pulley repairs and reduce the risk of failure in high-tension zones.

As surgical techniques and suture materials evolve, the ability to tailor repairs to the specific biomechanical demands of hand structures will lead to better outcomes. However, the success of these repairs ultimately depends on a thorough understanding of the biomechanical properties of sutures and their interaction with the repaired tissues.

Conclusion

Sutures are essential in hand surgery, promoting effective healing in a region characterized by intricate anatomy and demanding functional capabilities. The evolution of suturing materials, from traditional natural threads to cutting-edge synthetic and bioengineered options, reflects advancements in surgical methodologies and highlights the increasing emphasis on improving patient outcomes.

Suturing precision is essential in hand surgery, and selecting the right suture material is crucial in the healing process. Various sutures possess unique biomechanical properties, including tensile strength, elasticity, and tissue reactivity, which can significantly impact recovery outcomes. Because of their durability and strength, nonabsorbable sutures are typically utilized for internal repairs, such as tendon and ligament reconstructions. In contrast, absorbable sutures are often preferred for external use, particularly in pediatric patients under five, as they tend to minimize scarring and alleviate the anxiety associated with suture removal. Additionally, absorbable sutures like chromic sutures are commonly employed to prevent complications related to nail growth.

Furthermore, understanding the various suturing techniques—such as interrupted, continuous, and mattress sutures—enables surgeons to optimize tissue apposition and distribution of tension across wound edges. This knowledge is vital in achieving the best surgical repairs while minimizing complications like infection or dehiscence.

Hand surgeons must comprehensively grasp the mechanical characteristics of diverse suture types and the intricacies of suturing techniques.

References

1. Chittoria RK, Reddy BP. Suture materials—recent advances. CosmoDerma. 2023;3:175. https://doi.org/10.25259/CSDM_176_2023.
2. Faris A, Khalid L, Hashim M, Yaghi S, Magde T, Bouresly W, Hamdoon Z, Uthman AT, Marei H, Al-Rawi N. Characteristics of suture materials used in oral surgery: systematic review. Int Dent J. 2022;72(3):278–87. https://doi.org/10.1016/j.identj.2022.02.005. ISSN 0020-6539. https://www.sciencedirect.com/science/article/pii/S0020653922000302.
3. Hochberg J, Meyer KM, Marion MD. Suture choice and other methods of skin closure. Surg Clin N Am. 2009;89(3):627–41., ISSN 0039-6109. https://doi.org/10.1016/j.suc.2009.03.001.
4. Irwin CS, Parks BG, Means KR. Biomechanical analysis of novel flexor tendon coupler versus suture repair. Poster presented at AAHS. 2018.
5. Janiga P, Elayarajah B, Rajendran R, Rammohan R, Venkatrajah B, Asa S. Drug-eluting silk sutures to retard post-operative surgical site infections. J Ind Text. 2012;42(2):176–90. https://doi.org/10.1177/1528083711432948.
6. Jordan MC, et al. Suture material for flexor tendon repair: 3-0 V:Loc versus 3-0 Stratafix in a biomechanical comparison *ex* vivo. J Orthop Surg Res. 2014;9:72. https://doi.org/10.1186/s13018-014-0072-9.
7. Kalidasan V, Yang X, Xiong Z, et al. Wirelessly operated bioelectronic sutures for the monitoring of deep surgical wounds. Nat Biomed Eng. 2021;5:1217–27. https://doi.org/10.1038/s41551-021-00802-0.
8. Koppa M, House R, Tobin V, et al. Suture material choice can increase the risk of hypersensitivity in hand trauma patients. Eur J Plast Surg. 2023;46:239–43. https://doi.org/10.1007/s00238-022-01986-7.
9. Lawrence TM, et al. A biomechanical analysis of suture materials and their influence on a four-strand flexor tendon repair. J Hand Surg. 2005;30(4):836–41.
10. Lekic N, Dodds SD. Suture materials, needles, and methods of skin closure: what every hand surgeon should know. J Hand Surg Am. 2022;47(2):160–171.e1. https://doi.org/10.1016/j.jhsa.2021.09.019. ISSN 0363-5023. https://www.sciencedirect.com/science/article/pii/S0363502321006146.
11. McKernan MJ, Diaz MA, Kucharik M, Coutelle N, Simon P, Doarn MC. Biomechanical comparison of surgical steel wiring and suture tape tension band techniques for arthrodesis of the metacarpophalangeal and proximal interphalangeal joint. J Hand Surg Glob Online. 2024;6(6):898–902. https://doi.org/10.1016/j.jhsg.2024.08.014. ISSN 2589-5141. https://www.sciencedirect.com/science/article/pii/S2589514124001804.
12. Miller B, Dodds SD, deMars A, Zagoreas N, Waitayawinyu T, Trumble TE. Flexor tendon repairs: the impact of fiberwire on grasping and locking core sutures. J Hand Surg Am. 2007;32(5):591–6. https://doi.org/10.1016/j.jhsa.2007.03.003.
13. Muslu Ü, Alıç T, Demir E, Gölpınar M, Bilici İ, Demirezen MO. Biomechanical comparison of different suture materials with different techniques in tendon repair: an ex-vivo study. Hand Surg Rehabil. 2023;42(5):435–41. https://doi.org/10.1016/j.hansur.2023.06.010. ISSN 2468-1229. https://www.sciencedirect.com/science/article/pii/S2468122923001202.
14. Nemirov D, Dentcheva E, Thurmond T, Bachoura A, Hirsch D, Tosti R. A retrospective comparison of absorbable versus non-absorbable sutures

for elective hand surgery wound closures. J Hand Microsurg. 2024;100178. https://doi.org/10.1016/j.jham.2024.100178. ISSN 0974-3227. https://www.sciencedirect.com/science/article/pii/S0974322724005404.
15. Pastor T, Zderic I, van Knegsel KP, et al. How many knots are necessary to secure two high-strength suture tapes? A biomechanical comparative analysis. Arch Orthop Trauma Surg. 2024;145:43. https://doi.org/10.1007/s00402-024-05638-2.
16. Schöffl I, Meisel J, Lutter C, Schöffl V. Feasibility of a new pulley repair: a cadaver study. J Hand Surg Am. 2018;43(4):380.e1–7. https://doi.org/10.1016/j.jhsa.2017.08.028. ISSN 0363-5023. https://www.sciencedirect.com/science/article/pii/S0363502317315514.
17. Senju T, Takeuchi N, Kozono N, Higaki H, Shimoto T, Nakashima Y. Biomechanical comparison of a horizontal mattress, cross suture and vertical mattress for repair of a tendon weave in a porcine model. J Hand Surg Asian Pac Vol. 2022;6(3). https://doi.org/10.1142/S2424835522500436.
18. Su BW, Solomons M, Barrow A, et al. A device for zone-II flexor tendon repair. Surgical technique. J Bone Joint Surg Am. 2006;88(Suppl 1 Pt 1):37–49. https://doi.org/10.2106/JBJS.E.00978.
19. Suleiman AS, Abbass M, Hossain M, Choudhary P, Bhattacharya P, Islam MA. Impact of antibiotic-coated sutures on surgical site infections: a second-order meta-analysis. Int J Surg. 2024;110(1):507–19. https://doi.org/10.1097/JS9.0000000000000822.
20. Theopold C, Potter S, Dempsey M, O'Shaughnessy M. A randomized controlled trial of absorbable versus non-absorbable sutures for skin closure after open carpal tunnel release. J Hand Surg Eur Vol. 2012;37(4):350–3. https://doi.org/10.1177/1753193411422334.
21. Viinikainen A, Göransson H, Huovinen K, Kellomäki M, Rokkanen P. A comparative analysis of the biomechanical behaviour of five flexor tendon core sutures. J Hand Surg Br. 2004;29(6):536–43. https://doi.org/10.1016/j.jhsb.2004.05.007.
22. Visscher LE, Jeffery C, Gilmour T, Anderson L, Couzens G. The history of suture anchors in orthopaedic surgery. Clin Biomech (Bristol). 2019;61:70–8. https://doi.org/10.1016/j.clinbiomech.2018.11.008. Epub 2018 Nov 23.
23. Wormald JCR, Rodrigues JN, Cook JA, Prieto-Alhambra D, Costa ML. Hand and wrist trauma: antimicrobials and infection (HAWAII): a protocol for a multicentre, feasibility study of antimicrobial sutures in hand and wrist trauma surgery. Bone Jt Open. 2022;3(7):529–35. https://doi.org/10.1302/2633-1462.37.BJO-2022-0048.
24. Wu K, Tang P. The history of knots and surgical suturing. In: Tang P, Wu K, Fu Z, Chen H, Zhang Y, editors. Tutorials in suturing techniques for orthopedics. Singapore: Springer; 2021. https://doi.org/10.1007/978-981-33-6330-4_1.

Artificial Hands

26

Dmitriy Peresada, Majd Mzeihem, and Danil Rybalko

Introduction

We rely on our hands to experience and manipulate the world around us. Upper extremity amputations dramatically change a patient's ability to interact with the surrounding environment. Artificial prosthetics help bridge the deficit created by the anatomical changes following an amputation.

Patients acquire upper extremity amputations due to several causes. These include trauma, infection, burns, electrocution, vascular complications, oncological processes, contractures, and congenital differences. The leading cause of all upper extremity amputations in men ages 15–45 is trauma, accounting for 80% of acquired amputations. Of these injuries, 78% are transphalangeal amputations [1].

Incidence of major trauma-related upper extremity amputations, defined as injuries at the level of the palm and proximal, is 46 per 100,000 from 2009 to 2012, according to the National Trauma Database. The most common definitive surgical procedures performed for such injuries were at the levels of the humerus (35%), forearm (30%), and hand (14%) [2].

Depending on the level of amputation, the resulting disability is often significant. Even in cases of transphalangeal amputations, patients suffer substantial setbacks. Sagiv et al. reported a return to normal grip strength in only 43% of manual laborers with proximal finger amputations. In their study, only 32 out of 50 patients returned to the same occupation at an average of 7.5 months following injury [3]. Patients with major traumatic upper extremity amputations, including those proximal to the carpus, experience severe disability as indicated by patient-reported outcome measures. The median Disabilities of the Arm, Shoulder and Hand (DASH) score in one study was 33 for revision amputations. Many of the study's patients reported cold intolerance, low hand function, and aesthetic dissatisfaction—nearly all used prostheses [4].

The primary goal of surgical innovation and technological development of hand prosthetics is to improve the functional outcomes and quality of life. An ideal prosthesis offers restoration of motor functions, including fine motor control, provides sensory feedback, and appears lifelike. Sensory feedback includes the ability to perceive touch, pressure, position, and temperature, which enhances patients' control of the prosthesis and simulates the experience of a native extremity. Ultimately, prosthetics must also aid in the patients' psychological recovery and social integration.

D. Peresada (✉) · M. Mzeihem · D. Rybalko
Department of Orthopaedic Surgery, University of Illinois at Chicago, Chicago, IL, USA
e-mail: dperes2@uic.edu

F. Amirouche et al. (eds.), *Biomechanics of the Hand*,
https://doi.org/10.1007/978-3-032-04688-8_26

A number of additional requirements remain. Device comfort, weight, and intuitive control are reported as key factors in reducing device abandonment [need a source]. Biddiss et al. reported prosthetic abandonment of up to 20% in a 25-year survey [5]. Shortcomings in these areas may be responsible for the inferior patient outcomes reported with upper extremity amputations compared to limb salvage, despite increased complications associated with salvage [6].

To address these challenges, a variety of prosthetic devices and surgical techniques are being developed and employed. All are associated with unique benefits and advantages, while many are not yet commercially available. The following text provides an overview of commonly utilized hand prosthetic methods and emerging techniques.

Prosthetic Designs

Prosthetic designs can be categorized based on the level of amputation. These include wrist disarticulation/trans-radial amputations, elbow disarticulation and transhumeral amputations, and interscapulothoracic amputations. Alternatively, prosthetic devices are classified based on the mechanisms that power them. This classification includes passive, body-powered, or externally powered units [7].

Passive Prosthetics

Passive Cosmetic Prosthetics

Active components, such as proximal joint movement or motors, do not power passive devices. The most mechanically simplistic devices, such as passive silicone restorations, are purely cosmetic. These mimic the amputated parts using silicone custom-painted with high definition.

Kuret et al. found that cosmetic silicone finger prostheses provide patients with aesthetically satisfying results, improve social interaction, and enhance quality of life [8]. Additionally, cosmetic prosthetics were found to aid patients in overcoming the initial psychological trauma of amputation and adjusting to the associated physical limitations [9]. Despite the nomenclature, cosmetic prostheses sometimes improve appearance and enhance physical hand functions. This occurs via increasing the hand's span, providing an opposable post, and/or creating a palmar base for holding objects, helping facilitate bimanual activities [10]. Cosmetic prostheses are typically lightweight as compared to mechanically sophisticated alternatives. However, cosmetic devices were found to result in inferior function compared to active designs, which will be discussed later in this chapter. A large group of veterans with primarily forearm and distal amputations and patients using cosmetic devices reported more incredible difficulty in performing manual tasks compared to those using body-powered or myoelectric prostheses [11].

Passive Positional Prosthetics

Passive positional prosthetics are designed to surpass the functionality of purely passive devices, thus allowing the patient to perform a broader range of manual tasks. Positional functionality permits manually adjusting the position of one or multiple device joints. The devices typically utilize a soft inner socket and may be made of carbon fiber to decrease weight. The adjustable positional articulations may simulate larger joints, such as the elbow, or smaller joints, such as those of the thumb or the digits, allowing the degree of flexion and extension to be adjusted for the performance of specific tasks. Additionally, in a study by Weiss et al., passive positional prosthetics were found to reduce phantom limb pain by increasing the use of the amputation stump. This effect is not typically observed with passive cosmetic prosthetics, which do not encourage the active use of the residual limb [12]. Recent iterations of passive positional hand prostheses include the mHand Adapt (Moveable, Netherlands), which allows the patients to intuitively position all digits of the prosthetic hand, control the grasping force, and quickly release the grasp. Krinis et al. report that the mHand Adapt device achieves higher functionality scores and decreases contralateral hand interaction time compared to the prior generation of passive adaptable prostheses, allowing users to complete a broader range of tasks with greater ease [13] (Fig. 26.1).

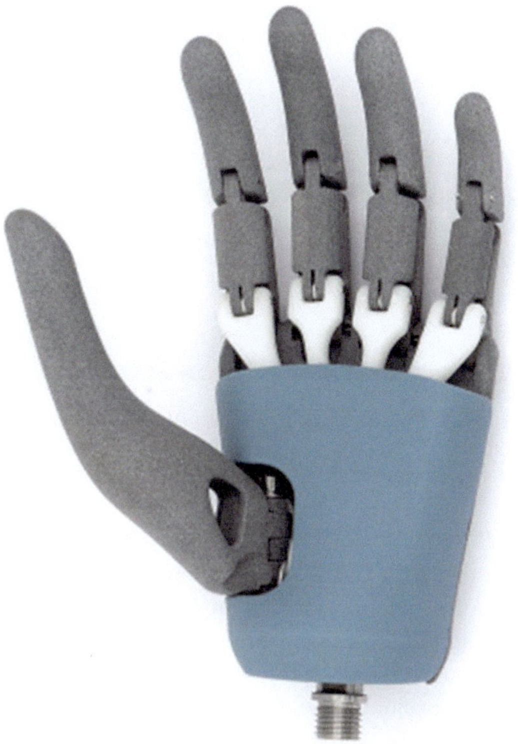

Fig. 26.1 mHand Adapt prosthesis. (Adapted from Krinis et al. [13], licensed under CC-BY 4.0)

Active Prosthetics

Body-Powered Prostheses

Body-powered prostheses allow patients to dynamically move and control their device by harnessing movement from functional native joints above the level of the amputation. The shoulder, elbow, or forearm movement may be harnessed via a control cable mechanism to control the distally attached prosthesis. This, in turn, actuates the prosthetic mechanism, such as the gripping motion of a prosthetic hand. For shoulder-powered hand prostheses, for example, shoulder flexion is typically used to control the prosthetic hand. The cable configuration must balance short enough to allow effective control and long enough to provide a sufficient range of motion [14] (Fig. 26.2).

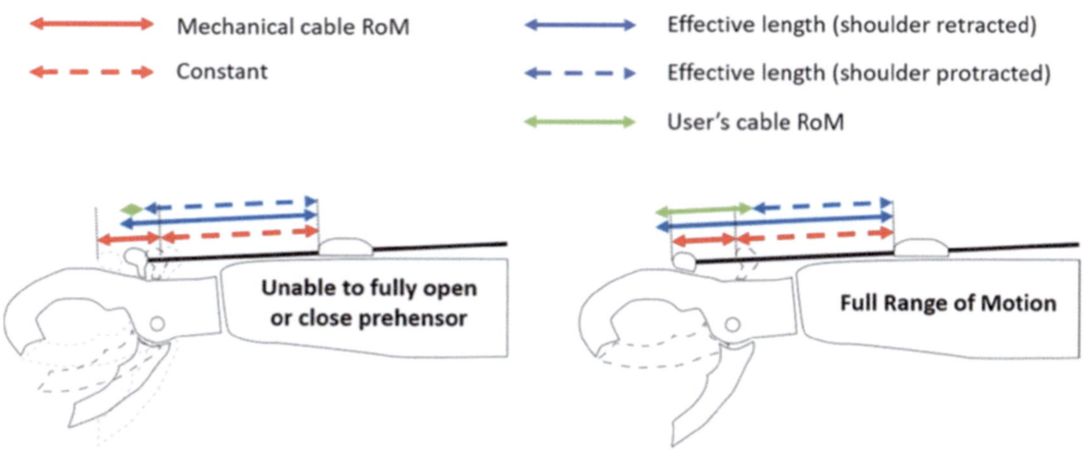

Fig. 26.2 Effective cable length and range of motion overlap. (Adapted from Chadwell et al.[14])

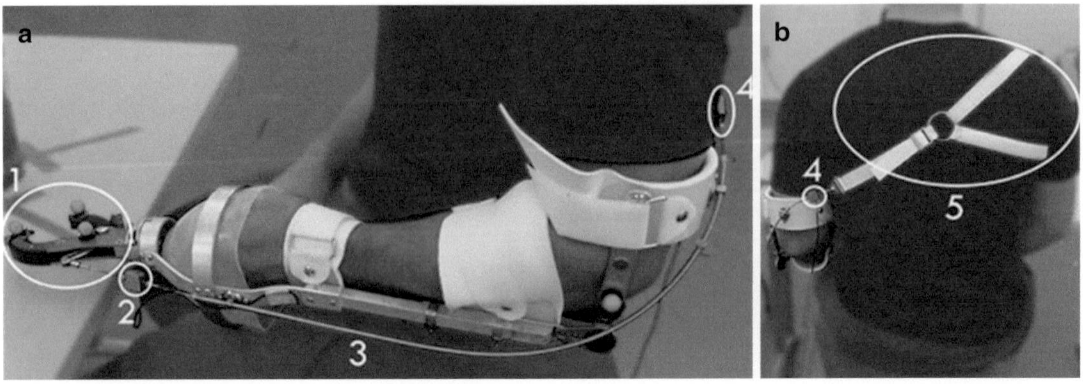

Fig. 26.3 Side view (**a**) and back view (**b**) of a participant wearing the customized voluntary-closing hand prostheses. The prehensor (1) is attached to the fitting and linked to the "figure-of-nine" harness (5) through a Bowden cable (3). Force sensor 1 (2) and force sensor 2 (4). (Adapted from [18], licensed under CC-BY 4.0)

Body-powered prostheses may be activity-specific, with varied terminal devices to interact with the patient's surroundings. The most common terminal devices include hooks and posts. These devices are reportedly the most commonly used prosthetics for below-the-elbow amputations. In one study, 97% of pediatric unilateral below-the-elbow amputees were fitted with a body-powered voluntary closing terminal device, and 41% of patients selected it as their prosthesis of choice in the long term [15].

There are several advantages of body-powered prostheses when compared to passive prostheses. Body-powered devices are mechanically simple, which makes them reliable and durable, especially in demanding environments. Patients typically require less time to learn how to use body-powered devices, as the intuitive control system utilizes simple body movements, facilitating quicker adaptation [16]. Body-powered prosthetics also provide rudimentary haptic feedback, allowing patients to sense the force they are applying through the prosthesis, which can improve the control and functionality of the device [17]. The wide range of activity-specific terminal devices, such as hooks and pegs, increase functionality, especially in performing activity- and labor-specific manual tasks. Resnik et al. found that body-powered prostheses were associated with less difficulty performing tasks and higher functionality when compared to cosmetic devices [11]. In contrast to externally powered myoelectric prostheses, body-powered prostheses are less expensive and, thus, more accessible to patients.

Despite their widespread use, body-powered prostheses present the patient with certain limitations. The physical demands of using a body-powered prosthesis, including the forces and postures required, can lead to patient discomfort and fatigue [17]. Higher forces that are typically needed to produce sufficient pinch force also result in decreased control of fine motor movements. This results in difficulty with precise manipulation of objects, particularly in voluntary-closing hand prostheses [18] (Fig. 26.3).

The harness and control cable mechanism can restrict the patient's reachable workspace. This results in a limited range of motion and decreased ability to perform tasks that require reaching across the body or overhead [14]. The overall quality of movements, including dexterity and speed, is reported to be inferior to more sophisticated externally powered devices, such as myoelectric prostheses [19].

Externally Powered Prostheses

Myoelectric Prostheses

Externally powered upper extremity prostheses utilize batteries, pneumatics, hydraulics, and/or motors to power the device. A commonly used type of externally powered device is a myoelec-

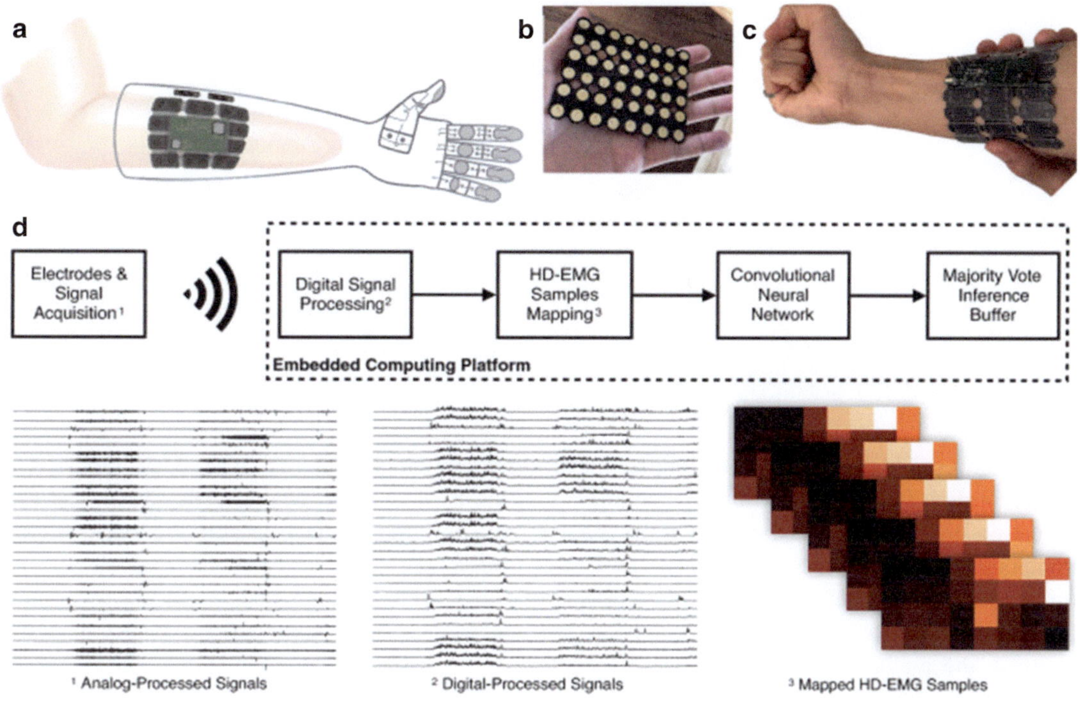

Fig. 26.4 Prosthesis system concept and hardware components. (**a**) Integrated hand prosthesis concept. (**b**, **c**) Bottom and top views of the electrode array and data acquisition platform. (**d**) Block diagram of the myoelectric prosthesis control system. The user's muscle contractions are detected by a 32-channel electrode array, filtered, digitized, and transmitted wirelessly. Processed via an MAV filter, the signals are mapped into 4 × 8 matrices, forming muscle activation maps. These are analyzed by a convolutional neural network, with final gesture recognition determined by majority voting. (Adapted from [20], licensed under CC-BY 4.0)

tric prosthesis. Myoelectric technology relies on electromyographic (EMG) electrodes to detect signals from the voluntary contraction of residual surface musculature at the stump site, which are then processed by a computer to trigger electrically powered motion of the prosthesis [need source].

Computer processing algorithms range from simple threshold-based systems to complex pattern recognition systems that involve machine learning to interpret multiple degrees of freedom movements. In a recent article, Tam et al. present a novel device that uses high-density EMG and deep learning algorithms to adapt to the patient's specific muscle contraction patterns, allowing for intuitive and responsive prosthesis control [20]. The processed signals control motors in the prosthesis, creating functional movements such as opening and closing the hand and individual finger movements. Advanced prostheses may also provide sensory feedback via sensors that detect tactile information (Fig. 26.4).

A variety of techniques have been developed to control myoelectric prostheses. Research efforts aim to amplify individual signals and increase the number of discrete signals available for EMG detection to allow precise control of a more significant number of prosthetic articulations. Proximal amputations present a challenge for applying myoelectric prostheses, as few muscle groups are available to provide discrete detectable signals. Targeted muscle reinnervation (TMR) is a surgical solution that involves a transfer of motor nerves to muscle or to other motor nerves to increase the amplitude and number of signals that are detectable by surface electrodes. Consequently, TMR allows patients with proximal amputations to increase the number of controlled prosthetic articulations [21].

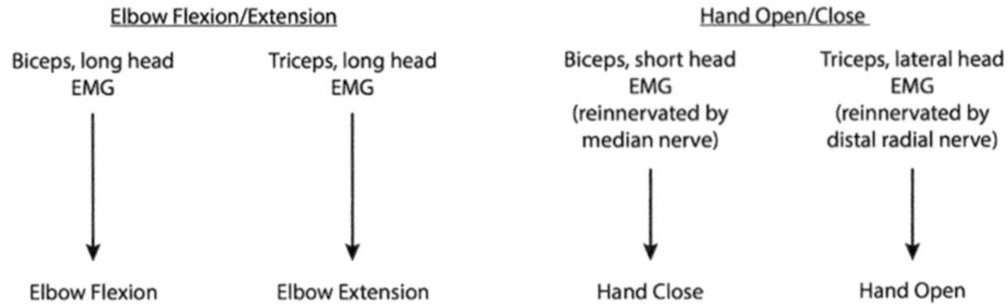

Fig. 26.5 Conventional myoelectric control for transhumeral amputation utilizes targeted muscle reinnervation, where native biceps and triceps control the elbow, while reinnervated biceps and triceps control the hand. Patients generate EMG signals by attempting to move the lost limb's corresponding joint, eliminating the need for mode switching in the same prosthesis. (Adapted from [27]) (permission granted from Thieme)

Numerous TMR operative techniques have been reported. TMR methods can be categorized based on the level of amputation. For transhumeral amputations, the radial nerve may be transferred to the motor nerve of the triceps' lateral head, preserving the triceps' long head. In contrast, the median nerve is transferred to the motor branch of the medial head of the biceps [22]. For transradial amputations, the radial nerve may be transferred to the motor branches of the brachioradialis [23] and the median nerve to the motor branches of the flexor muscles [24]. The starfish procedure has been developed for partial hand amputations as a variation of TMR. The technique involves the transfer of dorsal interossei muscles to subcutaneous positions that are advantageous in detectable signal strength and isolation. Early clinical reports indicate that the Starfish procedure resulted in reliable myoelectric signals detected from the transferred interossei muscles, which allowed for independent control of each digit using a myoelectric prosthesis. Patients enrolled in the study had improved outcomes and could perform precise and independent finger movements [25]. TMR has the additional benefit of decreasing phantom limb pain. The proposed mechanism provides motor targets to motor nerve stumps and reduces neuroma formation [26] (Fig. 26.5).

Myoelectric prostheses may be limited by the strength and number of detectable signals, even in conjunction with TMR [28], as well as the lack of proprioception and tactile feedback. Due to these and other issues, abandonment rates of myoelectric prostheses have been reported as high as 44% [29]. Myoelectric prostheses provide only rudimentary sensory feedback, leading to compensatory strategies that may cause musculoskeletal problems over time [30]. Most myoelectric devices require patients to rely on visual cues, which differs significantly from natural limbs and limits the efficiency of the prosthesis. Integration of tactile feedback mechanisms with myoelectric devices has improved object recognition accuracy, user confidence, and the ability to apply appropriate grip forces during object manipulation [31–34].

Peripheral Nerve Interfaces

To address the limitations of myoelectric prostheses, peripheral nerve interfaces (PNIs) have been developed. PNI techniques allow selectively recording nerve signals and/or stimulating nerves directly via nerve electrode implants. The ultimate goal of PNIs is to increase the number of controlled articulations, increase prosthetic dexterity, and provide improved sensory feedback [35].

PNIs can be categorized into intraneural, extraneural, and regenerative interfaces. Intraneural interfaces are electrodes inserted directly into the nerve fascicles, providing high selectivity but potentially causing more tissue damage. One example is the longitudinal intrafascicular electrode. Extraneural interfaces are electrodes that wrap around nerves, such as cuff electrodes. These devices are less invasive but may offer lower selectivity. Regenerative inter-

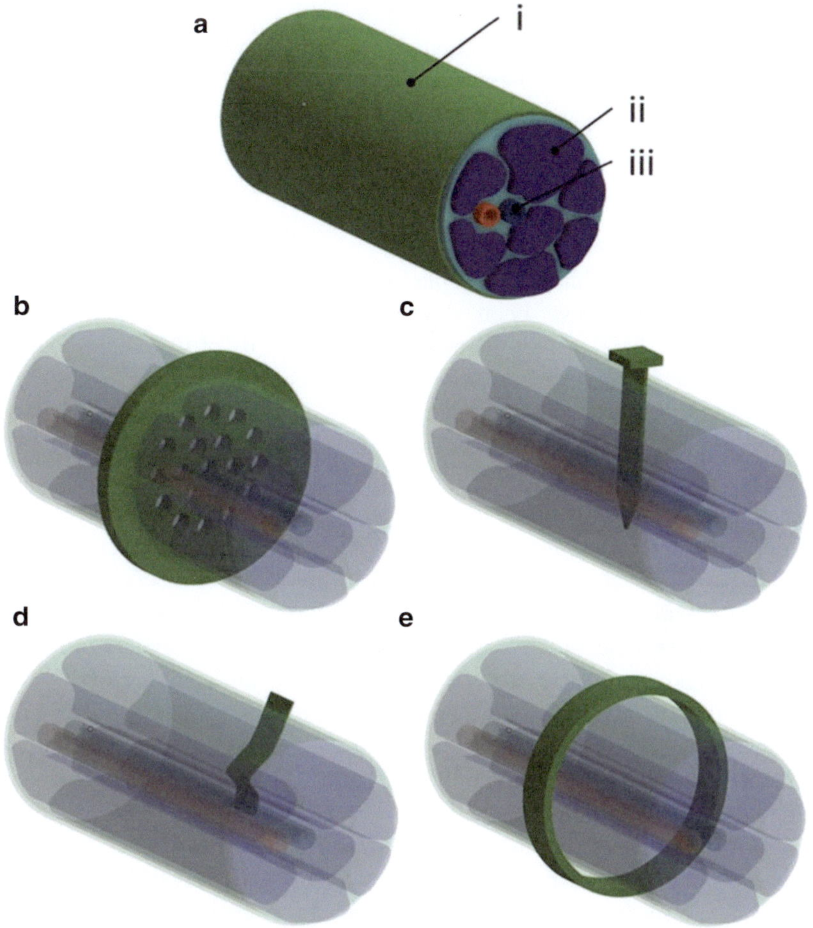

Fig. 26.6 (**a**) A simplified diagram of the peripheral nerve illustrating (i) the epineurium, (ii) a fascicle containing axons, and (iii) blood vessels. The schematic also depicts the representative positioning of electrodes for electrical interfacing with the PNS, including (**b**) regenerative, (**c**) intrafascicular, (**d**) interfascicular, and (**e**) extraneural placements (excluding electrical tethering in the illustrations). (Adapted from [36], licensed under CC-BY 4.0)

faces encourage nerve regeneration through the device, promoting close contact with the electrodes to potentially increase signal quality and selectivity [35] (Fig. 26.6).

Clinical applications of PNIs are currently limited to small case series. Davis et al. report on two patients with transradial amputations implanted with slanted electrode arrays (a type of intraneural nerve interface). Patients were able to utilize recorded neural activity to activate a prosthetic hand. The nerves were also stimulated to stimulate the touch sensation [37]. Vu et al. describe a technique combining regenerative PNIs and TMR principles. Fascicles of nerves were separated and implanted into small free muscle transfers to increase the specificity of efferent signals. This produced electromyographic signals with large signal-to-noise ratios with surface and intramuscular electrodes. Using these signals, subjects successfully controlled a hand prosthesis in real time up to 300 days postoperatively without control algorithm recalibration [31, 32, 34]. Multiple authors also achieved simulated tactile feedback utilizing PNIs [37–39].

Limitations of PNIs have emerged despite the small number of reported subjects. Reported complications of PNIs include foreign body reaction, fibrosis, and device micromotion [35]. Recent investigations reveal future directions of PNI applications, such as improving biocompatibility and mechanical compliance to reduce the body's foreign response, thus promoting long-term functionality. Soft, stretchable microelectrode arrays integrated with three-dimensional-printed microclips interfacing with small nerves were developed and studied in vivo by Rowan et al. for this purpose [40]. Moon et al. implemented soft-actuated cuff electrodes in vivo, designed to envelop nerves with minimal contact pressure, thus preserving nerve integrity and ensuring stable electrical performance [41].

Hybrid Techniques

Novel devices combining multiple aforementioned techniques have been developed. For instance, the Life Under Kinetic Evolution (LUKE) arm is an upper extremity prosthesis that combines several advanced techniques to provide motor and sensory functions to patients with transradial and transhumeral amputations. The device attaches to the residual limb via osseointegration rather than a conventional socket. Additional surgical procedures involved in preparing the patient for this prosthetic include targeted muscle reinnervation (TMR), implantation of epimysial electrodes to provide efferent motor signals to the prosthesis, and implantation of peripheral nerve interfaces to collect afferent tactile sensory signals [42]. The reported advantages of this technique are numerous because it relies on the patient's intuitive intent to perform movements and does not require formal training. Patients in a small case series used the device for 3–7 years, indicating significant durability. Precision and control of the prosthetic continued to improve in all patients throughout the midterm study period, and the detected myoelectric activity increased over time. All patients reported perceived tactile sensation with the use of the prosthesis. Patients who previously had phantom limb pain reported resolution. Some patients could return to recreational activities such as rally racing and fishing [43] (Fig. 26.7).

Central Nervous System Interfaces

Central nervous system interfaces (CNSIs), also known as neural machine interfaces or brain machine interfaces, are direct communication pathways between the brain and the prosthetic device. CNSIs function by recording neural activity, decoding the signals to interpret the patient's intentions, and then using the decoded signals to control a prosthetic limb. CNSIs may be invasive or noninvasive. Noninvasive CNSIs utilize electroencephalograms (EEGs) obtained via surface electrodes placed on the scalp. Schwarz et al. performed EEG on healthy subjects to decode three hand movements (palmar grasp, lateral grasp, and wrist supination) to control a robotic arm in a simulation environment. Participants achieved an average accuracy of 48% for these movements, suggesting the viability of EEG-based control of prosthetics [44].

Alternatively, invasive CNSIs rely on intracranial EEGs termed electrocorticograms (ECoGs). ECoGs are obtained via sensors placed directly on the brain surface, providing a superior intensity and higher-resolution signal than EEG [45]. Yanagisawa et al. developed a real-time system using ECoG signals to control a prosthetic hand. This system decoded hand movements from ECoG signals in a poststroke patient, achieving an accuracy of 69.2% in predicting movement types. The prosthetic hand mimicked the patient's movements with an average delay of 0.37 seconds [46]. Employing an ECoG device implanted 2 weeks prior, Willsey et al. demonstrated the ability to precisely control a virtual hand with 4 degrees of freedom by a tetraplegic patient. This allowed the patient to navigate a virtual quadcopter on a computer screen [47]. Such findings demonstrate future potential to control highly functional prosthetics using novel central CNSI techniques.

Hybrid techniques, combining CNSIs and PNIs, have been created to improve robotic hand control. By analyzing both cortical and peripheral signals, Tombini et al. enhanced the classification of motor commands, leading to improved prosthetic hand control. This approach also led to cortical reorganization and mitigation of phantom limb pain in studied subjects [48].

Cost of Externally Powered Prostheses

Externally powered prostheses utilizing myoelectric sensors, PNIs, and CNSIs represent significant advancements in the evolution of hand prosthetics. The proposed benefits of myoelectric and other externally powered devices include

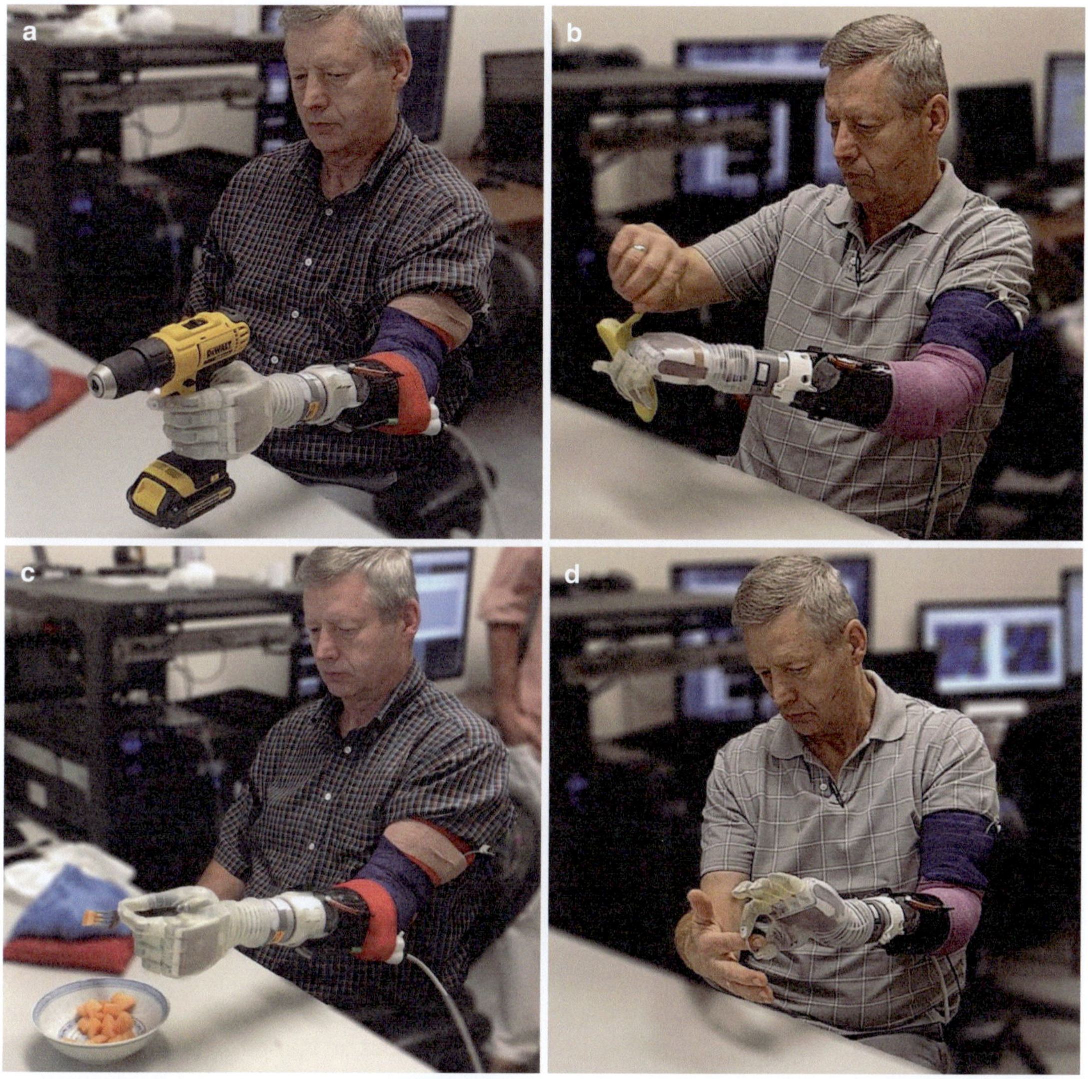

Fig. 26.7 Amputee participants utilized the modified Kalman filter to actively control all six degrees of freedom of the DEKA "LUKE" arm while performing various activities of daily living. Common tasks—such as using tools, eating, and dressing—were performed in a laboratory setting. The images depict amputee participant S6: (**a**) operating a drill, (**b**) peeling a banana, (**c**) eating with a fork, and (**d**) putting on or removing a ring. (Adapted from George et al. [42]) (permission granted)

superior prosthetic control, functionality, user satisfaction, and potential for sensory feedback. However, high surgical and prosthetic costs remain prohibitive for many patients. The cost of advanced prosthetics varies widely. Kerver et al. report mean annual costs for using standard commercially available myoelectric hand prostheses and the multigrip myoelectric hand prostheses of approximately $24,000 and $56,00, respectively. More advanced devices and techniques like CNSIs are only experimentally performed at academic institutions. Private entities are actively working on commercializing machine-brain interface technology, such as the Neuralink device, which utilizes thousands of implantable electrodes [49].

Sensory Feedback

Sensory feedback, including tactile sensation and proprioception, enables patients to control grip force and manipulate objects better. Lacking sensory feedback makes it difficult to achieve precise force application, such as when picking up fragile items. Sensory feedback alleviates the need for constant visual monitoring of the prosthetic extremity, decreasing the user's cognitive and physical load. Additionally, sensory input from the prosthesis contributes to a sense of embodiment, where the prosthetic limb is perceived as a natural part of the body [50].

Various techniques attempt to improve or restore sensation, including invasive and noninvasive methods. Noninvasive approaches strive to enhance sensory feedback in conventional prostheses, such as passive, passive positional, body-powered, and myoelectric prostheses. Several noninvasive strategies have been reported. Mechanical stimulation of the skin can be used to provide sensory information. This can be achieved through devices that apply pressure or deformation to the skin. A variation of mechanical stimulation involves vibratory motors placed on the skin to simulate different sensations. Electrical stimulation techniques have been developed to activate sensory nerves on the residual skin [51].

Invasive methods for providing sensory feedback include PNIs (such as extraneural electrodes, intraneural electrodes, and regenerative peripheral nerve interfaces), spinal cord stimulation, and CNSIs. Extraneural electrodes, such as cuff electrodes, are placed around nerves to provide sensory feedback by stimulating the nerve directly [51]. In a case series by Tan et al., electrical stimulation of upper extremity peripheral nerve interfaces (cuff electrodes) evoked tactile sensations in the prosthetic limb. This improved the patient's ability to manipulate objects using the prosthesis. Over a year after implantation of the peripheral nerve interfaces, the patients continued to experience touch sensations. Authors could simulate tapping, pressure, and vibration [38]. Unlike extraneural electrodes, intraneural electrodes are implanted within the nerve fascicles to collect and deliver electrical signals more precisely. Raspopovic et al. used intraneural electrodes to enable patients to effectively modulate the grasping force of hand prostheses with no visual or auditory feedback [52]. Regenerative PNIs involve nerve grafts to create a bidirectional interface that can provide both motor control and sensory feedback. In a case series by Gonzalez et al., RPNI stimulation consistently produced sensations such as tingling and kinesthesia, which were perceived in the amputated hand and forearm at a 54-month follow-up. One participant received RPNI stimulation proportional to prosthetic grip force to identify objects of different sizes and stiffness with 56% accuracy using stimulation alone, without visual feedback [30].

Epidural stimulation of the cervical spinal cord has been used to evoke tactile sensations perceived as coming from the missing limb, aiding in sensory discrimination tasks [53].

CNSIs interface directly with the brain and provide a unique opportunity to restore sensation in the prosthetic limb. Various authors have investigated this technique. O'Doherty et al. demonstrated that, in primates, a CNSI could control exploratory movements and provide artificial tactile feedback through ICMS (intracranial microstimulation), enabling monkeys to distinguish between different textures using a virtual-reality arm [54]. Armenta Salas et al. utilized ICMS to elicit cutaneous and proprioceptive sensations in a tetraplegic patient, enhancing the naturalistic perception of touch and limb position [55]. A recent study of human subjects demonstrated that multielectrode ICMS can produce stable and precise tactile sensations, which could significantly improve the functionality of brain-controlled bionic hands by providing more accurate and reliable sensory feedback. Authors utilized overlapping somatotropic cortical zones stimulated by multiple overlapping electrodes, rather than previously described approaches that utilized single electrodes for each sensory zone [56].

Hybrid techniques aim to restore proprioception and tactile feedback, thus enhancing the prostheses' functionality and user experience. D'Anna et al. used intraneural stimulation with myoelectric prosthesis to allow two patients with transradial amputations to effectively use the artificial hands to discriminate size and compliance with high proprioceptive acuity [57]. Marasco et al. demonstrated that combining tactile feedback, kinesthesia, and intuitive motor control through a complex neurorobotic interface significantly improves the performance of patients with upper extremity prostheses. Subjects using prosthetic devices that combine targeted motor and sensory reinnervation, a closed-loop neural-machine interface, and a noninvasive robotic architecture surpassed the performance of patients with body-powered prosthetics [58].

Future directions of prosthetic designs aim to improve further or restore sensory feedback. A recent novel design of a soft neuroprosthetic hand is described. Using a relatively affordable design and lightweight materials, including fingertip pressure sensors, Gue et al. demonstrated in one patient the ability to create primitive touch sensation and real-time closed-loop control. Patients using this device design outperformed a conventional rigid neuroprosthetic hand in speed and dexterity [59]. Other efforts are directed at creating synthetic skin substitutes to improve the sensory capabilities of hand prosthetics. Artificial skin prototypes integrate various sensors to detect pressure, strain, temperature, and other tactile stimuli, enabling sophisticated human-machine interactions and enhanced prosthetic functionality. Additionally, researchers have focused on creating flexible, stretchable materials capable of self-healing, which are essential for prolonged use and conformal attachment to irregular surfaces [28, 60].

Sockets

Sockets interface between the residual upper extremity and the prostheses. Sockets must be comfortable and durable to allow for daily use of prostheses. Sockets must fit firmly to provide a stable attachment site. Additionally, sockets ideally transfer sensory feedback to the residual limb [61].

Traditional Sockets

Traditional sockets are custom-molded from thermoplastics or laminated composites. Designed to fit the residual limb closely, traditional sockets provide a secure attachment and distribute pressure evenly. Some sockets are self-suspended, while others may require suspension via a harness [61]. Traditional sockets can be uncomfortable and unstable due to limb volume fluctuations, temperature changes, and sweating in the residual limb [62]. Three-dimensional (3D) scanning and printing techniques can be employed to manufacture sockets. Three-dimensional printing technology creates sockets that can be tailored to the exact shape of the residual limb, potentially improving fit and comfort [63].

Adaptive Sockets

Adaptive sockets incorporate dynamic adjustment mechanisms, such as air bladders, to accommodate changes in limb volume. This improves comfort and stability throughout a broader range of residual limb conditions and situations. Gu et al. report on novel self-adjusting air bladder sockets that continuously monitor and adjust the contact pressure, enhancing comfort during various activities [59].

Osseointegration

To address the limitations of traditional sockets, osseointegration techniques have been developed. Osseointegration refers to surgical and prosthetic methods that secure the prosthesis directly to the patient's residual bone. Typically performed in a staged fashion, the first stage of osseointegration consists of implanting a metal device into the bone of the residual limb. After bony ingrowth occurs around the prosthesis, the second stage connects an implant that protrudes through the skin and directly interfaces with the prosthesis [61].

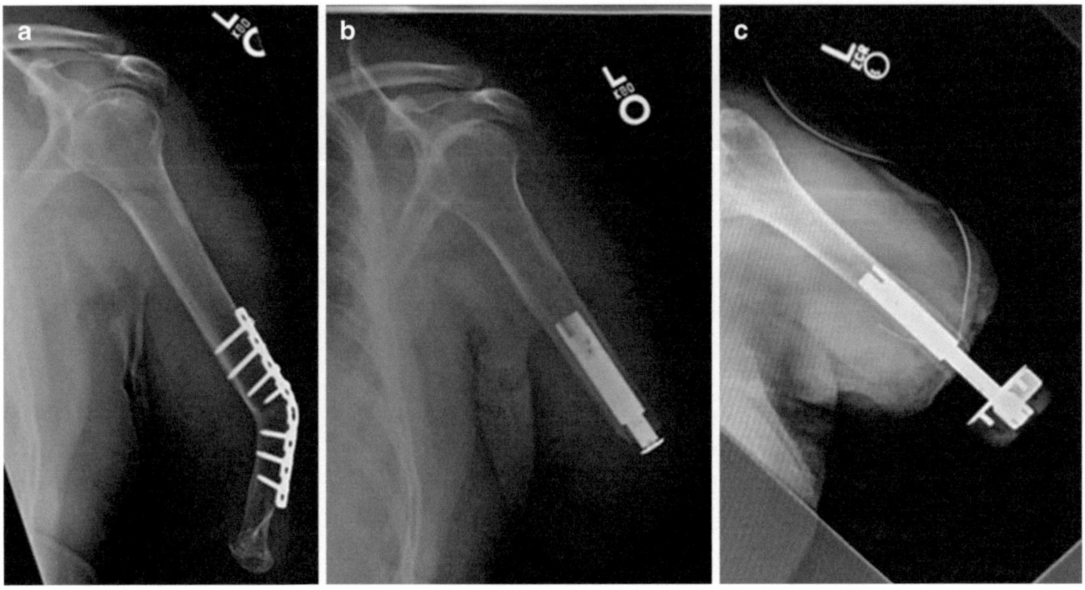

Fig. 26.8 From left to right: preoperative radiograph, postoperative radiographs after stage 1 and stage 2 of the OPRA Implant System procedure. (Adapted from Overmann et al. [67])

Various osseointegration surgical techniques and implants exist to create the implant-bone interface. Screw-type implants, such as the Osseointegrated Prostheses for the Rehabilitation of Amputees (OPRA) system, are commonly chosen. This type of implant involves a threaded design inserted into the bone [64]. Press-fit implants are also commercially available. These include the Endo-Exo Prosthesis (EEP) and the Osseointegration Prosthetic Limb (OPL) systems. Press-fit implants rely on an initial tight fit between the implant and the bone without the need for screws. Long-term stability is achieved via bony ingrowth into the implant [65]. Compressive devices were created for addressing massive bone loss in revision arthroplasty and complex reconstruction of bone tumors but are not available for osseointegration. The Compliant Pre-stress Device (CPS) uses compressive forces to enhance the integration of the implant with the bone. The primary advantage of compressive implants is immediate stability, allowing for early weight-bearing [66] (Fig. 26.8).

Osseointegration techniques allow more precise, intuitive prosthetic control, accomplished via a direct connection of the prosthetic to the bone. Additional benefits include reduced energy expenditure, increased range of motion, and enhanced sensory feedback [65]. In clinical comparisons, osseointegration-based prostheses were associated with improved prosthetic use, comfort, and quality of life compared to conventional socket prostheses. Patients reported superior control and functionality [64, 68] (Fig. 26.9).

Despite the reported advantages of osseointegration, there are significant challenges to its widespread use. Complications of this surgical technique occur both at the implant-bone and the implant-skin interface. In a series of 18 patients who underwent osseointegration procedures for transhumeral amputations, Tsikandylakis et al. reported a five-year survival of 83%. Superficial skin infections and superficial skin reactions were the most common complications, with 15 and 8 events reported, respectively. The majority of infections were effectively treated with antibiotics. Among the 17% of failed implants, all were caused by early loosening at the bone-implant interface. The most commonly reported radiographic findings are proximal trabecular buttressing, endosteal bone resorption and cancellization, cortical thinning, and distal bone resorption. Incomplete fractures after initial implantation into bone were noted in eight patients. None of

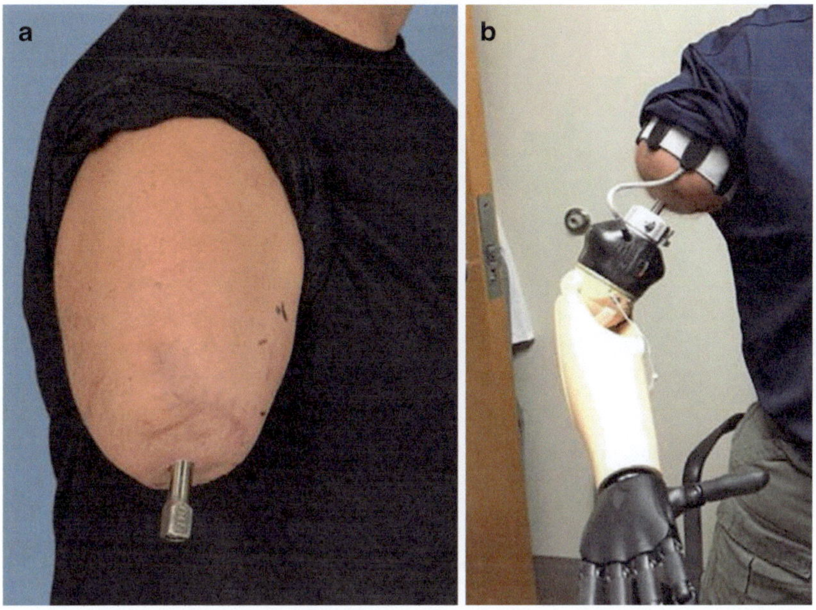

Fig. 26.9 Right transhumeral amputation following targeted muscle reinnervation and osseointegration using the Osseointegrated Prosthesis for the Rehabilitation of Amputees (OPRA) device (Integrum, Sweden). (**a**) The abutment extends through the skin at the distal end of the residual limb and is secured to the intramedullary fixture with an abutment screw. (**b**) The myoelectric prosthesis connects to the abutment via a white ring attachment. Without a traditional socket, the electrodes that detect muscle contractions and control the prosthesis are housed within a soft arm strap. Images courtesy of the Department of Defense Osseointegration Program. (Adapted from Bates et al. [69])

these required surgical treatment [70]. Taylor et al. report that the attached prosthetic device type can influence the loading at the bone-implant interface. More advanced prosthetic limbs, which are heavier, can increase the stress on the osseointegration implant, potentially affecting its long-term stability [71].

Rehabilitation

Regardless of the type of hand prosthesis, rehabilitative principles have emerged that guide patients' postoperative recovery. Preparation for the prosthesis involves preparing the residual limb and the patient for prosthetic fitting. This includes desensitization, range of motion exercises, and strengthening of the residual limb and surrounding muscles—the early fitting of the prosthesis, whether temporary or definitive, is recommended. Optimal outcomes have been reported with prosthetic fitting within 30 days of amputation [72]. Once the prosthesis is obtained and fitted, the patient undergoes a structured prosthetic training program. During the controls training stage, patients learn to operate the prosthetic device using the control mechanisms and practicing basic movements. Patients then progress to practicing repetitive drills. These exercises help reinforce the prosthesis' control through repetitive tasks, enhancing muscle memory and coordination. During the final bimanual functional skills training phase, patients focus on integrating the prosthetic hand into daily activities, emphasizing using both hands for functional tasks. A multidisciplinary approach is vital in effectively rehabilitating patients with amputations. Collaboration between various healthcare professionals, including certified prosthetists, occupational therapists, and physical therapists, allows the rehabilitation program to be tailored to the patient's needs [73] (Fig. 26.10).

Fig. 26.10 Before prosthetic fitting, a client participating in virtual training with a pattern recognition system (Coapt, Chicago, Illinois). During the session, the client practices bilateral hand movements while EMG signals are recorded, and prosthetic hand movements are simulated. In this image, the system learns her EMG pattern for hand opening. The training will progress to hand closing and may expand to wrist rotation, various grip patterns, and individual finger movements. Photo courtesy of Coapt, LLC. (Adapted from Bates et al. [69])

Future Directions

Current efforts are underway to create a highly functional artificial hand. Contemporary research involving implantable nervous system interfaces and peripheral nerve interfaces may lead to recreating both intuitive fine motor control and tactile sensation, which are critical functions of the human hand. Concurrently, neurosurgical and peripheral nerve operative techniques must be developed for efficient and effective surgical device implantation. Regarding prosthetic sockets, osseointegration methodologies are optimized to provide patients with a direct connection to the prosthesis, thus merging the prosthesis with the patient. Further efforts, however, are needed to minimize complications associated with osseointegration techniques. As devices employing interfaces with the nervous system and osseointegration are developed and prostheses become more lifelike, the lengthy rehabilitation courses and prosthetic training may become expedited or obsolete. Despite promising experimental results, a widely available artificial hand prosthesis has yet to be achieved. In summary, future efforts must focus on creating a highly functional prosthesis that is commercially accessible to a broad patient population.

Conclusion

The integration of sensory feedback into prosthetic technology represents a crucial advancement in enhancing modern prosthetics' functionality and user experience. Research has shown that noninvasive and invasive methods can significantly improve tactile sensation and proprioception, allowing users to manipulate objects with greater precision and confidence. These innovations help reduce the cognitive and physical demands placed on individuals relying on prosthetic limbs, fostering a more natural sense of embodiment and connection to their devices.

As we progress in understanding the complexities of sensory feedback mechanisms, hybrid techniques that combine multiple methods show promise in restoring the tactile sensations and the nuanced proprioceptive feedback essential for effective motor control. Studies highlighting successful applications, from peripheral nerve interfaces to cortical stimulation, underscore these technologies' potential to revolutionize amputees' lives.

Ultimately, the ongoing exploration of sensory feedback solutions is set to enhance the quality of life for individuals with limb loss, promoting independence and a more intuitive interaction with their environment. Future research and development in this field will continue bridging the gap between biological limbs and artificial replacements, paving the way for breakthroughs that redefine what it means to experience touch and manipulation in a prosthetic context.

References

1. Maduri P, Akhondi H. Upper limb amputation [Updated 2023 Aug 8]. In: StatPearls [Internet]. Treasure Island: StatPearls Publishing; 2025. Available from: https://www.ncbi.nlm.nih.gov/books/NBK540962/.

2. Inkellis E, Low EE, Langhammer C, Morshed S. Incidence and characterization of major upper-extremity amputations in the National Trauma Data Bank. JB JS Open Access. 2018;3(2):e0038. https://doi.org/10.2106/JBJS.OA.17.00038. PMID: 30280131; PMCID: PMC6145564.
3. Sagiv P, Shabat S, Mann M, Ashur H, Nyska M. Rehabilitation process and functional results of patients with amputated fingers. Plast Reconstr Surg. 2002;110(2):497–503; discussion 504–5. https://doi.org/10.1097/00006534-200208000-00020. PMID: 12142667.
4. Pyörny J, Karelson M, Sletten IN, Ukkola A, Jokihaara J. Patient-reported significant disability after major traumatic upper extremity amputation. J Hand Surg Eur Vol. 2024;49(8):1017–22. https://doi.org/10.1177/17531934231215791. Epub 2023 Nov 22. PMID: 37994012; PMCID: PMC11382439.
5. Biddiss EA, Chau TT. Upper limb prosthesis use and abandonment: a survey of the last 25 years. Prosthetics Orthot Int. 2007;31(3):236–57. https://doi.org/10.1080/03093640600994581. PMID: 17979010.
6. Hoftiezer YAJ, Lans J, van der Heijden BEPA, Chen NC, Eberlin KR, Lozano-Calderón SA. Long-term patient-reported outcome measures following limb salvage with complex reconstruction or amputation in the treatment of upper extremity sarcoma. J Surg Oncol. 2021;123(5):1328–35. https://doi.org/10.1002/jso.26426. Epub 2021 Mar 2. PMID: 33650694.
7. Kim GM, Powell JE, Lacey SA, Butkus JA, Smith DG. Current and emerging prostheses for partial hand amputation: a narrative review. PM R. 2023;15(3):392–401. https://doi.org/10.1002/pmrj.12764. Epub 2022 Feb 22. PMID: 35038226.
8. Kuret Z, Burger H, Vidmar G, Maver T. Adjustment to finger amputation and silicone finger prosthesis use. Disabil Rehabil. 2019;41(11):1307–12. https://doi.org/10.1080/09638288.2018.1424954. Epub 2018 Jan 11. PMID: 29322812.
9. Geffen S, Wilson S. New hands reduce handicap. Disabil Rehabil. 1998;20(11):428–30. https://doi.org/10.3109/09638289809166105. PMID: 9846243.
10. Leow ME, Kour AK, Pereira BP, Pho RW. Fitting a hand-glove prosthesis to enhance the reconstructed mutilated hand. J Hand Surg Am. 1998;23(2):328–33. https://doi.org/10.1016/S0363-5023(98)80135-4. PMID: 9556277.
11. Resnik L, Borgia M, Clark M. Function and quality of life of unilateral major upper limb amputees: effect of prosthesis use and type. Arch Phys Med Rehabil. 2020;101(8):1396–406. https://doi.org/10.1016/j.apmr.2020.04.003. Epub 2020 May 11. PMID: 32437692.
12. Weiss T, Miltner WH, Adler T, Brückner L, Taub E. Decrease in phantom limb pain associated with prosthesis-induced increased use of an amputation stump in humans. Neurosci Lett. 1999;272(2):131–4. https://doi.org/10.1016/s0304-3940(99)00595-9. PMID: 10507559.
13. Krinis SLL, Chadwell A, Kenney L, Smit G. A quantitative comparison between the mHand Adapt passive adjustable hand prosthesis and its predecessor, the Delft Self-Grasping Hand. PLoS One. 2024;19(3):e0300469. https://doi.org/10.1371/journal.pone.0300469. PMID: 38512857; PMCID: PMC10956796.
14. Chadwell A, Kenney L, Howard D, Ssekitoleko RT, Nakandi BT, Head J. Evaluating reachable workspace and user control over prehensor aperture for a body-powered prosthesis. IEEE Trans Neural Syst Rehabil Eng. 2020;28(9):2005–14. https://doi.org/10.1109/TNSRE.2020.3010625. Epub 2020 Jul 20. PMID: 32746324.
15. Crandall RC, Tomhave W. Pediatric unilateral below-elbow amputees: retrospective analysis of 34 patients given multiple prosthetic options. J Pediatr Orthop. 2002;22(3):380–3. PMID: 11961460.
16. Hashim NA, Abd Razak NA, Abu Osman NA, Gholizadeh H. Improvement on upper limb body-powered prostheses (1921-2016): a systematic review. Proc Inst Mech Eng H. 2018;232(1):3–11. https://doi.org/10.1177/0954411917744585. Epub 2017 Dec 4. PMID: 29199518.
17. Abbott ME, Stuart HS. Characterizing the force-motion tradeoff in body-powered transmission design. IEEE Trans Neural Syst Rehabil Eng. 2023;31:3064–74. https://doi.org/10.1109/TNSRE.2023.3297549. Epub 2023 Jul 28. PMID: 37471179.
18. Hichert M, Abbink DA, Kyberd PJ, Plettenburg DH. High cable forces deteriorate pinch force control in voluntary-closing body-powered prostheses. PLoS One. 2017;12(1):e0169996. https://doi.org/10.1371/journal.pone.0169996. PMID: 28099454; PMCID: PMC5242472.
19. Engdahl SM, Gates DH. Differences in quality of movements made with body-powered and myoelectric prostheses during activities of daily living. Clin Biomech (Bristol). 2021;84:105311. https://doi.org/10.1016/j.clinbiomech.2021.105311. Epub 2021 Mar 9. PMID: 33812199.
20. Tam S, Boukadoum M, Campeau-Lecours A, Gosselin B. Intuitive real-time control strategy for high-density myoelectric hand prosthesis using deep and transfer learning. Sci Rep. 2021;11(1):11275. https://doi.org/10.1038/s41598-021-90688-4. PMID: 34050220; PMCID: PMC8163779.
21. Gstoettner C, Laengle G, Harnoncourt L, Sassu P, Aszmann OC. Targeted muscle reinnervation in bionic upper limb reconstruction: current status and future directions. J Hand Surg Eur Vol. 2024;49(6):783–91. https://doi.org/10.1177/17531934241227795. Epub 2024 Feb 16. PMID: 38366374.
22. Dumanian GA, Ko JH, O'Shaughnessy KD, Kim PS, Wilson CJ, Kuiken TA. Targeted reinnervation for transhumeral amputees: current surgical technique and update on results. Plast Reconstr Surg. 2009;124(3):863–9. https://doi.org/10.1097/PRS.0b013e3181b038c9. PMID: 19730305.

23. Langeveld M, Bruin LL, Hundepool CA, Power D, Duraku LS, Zuidam JM. Anatomy of the superficial radial nerve and its target nerves for targeted muscle reinnervation: an anatomical cadaver study. Plast Reconstr Surg. 2024;153(1):95e–100e. https://doi.org/10.1097/PRS.0000000000010690. Epub 2023 May 16. PMID: 37189238.
24. Pierrie SN, Gaston RG, Loeffler BJ. Targeted muscle reinnervation for prosthesis optimization and neuroma management in the setting of transradial amputation. J Hand Surg Am. 2019;44(6):525.e1–8. https://doi.org/10.1016/j.jhsa.2018.11.019. Epub 2019 Feb 4. PMID: 30733097.
25. Gaston RG, Bracey JW, Tait MA, Loeffler BJ. A novel muscle transfer for independent digital control of a myoelectric prosthesis: the starfish procedure. J Hand Surg Am. 2019;44(2):163.e1–5. https://doi.org/10.1016/j.jhsa.2018.04.009. Epub 2018 Jun 19. PMID: 29908928.
26. Johnson CC, Loeffler BJ, Gaston RG. Targeted muscle reinnervation: a paradigm shift for neuroma management and improved prosthesis control in major limb amputees. J Am Acad Orthop Surg. 2021;29(7):288–96. https://doi.org/10.5435/JAAOS-D-20-00044. PMID: 33405489.
27. Cheesborough JE, Smith LH, Kuiken TA, Dumanian GA. Targeted muscle reinnervation and advanced prosthetic arms. Semin Plast Surg. 2015;29(1):62–72. https://doi.org/10.1055/s-0035-1544166. PMID: 25685105; PMCID: PMC4317279.
28. Yang D, Gu Y, Thakor NV, Liu H. Improving the functionality, robustness, and adaptability of myoelectric control for dexterous motion restoration. Exp Brain Res. 2019;237(2):291–311. https://doi.org/10.1007/s00221-018-5441-x. Epub 2018 Nov 30. PMID: 30506366.
29. Salminger S, Stino H, Pichler LH, Gstoettner C, Sturma A, Mayer JA, Szivak M, Aszmann OC. Current rates of prosthetic usage in upper-limb amputees - have innovations had an impact on device acceptance? Disabil Rehabil. 2022;44(14):3708–13. https://doi.org/10.1080/09638288.2020.1866684. Epub 2020 Dec 30. PMID: 33377803.
30. Gonzalez MA, Lee C, Kang J, Gillespie RB, Gates DH. Getting a grip on the impact of incidental feedback from body-powered and myoelectric prostheses. IEEE Trans Neural Syst Rehabil Eng. 2021;29:1905–12. https://doi.org/10.1109/TNSRE.2021.3111741. Epub 2021 Sep 21. PMID: 34516377.
31. Vu PP, Vaskov AK, Irwin ZT, Henning PT, Lueders DR, Laidlaw AT, Davis AJ, Nu CS, Gates DH, Gillespie RB, Kemp SWP, Kung TA, Chestek CA, Cederna PS. A regenerative peripheral nerve interface allows real-time control of an artificial hand in upper limb amputees. Sci Transl Med. 2020;12(533):eaay2857. https://doi.org/10.1126/scitranslmed.aay2857. PMID: 32132217; PMCID: PMC8082695.
32. Vu PP, Vaskov AK, Lee C, Jillala RR, Wallace DM, Davis AJ, Kung TA, Kemp SWP, Gates DH, Chestek CA, Cederna PS. Long-term upper-extremity prosthetic control using regenerative peripheral nerve interfaces and implanted EMG electrodes. J Neural Eng. 2023;20(2):026039. https://doi.org/10.1088/1741-2552/accb0c. PMID: 37023743; PMCID: PMC10126717.
33. Schiefer MA, Graczyk EL, Sidik SM, Tan DW, Tyler DJ. Artificial tactile and proprioceptive feedback improves performance and confidence on object identification tasks. PLoS One. 2018;13(12):e0207659. https://doi.org/10.1371/journal.pone.0207659. PMID: 30517154; PMCID: PMC6281191.
34. Vu PP, Lu CW, Vaskov AK, Gates DH, Gillespie RB, Kemp SWP, Patil PG, Chestek CA, Cederna PS, Kung TA. Restoration of proprioceptive and cutaneous sensation using regenerative peripheral nerve interfaces in humans with upper limb amputations. Plast Reconstr Surg. 2022;149(6):1149e–54e. https://doi.org/10.1097/PRS.0000000000009153. Epub 2022 Apr 11. PMID: 35404335; PMCID: PMC9133017
35. Paggi V, Akouissi O, Micera S, Lacour SP. Compliant peripheral nerve interfaces. J Neural Eng. 2021;18(3):031001. https://doi.org/10.1088/1741-2552/abcdbe. PMID: 33750743.
36. Russell C, Roche AD, Chakrabarty S. Peripheral nerve bionic interface: a review of electrodes. Int J Intell Robot Appl. 2019;3:11–8. https://doi.org/10.1007/s41315-019-00086-3.
37. Davis TS, Wark HA, Hutchinson DT, Warren DJ, O'Neill K, Scheinblum T, Clark GA, Normann RA, Greger B. Restoring motor control and sensory feedback in people with upper extremity amputations using arrays of 96 microelectrodes implanted in the median and ulnar nerves. J Neural Eng. 2016;13(3):036001. https://doi.org/10.1088/1741-2560/13/3/036001. Epub 2016 Mar 22. PMID: 27001946.
38. Tan DW, Schiefer MA, Keith MW, Anderson JR, Tyler J, Tyler DJ. A neural interface provides long-term stable natural touch perception. Sci Transl Med. 2014;6(257):257ra138. https://doi.org/10.1126/scitranslmed.3008669. PMID: 25298320; PMCID: PMC5517305.
39. Gonzalez MA, Nwokeabia C, Vaskov AK, Vu PP, Lu CW, Patil PG, Cederna PS, Chestek CA, Gates DH. Electrical stimulation of regenerative peripheral nerve interfaces (RPNIs) induces referred sensations in people with upper limb loss. IEEE Trans Neural Syst Rehabil Eng. 2024;32:339–49. https://doi.org/10.1109/TNSRE.2023.3345164. Epub 2024 Jan 16. PMID: 38145529; PMCID: PMC10938368.
40. Rowan CC, Graudejus O, Otchy TM. A microclip peripheral nerve interface (μcPNI) for bioelectronic interfacing with small nerves. Adv Sci (Weinh). 2022;9(3):e2102945. https://doi.org/10.1002/advs.202102945. Epub 2021 Nov 26. PMID: 34837353; PMCID: PMC8787429.
41. Moon H, Park B, Chou N, Park KS, Lee S, Kim S. Soft-actuated cuff electrodes with minimal contact for bidirectional peripheral interfaces. Adv Mater. 2024:e2409942. https://doi.org/10.1002/

adma.202409942. Epub ahead of print. PMID: 39523735.
42. George JA, Davis TS, Brinton MR, Clark GA. Intuitive neuromyoelectric control of a dexterous bionic arm using a modified Kalman filter. J Neurosci Methods. 2020;330:108462. https://doi.org/10.1016/j.jneumeth.2019.108462. Epub 2019 Nov 8. PMID: 31711883.
43. Ortiz-Catalan M, Mastinu E, Sassu P, Aszmann O, Brånemark R. Self-contained neuromusculoskeletal arm prostheses. N Engl J Med. 2020;382(18):1732–8. https://doi.org/10.1056/NEJMoa1917537. Erratum in: N Engl J Med. 2022 Nov 24;387(21):2008. doi: 10.1056/NEJMx220013. PMID: 32348644.
44. Schwarz A, Höller MK, Pereira J, Ofner P, Müller-Putz GR. Decoding hand movements from human EEG to control a robotic arm in a simulation environment. J Neural Eng. 2020;17(3):036010. https://doi.org/10.1088/1741-2552/ab882e. PMID: 32272464.
45. Caldwell DJ, Ojemann JG, Rao RPN. Direct electrical stimulation in electrocorticographic brain-computer interfaces: enabling technologies for input to cortex. Front Neurosci. 2019;13:804. https://doi.org/10.3389/fnins.2019.00804. PMID: 31440127; PMCID: PMC6692891.
46. Yanagisawa T, Hirata M, Saitoh Y, Goto T, Kishima H, Fukuma R, Yokoi H, Kamitani Y, Yoshimine T. Real-time control of a prosthetic hand using human electrocorticography signals. J Neurosurg. 2011;114(6):1715–22. https://doi.org/10.3171/2011.1.JNS101421. Epub 2011 Feb 11. PMID: 21314273.
47. Willsey MS, Shah NP, Avansino DT, Hahn NV, Jamiolkowski RM, Kamdar FB, Hochberg LR, Willett FR, Henderson JM. A high-performance brain-computer interface for finger decoding and quadcopter game control in an individual with paralysis. Nat Med. 2025;31(1):96–104. https://doi.org/10.1038/s41591-024-03341-8. Epub 2025 Jan 20. PMID: 39833405; PMCID: PMC11750708.
48. Tombini M, Rigosa J, Zappasodi F, Porcaro C, Citi L, Carpaneto J, Rossini PM, Micera S. Combined analysis of cortical (EEG) and nerve stump signals improves robotic hand control. Neurorehabil Neural Repair. 2012;26(3):275–81. https://doi.org/10.1177/1545968311408919. Epub 2011 Jul 5. PMID: 21730360.
49. Musk E, Neuralink. An integrated brain-machine Interface platform with thousands of channels. J Med Internet Res. 2019;21(10):e16194. https://doi.org/10.2196/16194. PMID: 31642810; PMCID: PMC6914248.
50. Thomas N, Fazlollahi F, Kuchenbecker KJ, Brown JD. The utility of synthetic reflexes and haptic feedback for upper-limb prostheses in a dexterous task without direct vision. IEEE Trans Neural Syst Rehabil Eng. 2023;31:169–79. https://doi.org/10.1109/TNSRE.2022.3217452. Epub 2023 Jan 30. PMID: 36346869.
51. Svensson P, Wijk U, Björkman A, Antfolk C. A review of invasive and non-invasive sensory feedback in upper limb prostheses. Expert Rev Med Devices. 2017;14(6):439–47. https://doi.org/10.1080/17434440.2017.1332989. PMID: 28532184.
52. Raspopovic S, Capogrosso M, Petrini FM, Bonizzato M, Rigosa J, Di Pino G, Carpaneto J, Controzzi M, Boretius T, Fernandez E, Granata G, Oddo CM, Citi L, Ciancio AL, Cipriani C, Carrozza MC, Jensen W, Guglielmelli E, Stieglitz T, Rossini PM, Micera S. Restoring natural sensory feedback in real-time bidirectional hand prostheses. Sci Transl Med. 2014;6(222):222ra19. https://doi.org/10.1126/scitranslmed.3006820. PMID: 24500407
53. Nanivadekar AC, Chandrasekaran S, Helm ER, Boninger ML, Collinger JL, Gaunt RA, Fisher LE. Closed-loop lateral cervical spinal cord stimulation in upper-limb amputees to enable sensory discrimination: a case study. Sci Rep. 2022;12(1):17002. https://doi.org/10.1038/s41598-022-21264-7. PMID: 36220864; PMCID: PMC9553970.
54. O'Doherty JE, Lebedev MA, Ifft PJ, Zhuang KZ, Shokur S, Bleuler H, Nicolelis MA. Active tactile exploration using a brain-machine-brain interface. Nature. 2011;479(7372):228–31. https://doi.org/10.1038/nature10489. PMID: 21976021; PMCID: PMC3236080.
55. Armenta Salas M, Bashford L, Kellis S, Jafari M, Jo H, Kramer D, Shanfield K, Pejsa K, Lee B, Liu CY, Andersen RA. Proprioceptive and cutaneous sensations in humans elicited by intracortical microstimulation. elife. 2018;7:e32904. https://doi.org/10.7554/eLife.32904. PMID: 29633714; PMCID: PMC5896877.
56. Greenspon CM, Valle G, Shelchkova ND, Hobbs TG, Verbaarschot C, Callier T, Berger-Wolf EI, Okorokova EV, Hutchison BC, Dogruoz E, Sobinov AR, Jordan PM, Weiss JM, Fitzgerald EE, Prasad D, Van Driesche A, He Q, Liu F, Kirsch RF, Miller JP, Lee RC, Satzer D, Gonzalez-Martinez J, Warnke PC, Ajiboye AB, Graczyk EL, Boninger ML, Collinger JL, Downey JE, Miller LE, Hatsopoulos NG, Gaunt RA, Bensmaia SJ. Evoking stable and precise tactile sensations via multi-electrode intracortical microstimulation of the somatosensory cortex. Nat Biomed Eng. 2024. https://doi.org/10.1038/s41551-024-01299-z. Epub ahead of print. PMID: 39643730.
57. D'Anna E, Valle G, Mazzoni A, Strauss I, Iberite F, Patton J, Petrini FM, Raspopovic S, Granata G, Di Iorio R, Controzzi M, Cipriani C, Stieglitz T, Rossini PM, Micera S. A closed-loop hand prosthesis with simultaneous intraneural tactile and position feedback. Sci Robot. 2019;4(27):eaau8892. https://doi.org/10.1126/scirobotics.aau8892. PMID: 33137741.
58. Marasco PD, Hebert JS, Sensinger JW, Beckler DT, Thumser ZC, Shehata AW, Williams HE, Wilson KR. Neurorobotic fusion of prosthetic touch, kinesthesia, and movement in bionic upper limbs promotes intrinsic brain behaviors. Sci Robot.

2021;6(58):eabf3368. https://doi.org/10.1126/scirobotics.abf3368.

59. Gu Y, Yang D, Osborn L, Candrea D, Liu H, Thakor N. An adaptive socket with auto-adjusting air bladders for interfacing transhumeral prosthesis: a pilot study. Proc Inst Mech Eng H. 2019;233(8):812–22. https://doi.org/10.1177/0954411919853960. Epub 2019 Jun 5. PMID: 31165676.
60. Yang JC, Mun J, Kwon SY, Park S, Bao Z, Park S. Electronic skin: recent progress and future prospects for skin-attachable devices for health monitoring, robotics, and prosthetics. Adv Mater. 2019;31(48):e1904765. https://doi.org/10.1002/adma.201904765. Epub 2019 Sep 19. PMID: 31538370.
61. Green DP, Hotchkiss RN, Pederson WC, Wolfe SW, Kozin SH, Cohen MS. Chapter 50. In: Green's operative hand surgery. 7th ed. Philadelphia: Elsevier; 2017.
62. Paterno L, Ibrahimi M, Gruppioni E, Menciassi A, Ricotti L. Sockets for limb prostheses: a review of existing technologies and open challenges. IEEE Trans Biomed Eng. 2018;65(9):1996–2010. https://doi.org/10.1109/TBME.2017.2775100. Epub 2018 Jan 23. PMID: 29993506.
63. Olsen J, Day S, Dupan S, Nazarpour K, Dyson M. 3D-printing and upper-limb prosthetic sockets: promises and pitfalls. IEEE Trans Neural Syst Rehabil Eng. 2021;29:527–35. https://doi.org/10.1109/TNSRE.2021.3057984. Epub 2021 Mar 3. PMID: 33587701.
64. Hoyt BW, Walsh SA, Forsberg JA. Osseointegrated prostheses for the rehabilitation of amputees (OPRA): results and clinical perspective. Expert Rev Med Devices. 2020;17(1):17–25. https://doi.org/10.1080/17434440.2020.1704623. Epub 2020 Jan 13. PMID: 31931635.
65. Li Y, Felländer-Tsai L. The bone anchored prostheses for amputees - historical development, current status, and future aspects. Biomaterials. 2021;273:120836. https://doi.org/10.1016/j.biomaterials.2021.120836. Epub 2021 Apr 19. PMID: 33894405.
66. Tropf JG, Potter BK. Osseointegration for amputees: current state of direct skeletal attachment of prostheses. Orthoplastic Surg. 2024. https://doi.org/10.1016/j.orthop.2023.05.004.
67. Overmann AL, Forsberg JA. The state of the art of osseointegration for limb prosthesis. Biomed Eng Lett. 2019;10(1):5–16. https://doi.org/10.1007/s13534-019-00133-9. PMID: 32175127; PMCID: PMC7046912.
68. Kunutsor SK, Gillatt D, Blom AW. Systematic review of the safety and efficacy of osseointegration prosthesis after limb amputation. Br J Surg. 2018;105(13):1731–41. https://doi.org/10.1002/bjs.11005. Epub 2018 Oct 11. PMID: 30307036.
69. Bates TJ, Fergason JR, Pierrie SN. Technological advances in prosthesis design and rehabilitation following upper extremity limb loss. Curr Rev Musculoskelet Med. 2020;13(4):485–93. https://doi.org/10.1007/s12178-020-09656-6.
70. Tsikandylakis G, Berlin Ö, Brånemark R. Implant survival, adverse events, and bone remodeling of osseointegrated percutaneous implants for transhumeral amputees. Clin Orthop Relat Res. 2014;472(10):2947–56. https://doi.org/10.1007/s11999-014-3695-6. PMID: 24879569; PMCID: PMC4160502.
71. Taylor CE, Drew AJ, Zhang Y, Qiu Y, Bachus KN, Foreman KB, Henninger HB. Upper extremity prosthetic selection influences loading of transhumeral osseointegrated systems. PLoS One. 2020;15(8):e0237179. https://doi.org/10.1371/journal.pone.0237179. PMID: 32760149; PMCID: PMC7410272.
72. Malone JM, Fleming LL, Roberson J, et al. Immediate, early, and late postsurgical management of upper-limb amputation. J Rehabil Res Dev. 1984;21(1):33–41.
73. Johnson SS, Mansfield E. Prosthetic training: upper limb. Phys Med Rehabil Clin N Am. 2014;25(1):133–51. https://doi.org/10.1016/j.pmr.2013.09.012. PMID: 24287244.

Virtual and Augmented Reality Applications for Hand Surgery

27

Jared Rubin and Farid Amirouche

History and Evolution of Virtual and Augmented Reality Applications

In recent years, virtual reality (VR) and augmented reality (AR) have become more integrated into orthopedic surgery through education, training, and procedural use. Initially, VR was an educational tool to train medical professionals by simulating complex surgical procedures in a low-stake environment [18]. As an effective education tool, VR was used to understand better the biomechanical properties of the musculoskeletal system, including the hand. In addition to VR, AR has risen in popularity due to its unique application in producing digital information in the surgical field. Both VR and AR have evolved in orthopedic surgery from just visual aids to now becoming safe and effective intraoperative tools for orthopedic surgeons to use. Some specific examples of these technologies include the HoloLens for AR and various VR simulators for training and preoperative planning [25].

Transitioning from being used as training devices to being tools incorporated during surgeries, VR and AR have proven essential for hand surgeries due to the intricate anatomy and biomechanics involved in such procedures. These technologies have allowed for better preoperative planning, increased intraoperative accuracy for procedures, and improved patient outcomes. For example, AR is effective in fracture management by helping guide the interlocking of intramedullary nails. The potential of VR and AR to enhance surgical precision reassures us about the future of orthopedic surgery and the confidence it can instill in surgeons and patients.

J. Rubin
Department of Orthopaedic Surgery, University of Illinois at Chicago, Chicago, IL, USA

F. Amirouche (✉)
Department of Orthopaedic Surgery, Orthopaedic and Spine Institute, Endeavor HealthSystem (Formerly Northshore University HealthSystem), an Affiliate of the University of Chicago Pritzker School of Medicine, Skokie, IL, USA

Department of Orthopaedic Surgery, University of Illinois at Chicago, Chicago, IL, USA
e-mail: famirouche@northshore.org

Preoperative Applications

Advances in virtual and augmented reality have significantly changed preoperative planning and training for orthopedic surgeons. Orthopedic residents have found that using such innovative technology helps them learn the anatomy and pertinent steps for procedures they will be performing [20]. Surgical simulations allow orthopedic residents to practice their technical skills

F. Amirouche et al. (eds.), *Biomechanics of the Hand*,
https://doi.org/10.1007/978-3-032-04688-8_27

and develop greater procedural accuracy in a risk-free environment.

More specifically, mobile-based simulation programs such as Touch Surgery have been implemented in programs across the country to better enhance the surgical training of residents and attending surgeons. Touch Surgery and similar applications allow participants to use their phones and simulate the procedure they will conduct. Tulipan et al. observed that when fellowship-trained hand surgeons, orthopedic residents, and medical students completed modules of carpal tunnel release on Touch Surgery, all cohorts of participants performed better on average with each subsequent attempt [38]. These findings validate that simulating hand procedures with Touch Surgery can improve the knowledge and technical skills required for surgical success. Other orthopedic surgeries have been similarly enhanced with Touch Surgery simulations, confirming its benefit in preoperative planning and practice for orthopedic trainees [35]. The use of VR and AR through mobile simulations such as Touch Surgery has immense potential for practical surgical training for any level of participant interested in orthopedic hand surgery.

In addition to Touch Surgery, other VR and AR simulators are being developed to help with preoperative planning and training. Leap Motion is a motion-tracking technology that tracks hand movements to allow participants to build the required surgical skills. The program can be used as an effective tool for practicing surgical hand motions needed for hand procedures [37]. Leap Motion has also been shown to enhance spatial recognition and improve procedural precision for surgical planning with patient data [39]. Similarly, gesture recognition tools are gaining popularity due to effective feedback while practicing surgical procedures. These tools can increase efficiency and preoperative planning time for orthopedic procedures [9]. Adding various VR and AR technologies into preoperative training can be valuable in orthopedic hand surgery, as intricate anatomy and biomechanics are involved in these procedures.

Biomechanical Analysis Using Virtual and Augmented Reality

Nuanced element analysis (FEA) is a computational modeling method to assess different forces in intricate anatomical structures. Previous research has demonstrated that FEA is an effective method to predict the biomechanical behavior of the upper extremity under various loading conditions [28]. Applying VR and AR in hand biomechanics has changed how FEA is implemented. The introduction of VR and AR with FEA has further advanced visualization and analysis of biomechanical data to help with the efficacy and success of hand surgeries. For patients undergoing wrist arthroplasty, FEA could accurately predict joint implant stability [3]. Similarly, for patients undergoing distal radius fractures, there is potential for enhanced fixation techniques via FEA and other computational models [23]. Furthermore, a study by Zhang et al. implemented FEA to identify methods to minimize stress concentrations while increasing the fixation stability for ulnar head fractures [42] (Fig. 27.1).

In addition to these promising findings for upper extremity pathologies, FEA has also been implemented for specific hand surgeries. In a study by Yu et al., the authors used FEA to determine a favorable approach for transverse carpal ligament transections in carpal tunnel release [40]. FEA and other computational models have also been applied to metacarpal shaft fractures, effectively analyzing stress distribution based on how many screws were used in the fixation procedure [43]. The results from studies assessing FEA and orthopedic surgeries are valuable for optimizing surgical techniques and minimizing biomechanical complications. Integrating VR and AR with FEA adds further value to its use for surgical success. Combining these technologies improves simulation performance in preparation for future surgeries [4]. Ultimately, surgeons can better understand stress distributions, fracture and fixation mechanics, and other biomechanical properties prevalent in hand surgeries.

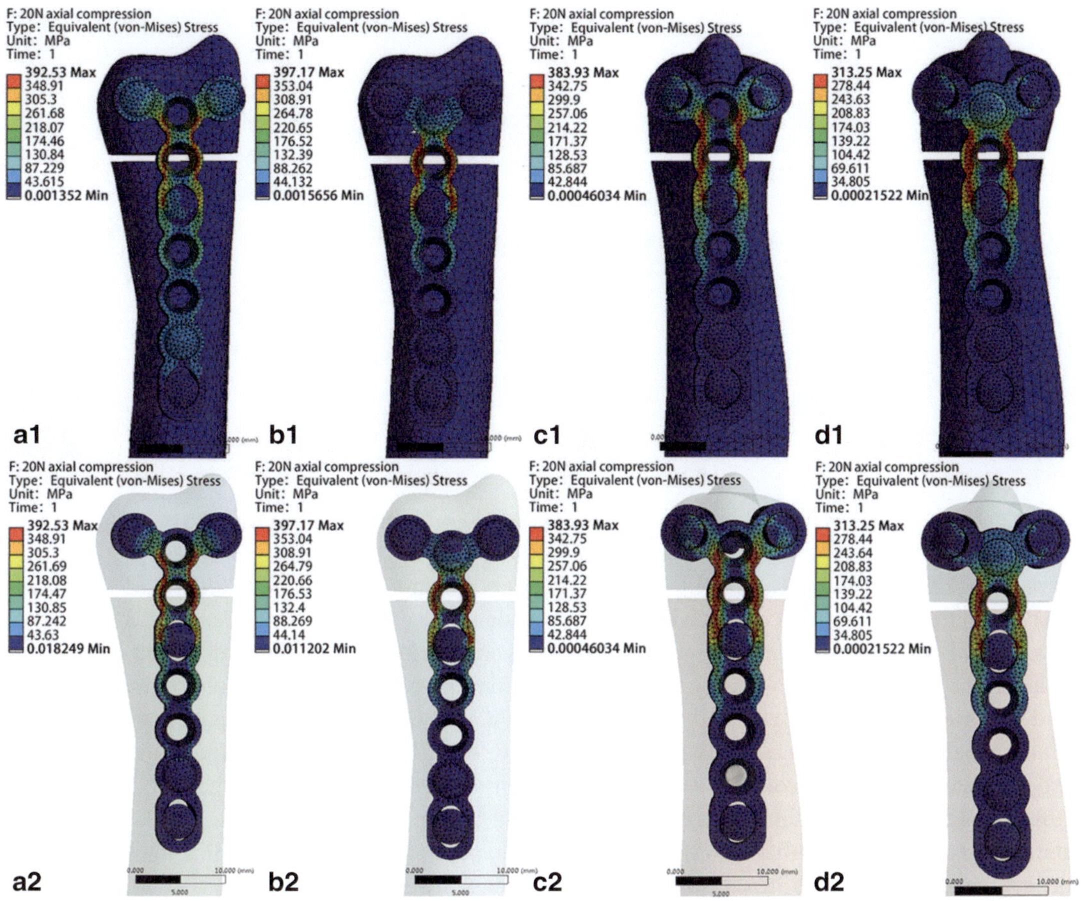

Fig. 27.1 Various stress distributions based on different fixation systems through nuanced element analysis [42]

Intraoperative Applications

VR and AR can be used for preoperative training and planning, and new technologies have been created that allow for intraoperative implementation. The Microsoft HoloLens is a wearable device that overlays three-dimensional anatomical images onto the surgeon's field of view during the procedure. Using devices like Microsoft HoloLens improves visualization of complex anatomy during hand surgeries while providing immediate performance feedback. By using speech, hand gestures, and specific eye movements, orthopedic surgeons can use this technology to focus more on the surgical field [12]. The HoloLens device uniquely aligns the surgeons' visual motor axis to improve performance outcomes during surgical procedures [1]. These applications can be particularly valuable in hand surgeries that require technical precision due to the complex biomechanics of the hand.

Furthermore, there has been an update from the first version of HoloLens, HoloLens 1, to a new and improved version, HoloLens 2. In a study conducted by Lastra et al., the authors observed that HoloLens 2 improved the comfortability, usability, and efficiency of the device in comparison to HoloLens 1 while also improving the AR projection accuracy [33]. HoloLens 2 allows for efficient and accurate registration

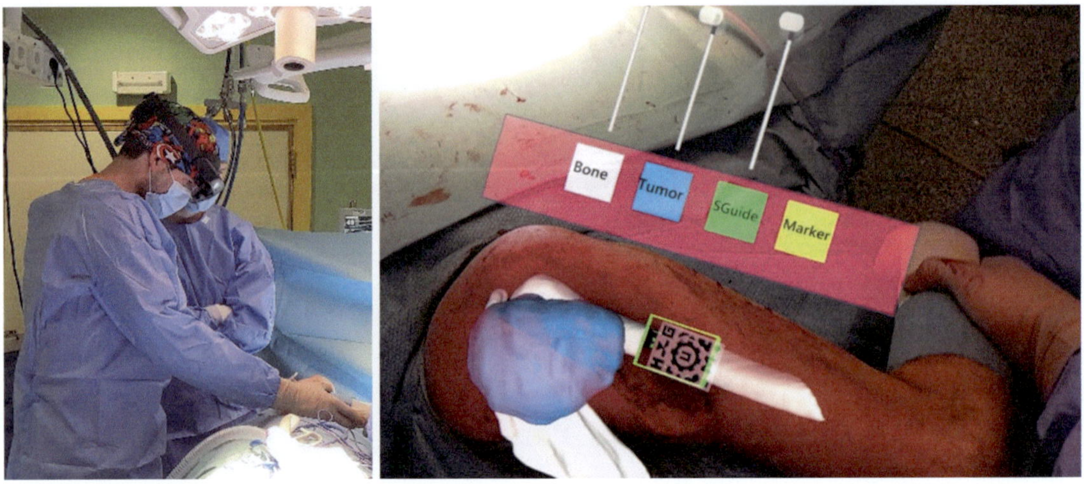

Fig. 27.2 Demonstration of the surgeon's perspective using HoloLens 2 technology in orthopedic surgery [33]

between real and virtually created objects, which can be used to improve surgical precision in hand surgeries [2]. VR and AR can facilitate real-time visualization and navigation during orthopedic hand surgeries. For instance, a study by Zemirline et al. revealed that AR can assist orthopedic surgeons with real-time feedback during wrist arthroscopies by making maneuvers easier, more efficient, and more comfortable [41]. HoloLens and similar devices can be incredibly beneficial for hand surgeons intraoperatively, including improved efficiency, real-time visualization and navigation, and greater accuracy in performing procedures (Fig. 27.2).

Intraoperative Benefits for Surgeons and Patients

Intraoperative applications of VR and AR not only improve accuracy and enhance the technical experience for surgeons but can also create a better experience for patients undergoing orthopedic surgery. The use of VR and AR during orthopedic surgeries has been evaluated in previous research. In a study by Chan et al., the researchers discovered that patients who underwent joint replacement surgery using immersive VR required less sedation than patients without VR use [7]. Furthermore, in such procedures, VR is well tolerated, reported to have highly satisfactory experiences, and does not have reports of significant side effects [17].

Promising findings have also been observed in patients specifically undergoing hand surgeries. In a study by Faruki et al., patients who offered VR or AR during hand surgeries required less intraoperative sedation than patients who did not use VR or AR [14]. In addition to sedative benefits, patients who undergo VR immersion during broad awake local anesthesia no tourniquet (WALANT) hand surgeries are more relaxed and report lower pain and anxiety scores in comparison to patients who do not receive VR immersion [16]. Not only can patients experience a significant anxiety reduction when undergoing WALANT carpal tunnel release with VR use, but they also have increased enjoyment of the overall experience [29] (Fig. 27.3).

In addition to sedative and patient comfort benefits, VR and AR can also improve the intraoperative experience for orthopedic surgeons. Prior literature has demonstrated that orthopedic surgeons appreciate the distraction the program creates for patients during awake procedures, providing more room for the physician to concentrate on the procedural tasks [27]. Overall, the intraoperative benefits of using VR and AR during hand surgeries go beyond provider satisfaction and enhanced accuracy of procedures. VR and AR during hand surgeries allow patients to receive anesthetic benefits, have decreased anxi-

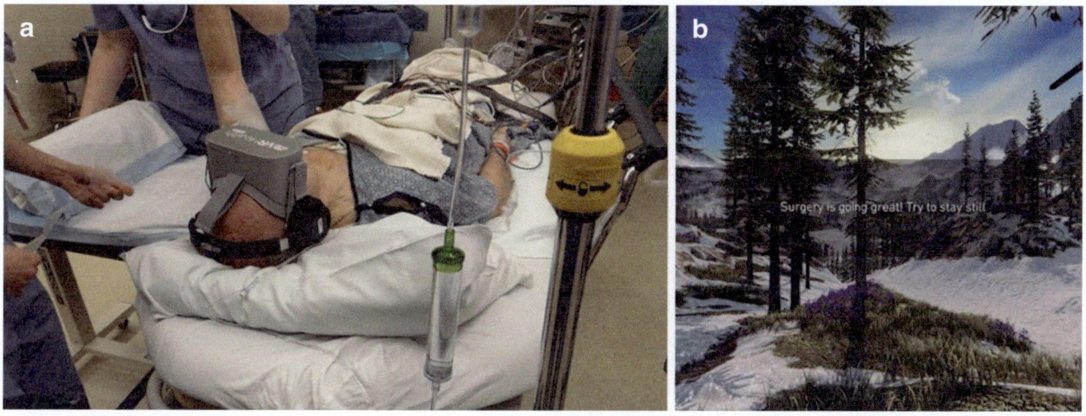

Fig. 27.3 (**a**) Participant using the VR headset during surgery. (**b**) Immersive environment seen by participants for comfort and communication [14]

ety during the procedure, and report overall increased satisfaction metrics. By benefitting both the surgeon and the patient during hand surgeries, the intraoperative use of VR and AR is changing the approaches toward conducting effective hand surgery.

Rehabilitation and Postoperative Outcomes

In addition to preoperative and intraoperative benefits, preliminary research has suggested that virtual and augmented reality can lead to better postoperative outcomes. For postupper extremity injuries, patients who use VR in addition to occupational therapy have better functional outcomes and quality of life compared to patients who do not have a VR intervention [34]. Furthermore, using VR or AR simulators such as Leap Motion controllers for upper extremity rehabilitation provides similar improvements for patients compared to those undergoing more conventional rehabilitation techniques [36]. Regarding hand rehabilitation, VR and AR interventions can improve hand strength, range of motion, and coordination in functional tasks [32]. AR has the potential to track specific finger range of motion changes after a patient has received hand surgery [22]. An additional benefit of using virtual and augmented technologies for recovery is the ability to deliver high-quality patient care while simultaneously reducing the cost for the patient [5]. While limited data and research have explored postoperative outcomes for patients who use VR and AR technology, there seems to be promising potential for its use in hand surgery.

Extended Reality and Mixed Reality

In addition to virtual and augmented reality, other forms of technology, such as mixed reality (MR), combine physical and virtual elements to enhance the surgical experience; these various immersive technologies fall under the umbrella of extended reality [15]. Extended reality considers the anatomical differences in individuals to promote practical training and surgical success [31]. Mixed reality combines explicit elements of virtual reality, interactive three-dimensional visualization, and augmented reality, projected images onto real environments [24]. Similar to VR and AR, MR can greatly benefit fields that require precise manipulation of complex anatomy, such as orthopedic hand surgery. MR allows for an efficient process to manipulate objects while integrating hand interactions [30]. Other benefits provided by MR during hand surgery include better depth perception, orientation, and spatial relationship of surgical tools about the objects they are working on [21]. MR can significantly supplement both training and surgical procedures. A final benefit that MR can also provide is enhanced intraoperative communication, as multiple surgeons can simultaneously observe identical visual information through their headsets during procedures [6].

Hand Microsurgery

Recent developments in VR and AR are also being integrated into microsurgery. While traditional microsurgical training courses and laboratories can be costly and difficult to access, numerous VR and AR microsurgery resources effectively increase the availability and feasibility of learning microsurgery [26]. These novel resources are an affordable and efficient method to develop the dexterity and hand-eye coordination required for successful microsurgery [8]. With the guidance of VR and AR, microsurgery can become easier to understand, and surgeons can develop critical microsurgical skills [13]. Specific parameters that VR and AR incorporate during training include tissue trauma, orientation angle, penetration angle, and motion smoothness while performing microsurgical tasks [19]. While these promising findings have been discovered in the context of other surgical fields, further studies can utilize such methods to determine if similar benefits occur in orthopedic hand procedures.

Limitations and Future Directions

VR and AR use in orthopedic surgery has already demonstrated significant value, enhancing preoperative and intraoperative education and training and improving patient outcomes [10]. Despite the early exciting findings to this date, there remain limitations in its use and challenges that must be addressed. Even when using the technology, orthopedic surgeons have a significant learning curve, often associated with high costs [11]. Future research should determine the long-term benefits of using VR and AR in hand surgery and optimize their implementation in orthopedic surgery. In addition, further validation of clinical patient outcomes needs to be analyzed before a definitive conclusion can be drawn on whether the new technology is significantly beneficial for the patient. Ultimately, future directions include optimizing the implementation of VR and AR programs in hand surgeries, finding solutions for the current limitations of the technology, and collecting more clinical data for determining long-term patient benefits.

References

1. Al Janabi HF, Aydin A, Palaneer S, et al. Effectiveness of the HoloLens mixed-reality headset in minimally invasive surgery: a simulation-based feasibility study. Surg Endosc. 2020;34:1143–9. https://doi.org/10.1007/s00464-019-06862-3.
2. Allen D, Peters T, Chen ECS. Enhancing surgical navigation: a robust hand-eye calibration method for the Microsoft HoloLens 2. Int J Comput Assist Radiol Surg. 2025;20:137–46. https://doi.org/10.1007/s11548-024-03250-8.
3. Bajuri MN, Abdul Kadir MR, Murali MR, Kamarul T. Biomechanical analysis of the wrist arthroplasty in rheumatoid arthritis: a finite element analysis. Med Biol Eng Comput. 2013;51:175–86. https://doi.org/10.1007/s11517-012-0982-9.
4. Berkley J, Turkiyyah G, Berg D, et al. Real-time finite element modeling for surgery simulation: an application to virtual suturing. IEEE Trans Vis Comput Graph. 2004;10:314–25. https://doi.org/10.1109/TVCG.2004.1272730.
5. Berton A, Longo UG, Candela V, et al. Virtual reality, augmented reality, gamification, and telerehabilitation: psychological impact on orthopedic patients' rehabilitation. J Clin Med. 2020;9:2567. https://doi.org/10.3390/jcm9082567.
6. Calem DB, Lubiatowski P, Trenhaile S, et al. Mixed reality applications in upper extremity surgery: the future is now. EFORT Open Rev. 2024;9:1034–46. https://doi.org/10.1530/EOR-24-0080.
7. Chan PY, Scharf S. Virtual reality as an adjunctive nonpharmacological sedative during orthopedic surgery under regional anesthesia: a pilot and feasibility study. Anesth Analg. 2017;125:1200–2. https://doi.org/10.1213/ANE.0000000000002169.
8. Choque-Velasquez J, Colasanti R, Collan J, et al. Virtual reality glasses and "eye-hands blind technique" for microsurgical training in neurosurgery. World Neurosurg. 2018;112:126–30. https://doi.org/10.1016/j.wneu.2018.01.067.
9. Chui EC-S, Mak KK-K, Ng RH-T, et al. Application of image recognition-based tracker-less augmented reality navigation system in a series of sawbones trials. Art. 2024;6:39. https://doi.org/10.1186/s42836-024-00263-1.

10. Combalia A, Sanchez-Vives MV, Donegan T. Immersive virtual reality in orthopaedics-a narrative review. Int Orthop. 2024;48(1):21–30. https://doi.org/10.1007/s00264-023-05911-w. Epub 2023 Aug 11. PMID: 37566225; PMCID: PMC10766717.
11. Dashtbozorg A, Heidarian E, Sagheb Ray Shirazi M, et al. Emerging technologies in hand orthopedic surgery: current trends and future directions. Galen Med J. 2024;13:1–10. https://doi.org/10.31661/gmj.v13i.3325.
12. De Oliveira ME, Debarba HG, Lädermann A, et al. A hand-eye calibration method for augmented reality applied to computer-assisted orthopedic surgery. Int J Med Robot. 2019;15:e1969. https://doi.org/10.1002/rcs.1969.
13. El Chemaly T, Athayde Neves C, Fu F, et al. From microscope to head-mounted display: integrating hand tracking into microsurgical augmented reality. Int J Comput Assist Radiol Surg. 2024;19:2023–9. https://doi.org/10.1007/s11548-024-03224-w.
14. Faruki AA, Nguyen TB, Gasangwa D-V, et al. Virtual reality immersion compared to monitored anesthesia care for hand surgery: a randomized controlled trial. PLoS One. 2022;17:e0272030. https://doi.org/10.1371/journal.pone.0272030.
15. Furman AA, Hsu WK. Augmented reality (AR) in orthopedics: current applications and future directions. Curr Rev Musculoskelet Med. 2021;14:397–405. https://doi.org/10.1007/s12178-021-09728-1.
16. Hoxhallari E, Behr IJ, Bradshaw JS, et al. Virtual reality improves the patient experience during wide-awake local anesthesia no tourniquet hand surgery: a single-blind, randomized, prospective study. Plast Reconstr Surg. 2019;144:408–14. https://doi.org/10.1097/PRS.0000000000005831.
17. Huang MY, Scharf S, Chan PY. Effects of immersive virtual reality therapy on intravenous patient-controlled sedation during orthopaedic surgery under regional anesthesia: a randomized controlled trial. PLoS One. 2020;15:e0229320. https://doi.org/10.1371/journal.pone.0229320.
18. Izard SG, Juanes JA, García Peñalvo FJ, et al. Virtual reality as an educational and training tool for medicine. J Med Syst. 2018;42:50. https://doi.org/10.1007/s10916-018-0900-2.
19. Kazemi H, Rappel JK, Poston T, et al. Assessing suturing techniques using a virtual reality surgical simulator. Microsurgery. 2010;30:479–86. https://doi.org/10.1002/micr.20766.
20. Kuhn AW, Yu JK, Gerull KM, et al. Virtual reality and surgical simulation training for orthopaedic surgery residents: a qualitative assessment of trainee perspectives. JB JS Open Access. 2024;9(e23):00142. https://doi.org/10.2106/JBJS.OA.23.00142.
21. Lee SC, Fuerst B, Tateno K, et al. Multi-modal imaging, model-based tracking, and mixed reality visualisation for orthopaedic surgery. Healthc Technol Lett. 2017;4:168–73. https://doi.org/10.1049/htl.2017.0066.
22. Létourneau SG, Jin H, Peters E, et al. Augmented reality-based finger joint range of motion measurement: assessment of reliability and concurrent validity. J Hand Surg Am. 2024;S0363-5023(24):00514–8. https://doi.org/10.1016/j.jhsa.2024.10.006.
23. Liu HC, Jiang J-S, Lin C-L. Biomechanical investigation of a novel hybrid dorsal double plating for distal radius fractures by integrating topology optimization and finite element analysis. Injury. 2020;51:1271–80. https://doi.org/10.1016/j.injury.2020.03.011.
24. Long AS, Almeida MN, Chong L, Prsic A. Live virtual surgery and virtual reality in surgery: potential applications in hand surgery education. J Hand Surg Am. 2023;48:499–505. https://doi.org/10.1016/j.jhsa.2023.01.004.
25. Longo UG, De Salvatore S, Candela V, et al. Augmented reality, virtual reality and artificial intelligence in orthopedic surgery: a systematic review. Appl Sci. 2021;11:3253. https://doi.org/10.3390/app11073253.
26. Margulies IG, Xu H, Henderson PW. Microsurgery training in the digital era: a systematic review of accessible digital resources. Ann Plast Surg. 2020;85:337–43. https://doi.org/10.1097/SAP.0000000000002214.
27. McCullough M, Osborne TF, Rawlins C, et al. The impact of virtual reality on the patients and providers experience in wide-awake, local-only hand surgery. J Hand Surg Glob Online. 2023;5:290–3. https://doi.org/10.1016/j.jhsg.2023.01.014.
28. Mena A, Wollstein R, Baus J, Yang J. Finite element modeling of the human wrist: a review. J Wrist Surg. 2023;12:478–87. https://doi.org/10.1055/s-0043-1768930.
29. Miller MB, Gabel SA, Gluf-Magar LC, et al. Virtual reality improves patient experience and anxiety during in-office carpal tunnel release. Plast Reconstr Surg Glob Open. 2022;10:e4426. https://doi.org/10.1097/GOX.0000000000004426.
30. Nguyen R, Gouin-Vallerand C, Amiri M. Hand interaction designs in mixed and augmented reality head-mounted display: a scoping review and classification. Front Virtual Real. 2023;4:1171230. https://doi.org/10.3389/frvir.2023.1171230.
31. Oyama S, Iwase H, Yoneda H, et al. Insights and trends review: use of extended reality (xR) in hand surgery. J Hand Surg Eur Vol. 2025;17531934241313208. https://doi.org/10.1177/17531934241313208.
32. Pereira MF, Prahm C, Kolbenschlag J, et al. Application of AR and VR in hand rehabilitation: a systematic review. J Biomed Inform. 2020;111:103584. https://doi.org/10.1016/j.jbi.2020.103584.
33. Pose-Díez-de-la-Lastra A, Moreta-Martinez R, García-Sevilla M, et al. HoloLens one vs. HoloLens 2: improvements in the new model for orthopedic oncological interventions. Sensors (Basel). 2022;22:4915. https://doi.org/10.3390/s22134915.
34. Shin J-H, Kim M-Y, Lee J-Y, et al. Effects of virtual reality-based rehabilitation on distal upper extremity

function and health-related quality of life: a single-blinded, randomized controlled trial. J Neuroeng Rehabil. 2016;13:17. https://doi.org/10.1186/s12984-016-0125-x.
35. Sugand K, Mawkin M, Gupte C. Validating Touch Surgery™: a cognitive task simulation and rehearsal app for intramedullary femoral nailing. Injury. 2015;46:2212–6. https://doi.org/10.1016/j.injury.2015.05.013.
36. Tarakci E, Arman N, Tarakci D, Kasapcopur O. Leap Motion Controller-based training for upper extremity rehabilitation in children and adolescents with physical disabilities: a randomized controlled trial. J Hand Ther. 2020;33:220–228.e1. https://doi.org/10.1016/j.jht.2019.03.012.
37. Tsuyuki S, Hoshina K, Miyahara K, et al. (2022) Motion analysis of suturing technique with Leap Motion Controller™. Proof-of-concept. Sci Prog 105:368504221103777. https://doi.org/10.1177/00368504221103777.
38. Tulipan J, Miller A, Park AG, et al. Touch surgery: analysis and assessment of validity of a hand surgery simulation "App.". Hand (N Y). 2019;14:311–6. https://doi.org/10.1177/1558944717751192.
39. Wright T, de Ribaupierre S, Eagleson R. Design and evaluation of an augmented reality simulator using leap motion. Healthc Technol Lett. 2017;4:210–5. https://doi.org/10.1049/htl.2017.0070.
40. Yu L, Jia J, Lakshminarayanan K, et al. A finite element analysis of the carpal arch with various locations of carpal tunnel release. Front Surg. 2023;10:1134129. https://doi.org/10.3389/fsurg.2023.1134129.
41. Zemirline A, Agnus V, Soler L, et al. Augmented reality-based navigation system for wrist arthroscopy: feasibility. J Wrist Surg. 2013;2:294–8. https://doi.org/10.1055/s-0033-1359321.
42. Zhang Y, Shao Q, Yang C, et al. Finite element analysis of different locking plate fixation methods for treating the ulnar head fracture. J Orthop Surg Res. 2021;16:191. https://doi.org/10.1186/s13018-021-02334-4.
43. Zhang J-F, Li Y, Huo Y-X, et al. Biomechanical analysis of locking plates for fixation of metacarpal shaft fractures: a finite element analysis. Orthop Traumatol Surg Res. 2022;108:103340. https://doi.org/10.1016/j.otsr.2022.103340.

28 Artificial Intelligence and Innovation: The New Frontier in the Future of Hand Surgery

Majd Mzeihem and Farid Amirouche

Introduction

Overview of AI and ML in Medicine

Artificial intelligence (AI), together with machine learning (ML), enhances medical diagnostics, treatment strategies, and patient recovery through extensive data analysis [7]. AI functions as a machine simulation of the human intellect, and ML represents a portion of AI that enables the training of programming frameworks to detect data patterns automatically [22]. AI applications in medicine exist across a broad spectrum that includes image analysis while simultaneously generating predictions and addressing all primary medical specialties, from radiology to oncology to surgery [7].

For example, AI diagnosis systems have shown several successes in diagnosing diseases, including cancers from medical images [19]. Thus, AI in hand surgery could further enhance diagnostic accuracy, surgical planning, and post-surgery management.

Innovation in Hand Surgery

Over the past century, hand surgery has been revolutionized by rapid technological developments and an increased understanding of the anatomy and function of the hand [51]. These innovations have greatly improved surgical outcomes and patient care, from the development of microsurgery to the integration of AI [18].

Historical Perspective on Technological Advances

Perhaps one of the most significant breakthroughs ever in hand surgery was the introduction of microsurgery in the 1960s [35]. Utilizing operating microscopes with specialized instruments, microsurgical techniques allow surgeons to repair delicate structures such as blood vessels and nerves [29]. This innovation thus changed the face of treating traumatic hand injuries, making possible the reattachment of amputated digits and limbs [12]. One example is that in 1968, Komatsu and Tamai reported the first successful replantation of a completely amputated thumb in a

M. Mzeihem
Department of Orthopaedic Surgery, University of Illinois at Chicago, Chicago, IL, USA

F. Amirouche (✉)
Department of Orthopaedic Surgery, Orthopaedic and Spine Institute, Endeavor HealthSystem (Formerly Northshore University HealthSystem), an Affiliate of the University of Chicago Pritzker School of Medicine, Skokie, IL, USA

Department of Orthopaedic Surgery, University of Illinois at Chicago, Chicago, IL, USA
e-mail: famirouche@northshore.org

F. Amirouche et al. (eds.), *Biomechanics of the Hand*,
https://doi.org/10.1007/978-3-032-04688-8_28

28-years old man, marking an important milestone in hand surgery [24].

The medical world witnessed the critical development of arthroscopy during the 1970s [40]. Implementing microscopic arthroscopy enables surgeons to examine and perform procedures on joints through small openings, thus minimizing surgical invasiveness [30]. The minimally invasive technique of hand arthroscopy provides adequate care to patients suffering from carpal tunnel syndrome, to treating triangular fibrocartilage complex (TFCC) injuries and wrist arthritis, as Zhou et al. [68] demonstrated. The research conducted by Zhou et al. [68] proved that patients achieved better postoperative function after TFCC repairs performed with arthroscopic assistance compared to standard open surgical procedures. Additionally, arthroscopic procedures led to quicker healing times.

The Growing Importance of AI in Hand Surgery

Healthcare professionals have been using AI and ML techniques to advance hand surgery boundaries since 2023 [32]. AI applications enhance medical diagnoses and create surgical designs and personalized postoperative therapy plans [67]. AI algorithms help medical teams determine bone fractures and analyze degenerative pathologies and ligamentous conditions with optimal precision through X-rays and MRI analysis [33, 34]. The software tools improve surgeons' diagnostic accuracy while decreasing their chances of incorrect analysis [33, 34].

Some modern examples of this in hand surgery include the role of three-dimensional modeling and surgical simulation in AI applications. AI-influenced three-dimensional models assist the surgeon with demarcating any complex anatomy and simulate different probable surgical outcomes well before actual operating room entry [50]. More specifically, research by Langlotz et al. [36] on AI-powered three-dimensional modeling showed that such methodologies could improve precision and reduce surgical time for different complex hand procedures.

Robotic-Assisted Surgery

Robotic systems are now used in hand surgery because they enable manipulations that traditional methods cannot achieve [56]. The da Vinci Surgical System received its first adaptation for thoracic and abdominal surgery before doctors modified it to perform hand procedures [53]. Outcomes of complex maneuvers such as nerve repair, microvascular anastomosis, and other tiny maneuvers become possible for surgeons through robotic assistance because of the system's dexterity and reduced tremor (Benedikt [57]). The results of complex hand surgery assisted by robotics proved positive, according to researchers such as Hendrick et al. [27].

Wearable Technology and Rehabilitation

Wearable technology in postoperative care plays a vital role [3]. Wearable devices embedded with sensors will track metrics related to grip strength, range of motion, and therapy compliance while simultaneously feeding that data back, in real time, to the patient and the healthcare professional [38]. Such wearables also come with personalized rehabilitation programs that the patient modifies [28]. Lanotte et al. [37] have established that employing wearable devices driven by AI enhances functional outcomes and increases the adherence rate in therapeutic protocols, compared to traditional methods, during hand rehabilitation.

AI and ML in Diagnostics

Artificial intelligence and machine learning have inspired a revolution in diagnostics within hand surgery. Advanced algorithms make those technologies useful to enhance diagnostic processes regarding speed and precision, creating conditions at early stages where personalized treatment can be offered crucially.

Enhancing Imaging Analysis

AI in Fracture Detection

According to Vassa et al. [65], medical imaging analysis with X-rays, MRI, and CT scans represents the most vital application of AI in hand surgery. Various orthopedic conditions, including fractures, ligament injuries, and degenerative conditions, receive successful diagnostic evaluations thanks to AI algorithms that use convolutional neural networks (CNNs) [32]. AI system diagnosis of wrist fractures in X-rays, according to Olczak et al. [48], reached accuracy rates above 90%, outperforming human radiologists in their assessments. Medical professionals require extreme precision during critical emergency procedures because precise and fast actions represent their primary priorities.

Automated Segmentation of Hand and Wrist Imaging

AI also automates complex hand and wrist structure segmentation and analyses in imaging studies [23]. Examples include the segmenting of bones, ligaments, and tendons from MRI using ML algorithms that can allow for the diagnosis of conditions related to carpal tunnel syndrome with an accuracy of 89.47% using anatomic measurements of the hands [66]. For instance, Thian et al. [63] and Parisa Karimi Darabi [15] presented the application of AI in automated wrist MRI segmentation, which showed high precision in identifying structures and pathological changes (Fig. 28.1). It reduces the time taken for manual analysis, minimizing the possibility of human errors [20].

Case Example: AI in Osteoarthritis Diagnosis

One such way that AI impacts the interpretation of imaging examinations is through the diagnosis it imparts to the evaluation of hand osteoarthritis. Osteoarthritis is a common enough condition that can cause pain and disability [13] unless detected early and treated. AI algorithm development is needed to analyze the early stages of changes in X-rays and MRIs of osteoarthritis, like joint space narrowing and osteophyte formation [44]. Norman et al. [47] demonstrated the excellent performance of a deep learning-based AI system to classify osteoarthritis severity from hand X-rays. They highlighted that it can be a critical support for treatment planning.

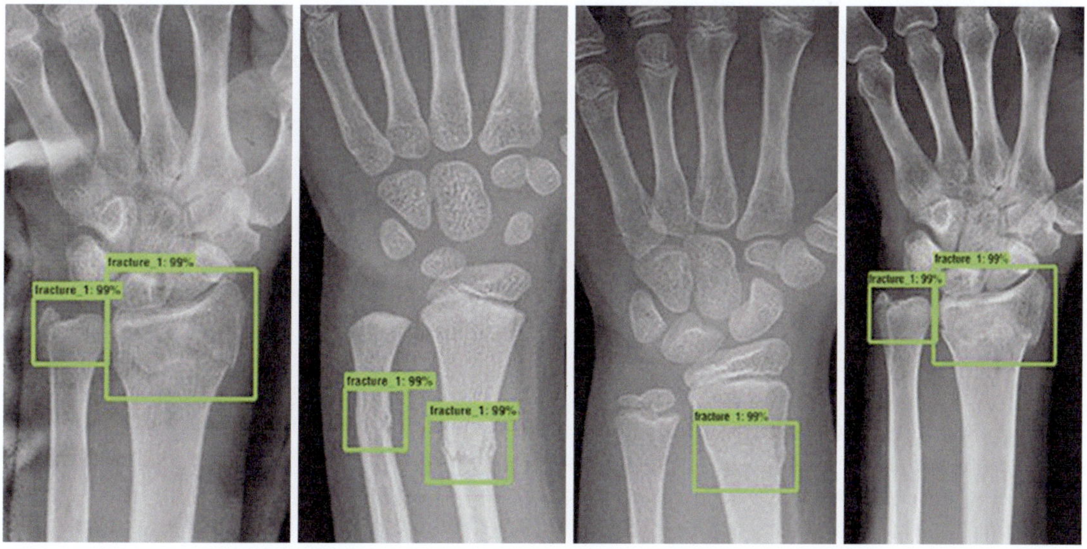

Fig. 28.1 Radiographs display true-positive examples of radius and ulna fractures. The green boxes highlight regions identified by the convolutional neural network (CNN) YOLOv8 model, which is trained to detect and localize fractures. The percentages inside each box indicate the network's confidence score in detecting a fracture within that area. (Adapted from Parisa Karimi Darabi [15])

Predictive Analytics

Predicting Disease Progression

Hand surgery has advanced because of predictive analytics, which runs on artificial intelligence. Algorithms process extensive patient data collections of medical records and images and genetic testing data to forecast disease advancements, such as rheumatoid arthritis and osteoarthritis [33, 34]. According to Momtazmanesh et al. [45], AI capabilities analyze both early imaging results and clinical data as the basis for anticipating the progression of rheumatoid arthritis. Through these predictions, surgeons can create specific treatment protocols that enhance patient results with fewer treatment complications [33, 34].

Risk Assessment for Surgical Complications

Decision support systems that utilize artificial intelligence technology enable healthcare providers to determine patient surgical complication risks, including postoperative infections, delayed recovery, and implant malfunctions [26]. The analysis tools evaluate individual variables from patients, such as their age, coexisting medical conditions, and the specifics of their surgical procedure, to establish risk scores that assist healthcare decision-making [59]. AI technology demonstrated its ability to predict surgical complications following major or minor procedures with outstanding results in identifying patients at high risk, according to research by Wissem Tafat et al. [61]. Knowledge acquired through this technology provides surgeons with essential information to take necessary safety precautions and enhance patient care management.

Integration of AI into Clinical Practice

Implementing AI technologies has brought disruptive changes to various diagnostic approaches developed and used by hand surgeons in a clinical treatment context. Applications of AI support integration with available electronic health records systems, simultaneously applying imaging techniques that provide direct clinical decisions to the relevant practitioners [32]. AI diagnosis assists hand surgeons by detecting aberrant medical image scans, subsequently encouraging further surgeon investigations [42].

AI and ML in Surgical Planning and Execution

Preoperative Planning

The field of preoperative planning in hand surgery changes through AI-backed three-dimensional modeling [14]. Hospital surgeons employ detailed three-dimensional body models to see complex tissue structures while performing preoperative simulations before surgical procedures to forecast results [36].

Through medical algorithms, surgeons obtain better surgical tactics by assessing individual patient data points [59]. These algorithms analyze recorded surgical data to identify the most suitable treatment paths for specific health conditions to enhance procedural outcomes [25].

Intraoperative Applications

In surgical procedures, AI navigation systems play a crucial role in enhancing precision and reducing the likelihood of errors. These systems can guide surgeons during procedures such as fracture fixation or tendon repair and provide real-time data from imaging and sensors, achieving greater accuracy and reducing complications. For instance, Figure 28.2 demonstrates how augmented reality can enhance surgical precision intraoperatively or better prepare the surgeon preoperatively.

Another area of innovation in hand surgery is robotic assistance. Adapting robotic systems, such as the da Vinci Surgical System, for delicate hand surgeries offers precision and control that surpasses traditional techniques [53]. According to Adegbesan et al. [1], these systems significantly reduce the risk of human error and improve outcomes, especially in complex cases, reassuring surgeons and patients (Fig. 28.3).

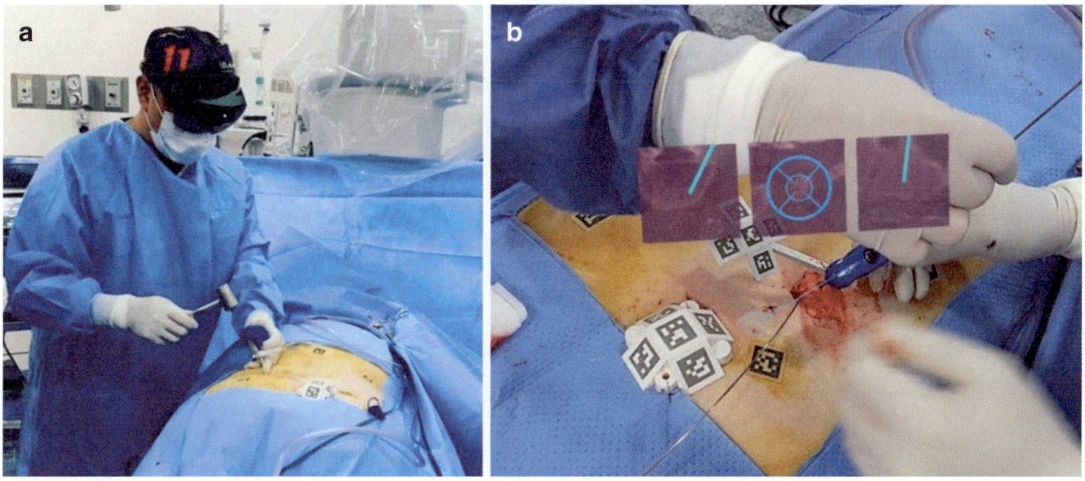

Fig. 28.2 Augmented reality in healthcare: Minimally invasive spine surgery (MISS) performed with augmented reality (AR) guidance using the Hololens 2 system. (**a**) AprilTags are attached to the patient's skin and also mounted on stable platforms secured with bone pins. The Jamshidi needle is aligned with a virtual trajectory and inserted through the skin. (**b**) Lateral and axial views show the ongoing MISS procedure guided by the Hololens. The Jamshidi needle has successfully entered the target pedicle, its core has been removed, and a K-wire has been placed to guide the insertion of a cannulated screw. An optical marker (AprilTag) is visible beneath the vertebra. (Adapted from De Jesus Encarnacion Ramirez et al. [17])

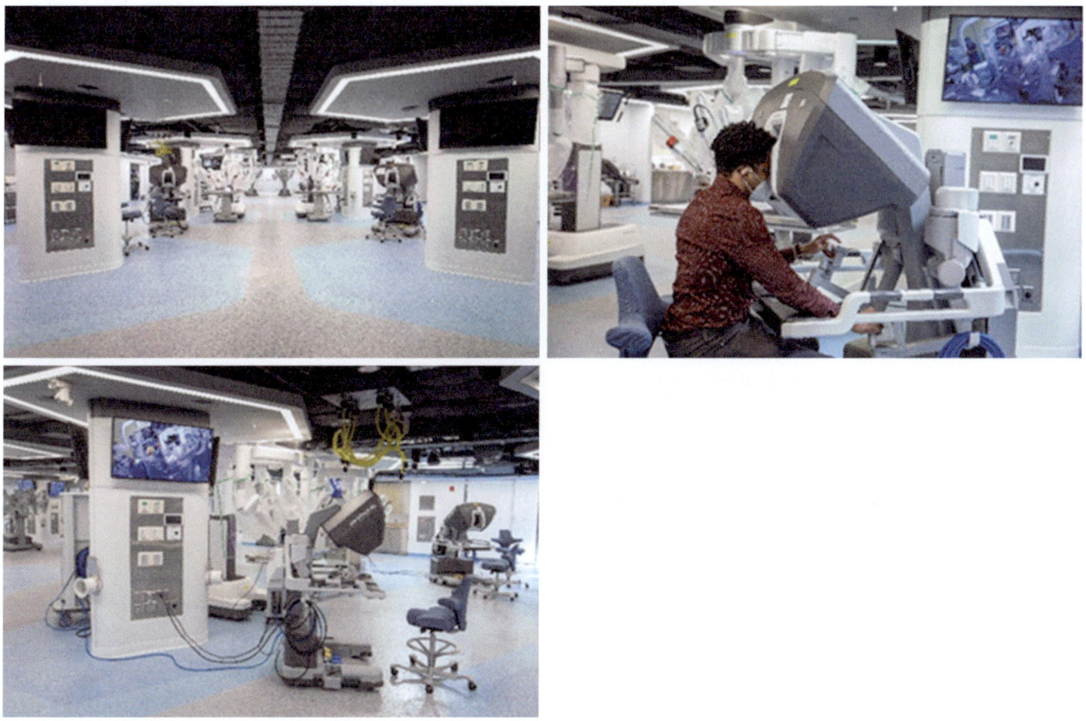

Fig. 28.3 Surgical solutions with multiarm robotic technologies: robotic systems enhance surgical precision with advanced visual feedback and controlled movements

AI in Rehabilitation and Postoperative Care

AI and ML provide rehabilitation and postoperative care, constituting the new phase for hand surgery patients [62]. AI facilitates the delivery of customized recovery regimens, direct patient progress observation, and predictively avoides further complications through informed analysis [26]. Improving the quality of postoperative therapy from wearable technology to AI-enabled treatment regimens improves quality while improving the functional outcome in cases of such kinds [37].

Personalized Recovery Plans

ML Algorithms for Predicting Recovery Timelines

The most important application of AI in postoperative care can be seen using ML algorithms, which predict an individual's recovery trajectory [5]. Using patient-specific factors such as age, comorbidities, the type of surgery, and functional preoperative status, AI thus provides personalized estimates of recovery times tailored for the needs of each patient [39]. For example, Keller et al. [32] discussed that AI can now anticipate patients' recovery timetable from flexor tendon repair. Using data from prior surgeries, it estimated the recovery time while recommending personalized therapy programs [32].

Recovery prediction of carpal tunnel release surgery was realized using ML [66]. Considering such data as patients' age, gender, and preoperative symptoms, the ML algorithm will give the chances of full recovery and recommend personalized rehabilitation plans [8]. According to these data, functional outcomes were comparatively better, and the recovery duration was also shorter in the patients who adhered to the AI-generated recovery plan than in cases where standard protocols were followed [60].

Customizable Therapy Programs

AI is also used to develop personalized therapy programs for patients undergoing hand surgery [52]. Such therapy programs use ML algorithms to readjust the therapy exercises in real time based on a patient's progress [54]. For example, an AI-powered therapy platform will be able to track key data from wearable devices, such as grip strength, and range of motion for the autoadjustment of the therapy program [49]. Research by Davids et al. [16] showed that AI-driven therapy programs increased adherence to rehabilitation protocols among patients and accelerated their recovery times (Fig. 28.4).

Monitoring and Feedback

Wearable Devices for Real-Time Monitoring

Postoperative monitoring depends more heavily on AI-enhanced wearables. These devices assess patient conditions by diagnosing grip strength and range of motion alongside therapy adherence measurements, generating real-time feedback to patients and healthcare providers [3]. These tracking devices track patient recovery while releasing feedback at an early stage to facilitate prompt treatment in delayed healing [6, 9, 55].

The research paper by Choi et al. [11] explains how wearable technology tracks the recovery process after fixations of distal radius fractures. The AI system uses tracked wearable data to produce personified feedback for each patient. Introducing AI-powered wearable technology improves patients' functional outcomes and better rehabilitation protocols following a medical procedure.

Predictive Analytics for Postoperative Complications

Predicting Surgical Site Infections

Predictive AI analytics are being carried out to estimate the risk of postoperative complications for surgical site infection [64]. AI algorithms can deliver the risk score that enables guidance for postoperative care through the analysis of patient-specific factors such as age, associated

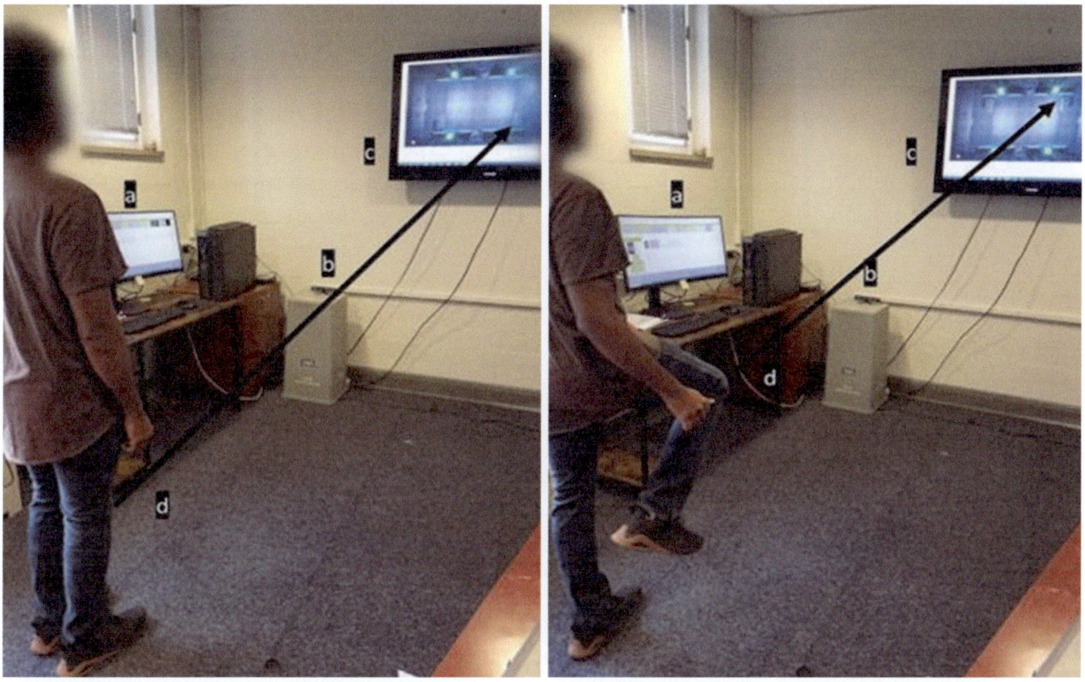

Fig. 28.4 Example of a VR system utilizing game-based exercises: Subject controlling a Pong paddle on the screen by raising and lowering their leg during gameplay. The VR system utilizes game-based exercises as part of the Serious Games for Therapy model to enhance patient engagement and satisfaction (**a**) Computer monitor used to launch the game, (**b**) desktop computer, (**c**) display screen showing the physical therapy game, and (**d**) knee elevation movement performed by the participant, which triggered the game light to turn green when the requested task was completed

comorbidities, and surgical technique [64]. The AI algorithm can, for example, analyze data to identify patients at a high risk of acquiring an infection and thus provide timely interventions to reduce postoperative infection [26].

Predicting Implant Failure

AI is also used to predict the risk of implant failure after wrist arthroplasty in patients. Since implant failure can be a significant concern in arthroplasties, reoperations are often required to revise the former surgery [52]. Such algorithms analyze information from previous operations and predict an implant failure concerning variables such as implant use, surgical techniques, and/or patient characteristics [58]. Shah et al. [58] established in their study that AI-powered predictive analytics increased the ability to predict implant failure and allowed surgeons to take necessary precautions and optimize patient care.

Challenges and Future Directions

Current Challenges

However, integrating AI into hand surgery is challenging [31]. Data privacy remains one of the major concerns because AI algorithms would require big data regarding patients' information to function effectively; hence, confidentiality and security of this data are key in maintaining patients' trust [46].

AI systems train on biased datasets, which generate biased results, as stated by Mittermaier et al. [43]. According to Carey et al. [10], the integration of diverse data collection alongside with continuous AI bias monitoring systems enables successful avoidance of this challenge.

AI implementation in hand surgery faces resistance because medical practitioners have limited knowledge about AI systems while simultaneously displaying mistrust toward them. Professional surgeons demonstrate little knowledge about AI, which leads them to refrain from using these systems when making essential clinical choices [2]. To successfully implement AI tools, clinicians need specific education programs alongside training that develop their trust [2].

Future Directions

The field of hand surgery is poised for a remarkable transformation, driven by a wave of innovative advancements that promise to revolutionize patient care and surgical outcomes. One of the most exciting developments lies in the integration of AI-enhanced tools, which are being disseminated to underserved regions lacking direct access to specialized surgical expertise. This initiative prioritizes expansion, enabling surgical teams in resource-constrained settings to benefit from sophisticated technologies that significantly enhance their capabilities [4]. Through its remarkable ability to simulate complex surgical experiences, AI is positioned to make substantial contributions to global efforts aimed at reducing mortality and morbidity rates [41].

In addition to surgical tools, groundbreaking innovations in the realm of prosthetics and orthoses are emerging. These AI-driven devices are designed to adapt seamlessly to the movements of their users, offering a sophisticated level of sensory feedback that far surpasses that of traditional prosthetics. As a result, individuals using these advanced systems enjoy enhanced functionality and comfort, completely redefining what is possible for those relying on assistive devices [21].

Furthermore, the landscape of surgical training is experiencing a paradigm shift with the advent of AI-driven simulated environments. These state-of-the-art platforms replicate the intricacies of complex procedures, providing aspiring surgeons with the opportunity for realistic, hands-on practice. By immersing trainees in these dynamic learning environments, the future generation of surgeons is being equipped to harness the full spectrum of benefits offered by AI-assisted hand surgeries, ultimately leading to improved patient outcomes and a new standard of care in surgical practice.

Conclusion

Artificial intelligence and machine learning are transforming the field of hand surgery by introducing advanced diagnostic tools, enhancing surgical planning, and improving postoperative care (Fig. 28.5). These technologies enable surgeons to enhance their accuracy, efficiency, and effectiveness, leading to better patient outcomes. However, leveraging the full potential of AI in hand surgery presents several challenges, such as concerns over data privacy, algorithm bias, and the adoption of these technologies by clinicians. Continued research and collaborative efforts are essential to fostering innovative breakthroughs in hand surgery through AI, ultimately aiming to provide superior care to patients on a global scale.

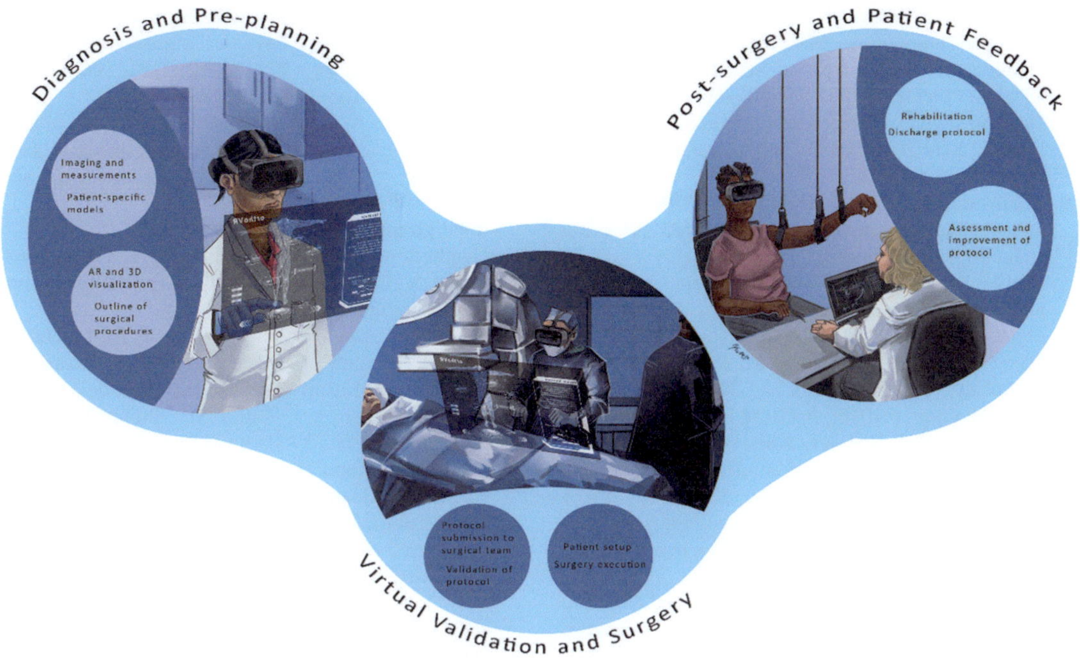

Fig. 28.5 The orthopedic future surgical platform: future platforms will integrate virtual visits, preplanning simulations, intraoperative AI assistance, and postoperative AI- and ML-driven rehabilitation to improve precision and outcomes

References

1. Adegbesan A, Akingbola A, Aremu O, Adewole O, Amamdikwa JC, Shagaya U. From scalpels to algorithms: the risk of dependence on artificial intelligence in surgery. J Med Surg Public Health. 2024;3:100140. https://doi.org/10.1016/j.glmedi.2024.100140.
2. Ahmed MI, Spooner B, Isherwood J, Lane MA, Orrock E, Dennison A. A systematic review of the barriers to implementing artificial intelligence in healthcare. Cureus. 2023;15(10):e46454. https://doi.org/10.7759/cureus.46454.
3. Amin T, Mobbs RJ, Mostafa N, Sy LW, Choy WJ. Wearable devices for patient monitoring in the early postoperative period: a literature review. Mhealth. 2021;7:50. https://doi.org/10.21037/mhealth-20-131.
4. Amin A, Cardoso SA, Suyambu J, Saboor HA, Cardoso RP, Husnain A, Isaac NV, Backing H, Mehmood D, Mehmood M, Maslamani ANJ, Amin A, Cardoso SA, Suyambu J, Saboor HA, Cardoso RP, Husnain A, Isaac NV, Backing H, Mehmood D. Future of artificial intelligence in surgery: a narrative review. Cureus. 2024;16(1):e51631. https://doi.org/10.7759/cureus.51631.
5. Apostolidis K, Kokkotis C, Moustakidis S, Karakasis E, Sakellari P, Koutra C, Tsiptsios D, Karatzetzou S, Vadikolias K, Aggelousis N. Machine learning algorithms for the prediction of language and cognition rehabilitation outcomes of post-stroke patients: a scoping review. Hum-Centric Intell Syst. 2023;4(1):147–60. https://doi.org/10.1007/s44230-023-00051-1.
6. Basu V, Cheng L, Zheng B. Technologies and sensors for artificial muscles in rehabilitation. Sensors. 2024;24(23):7532. https://doi.org/10.3390/s24237532.
7. Bitkina OV, Park J, Kim HK. Application of artificial intelligence in medical technologies: a systematic review of main trends. Digit Health. 2023;9. https://doi.org/10.1177/20552076231189331.

8. Boutry C, Moreau NN, Jaudet C, Lechippey L, Corroyer-Dulmont A. Machine learning and deep learning prediction of patient-specific quality assurance in breast IMRT radiotherapy plans using Halcyon-specific complexity indices. Radiother Oncol. 2024;200:110483. https://doi.org/10.1016/j.radonc.2024.110483.
9. Braun BJ, Grimm B, Hanflik AM, Richter PH, Sivananthan S, Yarboro SR, Marmor MT. Wearable technology in orthopedic trauma surgery - an AO trauma survey and review of current and future applications. Injury. 2022;53(6):1961–5. https://doi.org/10.1016/j.injury.2022.03.026.
10. Carey S, Pang A, de Kamps M. Fairness in AI for healthcare. Future Healthc J. 2024;11(3):100177. https://doi.org/10.1016/j.fhj.2024.100177.
11. Choi S, Lee J, Bae KJ, Kim DK, Lee YH, Lee Y. Clinical effect of rehabilitation after distal radius fracture surgery using a wearable device: a comparative prospective cohort study. Hand Surg Rehabil. 2024;43(6):101779. https://doi.org/10.1016/j.hansur.2024.101779.
12. Chung KC. Operative techniques: hand and wrist surgery. Elsevier Shop; 2021. https://shop.elsevier.com/books/operative-techniques-hand-and-wrist-surgery/chung/978-0-323-79415-2
13. Coaccioli S, Sarzi-Puttini P, Zis P, Rinonapoli G, Varrassi G. Osteoarthritis: new insight on its pathophysiology. J Clin Med. 2022;11(20):6013. https://doi.org/10.3390/jcm11206013.
14. Dababneh S, Colivas J, Dababneh N, Efanov JI. A review of artificial intelligence as an adjunctive tool in hand and wrist surgery. Artif Intell Surg. 2024;4(3):214–32. https://doi.org/10.20517/ais.2024.50.
15. Darabi PK. Bone fracture detection: computer vision project. ResearchGate. 2024. https://doi.org/10.13140/RG.2.2.14400.34569.
16. Davids J, Lidströmer N, Ashrafian H. Artificial intelligence for physiotherapy and rehabilitation. Springer EBooks; 2021. p. 1–19. https://doi.org/10.1007/978-3-030-58080-3_339-1.
17. De Jesus Encarnacion Ramirez M, Chmutin G, Nurmukhametov R, Soto GR, Kannan S, Piavchenko G, Nikolenko V, Efe IE, Romero AR, Mukengeshay JN, Simfukwe K, Mpoyi Cherubin T, Nicolosi F, Sharif S, Roa JC, Montemurro N. Integrating augmented reality in spine surgery: redefining precision with new technologies. Brain Sci. 2024;14(7):645. https://doi.org/10.3390/brainsci14070645.
18. Dy CJ. What's new in hand and wrist surgery. J Bone Joint Surg (Am Vol). 2020;102(6):453–9. https://doi.org/10.2106/jbjs.19.01333.
19. Esteva A, Kuprel B, Novoa RA, Ko J, Swetter SM, Blau HM, Thrun S. Dermatologist-level classification of skin cancer with deep neural networks. Nature. 2017;542(7639):115–8. https://doi.org/10.1038/nature21056.
20. Gassenmaier S, Küstner T, Nickel D, Herrmann J, Hoffmann R, Almansour H, Afat S, Nikolaou K, Othman AE. Deep learning applications in magnetic resonance imaging: has the future become present? Diagnostics. 2021;11(12):2181. https://doi.org/10.3390/diagnostics11122181.
21. Geethanjana A, & Hettige. Application of artificial intelligence in prosthetics: a review. ResearchGate; unknown. 2024. https://www.researchgate.net/publication/378907593_Application_of_Artificial_Intelligence_in_Prosthetics_A_Review.
22. Goodfellow I, Bengio Y, Courville A. Deep learning. MIT Press; 2016. https://mitpress.mit.edu/9780262035613/deep-learning/.
23. Gu Y, Xu Y, Shen Y, Huang H, Liu T, Jin L, Ren H, Wang J. A review of hand function rehabilitation systems based on hand motion recognition devices and artificial intelligence. Brain Sci. 2022;12(8):1079. https://doi.org/10.3390/brainsci12081079.
24. Hadley SR, Capo JT. Digit replantation the first 50 years. Bull Hosp Jt Dis (2013). 2015;73(2):148–55. https://pubmed.ncbi.nlm.nih.gov/26517169/
25. Hashimoto DA, Rosman G, Rus D, Meireles OR. Artificial intelligence in surgery. Ann Surg. 2018;268(1):70–6. https://doi.org/10.1097/sla.0000000000002693.
26. Hassan AM, Rajesh A, Asaad M, Nelson JA, Coert JH, Mehrara BJ, Butler CE. Artificial intelligence and machine learning in prediction of surgical complications: current state, applications, and implications. Am Surg. 2023;89(1):25–30. https://doi.org/10.1177/00031348221101488.
27. Hendrick RJ, Mitchell C, Duke Herrell S, Webster RJ. Hand-held transendoscopic robotic manipulators: a transurethral laser prostate surgery case study. Int J Rob Res. 2015;34(13):1559–72. https://doi.org/10.1177/0278364915585397.
28. Hoff T, Kitsakos A, Silva J. A scoping review of the patient experience with wearable technology. Digit Health. 2024;10:20552076241308439. https://doi.org/10.1177/20552076241308439.
29. Horta R, Silva P, Costa-Ferreira A, Amarante J, Silva Á. Microsurgical soft-tissue hand reconstruction: an algorithm for selection of the best procedure. J Hand Microsurg. 2011;03(02):73–7. https://doi.org/10.1007/s12593-011-0046-7.
30. Jackson RW. A history of arthroscopy. Arthroscopy. 2010;26(1):91–103. https://doi.org/10.1016/j.arthro.2009.10.005.
31. Jiang L, Wu Z, Xu X, Zhan Y, Jin X, Wang L, Qiu Y. Opportunities and challenges of artificial intelligence in the medical field: current application, emerging problems, and problem-solving strategies. J Int Med Res. 2021;49(3):030006052110001. Sagepub. https://doi.org/10.1177/03000605211000157.
32. Keller M, Guebeli A, Thieringer F, Honigmann P. Artificial intelligence in patient-specific hand surgery: a scoping review of the literature. Int J Comput Assist Radiol Surg. 2023;18(8):1393–403. https://doi.org/10.1007/s11548-023-02831-3.
33. Khalifa M, Albadawy M. AI in diagnostic imaging: revolutionising accuracy and efficiency.

Comput Methods Programs Biomed Update. 2024a;5(100146):100146. https://doi.org/10.1016/j.cmpbup.2024.100146.
34. Khalifa M, Albadawy M. Artificial intelligence for clinical prediction: exploring key domains and essential functions. Comput Methods Programs Biomed Update. 2024b;5:100148. https://doi.org/10.1016/j.cmpbup.2024.100148.
35. Kleinert HE, Serafín D, Daniel RK. The place of microsurgery in hand surgery. Orthop Clin N Am. 2021;4(4):929–44. https://doi.org/10.1016/S0030-5898(20)30828-2.
36. Langlotz CP, Allen B, Erickson BJ, Kalpathy-Cramer J, Bigelow K, Cook TS, Flanders AE, Lungren MP, Mendelson DS, Rudie JD, Wang G, Kandarpa K. A roadmap for foundational research on artificial intelligence in medical imaging: from the 2018 NIH/RSNA/ACR/the academy workshop. Radiology. 2019;291(3):781–91. https://doi.org/10.1148/radiol.2019190613.
37. Lanotte F, O'Brien MK, Jayaraman A. AI in rehabilitation medicine: opportunities and challenges. Ann Rehabil Med. 2023;47(6):444–58. https://doi.org/10.5535/arm.23131.
38. Lebleu J, Daniels K, Pauwels A, Dekimpe L, Mapinduzi J, Poilvache H, Bonnechère B. Incorporating wearable technology for enhanced rehabilitation monitoring after hip and knee replacement. Sensors. 2024;24(4):1163. https://doi.org/10.3390/s24041163.
39. Lee HJ, Kim JK, Park D, Kim JH, Kim CR, Choo YJ. The use of artificial intelligence to predict the prognosis of patients undergoing central nervous system rehabilitation: a narrative review. Healthcare. 2023;11(19):2687. https://doi.org/10.3390/healthcare11192687.
40. Lubowitz JH, Perry NPJ, Matzkin E, Rossi MJ. Introducing foundations of arthroscopy techniques. Arthroscopy. 2024;40(12):2777–9. https://doi.org/10.1016/j.arthro.2024.09.002.
41. Malhotra K, Ngie B, Lee S, Franco H, Singh C, Silva L, Iraqi HA, Sinha A, Burger S, Breedt DS, Goyal K, Dagli MM, Bawa A. Role of artificial intelligence in global surgery: a review of opportunities and challenges. Cureus. 2023. https://doi.org/10.7759/cureus.43192.
42. Miller R, Farnebo S, Horwitz MD. Insights and trends review: artificial intelligence in hand surgery. J Hand Surg (Eur Vol). 2023:175319342311525. https://doi.org/10.1177/17531934231152592.
43. Mittermaier M, Raza MM, Kvedar JC. Bias in AI-based models for medical applications: challenges and mitigation strategies. NPJ Digit Med. 2023;6(1):1–3. https://doi.org/10.1038/s41746-023-00858-z.
44. Mohammadi S, Salehi MA, Jahanshahi A, Shahrabi Farahani M, Zakavi SS, Behrouzieh S, Gouravani M, Guermazi A. Artificial intelligence in osteoarthritis detection: a systematic review and meta-analysis. Osteoarthr Cartil. 2024;32(3):241–53. https://doi.org/10.1016/j.joca.2023.09.011.
45. Momtazmanesh S, Nowroozi A, Rezaei N. Artificial intelligence in rheumatoid arthritis: current status and future perspectives: a state-of-the-art review. Rheumatol Ther. 2022. https://doi.org/10.1007/s40744-022-00475-4.
46. Murdoch B. Privacy and artificial intelligence: challenges for protecting health information in a new era. BMC Med Ethics. 2021;22(1). https://doi.org/10.1186/s12910-021-00687-3.
47. Norman B, Pedoia V, Majumdar S. Use of 2D U-net convolutional neural networks for automated cartilage and meniscus segmentation of knee MR imaging data to determine relaxometry and morphometry. Radiology. 2018;288(1):177–85. https://doi.org/10.1148/radiol.2018172322.
48. Olczak J, Fahlberg N, Maki A, Razavian AS, Jilert A, Stark A, Sköldenberg O, Gordon M. Artificial intelligence for analyzing orthopedic trauma radiographs. Acta Orthop. 2017;88(6):581–6. https://doi.org/10.1080/17453674.2017.1344459.
49. Olyanasab A, Annabestani M. Leveraging machine learning for personalized wearable biomedical devices: a review. J Pers Med. 2024;14(2):203. https://doi.org/10.3390/jpm14020203.
50. Park JJ, Tiefenbach J, Demetriades AK. The role of artificial intelligence in surgical simulation. Front Med Technol. 2022;4. https://doi.org/10.3389/fmedt.2022.1076755.
51. Pederson WC. Hand surgery as a paradigm for orthoplastic surgery. Orthoplastic Surg. 2024;18:8–9. https://doi.org/10.1016/j.orthop.2024.11.001.
52. Pressman SM, Borna S, Gomez-Cabello CA, Haider SA, Forte AJ. AI in hand surgery: assessing large language models in the classification and management of hand injuries. J Clin Med. 2024;13(10):2832. https://doi.org/10.3390/jcm13102832.
53. Probst P. A review of the role of robotics in surgery: to DaVinci and beyond! Mo Med. 2023;120(5):389. https://pmc.ncbi.nlm.nih.gov/articles/PMC10569391/
54. Rasa AR. Artificial intelligence and its revolutionary role in physical and mental rehabilitation: a review of recent advancements. Biomed Res Int. 2024;2024(1). https://doi.org/10.1155/bmri/9554590.
55. Ratcliffe AM, Zhai B, Guan Y, Jackson DG, South West Anaesthesia Research Matrix (SWARM), Sneyd JR. Patient-centred measurement of recovery from day-case surgery using wrist worn accelerometers: a pilot and feasibility study. Anaesthesia. 2021;76(6):785–97. https://doi.org/10.1111/anae.15267.
56. Rivero-Moreno Y, Echevarria S, Vidal-Valderrama C, Pianetti L, Cordova-Guilarte J, Navarro-Gonzalez J, Acevedo-Rodríguez J, Dorado-Avila G, Osorio-Romero L, Chavez-Campos C, Acero-Alvarracín K, Rivero Y, Echevarria S, Vidal-Valderrama C, Pianetti L, Guilarte JC, Navarro-Gonzalez J, Acevedo-Rodríguez J, Avila GLD, Osorio-Romero L. Robotic surgery: a comprehensive review of the literature and current trends. Cureus. 2023;15(7). https://doi.org/10.7759/cureus.42370.

57. Schäfer B, Bahm J, Beier JP. Nerve transfers using a dedicated microsurgical robotic system. Plast Reconstr Surg Glob Open. 2023;11(8):e5192. https://doi.org/10.1097/gox.0000000000005192.
58. Shah AK, Lavu MS, Hecht CJ, Burkhart RJ, Kamath AF. Understanding the use of artificial intelligence for implant analysis in total joint arthroplasty: a systematic review. Arthroplasty. 2023;5(1):54. https://doi.org/10.1186/s42836-023-00209-z.
59. Stam WT, Goedknegt LK, Ingwersen EW, Schoonmade LJ, Bruns ERJ, Daams F. The prediction of surgical complications using artificial intelligence in patients undergoing major abdominal surgery: a systematic review. Surgery. 2021; https://doi.org/10.1016/j.surg.2021.10.002.
60. Swarnakar R, Yadav SL. Artificial intelligence and machine learning in motor recovery: a rehabilitation medicine perspective. World J Clin Cases. 2023;11(29):7258–60. https://doi.org/10.12998/wjcc.v11.i29.7258.
61. Tafat W, Budka M, McDonald D, Wainwright TW. Artificial intelligence in orthopaedic surgery: a comprehensive review of current innovations and future directions. Comput Struct Biotechnol Rep. 2024;1. https://doi.org/10.1016/j.csbr.2024.100006.
62. Takeuchi M, Kitagawa Y. Artificial intelligence and surgery. Ann Gastroenterol Surg. 2023. https://doi.org/10.1002/ags3.12766.
63. Thian YL, Li Y, Jagmohan P, Sia D, Chan VEY, Tan RT. Convolutional neural networks for automated fracture detection and localization on wrist radiographs. Radiol Artif Intell. 2019;1(1). https://doi.org/10.1148/ryai.2019180001.
64. van der Meijden SL, van Boekel AM, Schinkelshoek LJ, van Goor H, Steyerberg EW, Nelissen RGHH, Mesotten D, Geerts BF, de Boer MGJ, Arbous MS. Development and validation of artificial intelligence models for early detection of postoperative infections (PERISCOPE): a multicentre study using electronic health record data. Lancet Reg Health Eur. 2024;49:101163. https://doi.org/10.1016/j.lanepe.2024.101163.
65. Vassa R, Garg A, Omar IM. Magnetic resonance imaging of the wrist and hand. Pol J Radiol. 2020;85:e461–88. https://doi.org/10.5114/pjr.2020.99034.
66. Yetiş M, Kocaman H, Canlı M, Yıldırım H, Yetiş A, Ceylan İ. Carpal tunnel syndrome prediction with machine learning algorithms using anthropometric and strength-based measurement. PLoS One. 2024;19(4):e0300044. https://doi.org/10.1371/journal.pone.0300044.
67. Zeltzer D, Herzog L, Pickman Y, Steuerman Y, Ber RI, Kugler Z, Shaul R, Ebbert JO. Diagnostic accuracy of artificial intelligence in virtual primary care. Mayo Clin Proc Digit Health. 2023;1(4):480–9. https://doi.org/10.1016/j.mcpdig.2023.08.002.
68. Zhou JY, Tuyishime H, Yao J. Arthroscopic-assisted repair of the triangular fibrocartilage complex. J Hand Surg Glob Online. 2024. https://doi.org/10.1016/j.jhsg.2024.03.011.

Correction to: Motion Analysis of Hand Movement

Yiru Wang and Farid Amirouche

Correction to:
Chapter 7 in: F. Amirouche et al. (eds.), *Biomechanics of the Hand*, https://doi.org/10.1007/978-3-032-04688-8_7

In Chapter 7 "Motion Analysis of Hand Movement," the author's name was incorrectly printed as Jesse Wang. This has been corrected to Yiru Wang in Chapter 7 (page 65), Table of Contents (page VII), and Contributors List.

The updated version of this chapter can be found at https://doi.org/10.1007/978-3-032-04688-8_7

F. Amirouche et al. (eds.), *Biomechanics of the Hand*,
https://doi.org/10.1007/978-3-032-04688-8_29

Index

F. Amirouche et al. (eds.), *Biomechanics of the Hand*, https://doi.org/10.1007/978-3-032-04688-8

F

G

MIX
Papier aus verantwortungsvollen Quellen
Paper from responsible sources
FSC® C105338

If you have any concerns about our products,
you can contact us on
ProductSafety@springernature.com

In case Publisher is established outside the EU,
the EU authorized representative is:
Springer Nature Customer Service Center GmbH
Europaplatz 3, 69115 Heidelberg, Germany

Printed by Libri Plureos GmbH
in Hamburg, Germany